QD 1 .A355 n.207

The Chemistry of solid wood

ADVANCES IN CHEMISTRY SERIES **207**

The Chemistry of Solid Wood

Roger Rowell, Editor
U.S. Department of Agriculture

Based on a short course and symposium
sponsored by the Division
of Cellulose, Paper, and Textile Chemistry
at the 185th Meeting
of the American Chemical Society,
Seattle, Washington,
March 20–25, 1983

American Chemical Society, Washington, D.C. 1984

Library of Congress Cataloging in Publication Data

The chemistry of solid wood.
 (Advances in chemistry series, ISSN 0065-2393; 207)

 "Based on a symposium sponsored by the Division of Cellulose, Paper, and Textile at the 185th Meeting of the American Chemical Society, Seattle, Washington, March 20-25, 1983."

 Includes bibliographies and indexes.

 1. Wood—Composition—Congresses.

 I. Rowell, Roger M. II. American Chemical Society. Cellulose, Paper, and Textile Division. III. Series.

QD1.A355   no. 207   [TA419.A1]
540s   [674'.134]   83-22451
ISBN 0-8412-0796-8

Copyright © 1984

American Chemical Society

All Rights Reserved. The appearance of the code at the bottom of the first page of each chapter in this volume indicates the copyright owner's consent that reprographic copies of the chapter may be made for personal or internal use or for the personal or internal use of specific clients. This consent is given on the condition, however, that the copier pay the stated per copy fee through the Copyright Clearance Center, Inc., 21 Congress Street, Salem, MA 01970, for copying beyond that permitted by Sections 107 or 108 of the U.S. Copyright Law. This consent does not extend to copying or transmission by any means— graphic or electronic—for any other purpose, such as for general distribution, for advertising or promotional purposes, for creating a new collective work, for resale, or for information storage and retrieval systems. The copying fee for each chapter is indicated in the code at the bottom of the first page of the chapter.

The citation of trade names and/or names of manufacturers in this publication is not to be construed as an endorsement or as approval by ACS of the commercial products or services referenced herein; nor should the mere reference herein to any drawing, specification, chemical process, or other data be regarded as a license or as a conveyance of any right or permission, to the holder, reader, or any other person or corporation, to manufacture, reproduce, use, or sell any patented invention or copyrighted work that may in any way be related thereto. Registered names, trademarks, etc., used in this publication, even without specific indication thereof, are not to be considered unprotected by law.

PRINTED IN THE UNITED STATES OF AMERICA

Advances in Chemistry Series

M. Joan Comstock, *Series Editor*

Advisory Board

Robert Baker
U.S. Geological Survey

Martin L. Gorbaty
Exxon Research and Engineering Co.

Herbert D. Kaesz
University of California—Los Angeles

Rudolph J. Marcus
Office of Naval Research

Marvin Margoshes
Technicon Instruments Corporation

Donald E. Moreland
USDA, Agricultural Research Service

W. H. Norton
J. T. Baker Chemical Company

Robert Ory
USDA, Southern Regional
   Research Center

Geoffrey D. Parfitt
Carnegie-Mellon University

Theodore Provder
Glidden Coatings and Resins

James C. Randall
Phillips Petroleum Company

Charles N. Satterfield
Massachusetts Institute of Technology

Dennis Schuetzle
Ford Motor Company
   Research Laboratory

Davis L. Temple, Jr.
Mead Johnson

Charles S. Tuesday
General Motors Research Laboratory

C. Grant Willson
IBM Research Department

FOREWORD

ADVANCES IN CHEMISTRY SERIES was founded in 1949 by the American Chemical Society as an outlet for symposia and collections of data in special areas of topical interest that could not be accommodated in the Society's journals. It provides a medium for symposia that would otherwise be fragmented, their papers distributed among several journals or not published at all. Papers are reviewed critically according to ACS editorial standards and receive the careful attention and processing characteristic of ACS publications. Volumes in the ADVANCES IN CHEMISTRY SERIES maintain the integrity of the symposia on which they are based; however, verbatim reproductions of previously published papers are not accepted. Papers may include reports of research as well as reviews since symposia may embrace both types of presentation.

ABOUT THE EDITOR

ROGER M. ROWELL is a carbohydrate chemist at the Forest Products Laboratory, U.S. Department of Agriculture, Forest Service in Madison, Wisconsin. His research involves the enhancement of wood properties through the chemical modification of wood cell walls. His research has included cooperative studies in the United States and in Australia, China, Great Britain, and Sweden. In 1967 he was appointed an Honorary Research Fellow at the University of Birmingham, England. In September 1983, he was invited to spend four months at Sweden's Chalmers University of Technology to establish a research program on the chemical modification of wood. He currently serves as an adjunct professor in the forestry department at the University of Wisconsin–Madison.

In addition to authoring more than 60 publications and receiving two patents, Rowell has presented numerous papers at national and international scientific meetings, organized national and international symposia, and has been active in consulting and technology transfer of his research.

Rowell is a member of the Sigma Xi Honorary Research Society and the American Chemical Society where he served as chairman of the Cellulose, Paper, and Textile Division in 1980.

Rowell received his B.S. degree in math and chemistry in 1961 from Southwestern College in Kansas. He received his M.S. and Ph.D. in biochemistry from Purdue University in 1963 and 1965, respectively.

CONTENTS

Preface ix

STRUCTURE AND BASIC CHEMISTRY

1. Formation and Structure of Wood 3
   Russell A. Parham and Richard L. Gray

2. The Chemical Composition of Wood 57
   Roger C. Pettersen

3. Wood–Water Relationships 127
   C. Skaar

PROPERTIES AND REACTIVITY

4. Penetration and Reactivity of Cell Wall Components 175
   Roger M. Rowell

5. The Chemistry of Wood Strength 211
   Jerrold E. Winandy and Roger M. Rowell

6. Wood–Polymer Materials 257
   John A. Meyer

7. Bioactive Wood–Polymer Composites 291
   R. V. Subramanian

8. Interaction of Preservatives with Wood 307
   Darrel D. Nicholas and Alan F. Preston

SURFACE CHEMISTRY

9. Chemistry of Adhesion 323
   R. V. Subramanian

10. Activation of Wood Surface and Nonconventional Bonding 349
    Eugene Zavarin

11. Chemistry of Weathering and Protection 401
    William C. Feist and David N.-S. Hon

DEGRADATION CHEMISTRY

12. **Biological Decomposition of Solid Wood** 455
    T. Kent Kirk and Ellis B. Cowling

13. **The Chemistry of Pyrolysis and Combustion** 489
    Fred Shafizadeh

14. **Chemistry of Fire Retardancy** 531
    Susan L. LeVan

15. **Degradation of Wood by Chemicals** 575
    Irving S. Goldstein

Abbreviations 587
Index 591

PREFACE

Most books published under the title "wood chemistry" could more appropriately be titled "The Chemistry of Wood Components" because they deal mainly with the chemistry of cellulose, hemicelluloses, lignin, extractives, bark, and pulping. Understanding the chemistry of the components is essential, but we use the total composite as a material. The title of this book may seem somewhat redundant, but it was chosen to make the point that this text concentrates on the chemistry of solid wood and deals with component chemistry only to the extent necessary to explain the composite.

Until the 1920s, wood chemistry was a significant scientific discipline. The advent of petroleum chemistry took wood chemistry out of the limelight. Over the past 20 years, the steady decline of trained people in carbohydrate, lignin extractives, and wood chemistry has reduced drastically the available work force, and these workers are scattered around the world. The "easy" wood chemistry questions have been answered; the more difficult ones will require a well-directed, cooperative research effort. We may not be able to answer the difficult questions, but through multidisciplinary research we will, at least, be able to ask more intelligent ones.

There almost seems to be a stigma associated with wood chemistry. Some members of the scientific community feel that because wood has been used for so long, we must surely know all there is to know about it. And besides, it is such a variable material with inconsistent properties that it does not require a high level of sophistication in its research approach. Using a material and understanding it are a world apart! Wood will never reach its highest use potential until we fully describe it, understand the mechanisms that control its properties, and, finally, are able to manipulate those properties to suit our needs. Because the properties of wood are the results of the chemistry of wood, this manipulation is possible only through wood chemistry research.

Perhaps what is needed is a little modern "hype." Instead of calling this book "The Chemistry of Solid Wood," the title "The Chemistry of Three-Dimensional Biopolymer Composites" might attract more attention. This approach could bring new people into a "new" science and reverse the decline in the number of trained wood chemists. At any rate, the time is right for a renewed consideration of this science with national and international attention focused on renewable resources. Wood will receive a fresh look by scientists, politicians, and economists because of its unique properties, aesthetics, low cost, low conversion energy, abundance, availability, and renewability. The large attendance at the Seattle short course from which this book was derived indicates a more active future for wood chemistry research.

This book is a result of many hours of effort by many people. I thank all of them for their time and concern for this project. I would especially like to thank my wife, Judy, for her efforts as the secretary to the short course and to Kathy Walker who has been the secretary for this book.

The book is dedicated to all past, present, and future scientists who proudly consider themselves wood chemists.

ROGER M. ROWELL
U.S. Department of Agriculture
Madison, WI 53705

August 1983

STRUCTURE AND BASIC CHEMISTRY

Formation and Structure of Wood

RUSSELL A. PARHAM and RICHARD L. GRAY
ITT Rayonier, Inc., Shelton, WA 98584

> *Wood—a natural, cellular, composite material of botanical origin—possesses unique structural and chemical characteristics that render it desirable for a broad variety of end uses. The level of suitability for a given end use (i.e., wood quality) is frequently determined by the wood's response to imposed physical and chemical treatments. However, in addition to these criteria, wood quality is also often based on the behavior of wood when subjected to the natural forces of the environment (e.g., weather, fire, and decay). All of these performance criteria are related either directly or indirectly to wood chemistry together with the wood's organizational architecture at the macroscopic and microscopic levels. This chapter reviews both of these structural domains, the development of wood characteristics in a growing tree, and why these characteristics can govern the behavior of wood under various treatment regimes. Emphasis is also given to the potential sources of wood variability, including wood type (softwood or hardwood), tree genus or species, and the variability present even within a single tree.*

Wood Sources

Botanical Origin. Wood is a natural material familiar in at least some way to everyone. Wood is obtained from two broad categories of plants known commercially as *softwoods* and *hardwoods*. These general names cannot be used universally to refer to the actual physical hardness or density of all woods because some softwoods are quite hard (e.g., Douglas-fir and southern yellow pines) and some hardwoods are soft (e.g., yellow buckeye, aspen, and cottonwood). Nevertheless, the names do accurately apply to many woods within these two categories and thus can be used as practical designations for the two general classes of commercial timbers.

From a more scientific perspective, softwoods are tree species of a class of plants called *gymnosperms* (seeds are borne naked), and hardwoods are woody, dicotyledenous (two seed leaves) *angiosperms* (seeds are borne in a fruit structure) (*see* Scheme I). The softwoods are also referred to as *conifers* because many produce seed cones, pollen cones, or both. The conifers have needlelike (e.g., pine) or scalelike (e.g., cedar) leaves and appear to be evergreen in that they retain new leaves for up to several years (with the exception of larches and baldcypress).

Hardwoods have leaves that are generally broad or bladelike, and most commercial species—at least in temperate climates—are deciduous, which means they commonly shed their leaves each fall at the end of the tree's growing season.

The selection of a particular tree species for various end uses is, in most cases, based on the physical and chemical characteristics of the wood. However, in many situations proximity to the wood source and total wood procurement cost can significantly affect species selection.

A comprehensive listing of commercial softwoods and hardwoods including domestic and foreign species, and their scientific (or Latin) names, is presented in Chapter 2.

Technical Nature. Wood is a complex plant tissue composed of several distinct types of cells. In the trees discussed in this text, wood is easily recognized as that tissue located to the inside of the tree bark and forming the interior bulk of all major stems, branches, and roots. Technically, wood is the main conductive and mechanical (or supportive) tissue of the tree, and is largely responsible for the upward translocation of water and dissolved minerals from the root system to the active tree crown (buds and functioning leaves) (*1*). The histological and cellular structure of wood and how it serves conductive, mechanical, and storage functions in the tree will be dealt with later in this chapter.

In their fully mature state, the vast majority of wood cells are dead and hollow, and the resulting tissue—known technically as secondary xylem—is composed essentially of only cell walls and voids, the voids being the hollow interiors of the cells (or lumens) (Figure 1). In softwoods, the cells making up 90–95% of the wood volume are fibrous in form (morphology) and are thus termed *fibers*. Hardwoods, on the other hand, are composed largely of fibers and much wider cells called *vessel elements*. The vessel elements are joined end-wise to form tubes or vessels along the stem, branch, or root axis and are seen as *pores* on the wood cross section (Figure 1). Among various hardwoods, fiber volume will vary over a wide range but averages about 50% for commercial species (*2*). Both softwoods

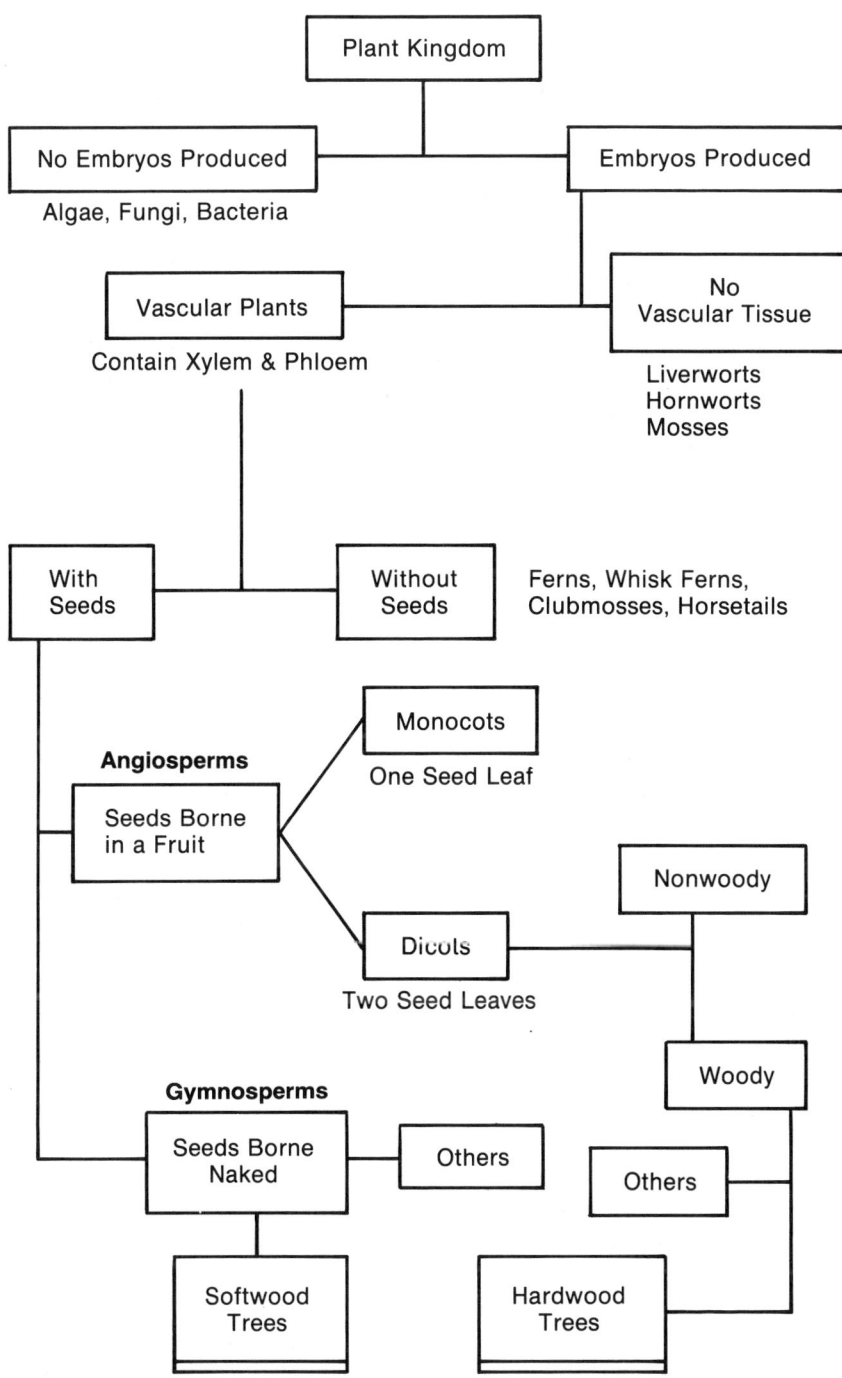

Scheme I. Botanical origin of commercial woods.

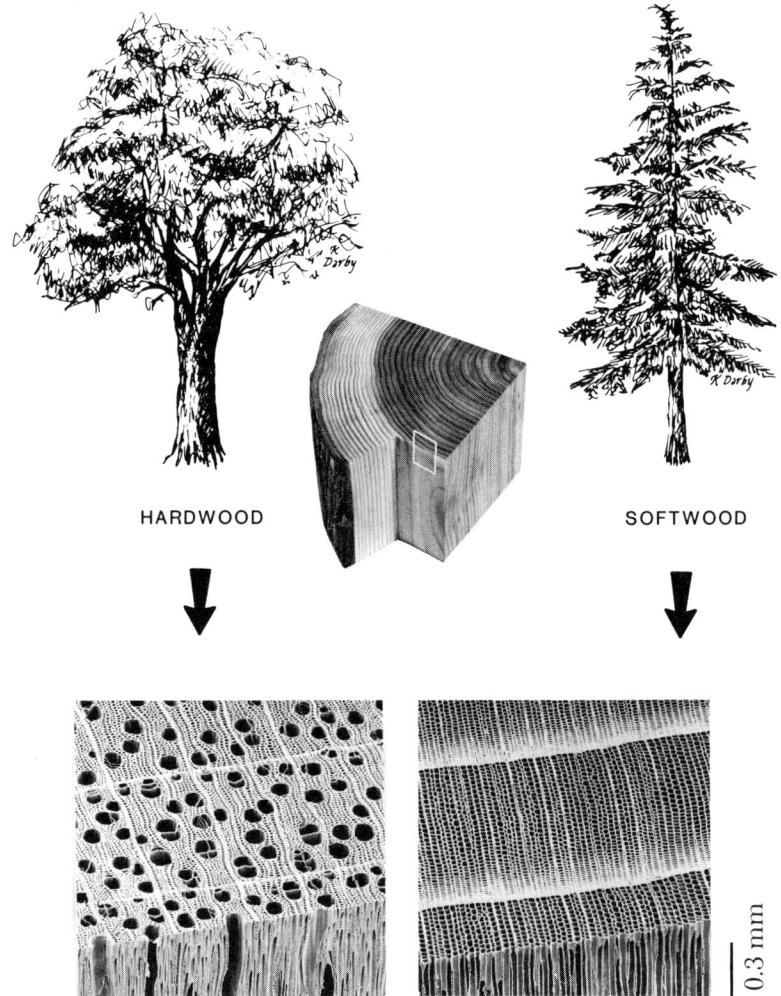

Figure 1. Commercial timber is obtained from angiosperms (hardwoods) or from gymnosperms (softwoods). The wood tissue is made up largely of dead, hollow plant cells that are arranged to form a composite material of substantial void volume.

and hardwoods contain various other nonfiber cells which will be given appropriate attention in the "Wood Anatomy" section.

From a chemical perspective, wood tissue (including cells and intercellular substance) is a composite material constructed from a variety of organic polymers—molecules that are made of many repeating subunits or monomers. The basic structural or skeletal material of all wood cell walls is cellulose, a long-chain, linear sugar molecule or polysaccharide (and carbohydrate) composed of glucose monomers. Glucose is a hexose or six-carbon ring sugar, and as a cellulose polymer it accounts for about 40–45% of the dry weight of normal wood tissue (3).

Serving as a matrix substance for the cellulose superstructure are other, lower molecular weight polysaccharides that contain short side chains. These carbohydrates represent mostly combinations of various five-carbon sugars (xylose and arabinose) and six-carbon sugars (glucose, mannose, and galactose). They are different in several respects from cellulose (mainly in conformation and molecular weight), but they are similar enough in form to warrant the name *hemicelluloses*.

The third major constituent of wood chemistry is *lignin*, a three-dimensional, highly branched, and polyphenolic molecule of complex structure and high molecular weight. Lignin is frequently compared to an incrustant substance because it enjoys an essentially ubiquitous distribution in fully mature wood tissue. It permeates both cell walls and intercellular regions, or *middle lamella*, and renders wood a hard, rigid material able to withstand considerable mechanical stress. The middle lamella region (Figure 2, A and B) is about 70–80% (or more) lignin by weight and is the cementing material that helps bind all wood cells together. Although the middle lamella region has a very high lignin content, because the cell walls are also lignified and occupy most of the wood's solid volume, about 70% or more of the total wood lignin is located in the cell walls themselves (4, 5) (Figure 2C).

The wood cell wall polysaccharides—cellulose and the hemicelluloses—have a strong affinity for water molecules in either their liquid or vapor state. Lignin, on the other hand, is almost water repelling. In fact, within the wood's internal architecture or ultrastructure (*see* later sections) lignin acts to block potential sites for water absorption.

It follows that the gain or loss of water or other liquids or vapors into or out of wood tissue can be influenced greatly by the nature, amount, and distribution of wood polysaccharides. As will be discussed in Chapter 3, these same sorption phenomena (i.e., adsorption and desorption), together with the architectural arrangement of wood cells in the tree, are responsible for particular patterns in wood swelling and shrinkage when wood is subjected to various environments. The arrangement of polysaccharide molecules within the cell wall itself, especially that of the cellulose, also has a profound effect on the physical and mechanical properties of individual cells and wood as a whole.

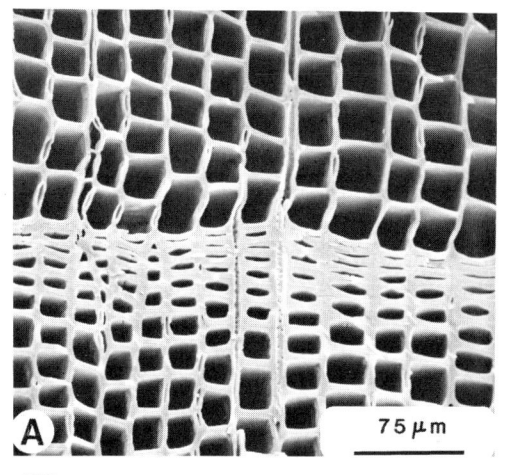

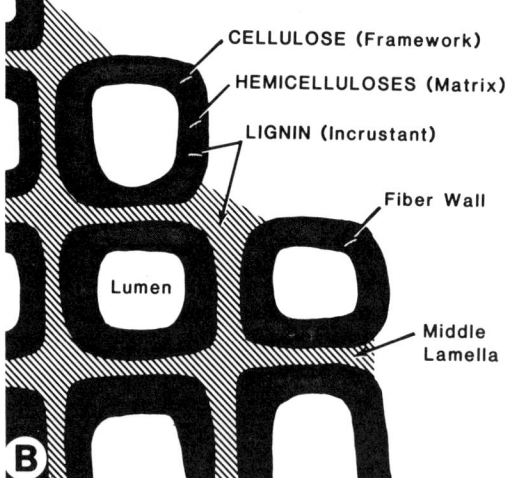

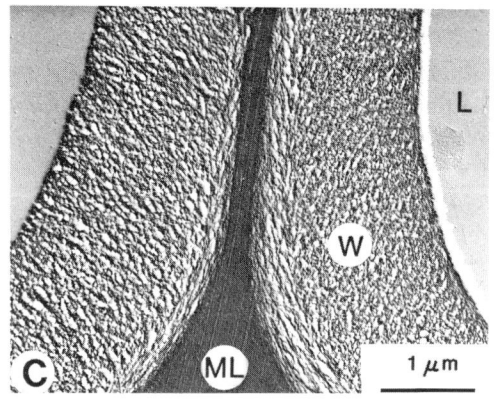

Wood Formation in Trees

Meristems. The formation of wood in trees is an integral aspect of total tree growth and includes not only increases in the diameters of the stem, branches, and roots, but also the elongation of these major tree parts. All of this visible or macroscopic growth is the result of activity by special cell zones called *meristems*, derived from the Greek, meristos, meaning divisible (1). Thus, meristematic cells are those in the tree that retain a continued ability to divide, eventually producing many cells (derivatives or daughter cells) of a given type (e.g., wood and bark) from only one cell. The latter cell is called an *initial* and remains in the meristem.

Developing trees contain two major types of meristems: (1) terminal or apical meristems and (2) lateral meristems. *Apical meristems* are located at the tips of all stems and branches (both termed shoots) where they are contained within terminal buds; they are also located within the tip regions of all roots. In the tip regions, the meristematic zone is usually protected by another zone of cells called the *root cap*. Root hairs, or microscopic roots, have no apical meristems, but these minute structures are lateral projections of roots that do have apical meristems.

In very young trees or seedlings, apical meristems are responsible for essentially all shoot and root growth—called primary growth. During the first year of seedling growth, a lateral meristem is initiated throughout most of the main shoots and roots. This lateral meristem—the *vascular cambium*—is responsible for the manufacture of all initial and subsequent wood tissue. The cambium is a thin, circumferential sheath of cells that produces wood or secondary xylem to the inside (i.e., toward the tree center) and *phloem* or inner bark tissue to the outside (Figure 3A).

Even in its first few months of growth, a seedling already has well-developed tissues that function in translocation and physical support. These early but critically important tissues are called *primary xylem* and *primary phloem* and are produced by derivatives of apical meristems; they are longitudinally arranged into tissue zones called *vascular bundles* (1, 2). These structures are the only circulatory

Figure 2. The technical nature of wood tissue. (A) Scanning electron micrograph (SEM) of the surface of a softwood resulting from a cut directed perpendicular to the tree axis. This view is known technically as the transverse or cross-sectional plane. (B) Schematic showing the location of the major constituents of wood chemistry. (C) Transmission electron micrograph (TEM) of the transverse section of a white spruce wood lignin skeleton that was prepared by removing cellulose and hemicelluloses with hydrofluoric acid. Key: L, lumen; W, wall; and ML, middle lamella. (Reproduced with permission from Ref. 38. Copyright 1974, Forest Products Research Society.)

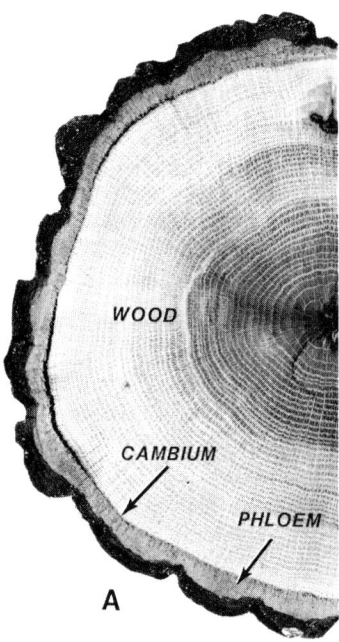

Figure 3A. The location and manufacture of wood cells in an oak tree cross section illustrating various tissue systems. (Reproduced from Ref. 39. Copyright 1982, American Chemical Society.)

Figure 3B. Light micrograph of the location and manufacture of wood cells in the cross section of a young pine (softwood) stem. The cambial zone (CZ) produces phloem or inner bark cells to the outside and wood cells to the inside.

tissues in the tree until they are complemented in older tree parts by *secondary xylem* and *secondary phloem* from the vascular cambium. Whether primary or secondary in origin, xylem functions largely in upward conduction and mechanical support, and phloem acts as a conduit for downward movement of photosynthates (manufactured foodstuffs) and hormones from the leaves and buds. Both xylem and phloem also function in a storage capacity, which takes place largely in *parenchyma* cells (*see* later sections).

Wood Cell Production. The site of wood cell production, the vascular cambium, is illustrated in Figure 3B. Technically, it is a microscopic sheath of meristematic cells. However, the exact circumferential line of cambial cells is very difficult to locate precisely, particularly during the tree's growing season, because of the presence of recent xylem and phloem derivatives. Therefore, it is more common to reference this lateral meristem as the *cambial zone* (2).

All cells in the cambial zone are living. However, as xylem derivatives (i.e., developing wood cells) begin a sequence of transformations that will convert them into mature wood elements, they embark on a path of cell specialization or *differentiation* that will lead eventually (for fibers, vessel elements, and certain other cells) to cell death.

Recent xylem derivatives may function for a period of time as *mother cells*, dividing to form still other derivatives. Nevertheless, the ultimate fate of most xylem derivatives is self destruction, *autolysis*, of their living contents, *protoplast*, and the eventual products are fully differentiated, or specialized, wood cells possessing rather elaborate walls and hollow centers, *lumens*. Only a relatively small number of cells in wood—called *parenchyma*—retain a viable protoplast after exiting the cambial and differentiation zones. Parenchyma are small, nonfibrous cells that have special storage or secretory functions.

In hardwoods, parenchyma cells may be organized into numerous vertical strands and can occupy up to 25–50% of the wood volume (2). However, such strands usually occupy a relatively small percentage of the total wood volume in conifers (1–2%). In many softwoods and hardwoods the vertical strand parenchyma are essentially absent (2).

Parenchyma also compose most (or all in hardwoods) of the wood tissue regions referred to collectively as wood *rays*. These structures are narrow ribbons of cells that are oriented lengthwise along the tree radius and perpendicular to the stem axis (Figure 4). Ray height is measured along the stem axis, and ray width is measured parallel (tangent) to the tree circumference. In addition to their storage or secretory functions (*see* later sections), ray parenchyma

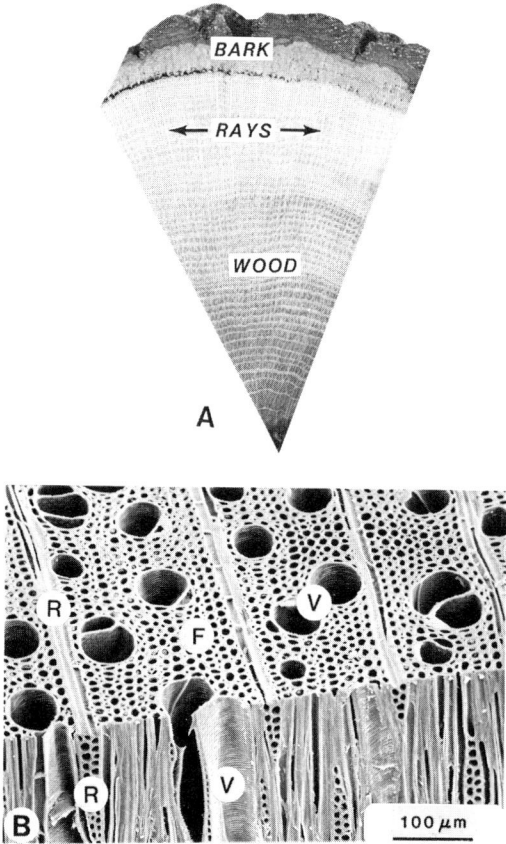

Figure 4. The location of wood rays. (A) Portion of an oak tree cross section showing that rays are oriented normal to tree growth rings and along the tree radius. (Photo by W. J. McCleary.) (B) SEM of wood rays (R), vessels or pores (V), and wood fibers (F) as seen in a two-plane view of soft maple.

cells also function in radial transport of tree photosynthates and biochemicals inward from the phloem and cambial zone.

During the division and enlargement phases of wood cell development, the cell wall is a thin, deformable, and extensible envelope of material referred to as the *primary wall*. Near the cessation of cell enlargement, however, a *secondary wall* may begin to be manufactured to the lumen side of the primary wall. Wood fibers, vessel elements, and certain other xylem or phloem elements that function in passive conduction and/or support normally develop a secondary wall (Figure 5).

Perhaps as early as the manufacture of the secondary wall, differentiating wood fibers (and most other wood cells) and the regions

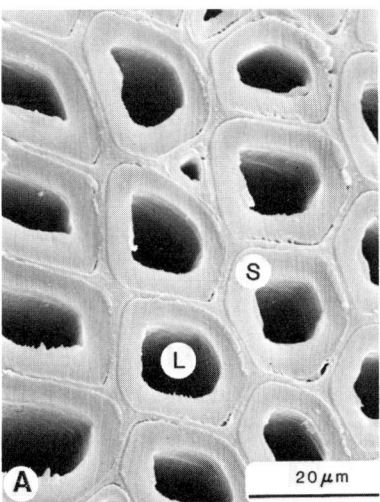

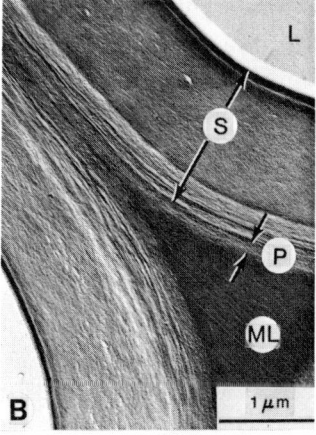

Figure 5. Cross-sectional view of fully differentiated wood fibers. (A) SEM of a Douglas-fir. Note the thick secondary walls (S) and the fiber lumen (L). (Reproduced from Ref. 39. Copyright 1982, American Chemical Society.) (B) TEM of two adjacent fibers in white spruce. Key: P, primary wall; S, secondary wall; L, lumen; and ML, middle lamella. (Reproduced with permission from Ref. 38. Copyright 1974, Forest Products Research Society.)

between the cells (middle lamella) start to become *lignified*. That is, lignin manufactured by the cells starts to infiltrate and incrust the entire wood tissue, and most of the process comes near the end of secondary wall deposition. In fully differentiated wood tissue, lignin composes about 25–35% of the total wood weight (moisture-free basis) and plays a major role in imparting rigidity to the polysaccharide wall substances (3).

The overall process of wood cell production is influenced by numerous factors, including genetics, climate, photoperiod or day length, and forest soil conditions. However, these factors are only indirectly involved in the control of wood cell development, but they directly affect the tree crown—the leaves and buds—and the crown then exerts a directing influence on cambial zone activities, including the form and number of derivatives produced.

Wood Tissues and Growth Increments. Wood cell production occurs during only part of each year (in the spring and summer months) in temperate zones. Wood production occurs on a more irregular time schedule in tropical regions of the world and, to some extent, in subtropical areas. In tropical and subtropical regions, the cambium has no true seasonal dormancy, and tree growth is related more to variations in local climate, particularly to the supply of available water (6). In tropical trees, zones of wood tissue that can be attributed to specified periods of growth are not usually decipherable because the trees can grow essentially on a continuous basis.

In the subtropics, or in highly elevated tropical areas, some trees can lose their leaves during part of the year. This situation can then be reflected in a visible change in the type of wood produced. The general result is a detectable *growth increment* that appears as a *growth ring* on a cross-section of the tree stem.

Growth increments reach their most advanced form in temperate-zone trees where the norm is a single, generally distinct, and circumferential band of wood production each year (Figure 6). The appearance of these so-called annual rings varies between hardwoods and softwoods with species, tree age, and growing conditions. These factors, together with certain other environmental effects, can also

Figure 6. Stem cross section of Sitka spruce showing about 24 annual growth increments. (Photo by W. J. McCleary.)

lead to aberrations in growth known as *multiple rings* or *discontinuous rings* (2).

Growth rings and the types and distribution of cells therein give rise to the situation whereby the general structure of wood tissue can be referenced by three major planes or surfaces. Each surface has a unique appearance that stems from the manner in which cells of specific orientations are cut by planes directed either parallel or perpendicular to the tree's longitudinal axis. These three planes are the cross-sectional or transverse plane, the radial plane, and the tangential plane. They are illustrated and defined in Figure 7.

Because trees grow from the tips of stems and roots, growth rings are laid onto the tree in the form of inverted hollow cones, and each ring (except the most recent) extends only part way up the tree and tapers to a point in the direction of the stem tip (Figure 8). Therefore, counting growth increments on a tree cross section will almost always not reveal the total age of a tree but only the age of that particular vertical level in the stem. Other aspects of tree growth, some being species-dependent, preclude the accurate esti-

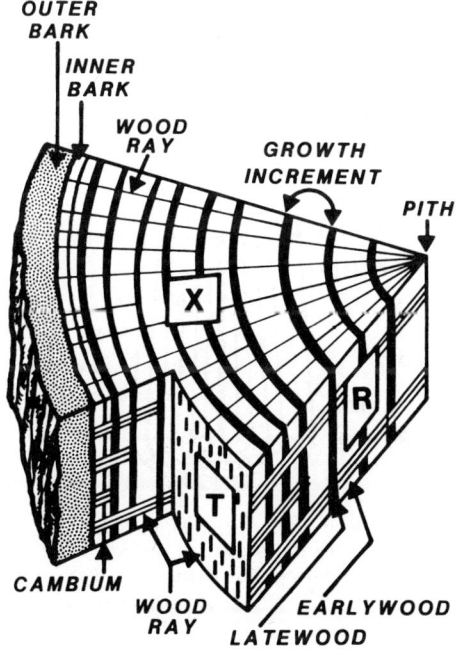

Figure 7. Representation of a tree cross section cut to reveal the three major structural planes of wood. This particular stem was cut in the spring of its 10th growing season. Key: X, cross-sectional or transverse plane/surface; R, radial surface; T, tangential surface. (Reproduced with permission from Ref. 40. Copyright 1982, Technical Association of the Pulp and Paper Industry Press.)

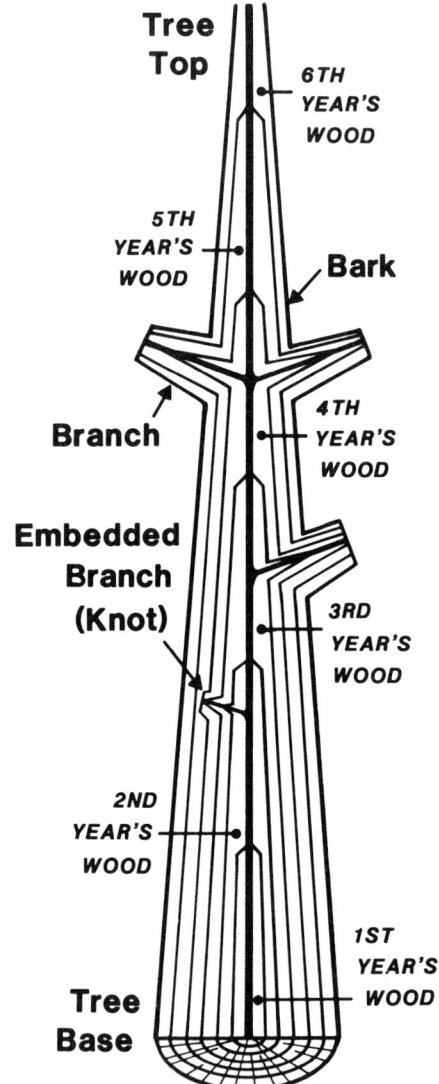

Figure 8. Representation of growth increments in a 6-year-old tree stem. Note that the increments are deposited as inverted hollow cones. (Reproduced with permission from Ref. 40. Copyright 1982, Technical Association of the Pulp and Paper Industry Press.)

mate of tree age via ring counting. These topics are dealt with in other texts (2, 7).

The initiation of a growth increment and the production of new wood cells each year in a given tree are governed by a complex system of tree physiology involving interaction of the tree crown with climate, soil, and other environmental factors, including imposed forest management (8). Nevertheless, in temperate trees it is normally possible to separate visibly each "annual" ring into two distinct tissues. These two tissue zones are known technically as *earlywood* (also springwood or initial wood) and *latewood* (also summerwood or terminal wood). They are manufactured during approximately the spring and summer months, respectively, of a given geographical region. Growth rings are normally readily detected in both softwoods and most hardwoods because of differences in the diameter and wall thickness of cells composing the earlywood and latewood zones.

SOFTWOODS. Earlywood fibers in softwoods are thin-walled and have wide radial diameters. As spring passes into summer, the fibers develop thicker walls, either abruptly or gradually, depending on the species. At the same time fiber radial diameter is reduced (Figure 9). These changes in fiber structure and morphology provide a latewood tissue that is normally several times denser (weight/volume) and considerably harder than earlywood of the same growth increment (2). The latewood zone appears darker than earlywood on the tree cross section as well as on the radial and tangential surfaces, and this difference in appearance between earlywood and latewood, together with the angle at which the tree or wood is cut or machined, helps give rise to *figure* or *grain* in softwood timber.

HARDWOODS. For most hardwoods growth rings are distinguishable but not because of a significant change in the fiber cells from earlywood to latewood. Fiber morphology and wall thickness are relatively similar within a given ring. In hardwoods, growth increments are most often distinguished by changes in vessel or pore diameters—changes that range from subtle to very distinct.

Figure 10 shows that commercial temperate hardwoods can be classified as having abrupt transition, gradual transition, or little if any transition in pore size from earlywood to latewood. These three types of woods are commonly known as *ring-porous*, *semi-ring-porous* (or *semi-diffuse-porous*), and *diffuse-porous*, respectively. Here, as in the softwoods, the structural differences, or lack thereof, between earlywood and latewood tissue give rise to the aesthetic property of wood grain. Because of the more complex or varied anatomy of hardwoods, grains here show much more diversity than in the softwoods.

Wood Quality. The relative number of thin-walled and thick-walled fibers in softwoods and hardwoods, and the vessel size and number in hardwoods, have a major impact on wood characteristics that often dictate the wood's end use. Such characteristics include the mechanical properties of wood, the type of surface resulting from wood machining, the permeability of wood to liquids and gases, wood weatherability, and perhaps others. In essence, the types and pro-

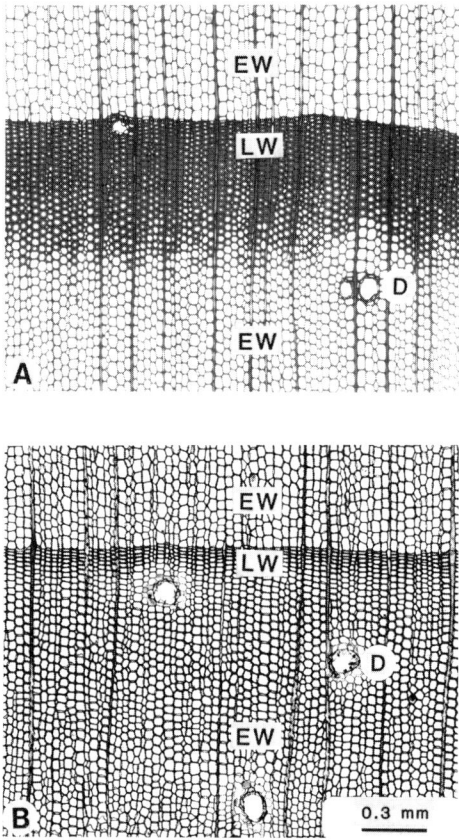

Figure 9. Light micrograph of the earlywood-to-latewood transition within softwood growth increments as viewed in cross section. Key: A, abrupt transition in eastern larch, with thick-walled latewood fibers (D = resin duct); and B, gradual transition in eastern white pine, with relatively thin-walled latewood fibers.

portions of various cell types in a given piece of wood, tree, or species are responsible for the general concept of what is widely referred to as wood quality. However, wood quality is such an arbitrary term because it is employed to describe the general suitability of a particular wood source for a very specific end use application (8). That specific use might be furniture parts, baseball bats, plywood, paneling, railroad ties, toothpicks, firewood, pencils, fence posts, etc. The point is that a wood source rated as excellent for one application may not be suitable for certain other uses, and vice versa.

This discussion may seem to imply that one of the most influential factors controlling wood quality, i.e., wood variability, is not desirable and that a major goal of forest management should be to eliminate or at least minimize this variability. This is often indeed the objective in some cases, particularly for trees of a given forest stand or for the same species between stands. However, among species or tree types, it is this same natural variability that gives rise to the diversity of available timber resources and the specific or even unique wood properties upon which many end products are based.

Wood Anatomy

Softwoods. The wood or secondary xylem of gymnosperms is composed of relatively few cell types (*see* box).

The predominant cell type in softwoods is the vertically oriented (along the stem axis) *longitudinal tracheid*. More commonly known as fibers, these tracheids are hollow, square to rectangular in cross section, have closed and tapering ends, and are arranged so that their

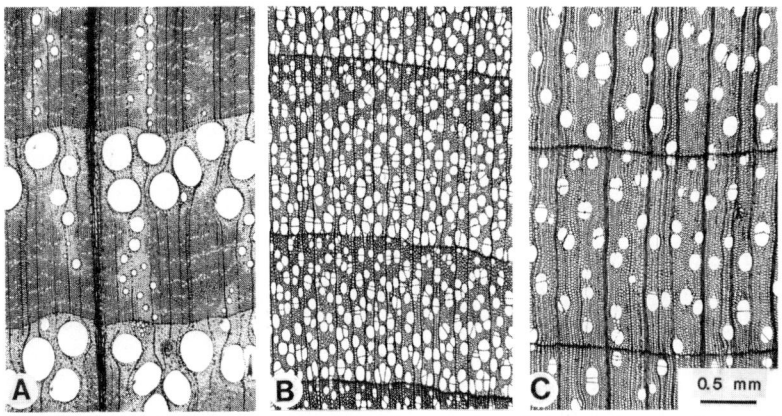

Figure 10. Light micrograph of the three types of pore patterns of growth increments in hardwoods as seen in cross section. Key: A, ring-porous (red oak); B, semi-ring-porous (aspen); and C, diffuse-porous (yellow birch). (Reproduced with permission from Ref. 40. Copyright 1982, Technical Association of the Pulp and Paper Industry Press.)

Major Types of Cells in Softwoods	
Vertically Oriented	*Horizontally Oriented*
A. Longitudinal tracheids B. Axial parenchyma C. Epithelial cells[b]	A. Ray tracheids[a] B. Ray parenchyma C. Epithelial cells[b] Homocellular rays: A or B Heterocellular rays: A + B (Fusiform ray: A + B + C)

[a] Presence is species-dependent.
[b] Surround normal resin ducts in pines, spruces, larches, and Douglas-fir, and traumatic resin ducts (formed as a result of tree injury) in these and other softwoods (2).

ends overlap adjacent fibers. They are also arranged into well-aligned radial rows (*see* Figure 9).

Widths of longitudinal tracheids generally range from 35 to 50 μm and lengths average between 3 and 5 mm. These cells serve a dual role of providing strength and mechanical support as well as being the pathway by which water and dissolved minerals are translocated from the tree's root system upward to the tree crown.

Arranged horizontally or radially in the tree are the wood rays, which, as mentioned earlier, are composed predominantly of small, bricklike, and often living cells called parenchyma (*see* Figures 4, 7, and 11). These cells function in radial translocation but have a major role as a storage receptacle, and frequently contain extraneous materials such as starch, fats, oils, various sugars, and inorganic depositions such as calcium oxalate crystals or silica (Figure 12).

The rays in some species also contain cells known as *ray tracheids*, which are similar in size to parenchyma but are dead at maturity (*see* Figure 11). Presence and/or type of ray tracheid is sometimes a very useful diagnostic feature in identification of a particular wood genus or species. With or without ray tracheids, rays are usually several cells high, but in softwoods they are generally only one cell wide (uniseriate) (Figure 13), except in special cases where they can be up to several cells wide (multiseriate).

Resin ducts or *resin canals* are tubelike voids that are both longitudinally and radially oriented throughout the xylem of some softwoods (*see* Figure 9). These ducts are lined with specialized parenchyma, called *epithelial cells*, that secrete into the duct a substance called oleoresin (Figure 14A) (*3*).

Vertical and horizontal resin ducts are natural and constant features of four domestic genera: pines (*Pinus*), spruces (*Picea*), larches (*Larix*), and Douglas-fir (*Pseudotsuga*). Resin ducts also develop as a response to injury or trauma in other genera, as well as in the four listed above (*2*). Horizontal resin ducts are contained in special multiseriate rays, called *fusiform rays* because of their spindle shape in tangential view (Figure 14B).

A few softwoods, notably the cedars and baldcypress, contain a noticeable amount of longitudinal (strand) parenchyma, but as mentioned previously, these axially oriented parenchyma are generally sparse in most conifers.

Hardwoods. Various cell types are found in hardwoods (*see* box). Hardwood anatomy is more varied or complicated than that of the softwoods, but most structural concepts are analogous.

Hardwoods contain a substantial volume of fiber cells, but the distinguishing feature of angiosperm xylem is the occurrence of *vessels*. The vessels are seen on the wood cross section as holes or pores

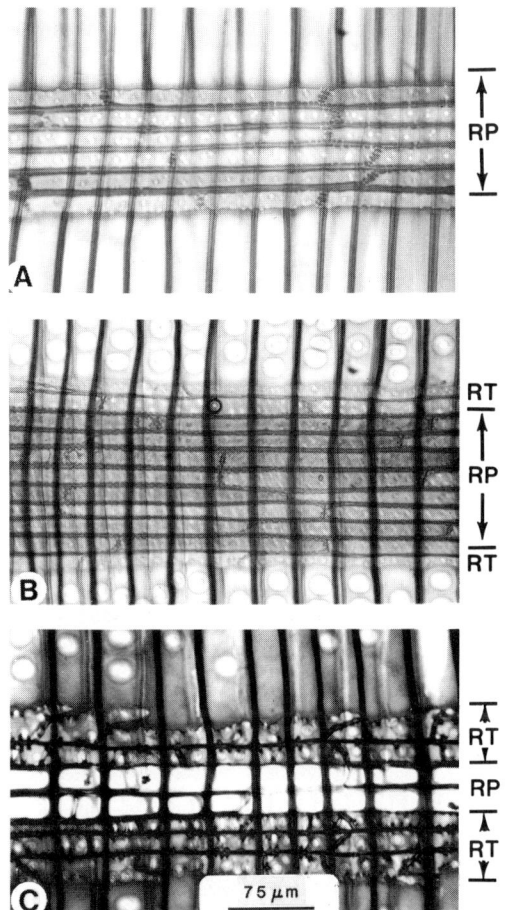

Figure 11. Light micrograph of ray structure in softwoods as seen in radial section. Key: A, homocellular ray composed of procumbent parenchyma (RP) in white fir; B, heterocellular ray with nondentate ray tracheids (RT) in black spruce; and C, heterocellular ray with dentate ray tracheids in red pine. (Reproduced with permission from Ref. 40. Copyright 1982, Technical Association of the Pulp and Paper Industry Press.)

in various patterns. Thus, all hardwoods are also referred to as porous woods, in contrast to the softwoods that are technically nonporous (*compare* Figures 9 *and* 10).

An individual vessel or pore consists of a vertical series of short (0.02–0.5 mm) vessel segments, which are joined end-to-end along the grain. Individual vessels can meander to a limited extent in the radial or tangential direction to join, terminate in, or depart from other vessels, but their major function is the vertical translocation of

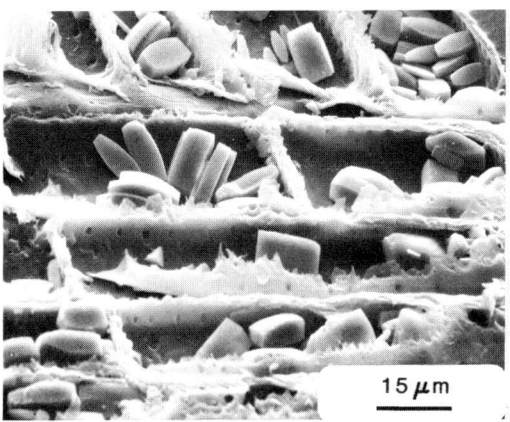

Figure 12. SEM of calcium oxalate crystals in the ray parenchyma of the wood radial surface of a tropical hardwood. (Reproduced from Ref. 39. Copyright 1982, American Chemical Society.)

sap. To facilitate this translocation, the ends of all vessel segments are perforate; that is, the ends are open for free flow of liquids between cells, in contrast to the situation of completely imperforate ends of all wood fibers. In some hardwoods the vessel segment ends are entirely open (simple), while in others the ends contain a series of parallel crossbars (scalariform) or some other design (e.g., reticulate). The particular type of opening here is of considerable value in wood

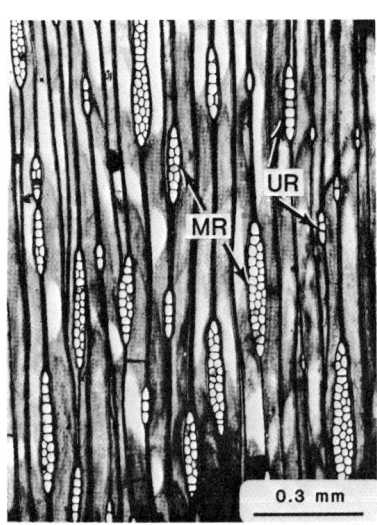

Figure 13. Light micrograph of tangential section of redwood showing uniseriate rays (UR), the most common type of ray in softwoods, together with multiseriate rays (MR). The latter, if not containing a horizontal resin duct, are rare in softwoods.

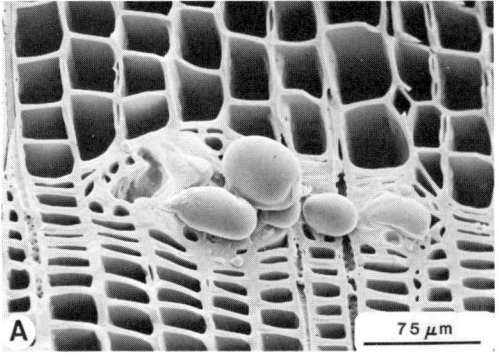

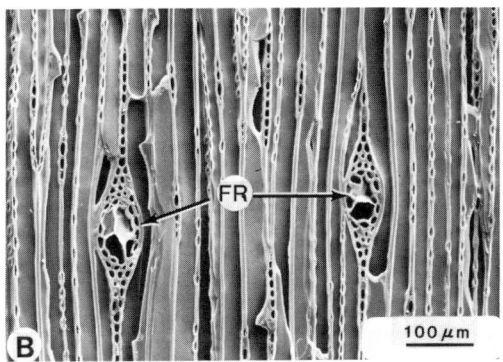

Figure 14. SEM of resin ducts in spruce wood. (A) Vertical duct with exuded resin droplets. (Reproduced from Ref. 39. Copyright 1982, American Chemical Society.) (B) Horizontal ducts contained in fusiform rays (FR) of the wood tangential surface.

species identification (2). (*See discussion on* "Vessel Elements: Architecture," page 34).

Hardwood fibers, because of the presence of vessels, occupy a proportionally smaller volume of wood tissue than softwood fibers

Major Types of Cells in Hardwoods	
Vertically Oriented	*Horizontally Oriented*
A. Fibers	Ray parenchyma
1. Libriform fibers	1. Procumbent cells
2. Fiber tracheids	2. Upright cells[a]
3. Vasicentric tracheids	Homocellular rays: 1 or 2
B. Axial parenchyma	Heterocellular rays: 1 and 2
C. Vessel elements	

[a] Upright cells are actually oriented parallel to the fiber axis (i.e., axially), but within a ray they are arranged in radial or horizontal lines (with or without procumbent cells).

do. The fibers themselves are also smaller, and average about one-half the width and one-third the length of softwood tracheids (2). The sculpturing of hardwood fiber walls differs in several respects from that of softwood fibers, and these details, in addition to structural information on softwood fibers, are discussed in the section on "Wood Cell Walls."

The parenchyma-cell content of hardwoods is, on the average, much greater than that of softwoods. This situation is a result of the wider rays (1–50 cells) and greater ray volume of hardwoods, and also the relatively high proportion of longitudinal parenchyma (2). Additionally, the rays are all parenchyma—no ray tracheids.

The volume ratio of vessels to fibers and fiber wall thickness are two important factors influencing the hardness and density of different hardwood species and the permeability of these woods to liquids and gases. Wood grain is also partly a function of these two parameters.

Research on the chemical nature of hardwood xylem has revealed that the walls of fibers and ray cells contain lignin of one type, *syringyl,* while vessel walls together with the surrounding middle lamella are rich in lignin of a second type, *guaiacyl,* which is the same type found in softwood xylem (9, 10). This basic chemical difference between hardwoods and softwoods could, in some instances, potentially influence phenomena that depend on the chemical nature or reactivity of wood tissue at the cellular level. Further details on the nature and distribution of wood chemical constituents are found in Chapter 2.

Wood Cell Walls

Fibers. ARCHITECTURE. The basic skeletal substance of the wood cell wall—cellulose—is, in the mature cell, aggregated into larger units of structure called *elementary fibrils* that, in turn, are aggregated to form threadlike entities known as *microfibrils.* The latter are readily observed in the electron microscope and are usually found together in still larger entities that could be labeled *macrofibrils,* although they are still microscopic (Figure 15). Considerable controversy still reigns over the size, morphology, or even existence of elementary fibrils [or subelementary fibrils in cambial tissue (11)] and the architecture of microfibrils (12–14). However, because at least some type of threadlike entity is readily distinguishable, we will refer to that element of structure as a microfibril and recognize its aggregation into macrofibrils. Microfibrils and macrofibrils combine to form sheets of wall substance, and ultimately these sheets or

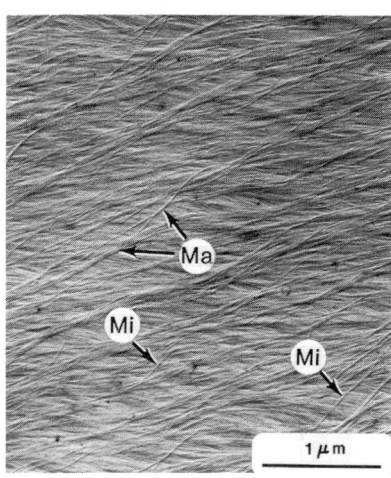

Figure 15. TEM of cellulose microfibrils (Mi) and macrofibrils (Ma) within the secondary fiber wall of balsam fir; direct carbon replica.

lamellae form relatively discrete wall layers. This architectural scheme is diagramed in Figure 16 for a typical, normal wood fiber.

As mentioned earlier, the initial portion of a fiber cell wall is manufactured in the cambial zone and is referred to as the *primary wall*. Here, cellulose microfibrils form a random, irregular, and interwoven network (Figure 17) to facilitate cell expansion during the enlargement phase of fiber development. In addition to cellulose, the primary wall contains a large proportion of matrix carbohydrates, particularly pectic materials and hemicelluloses (*see* Chapter 2). The combination of two adjacent primary walls and the interdisposed true middle lamella zone is collectively referred to as the *compound middle lamella*. Microscopically, it is difficult to separate wall substance here from the interfiber substance.

At the innermost region of the primary wall, microfibrils begin to exhibit a preferred orientation and show a tendency to align themselves about the fiber axis as a helix and to have a specific *microfibril angle* (measured as the angular displacement from the fiber axis). From this point inward to the fiber lumen, the wall is referred to as *secondary wall*. This portion of the wall is initiated at the end of the cell enlargement phase and is composed of many lamellae, each with a specific orientation. The lamellae are organized into distinguishable wall layers (Figure 18A).

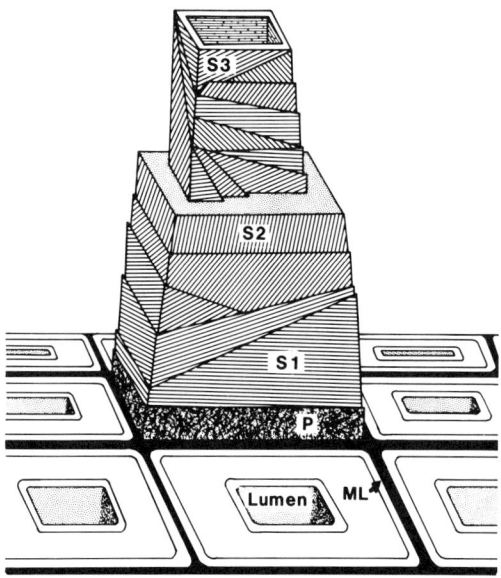

Figure 16. Schematic of what is widely considered to be (at least in principle) the general wall architecture of normal wood fibers. Key: ML, middle lamella; P, primary wall; and S_1, S_2, and S_3, layers of the secondary wall. (Adapted from Ref. 15.)

In normal wood tissue, the fiber secondary wall consists of three fairly distinct layers. The outermost layer or S_1 is very thin (0.1–0.2 μm) and exhibits an average microfibril angle (for the layer as a whole) of about 50–70° (2). The bulk of the secondary wall is made up of the S_2 layer, which is typically several micrometers thick (Figure 18). Here the microfibrils are usually oriented to the fiber axis at a relatively small angle (5–20°). The thickness and small microfibril angle

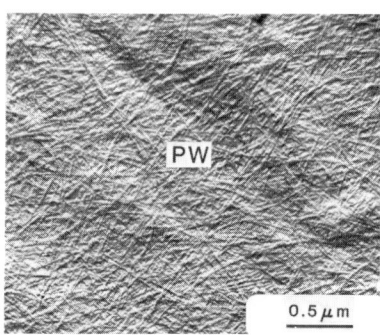

Figure 17. TEM of randomly arranged microfibrils in the fiber primary wall (PW) of balsam fir; direct carbon replica.

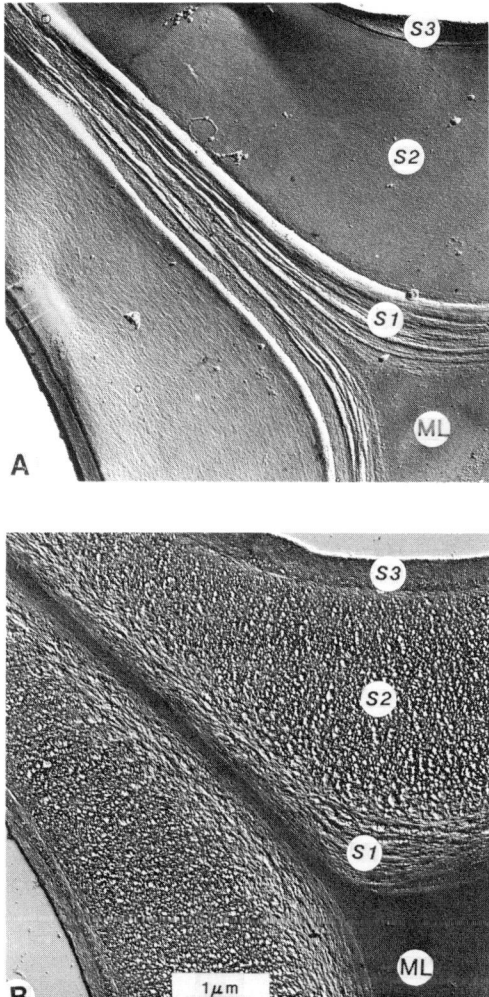

Figure 18. TEM of normal layering ($S_1/S_2/S_3$) in the fiber secondary wall of loblolly pine; transverse sections. (Reproduced from Ref. 41. Copyright 1971, Springer–Verlag.) Key: A, portion of two earlywood fibers (ML = middle lamella); and B, lignin skeleton of region similar to that shown in A.

of the S_2 contribute significantly to the high strength properties of wood parallel to grain (tensile and compression, see Chapter 5) (2).

The innermost layer of normal wood fibers, the S_3, is generally similar to the S_1, although it is perhaps a little thinner, and has sublamellae averaging 60–90° to the fiber axis (15).

Figure 16 shows that the various sublayers or lamellae in the S_1, S_2, and S_3 can exhibit left-handed (parallel to middle bar of an S) and/or right-handed (parallel to middle bar of a Z) helices. This particular variation in fiber-wall architecture has been investigated in very few wood species, and generalization to include all softwood and hardwood fibers may be a bit risky. However, available data imply that all normal wood fibers may be constructed in principle from a similar blueprint.

SCULPTURING. *Pits.* Softwood and hardwood fibers have closed ends, but a special wall feature facilitates movement of the tree's sap stream from one fiber to another, from fibers to vessel elements, and from fibers to ray cells. This special feature is a small opening or recess in the fiber secondary wall known technically as a *pit*.

The most obvious fiber pits are those that occur between contiguous softwood fibers. They are donut-shaped in face view with a circular ridge of wall material overarching and bordering the actual aperture in the wall. These pits are known technically as *interfiber bordered pits* and are located predominantly on radial fiber walls (Figure 19A). Actually, pits in contiguous cells usually occur in matched pairs; each of the two participating cells contributes one-half of a pit pair (Figure 19, B and C).

Within a growth ring, softwood interfiber pits are larger and more abundant in earlywood. In latewood they are fewer, smaller, and often appear slitlike in very thick-walled fibers (2). This same type of pit in hardwood fibers varies morphologically with the fiber type, changing from an obviously bordered pit in thin-walled cells to only a slitlike aperture in fibers with thick walls.

As all pits develop in softwoods and hardwoods, a specialized pit membrane remains within the pit complex (Figure 19, D and E). This membrane is initially constructed from the compound middle lamella in all cases, but in its fully differentiated state the membrane can differ considerably between various cell types, between softwoods and hardwoods, and to some extent even between different species (3). In hardwoods, pit membranes are observed to be thin and generally nonporous partitions of microfibrils, matrix materials, and lignin (Figure 20). Movement of liquids through the pit complex to an adjacent cell must therefore occur largely by diffusion rather than by free liquid translocation. Fortunately, hardwoods have an effective alternate mechanism for liquid movement, at least in the vertical direction, and that mechanism is the vessel system.

The interfiber-pit membranes in softwoods are substantially different from the pit membranes in hardwoods. Much of the membrane periphery is quite open, with only the central portion being nonpo-

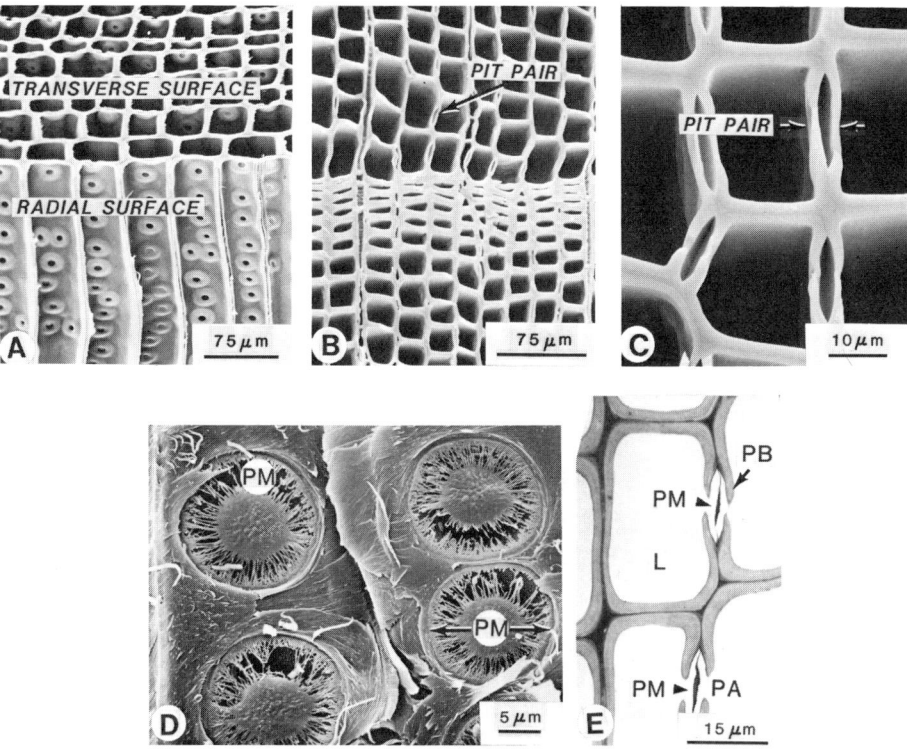

Figure 19. Anatomy of softwood interfiber pits. (Reproduced from Ref. 39. Copyright 1982, American Chemical Society.) (A) SEM of interfiber pits in earlywood as seen on the wood radial face. Note the donut-shaped borders. (B and C) SEM of pit pairs between adjacent fibers; cross-sectional surface. (D) SEM of bordered-pit membranes (PM) in face view of a split wood radial surface. (E) Light micrograph of pit pairs as seen in cross section with a light microscope. Key: PM, pit membranes; PB, pit border; and PA, pit aperture.

rous (Figure 21A). The exact morphology of the membrane is species-dependent (2).

In the standing, living tree the bordered-pit membranes between softwood fibers act as *valves* to prevent the spread of air or bubbles into sap-filled cells in the event of tree injury and potential rupture to vertical water columns. Unfortunately, they perform a similar function in the processing of wood into commercial products. For example, during wood drying, substantial capillary and surface tension forces are developed upon water retreat from the fiber lumens through the pits, and the membranes move effectively (particularly in earlywood) to seal the apertures in the direction of water

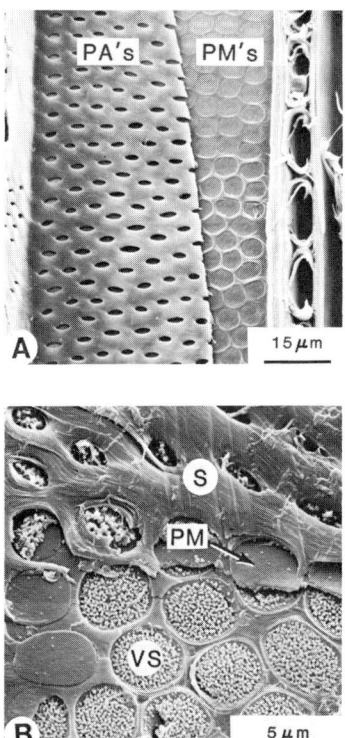

Figure 20. SEM of intervessel-pit membranes in hardwoods. (A) Vessel in western red alder. The cell wall at the lumen is partly torn away to reveal the nonporous nature of the pit membranes (PM) (PA = pit aperture). (B) High magnification of pits in Anthocephalus, a tropical species. The secondary wall (S) at the lumen has been removed to expose the nonporous pit membranes (PM) and a special structure known as vestures. The latter can be found in the pit complex of various hardwoods (2). (Reproduced with permission from Ref. 40. Copyright 1982, Technical Association of the Pulp and Paper Industry Press.)

movement (Figure 21, B–D). This condition, known as *pit aspiration*, greatly impedes the subsequent movement of fluids or gases through the wood tissue in question (2). Pit aspiration occurs naturally in conifers upon conversion of sapwood to heartwood (discussed later). However, aspiration is an almost unavoidable consequence of any circumstance that promotes wood drying, i.e., the creation of water/air interfaces within the wood structure.

Pits with some form of border also occur at irregular intervals along softwood fibers where the fibers contact ray cells (Figure 22). Such pits are very rare in hardwood fibers, but in softwoods they are abundant and conspicuous, especially in earlywood. These pits are known technically as ray *cross-field pits* (*see* Figure 11), and they

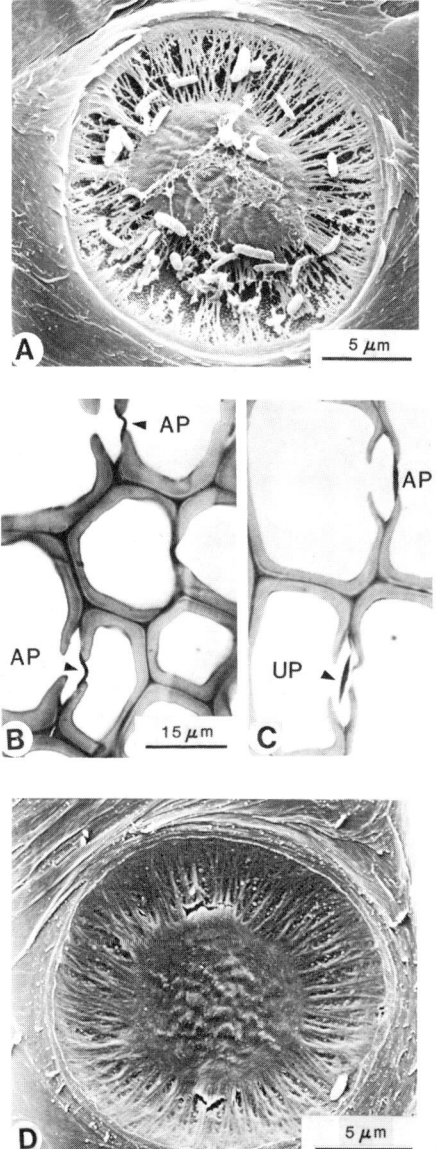

Figure 21. Softwood bordered-pit membranes of western hemlock. (Reproduced from Ref. 39. Copyright 1982, American Chemical Society.) (A) SEM of unaspirated pit in earlywood. Note porous periphery of the membrane. Rodlike bacteria are also present here, apparently filtered out onto the membrane during sample preparation; split wood radial surface. (B) Light micrograph of aspirated pits (AP) in latewood; cross section. (C) Light micrograph of aspirated pit and unaspirated pit (UP) in earlywood; cross section. (D) Fully aspirated pit in earlywood. Note the reduction in porosity upon aspiration. (Compare to A above.)

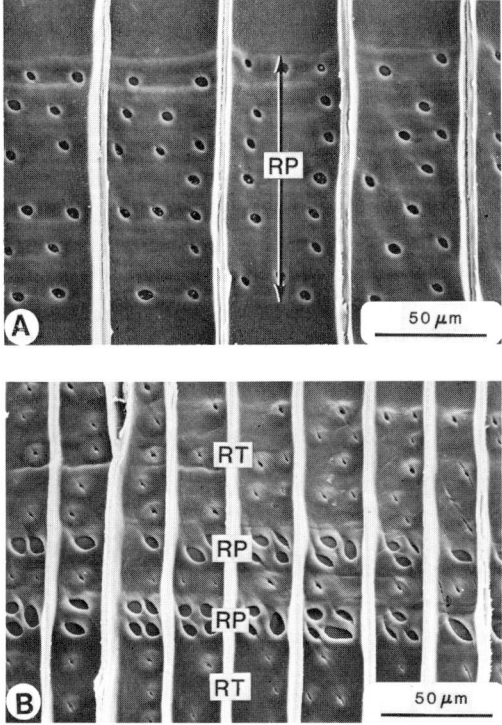

Figure 22. SEM of ray cross-field pits in softwoods as seen on the wood radial surface. Key: A, pits to ray parenchyma (RP) in western white fir; and B, pits to ray parenchyma and ray tracheids (RT) in lodgepole pine.

provide much of the information needed for identification of softwoods (2).

Membranes within ray cross-field pits are essentially nonporous partitions of *compound middle lamella,* although the membrane architecture varies to some extent, depending on whether fiber contact is made to ray parenchyma or ray tracheids. In either case, cross-field pits do provide a path of communication between fibers and rays that is at least somewhat more easily traversed than one requiring passage of materials through entire cell walls. Radial transport of liquids is facilitated by diffusion through simple pits (i.e., without borders) located in the side and end walls of contiguous parenchyma cells and by small bordered pits between contiguous ray tracheids (2).

Spiral Thickenings. The fibers of a very few species of softwoods and hardwoods are lined with helically oriented ridges of wall material (Figure 23). These ridges, which can be ropelike in certain woods, are associated with, and are an integral part of, the S_3 wall

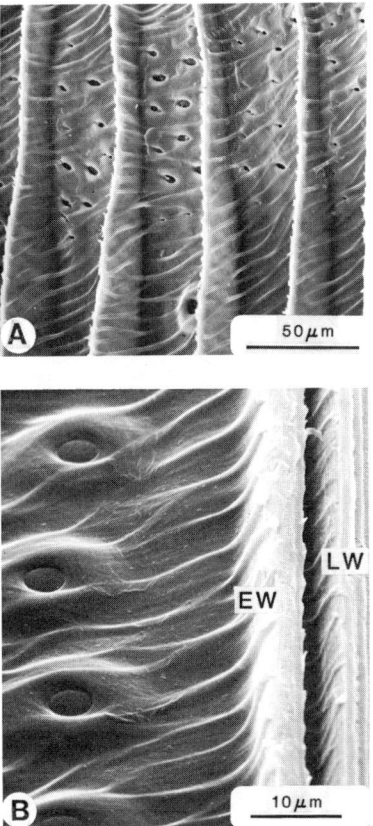

Figure 23. SEM of spiral thickenings in the fibers of Douglas-fir; wood radial surfaces. Key: A, spirals in the vicinity of ray cross-field pits in earlywood; and B, high magnification of spirals in the last latewood fiber of one year and the first earlywood fiber of the next year. Pits shown in B are interfiber-bordered pits.

layer (2, 16). The exact morphology of such spirals varies with the species in question, but in all cases they probably do not have a detectable influence on wall physics or chemical reactivity. They are, however, a feature of diagnostic value. On this basis, Douglas-fir is readily distinguished from other commercial, domestic softwood timbers.

Warts. Warts are conelike or droplike protuberances, sometimes found covered with an amorphous deposition, that are scattered in a random pattern on the inner fiber-wall surface in most softwoods and the fibers of some hardwood species (Figure 24). The wart structure, known collectively as the *warty layer* (17), is manufactured by the living cell protoplast before cell autolysis (18, 19). The warty layer is ligninlike in nature but has no apparent physiological role; it prob-

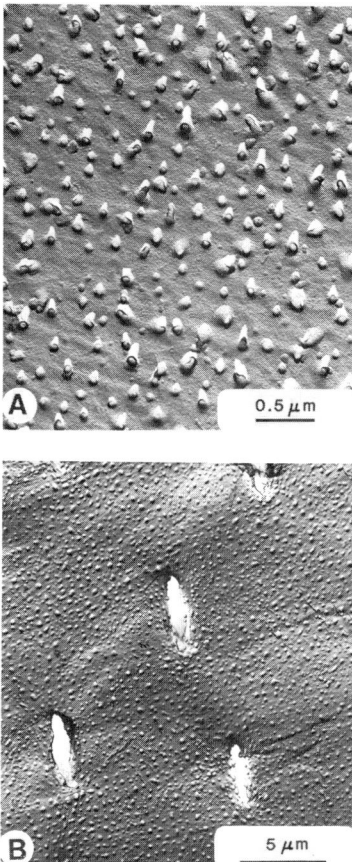

Figure 24. TEM of warty layer in softwood (A) and hardwood (B) fibers; direct carbon replicas. (A) Lumen surface of fiber in balsam fir. Note the amorphous substance masking the S_3. (Reproduced with permission from Ref. 20. Copyright 1974, Society of Wood Science and Technology.) (B) Lumen surface of a warty fiber in sycamore. The slits shown here are pit apertures. (Reproduced with permission from Ref. 23. Copyright 1974, Springer–Verlag.)

ably also has little or no effect on wood physical behavior. However, lignin in the warty layer does seem to be more highly condensed (or bonded three-dimensionally) than the lignin in the rest of the fiber wall (20, 21), and perhaps this situation could influence wall chemical reactivity or penetrability at the cellular level.

Vessel Elements. ARCHITECTURE. The cell wall of vessel elements appears to be constructed along the same general scheme as wood fibers. However, the layering is generally more complicated, and the presence in many species of numerous intervessel bordered pits causes marked deviations in the wall's internal microfibrillar arrangement (22).

SCULPTURING. *Pits.* The regions on vessel elements where contact is made with adjacent vessels, fibers, and parenchyma are distinctly pitted (Figure 25). The arrangement, size, and shape of these pits are often species-dependent and are thus valuable in wood identification efforts (2). However, hardwood pits are not an open route for rapid intercellular transport (Figure 20); consequently, there is a need for vessel element perforations as a more effective route for fluid translocation.

Perforations. The open areas at the ends of vessel elements are called *perforation plates.* The form of these end-wall regions varies between species and sometimes between earlywood and late-

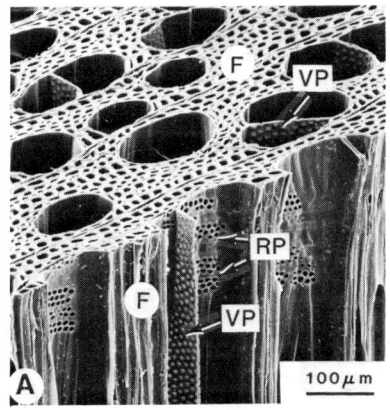

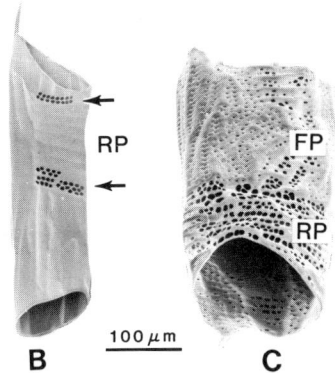

Figure 25. SEM of vessel pitting in hardwoods. (A) Vessel/vessel pitting (VP) and vessel/ray parenchyma pitting (RP) as seen from a transverse/ radial perspective in cottonwood (F = fibers). (B) Individual vessel element of cottonwood isolated by chemical pulping. (C) Isolated earlywood vessel element of white oak (FP = fiber/vessel pitting).

wood of the same species (2). However, in all cases the perforation plates allow the free flow of liquids or gases between vertically contiguous vessel elements along the stem axis. Major types of perforation plates are illustrated in Figure 26. Their variability among different wood species is beneficial in efforts to identify a particular wood source.

Spiral Thickenings. These structures are common in the vessels of many hardwoods (e.g., maple, cherry, basswood, buckeye, southern magnolia, and madrone). The particular form is valuable to the wood anatomist for species identification, but the presence or absence of spirals has no apparent effect on wood behavior (2).

Warts. The presence of a vessel element warty layer is species-dependent, and where it is found, it tends to be associated more with the occurrence of scalariform-type perforation plates (1, 23). As in the case of fibers, vessel warts have no obvious role at the cellular

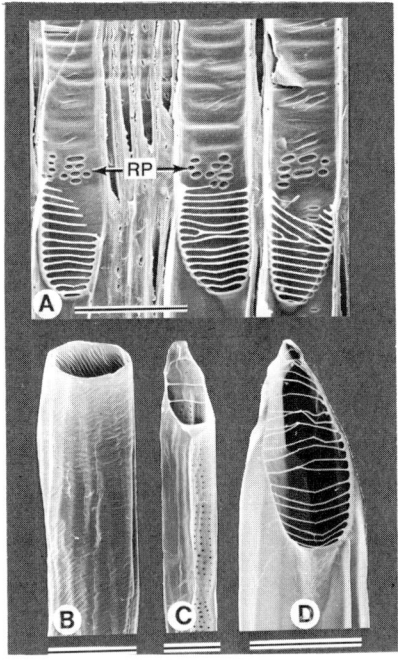

Figure 26. SEM of major types of perforation plates in hardwood vessel elements. (A) Wood radial section of redgum showing scalariform (ladderlike) perforation plates in three vessel elements (RP = ray cross-field pits). (B) Portion of a maple vessel element containing a simple perforation plate; chemically pulped wood. (C and D) Portions of vessel elements from yellow poplar (C) and paper birch (D) showing scalariform plates with very few bars and with numerous bars; chemically pulped wood.

level and are likely to have no decipherable influence on wood properties.

Other Cells. The wall architecture/sculpturing of ray tracheids in softwoods and of the ray and axial parenchyma cells in both softwoods and hardwoods is discussed at length in other texts (2 and references cited therein). Details showing variability include pit type, wall layering, degree of lignification, and wall sculpturing peculiar to a few species (e.g., dentations in ray tracheids of the hard pines). Although these details are of botanical and academic interest, their exact form and variability are of little relevance to the remainder of this text and the major aspects of wood chemistry and wood behavior.

Wood Physical Characteristics

The weight and strength properties of wood, together with the behavior of wood in response to weather, chemical treatment, fire, or microbial organisms, are influenced greatly by the wood's water content and the mass of wood tissue per unit volume (its density). The physical characteristics of wood tissue at given levels of moisture and density will be given specific attention in Chapter 3. However, some basic aspects of wood physics are included here to provide a more comprehensive introduction to the technical nature of solid wood.

Hygroscopicity. The chemical nature of wood substance, particularly that of the polysaccharides, renders wood cell walls hygroscopic (or hydrophilic). The hydroxyl groups on the cellulose and hemicellulose molecules are responsible for this great affinity for water and have a very strong propensity to form hydrogen bonds (2). Lignin, on the other hand, possesses comparatively few free hydroxyls, and as a result is much less hygroscopic. In fact, for all practical purposes, lignin is generally considered to be essentially hydrophobic (or lipophilic).

As a consequence of its hydrophilicity, wood tissue will seek to maintain, through either gain or loss of moisture, an equilibrium moisture content with the surrounding atmosphere. If the wood takes on water, the cell walls proceed to swell until the cell walls become water-saturated. The latter moisture content is called the wood's fiber saturation point. In contrast, loss of wood water (below the fiber saturation point), due to diffusion and evaporation, results in wood shrinkage.

The water content of the wood cell wall has a strong influence on the wood's mechanical properties, and a higher moisture content, at least below the fiber saturation point, and normally is inversely related to most strength properties (*see* Chapter 5). This situation is easily reconciled if one considers that the takeup of water below the fiber saturation point results in a pushing apart of polysaccharide molecules as water is imbibed, and that the drying of wet wood, if resulting in wood shrinkage, promotes coalescence and bonding of the wood's internal architecture or ultrastructure.

Other ways in which the presence of water influences wood behavior will be discussed later.

Density and Specific Gravity. DEFINITION AND CALCULATION. A wood property that furnishes one of the most useful indices to the predicted behavior and treatability of wood is wood's *bulk density*—the mass or weight of wood substance per unit volume, usually expressed as grams per cubic centimeter or kilograms per cubic meter (2, 24). Unfortunately, as defined above, bulk wood density has two major problems associated with its measurement. The first is the constant question: at exactly which moisture content should the weight and volume of the sample be determined? This is always a consideration because the volume (at least below the fiber saturation point) and weight of wood tissue change with its moisture content. These measurements can be made at a given moisture content, but accurately achieving and maintaining a particular moisture content is not straightforward and certainly not convenient.

The second problem associated with the determination of wood density is strictly one of logistics; that is, exactly how should the necessary measurements be made, particularly that of volume? If all wood samples of interest could be easily dressed to perfect geometrical shapes, volume measurement would not be a problem. However, this is not the case, and density information is often desired for large specimens, irregularly shaped specimens, or small samples of earlywood or latewood of a single growth increment.

To circumvent these two problems, the wood technologist uses a special or artificial parameter known as *basic density*. It is an artificial parameter because the weight and volume measurements required for its calculation are made on extremely different wood conditions—the completely dry [or oven-dry (OD)] state for weight measurements and the completely wet (or water-saturated) state for volume measurements. In this way, even though wood changes in weight and volume with changes in moisture content, measurements of weight and volume of a given sample are possible at conditions that are as nearly constant and reproducible as can be obtained with wood tissue (2). An additional advantage of using water-swollen or green wood volume is that the simple technique of water displacement can be used to measure the volume of large, small, and/or irregularly shaped samples. This convenience, together with the fact that 1 cm^3 (or 1 mL) of water weighs 1 g, permits the conversion of basic density expressed as OD wt/green vol to an equivalent fraction expressed as OD wt/wt water displaced. The latter fraction is a pure number (i.e., without units), and as such, is equivalent to a related term—*specific gravity*.

The specific gravity of a material is defined as the ratio of the density of the material to the density of water (2). If this concept is broadened to incorporate the parameter of *basic density*, another term is obtained: *basic* specific gravity, which is defined as:

$$\text{Basic specific gravity} = \frac{\text{basic density}}{\text{density of water}} = \frac{\frac{\text{OD wt}}{\text{green vol}}}{1} = \frac{\text{OD wt}}{\text{wt displaced water}}$$

Consequently, the terms basic density and basic specific gravity give the same information, and they are different only in the fundamental sense that basic specific gravity is a pure number and basic density is not. The choice of one term over the other for descriptions of wood quality is a matter of preference and varies with particular authors or investigators. However, the assignment of units to these terms is appropriate only for basic density.

Available techniques make it possible to measure the true density of wood (at a given moisture content). These include special procedures to obtain dry-wood volumes (25), as well as special methods using β-ray or X-ray technology (26). However, these procedures are not easily applied in a routine fashion, on a large scale, or on large wood specimens.

Any procedure for wood specific gravity or density based on measuring weight and volume can be considered accurate only if the wood sample has been first extracted with suitable organic solvents to remove extraneous resins, oils, fats, gums, etc. (27). These materials bulk cell walls, block potential sites for water adsorption/absorption, alter potential wood swelling/shrinkage, and thereby interfere with the accurate characterization of wood tissue.

SIGNIFICANCE. The basic density of wood varies with cell size, cell wall thickness, and the volume proportion of cells of a given type. It affects wood shrinkage and swelling, machinability, surface texture and microsmoothness, gluability, penetrability of fluids and gases, and in other respects, governs the degradation of wood by chemicals, fire, and microorganisms. In particular, the strength of wood and its stiffness closely parallel changes in the basic density. Basic density is not an independent predictor of wood physical performance because total wood behavior is profoundly affected by its moisture content (2). Nevertheless, it may be possible to learn more about the nature of a given wood sample by determining its basic density than by any other single measurement (24).

In softwoods, basic density is strongly related to the volume proportion of latewood and its average fiber wall thickness. However, hardwood basic density depends not only on fiber wall thickness but also involves the volume ratio of fibers to vessels. Native commercial woods fall mostly in the basic density range of 0.35–0.65 g/cm^3, although native species can be as low as 0.21 g/cm^3 (corkwood) and as high as 1.04 g/cm^3 (black ironwood) (2).

Woods with basic density values (means for the species) that fall in the range of <0.36, 0.36–0.50, and >0.50 g/cm^3 are considered light, moderately light to moderately heavy, and heavy, respectively, and include both temperate and tropical woods (2). However, for a given species, there is considerable variability about any published and accepted mean. Specifically, at least for most North American woods, the expected coefficient of variation (i.e., standard deviation divided by the mean) is about 10% (24). Thus, if the 95% probability level is to be considered, a reasonable estimate of the total expected range of variability would be the mean basic density ± (10% × 1.96 × mean basic density). Table I presents the ranges of basic density that might be anticipated for several important U.S. woods.

At the cellular level, the true density of dry cell wall substance (i.e., within the cell wall) has been determined to be about 1.5 g/cm^3, varying to some extent with the method of measurement and species (2). There are voids within the dry wood cell wall, but the void volume here (i.e., micropores) is reported to be only about 2–4%. However, this figure would be expected to increase as wood moisture content is increased to the fiber saturation point (28).

The importance of wood moisture and basic density in determining wood behavior will become more evident in subsequent chapters. Suffice it here to say that variation in the amount of cell wall substance at a given moisture content that must be traversed by a penetrating liquid or chemical, microbe, etc., can determine the rate of reaction as well as the extent of reaction or the change in the character of the wood in question.

Wood Variability

Causes. Different specimens of wood even from the same tree are never identical and are similar only within broad limits (2). Within the larger categories of softwoods and hardwoods, such variability is extended to different trees of the same species and to different genera and families. All of this variability occurs naturally and is the combined result of tree genetics, the environment, and the age of the vascular cambium (i.e., tree age). Observed consequences are changes in the type, number, and form of wood cells. Additionally, such changes are not infrequently accompanied by varying wood chemistry and cell wall ultrastructure.

Table I. Basic Density (BD) of Some Important U.S. Woods and the Expected Range of Variability

Species	Mean BD[a] (g/cm^3)	Expected Range[b]
Softwoods		
Slash pine	0.54	0.43–0.65
Longleaf pine	0.54	0.43–0.65
Loblolly pine	0.47	0.38–0.56
Douglas-fir	0.45	0.36–0.54
Western hemlock	0.42	0.34–0.50
Ponderosa pine	0.38	0.31–0.45
White fir	0.37	0.30–0.44
W. white pine	0.35	0.28–0.42
Hardwoods		
Shagbark hickory	0.64	0.51–0.77
N. red oak	0.56	0.45–0.67
Sugar maple	0.56	0.45–0.67
White ash	0.55	0.44–0.66
S. red oak	0.52	0.42–0.62
Redgum	0.46	0.37–0.55
Blackgum	0.46	0.37–0.55
Yellow poplar	0.42	0.34–0.50
W. red alder	0.37	0.30–0.44
Quaking aspen	0.35	0.28–0.42

[a] Based on oven-dry weight and green volume (g/cm^3).
[b] Assumes a coefficient of variation of 10% (see Ref. 24). Additional information was calculated from data in Ref. 2.

Superimposed on natural wood variability is that variability induced by the professional forester, who is charged with the task of *silviculture*—the science of producing and maintaining a forest. The practices of scientific forest management are usually aimed at increasing tree growth rate and wood production, increasing the average length of the branch-free bole, and simultaneously maintaining a suitable, if not improved, level of wood quality. Well-established procedures include, but are not limited to, regulation of the number of stems per acre (stand density), pruning the lower branches of individual stems, fertilization, and irrigation (8). As in the case of natural tree variability, imposed treatments cause changes in wood physical characteristics through alteration of the form and number of various cell types. Wood chemistry may also be affected by such treatments.

Within a given tree, certain aspects of wood variability are related to the physical position of wood tissue in the bole. One major aspect of wood variability is found along the tree radius; wood in the center of the tree is different from the wood much further out. This situation then leads directly to a second major aspect. That is, because wood is laid onto the developing stem in the form of inverted hollow cones (*see* Figure 8) variation in the radial direction is responsible for a gradual change in wood characteristics with increasing tree height or position up the tree bole.

Changes in wood properties resulting from natural or induced tree variability often include a change in wood basic density. Associated changes in wood anatomy include altered growth rates (ring width or number of rings per inch), fiber wall thickness, earlywood to latewood transition, percent latewood, and vessel:fiber ratio. The ensuing changes in behavior or reactivity of wood in the affected trees will vary accordingly, following the same trends for changing basic density discussed previously.

Major Types. HEARTWOOD AND SAPWOOD. When a tree is young (a relative term, however), upward conduction of sap in the xylem is possible through the entire cross-sectional plane. (Upward conduction may not always involve the entire bole cross section, which is dependent on tree type, but at least it is possible.) Additionally, in the four softwood genera mentioned earlier, resin canals are normally functional, and in all trees, most storage parenchyma are still living. At this point, the xylem is known appropriately as *sapwood*, which is normally light-colored (Figure 27). As trees grow older, the central portion of most trees is eventually altered, primarily chemically but also structurally to some extent, to the point where it must be distinguished from normal sapwood. In its altered state this central region of the tree bole, variable in size and other characteristics, is termed *heartwood* and continues to enlarge with time (2).

The characteristics of heartwood that serve to distinguish it from normal sapwood are the following:

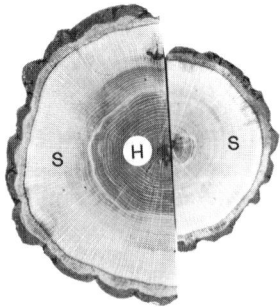

Figure 27. Heartwood distribution (H) as seen in tree cross sections from two blackjack oaks. Note that the heartwood zone increases with tree age. (Photo by W. J. McCleary.)

1. The entire region is normally infiltrated and incrusted with organic extractives that are largely polyphenolic in nature. These materials are derived from the conversion of starch, sugars, and organic extractives present in the sapwood parenchyma. Upon heartwood formation, the parenchyma cells usually die, and their contents then proceed to infiltrate cell walls, incrust pit membranes (Figure 28), and can even plug vessels in the affected areas (2).
2. The moisture content of heartwood in softwood trees is reduced to a level much lower than that of normal sapwood (2, 24). During the moisture reduction period, the membranes of bordered pits in sapwood fibers have a strong tendency to become aspirated. This situation, together with that of pit membrane incrustation, greatly reduces the natural permeability of heartwood tissue to liquids and gases.
3. The vessels in many hardwood species (e.g., white oaks and hickories) become plugged with bubblelike intrusions from adjacent parenchyma cells (Figure 29). These structures, termed *tyloses* (plural), invade the vessel lumen through vessel element/parenchyma pits and represent eruptive outgrowths of the parenchyma's living contents (29). Tyloses greatly reduce wood permeability.
4. Heartwood may take on a distinctive color, including shades of brown, yellow, orange, or red (e.g., southern pine, Douglas-fir, redwood, oak, black walnut, and

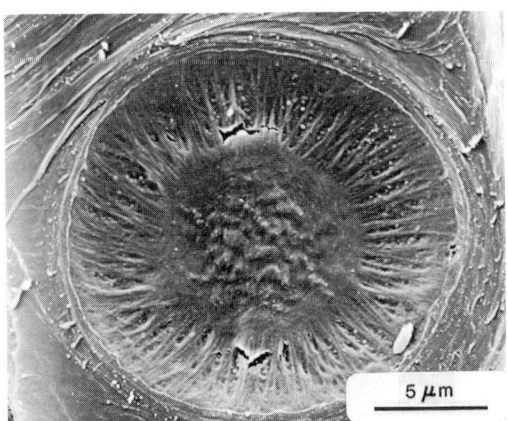

Figure 28. SEM of an incrusted, aspirated-pit membrane in western hemlock heartwood. (Reproduced from Ref. 39. Copyright 1982, American Chemical Society.)

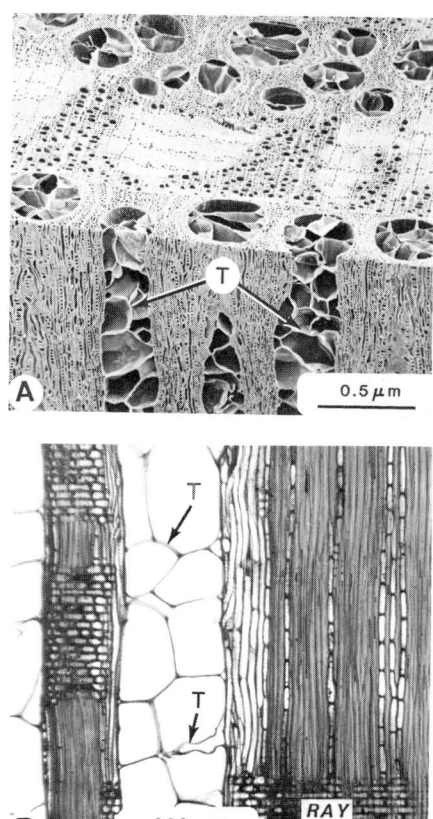

Figure 29. Tyloses in white oak heartwood. (A) SEM of transverse/tangential view showing vessels plugged with numerous tyloses (T). (Reproduced from Ref. 39. Copyright 1982, American Chemical Society); (B) Light micrograph of radial section of earlywood showing a plugged vessel.

black cherry). However, in other species, this zone is not noticeably changed in color (e.g., spruce, fir, hemlock, aspen, and cottonwood) (2). Nevertheless, the heartwood even in these species still contains an abnormally high extractives content as well as exhibiting the physical alterations described in characteristics 1–3. Just the extractives themselves, whether colored or not, can have an effect on wood reactivity or treatment processes such as finishing, preservation, gluing, and the production of polymer/wood composites.

5. The heartwood extractives in certain species (e.g., redwood and cedars) are toxic, to some extent, and render the wood more resistant to decay microorganisms and/

or insects (2). However, wood durability is due to the specific toxicity and amount of the extractives, rather than to heartwood color per se.

REACTION WOOD. When a tree stem or branch is brought out of its normal equilibrium position in space by an outside force, an accelerated radial growth is promoted on either the lower or upper side of the stem or branch in question. Such growth is initiated by the affected tree part as a means to reachieve its natural or equilibrium position. In doing so, the stem or branch may develop eccentric or elliptical growth rings (as seen in cross section), and the widest side of the eccentricity is composed of cells altered both structurally and chemically from normal wood, adjacent or side wood, and opposite wood. The atypical cells or tissues generated in the regions of accelerated growth are known collectively as *reaction wood* because the tissues reflect the tree's response or reaction to an alteration in its usual environment (30). Gravitational and hormonal stimuli are involved, but the precise physiology governing reaction wood formation remains to be clarified.

Reaction wood in softwoods is normally concentrated on the underside or lower side of the affected tree or branch. Because the wood in such regions appears to be subjected to compressional forces, this wood is given the nontechnical designation of *compression wood* (Figure 30). In hardwoods the location and concentration of reaction wood tissue is normally opposite that in softwoods, i.e., on the upper side of branches and leaning stems, and the tissue is called *tension wood* (Figure 31). Reaction wood in softwoods and hardwoods is also located in the vicinity of points on any tree bole where branches originate.

Although concentrated in and around branches and in obviously leaning tree boles, reaction wood zones can be frequently scattered

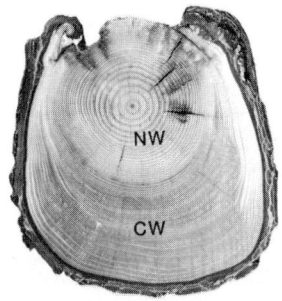

Figure 30. Compression wood in the stem cross section of a severely leaning Douglas-fir. Note the growth ring eccentricity (NW = normal wood). (Reproduced from Ref. 39. Copyright 1982, American Chemical Society.)

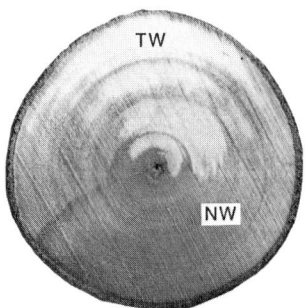

Figure 31. Cross-sectional surface of a young quaking aspen stem. Note the white arcs of tension wood (TW) (NW = normal wood). (Reproduced from Ref. 39. Copyright 1982, American Chemical Society.)

throughout any stem cross-section in short arcs, reflecting the variation in stresses imposed on a tree with time.

COMPRESSION WOOD. The formation, structure, and chemistry of compression wood have been given considerable attention (*see* Reference 30 *and* references cited therein). Compression wood tissue exhibits both major and subtle differences from normal softwood xylem. These differences (Table II) are either directly or indirectly responsible for any distinguishable changes in the behavior and treatability of compression wood versus the behavior generally noted for normal wood.

Perhaps the most influential properties of compression wood are its high lignin and low cellulose contents, its thick and ultrastructurally modified fiber walls, and the reduction of total middle lamella substance (and intercellular adhesion) due to rounded fiber outlines and intercellular spaces (Figure 32). Attendant wood features include (usually) higher-than-normal basic density and wood hardness, extremely high longitudinal shrinkage, lower moisture holding capacity, brashness (abruptness) in mechanical fracture, and altered strength properties (usually lower than normal).

In severely leaning softwood trees, compression wood is formed in very large proportions and mainly unilaterally, giving rise to the aforementioned eccentric growth rings. In relatively straight or erect trees, compression wood is found in smaller, scattered arcs or lunes, particularly near the pith or tree center. Its dark red color in domestic softwoods readily distinguishes compression wood along the tree bole and at the base of all branches and around knots. It also appears to be a subtle but constant feature of young, rapidly growing conifers and may be promoted in young or older trees remaining after heavy thinning of a crowded forest stand (*2, 31*).

TENSION WOOD. Reaction wood is not found as consistently in

hardwoods as in softwoods. That is, tension wood is not detected in some species, at least in a definitive manner, yet it can manifest itself differently in different species, within the same tree, or even within the same growth ring (32, 33). It may consist of large bands similar to those of compression wood, but tension wood is commonly distributed (particularly in fast-growing, young, and/or relatively straight trees) in a more diffuse pattern of smaller areas (33). In some of these latter cases, it may be so diffuse as to be detectable only by microscopical examination (34).

The salient features of tension wood are summarized in Table II. Most notable are the increased volume of fibers, the high cellulose content, low lignin content, and the special wall architecture of tension wood fibers.

Depending on species, growth rate, and severity of tree lean, tension wood fibers can exhibit one of several forms. In most cases the formation of tension wood involves the deposition of a loosely attached layer at (normally) the fiber lumen (Figure 33), a layer that is about 98% cellulose and whose microfibrils are aligned essentially parallel to the fiber axis. As seen in unembedded, microtomed wood cross sections stained for light microscopy, this special wall layer is almost always distorted or convoluted and separated from the rest of the secondary wall to give the impression of being soft or perhaps gelatinous (Figure 33). Based on these observations, tension wood fibers have been referred to as *gelatinous fibers* or simply *G-fibers* (meaning they contain a gelatinous or G-layer) and have been used as a means by which to define the existence and location of tension wood. In some species the G-layers are missing in the tension wood zones, but such zones still tend to have reduced lignin contents (2, 35).

The extra cellulose content of tension wood tissue is most commonly due to the presence of fiber G-layers. However, the layers themselves are not really gelatinous. On the contrary, they are quite highly crystalline, and this fact, together with the axial orientation of their microfibrils, renders this layer easily distorted in the horizontal plane (i.e., normal to the fiber axis).

The other significant structural feature of tension wood fibers is the nature of the rest of the secondary wall, which may lack an S_3 or S_3 and S_2 (36) (Figure 34).

The reduced vessel volume of tension wood, together with thickened fiber walls, can lead to a higher than normal basic density. This general situation, coupled with a difference in wood chemistry, could cause a variable response of such tissue to both chemical and physical treatments or to microbial degradation when compared to normal hardwood xylem.

Although much is known about reaction wood, the specific effects of its characteristics on such processes as wood preservation,

Table II. Major Structural and Chemical Characteristics of Reaction Wood

Characteristics	Compression Wood	Tension Wood
Macroscopic features	1. More prevalent in middle or at end of growth increments. 2. Abrupt earlywood/latewood transitions become more gradual, and vice versa. 3. Tissue is reddish color and darker than normal wood (in northern hemisphere).	1. Develops more readily in earlywood. 2. If concentrated into large areas, is whitish in color compared to normal wood. 3. Wooly surface of boards sawn while green, due to fiber pull-out.
Anatomical features	1. Fiber cross-sectional outlines tend to be rounded, particularly in latewood areas. 2. Intercellular spaces between fibers and between fibers and ray cells are common, varying with compression wood severity. 3. The total amount of middle lamella is less than in normal wood. 4. Interfiber pits are fewer and slitlike.	1. Volume ratio of fibers to vessels is increased and vessels may be smaller in diameter. 2. May occur as large bands, small and diffusely scattered areas, and/or as individual fibers or fiber clusters detectable only with a microscope.

Ultrastructure	1. S_3 layer is missing; S_1 layer is considerably thicker than normal.
2. S_2 layer in vast majority of species contains helical cavities in the same direction as S_2 microfibrils.
3. The S_2 orientation is commonly 30–50°, greater than in normal wood.
4. The warty layer is still present.	1. Tension wood fiber layering varies with species and tension wood severity, and may consist of $S_1 + S_2 + S_3 + G$, $S_1 + S_2 + G$, or $S_1 + G$, where G is the gelatinous layer.
2. G-layers are oriented essentially parallel to the fiber axis.
3. G-fibers may contain minute wall dislocations or axial compression failures.
Chemistry	1. Very heavily lignified (about 30–40% more than normal), but middle lamella contains less lignin than in normal wood tissue.
2. Lignin itself is more highly condensed than in normal wood.
3. Cellulose content is 20–25% less than in normal wood and is less crystalline.
4. Less galactoglucomannan and more galactan than normal; other polysaccharides are similar to those in normal wood.
5. Possibly higher extractives content, but this is species-dependent.	1. As a percentage of total wood, tension wood has more cellulose, less lignin, and fewer xylose residues than normal wood.
2. Fiber secondary wall exterior to the G-layer is just as lignified, if not more so, than in normal wood fibers.
3. Amounts of stored sugars, including starch, are lower than for normal wood tissue.

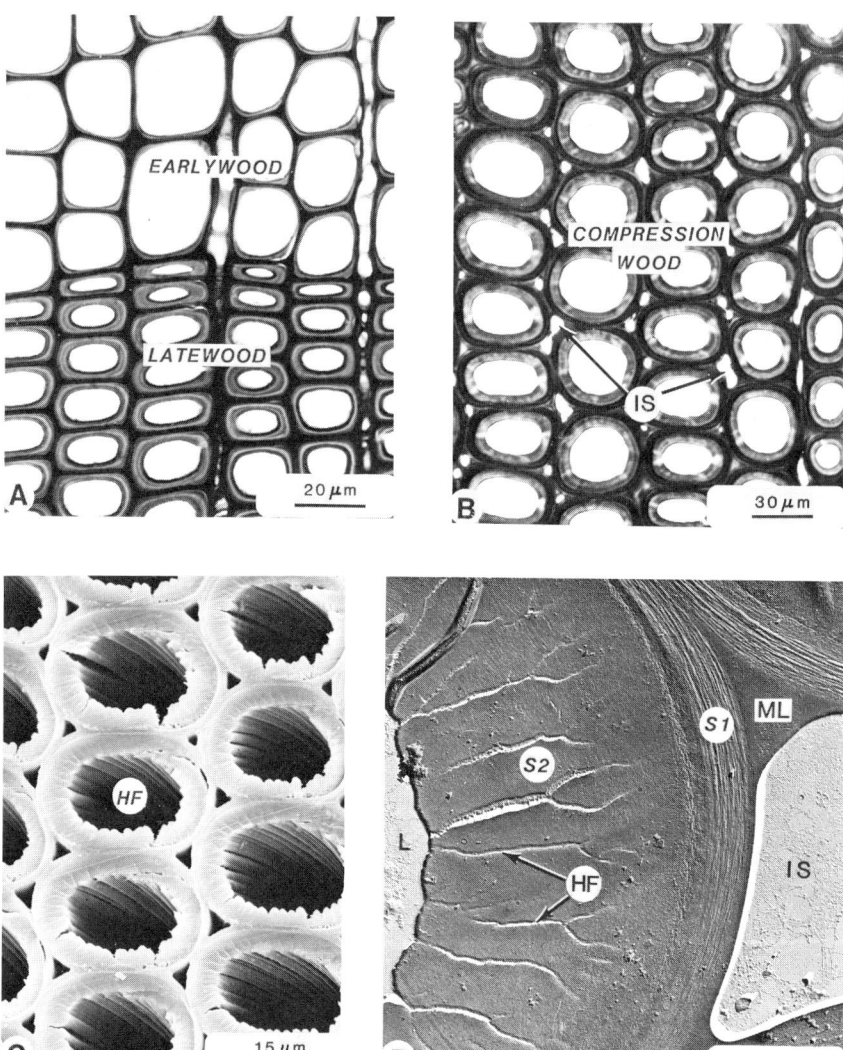

Figure 32. Cross-sectional views of normal and compression wood. (A) Light micrograph of normal wood of loblolly pine at a growth ring boundary. Key: EW, earlywood; and LW, latewood. (Reproduced with permission from Ref. 41. Copyright 1971, Springer–Verlag.) (B) Light micrograph of compression wood of loblolly pine. Note the rounded fiber outlines and the presence of intercellular spaces (IS). (Reproduced with permission from Ref. 41. Copyright 1971, Springer–Verlag.) (C) SEM of Douglas-fir compression wood. Note the helical fissures (HF) extending radially within the fiber wall. (Reproduced from Ref. 39. Copyright 1982, American Chemical Society.) (D) High magnification (TEM) of a compression wood fiber in loblolly pine. Note the absence of an S_3 layer. The helical fissures here terminate in the outer region of the S_2 layer. Key: ML, middle lamella; IS, intercellular space; and L, lumen.

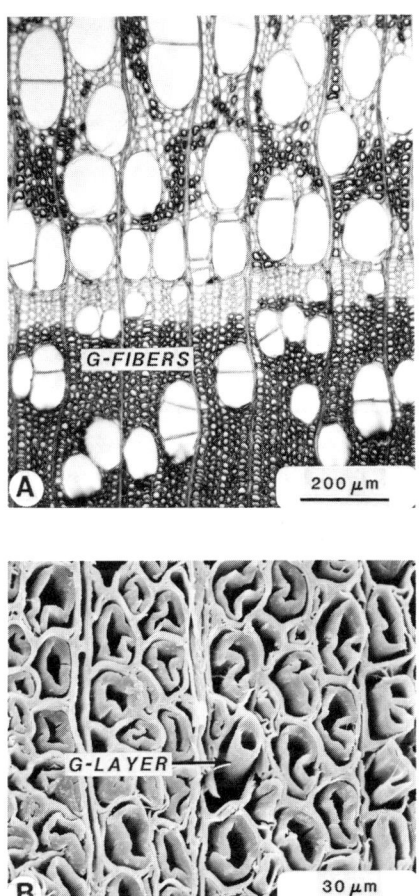

Figure 33. Cross-sectional views of tension wood in a young quaking aspen stem. (Reproduced from Ref. 39. Copyright 1982, American Chemical Society.) (A) Light micrograph of a section that was selectively stained to differentiate the gelatinous layers in G-fibers. (B) SEM of a surface of tension wood fiber zone. The G-layers, which are loosely attached to the rest of the fiber wall, were dislodged during specimen preparation and drying.

finishing, gluing, or its resistance to fire, weathering, or decay are not clearly documented yet. However, some trends in the general behavior of reaction wood can be predicted from a broad knowledge of the natural variability and behavior of normal wood. One fact is clear; the occurrence of reaction wood is a major contribution to xylem nonuniformity, a factor that always tends to deter the accurate prediction of wood quality.

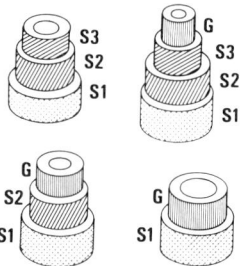

Figure 34. Possible types of tension wood fibers found among different hardwood species. (Adapted from Ref. 37.)

JUVENILE AND MATURE WOOD. The wood produced in young trees, the wood formed during the early or juvenile years of older trees (including the tree top and entire tree core), and most branches differ in several respects from the wood formed in the outer trunk of the same tree when the tree is more mature (Figure 35). These dissimilarities include cell size, wall thickness and microfibril orientations, varying proportions of particular cell types, growth rate and ring width, knot volume, in situ moisture content (varying inversely with percent heartwood), and wood chemistry. The differences are responsible for the often somewhat peculiar physical behavior and chemical nature of juvenile wood including its basic density, bending strength, elasticity, and shrinkage (2).

A complete explanation of why wood formed near the pith during the early life of a tree, also called *corewood*, is not the exact type

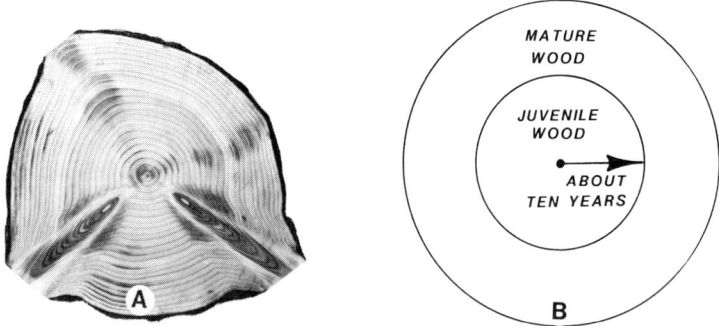

Figure 35. The general location of juvenile wood and knotwood in tree stems. (A) Cross-section of a Sitka spruce showing portions of two branch traces (knots). (Reproduced from Ref. 39. Copyright 1982, American Chemical Society.) (B) Schematic indicating the typical location of juvenile wood or corewood in softwood trees.

produced when the tree is more mature, also called *outerwood*, remains to be specified. However, the situation appears to be related in some ways to the sexual maturity of the tree or species in question, maturity being when the tree is physiologically capable of producing flowers, fruits, and seeds. Additionally, the proximity of the vascular cambium to the tree crown (leaves and buds) also seems to exert a particularly strong influence on wood cell characteristics during the first 5–20 years of tree growth (2). Thus, the changes in wood characteristics in a given tree from pith to bark (i.e., with increasing tree age) are fundamentally related to the age or maturity of the cambium itself, as well as to the effects on cambial activity caused by natural and induced changes in the environment. Both softwoods and hardwoods are affected, but juvenility is especially well marked in the xylem of softwoods.

In either softwoods or hardwoods the trends in wood variability from pith to bark impose a progressive but inverse change of similar type with increasing distance up the tree, because new wood is laid onto a developing tree bole in the form of inverted hollow cones (*see* Figure 8). In other words, juvenile wood features diminish outwardly along the tree radius at a given level in the tree but are accentuated with increasing height in the tree bole (Figure 35).

Physical Properties. Juvenile wood typically (but not always) exhibits wider growth rings, a higher earlywood/latewood ratio, lower basic density (variable with species), higher moisture content, and much greater longitudinal shrinkage than mature, normal sapwood of the same tree (2). About two-thirds of commercial softwoods and hardwoods show a rapidly increasing basic density in the radial direction, leveling out in mature wood or increasing gradually for many years. Other tree species show some kind of parabolic trend, starting out at the pith either higher or lower than in outerwood (2).

In softwoods, the generally higher basic density (10–15%) of outerwood is due to thicker walled fibers and an increased proportion of latewood. However, in hardwoods, juvenile wood and mature wood densities vary not only with fiber wall thickness but also with the ratio of fiber/vessel volume. Radial patterns in basic density of hardwoods are also less consistent than in the softwoods, apparently dependent to a large extent on whether the species is ring-, semi-ring-, or diffuse-porous, the tree's growth rate and percent of latewood, and the vertical position in the trunk (2). Wood tissue in which the fiber/vessel ratio or fiber wall thickness is greater than surrounding areas will, in general, have a higher basic density.

At the ultrastructural level, juvenile wood fibers have a much greater S_2 microfibril angle than normal mature wood fibers. The net result of this situation is the greater than usual wood shrinkage along the grain. The high microfibril angle of juvenile wood can in part be attributed to the common occurrence of compression wood and tension wood near the pith. However, even the normal wood fibers in juvenile wood have an inherently high S_2 angle, particularly in softwoods (2).

Chemistry. For both softwoods and hardwoods, the interpretation of general trends in changing wood chemistry along the tree radius is complicated by the potential occurrence and distribution of reaction wood. Nevertheless, hardwoods show relatively little change in cellulose or lignin contents from pith to bark and from tree base to top (2). However, softwoods have tree cores that are about 3–20% lower in cellulose and up to several percent higher in lignin content (probably due at least in part to the presence of compression wood) (2).

There is limited information regarding potential differences in the nature and amount of juvenile wood and mature wood hemicelluloses (2). What data are available indicate that there can be limited changes in the relative amounts of some simple sugars, but there is apparently little or no significant difference for either softwoods or hardwoods.

The amount and nature of extractives in normal juvenile wood are generally similar to those of mature sapwood in the same tree, except that extractives near the tree center can be altered with time in many trees by the development of heartwood. Specifically, the amount and toxicity of juvenile wood extractives may both increase from the pith to the heartwood/sapwood boundary (2). However, resin extractives are a somewhat special case. In trees with or without heartwood, resin content is reported to be highest near the pith and at the tree base. It then decreases outward and upward in the stem (2).

KNOTWOOD. From the outerwood of older trees, lumber, veneer, or chips can be cut without the inclusion of natural wood defects called *knots*. Knots are residual, embedded portions of branches, or more specifically, branch bases. Although knots are usually concentrated in wood near the pith, i.e., in crown-formed wood, they are characteristic of wood in any stem region that is manufactured while in the proximity of branches (Figure 35A).

Knots represent what is left after the death and natural loss or pruning of tree branches, which takes place on the lower tree bole with increasing tree age. Natural pruning takes place relatively early in crowded forest stands (pole stands), and knot volume in these cases is minimal. Pruning is delayed in more widely spaced or open-grown stands (8). However, in open-grown stands, artificial pruning by the forest manager can be used to produce trees with a greater proportion of knot-free wood.

Branch bases become gradually embedded in any enlarging tree bole. Resulting knots may be designated as *tight* (red or intergrown) if growth of the stem around the branch base takes place while that portion of the branch is still living. Conversely, if the branch base is already dead from the time it becomes embedded, there will be a break in tissue continuity with the stem, and the result will be a *loose* (black or encased) knot (2). The loose knot is more likely to leave holes upon processing of the tree into wood products. The ultimate number, size, and type of knot will depend on the number and size of limbs from which they originate, limb age at death, and length of time the dead limb stubs persist or remain on the tree (2).

Knotwood is extremely dense and hard, contains a high percentage of reaction wood, and in some softwoods is very resinous (for species with normal resin ducts). The stem regions below and above knots are high in reaction wood, and these regions also exhibit severely distorted grain. The exact areas affected vary with knot size, but the grain normally passes in wide sweeps to either side. The presence of knots and associated differences in wood properties are generally considered detrimental to processes involving wood machining, finishing, gluing, and to most wood strength properties, particularly bending strength (2).

Literature Cited

1. Esau, K. "Anatomy of Seed Plants"; John Wiley & Sons: New York, 1977.
2. Panshin, A. J.; de Zeeuw, C. "Textbook of Wood Technology"; McGraw-Hill: New York, 1980.
3. Sjostrom, E, "Wood Chemistry—Fundamentals and Applications"; Academic Press: New York, 1981.
4. Berlyn, G.; Mark, R. E. *For. Prod. J.* **1965**, *16*(3), 140.
5. Fergus, B. J.; Proctor, A. R.; Scott, J.; Goring, D. A. I. *Wood Sci. Technol.* **1969**, *3*(2), 117.
6. Alvim, P. T. in "The Formation of Wood in Forest Trees," Zimmermann, M. H., Ed.; Academic Press: New York, 1964; p. 479.
7. Fritts, H. C. "Tree Rings and Climate"; Academic Press: New York, 1976.
8. Larson, P. R. "Wood Formation and the Concept of Wood Quality"; Bull. No. 74; Yale University Press: New Haven, Conn., 1969.
9. Fergus, B. J.; Goring, D. A. I. *Holzforschung* **1970**, *24*(4), 118.
10. Musha, Y.; Goring, D. A. I. *Wood Sci. Technol.* **1975**, *9*(1), 45.
11. Hanna, R. B.; Côté, W. A. *Cytobiologie* **1974**, *10*, 102.
12. Heyn, A. N. J. *Tappi* **1977**, *60*(11), 159.
13. Berlyn, G. *For. Prod. J.* **1964**, *14*(10), 467.
14. Gardner, K. H.; Blackwell, J. *Biopolymers* **1974**, *13*(10), 1975.
15. Dunning, C. E. *Wood Sci.* **1968**, *1*(2), 65.
16. Parham, R. A.; Kaustinen, H. *Bull. IAWA* **1973**, *1973/2*, 8.
17. Liese, W. in "Cellular Ultrastructure of Woody Plants," Côté, W. A., Ed.; Syracuse University Press: Syracuse, 1965; p. 251.
18. Cronshaw, J. *Protoplasma* **1965**, *60*, 233.
19. Kutscha, N. P.; Schwarzmann, J. M. *Holzforschung* **1975**, *29*(3), 80.
20. Baird, W. M.; Johnson, M. A.; Parham, R. A. *Wood Fiber* **1974**, *6*(3), 211.
21. Parameswaran, N.; Liese, W. *Wood Sci. Technol.* **1977**, *11*, 313.
22. Kishi, K.; Harada, H.; Saiki, H. *J. Jpn. Wood Res. Soc.* **1979**, *25*(8), 521.
23. Parham, R. A.; Baird, W. M. *Wood Sci. Technol.* **1974**, *8*(1), 1.

24. Haygreen, J. G.; Bowyer, J. L. "Forest Products and Wood Science"; Iowa State University Press: Ames, 1982.
25. Stamm, A. J. "Wood and Cellulose Science"; Ronald Press Co.: New York, 1964.
26. Polge, H. *Wood Sci. Technol.* **1978,** *12,* 187.
27. Taras, M. A.; Saucier, J. R. *For. Prod. J.* **1967,** *17*(9), 97.
28. Kellogg, R. M.; Wangaard, F. F. *Wood Fiber* **1969,** *1*(3), 180.
29. Meyer, R. W. *For. Prod. J.* **1967,** *17*(12), 50.
30. Timell, T. E. in "Chemistry and Morphology of Wood Components", Vol. 1; Proc. Int. Symp. Wood and Pulping Chemistry, Rep. 38; SPCI: Stockholm, 1981; p. 99.
31. Barger, R. L.; Ffolliott, P. F. *Wood Sci.* **1976,** *8*(3), 201.
32. Hughes, F. E. *For. Abstr.* **1965,** *26*(1,2), 2, 179.
33. Berlyn, G. *Iowa State J. Sci.* **1961,** *35*(3), 367.
34. Robards, A. W.; Purvis, M. J. *Stain Technol.* **1964,** *39,* 309.
35. Côté, W. A.; Day, A. C.; Timell, T. E. *Wood Sci. Technol.* **1969,** 3(4), 257.
36. Saki, H.; Ono, K. *Bull. Kyoto Univ. For.* **1971,** *42,* 210.
37. Dadswell, H. E.; Wardrop, A. B.; Watson, A. J. in "Fundamentals of Papermaking Fibers"; Bolam, F., Ed.; Tech. Sect. BPBMA: London, 1958; p. 187.
38. Parham, R. A. *Wood Sci.* **1974,** *6*(4), 308.
39. Gray, R. L.; Parham, R. A. *CHEMTECH* **1982,** *12*(4), 232.
40. Parham, R. A.; Gray, R. L. "The Practical Identification of Wood Pulp Fibers"; TAPPI Press: Atlanta, Ga.; 1982.
41. Parham, R. A.; Côté, W. A. *Wood Sci. Technol.* **1971,** *5*(1), 49.

NOTE: This is Contribution No. 236 from the Research Center of ITT Rayonier Inc.

RECEIVED for review May 2, 1983. ACCEPTED July 1, 1983.

2

The Chemical Composition of Wood

ROGER C. PETTERSEN

U.S. Department of Agriculture, Forest Service, Forest Products Laboratory, Madison, WI 53705

This chapter includes overall chemical composition of wood, methods of analysis, structure of hemicellulose components and degree of polymerization of carbohydrates. Tables of data are compiled for woods of several countries. Components include: cellulose (Cross and Bevan, holo-, and alpha-), lignin, pentosans, and ash. Solubilities in 1% sodium hydroxide, hot water, ethanol/benzene, and ether are reported. The data were collected at Forest Products Laboratory (Madison, Wisconsin) from 1927–68 and were previously unpublished. These data include both United States and foreign woods. Previously published data include compositions of woods from Borneo, Brazil, Cambodia, Chile, Colombia, Costa Rica, Ghana, Japan, Mexico, Mozambique, Papua New Guinea, the Philippines, Puerto Rico, Taiwan, and the USSR. Data from more detailed analyses are presented for common temperate-zone woods and include the individual sugar composition (as glucan, xylan, galactan, arabinan, and mannan), uronic anhydride, acetyl, lignin, and ash.

THE CHEMICAL COMPOSITION of wood cannot be defined precisely for a given tree species or even for a given tree. Chemical composition varies with tree part (root, stem, or branch), type of wood (i.e., normal, tension, or compression) geographic location, climate, and soil conditions. Analytical data accumulated from many years of work and from many different laboratories have helped to define average expected values for the chemical composition of wood. Ordinary chemical analysis can distinguish between hardwoods (angiosperms) and softwoods (gymnosperms). Unfortunately, such techniques cannot be used to identify individual tree species because of the variation within each species and the similarities among many species. Further identification is possible with detailed chemical analysis of extractives (chemotaxonomy). Chemotaxonomy is discussed fully elsewhere in the literature (1, 2).

There are two major chemical components in wood: lignin (18–35%) and carbohydrate (65–75%). Both are complex, polymeric materials. Minor amounts of extraneous materials, mostly in the form of organic extractives and inorganic minerals (ash), are also present in wood (usually 4–10%). Overall, wood has an elemental composition of about 50% carbon, 6% hydrogen, 44% oxygen, and trace amounts of several metal ions.

A complete chemical analysis accounts for all the components of the original wood sample. Thus, if wood is defined as part lignin, part carbohydrate, and part extraneous material, analyses for each of these components should sum to 100%. The procedure becomes more complex as the component parts are defined with greater detail. Summative data are frequently adjusted to 100% by introducing correction factors in the analytical calculations. Wise and coworkers (3) presented an interesting study on the summative analysis of wood and analyses of the carbohydrate fractions. The complete analytical report also includes details of the sample, such as species, age, and location of the tree, how the sample was obtained from the tree, and from what part of the tree. The type of wood analyzed is also important; i.e., compression, tension, or normal wood.

Vast amounts of data are available on the chemical composition of wood. Fengel and Grosser (4) made a compilation for temperate-zone woods. This chapter is a compilation of data for many different species from all parts of the world, and includes much of the data in Reference 4. The tables at the end of this chapter summarize these data.

Chemical Components

Carbohydrates. The carbohydrate portion of wood comprises cellulose and the hemicelluloses. Cellulose content ranges from 40 to 50% of the dry wood weight, and hemicelluloses range from 25 to 35%.

CELLULOSE. Cellulose is a glucan polymer consisting of linear chains of 1,4-β-bonded anhydroglucose units. (The notation 1,4-β describes the bond linkage and the configuration of the oxygen atom between adjacent glucose units.) Figure 1 shows a structural diagram of a portion of a glucan chain. The number of sugar units in one molecular chain is referred to as the degree of polymerization (DP). Even the most uniform sample has molecular chains with slightly different DP values. The average DP for the molecular chains in a given sample is designated by $\overline{\mathrm{DP}}$.

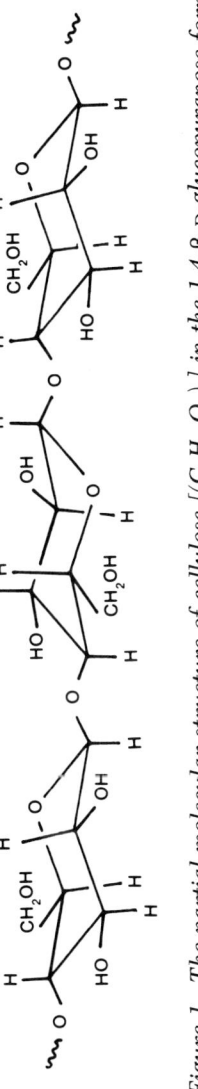

Figure 1. The partial molecular structure of cellulose $[(C_6H_{10}O_5)_x]$ in the 1,4-β-D-glucopyranose form.

Goring and Timell (5) determined the $\overline{DP}$ for native celluloses from several sources of plant material. They used a nitration isolation procedure that attempts to maximize the yield while minimizing the depolymerization of the cellulose. These molecular weight determinations, done by light-scattering experiments, indicate wood cellulose has a $\overline{DP}$ of at least 9,000–10,000, and possibly as high as 15,000. A DP of 10,000 would mean a linear chain length of approximately 5 μm in wood.

The $\overline{DP}$ obtained from light-scattering experiments is biased upward because light scattering increases exponentially with molecular size. The value obtained is usually referred to as the weighted $\overline{DP}$ or $\overline{DP}_w$. The number average degree of polymerization ($\overline{DP}_n$) is usually obtained from osmometry measurements. These measurements are linear with respect to molecular size and, therefore, a molecule is counted equally as one molecule regardless of its size. The ratio of $\overline{DP}_w$ to $\overline{DP}_n$ is a measure of the molecular weight distribution. This ratio is nearly one for native cellulose in secondary cell walls of plants (6). Therefore, this cellulose is monodisperse and contains molecules of only one size. Cellulose in the primary wall has a lower $\overline{DP}$ and is thought to be polydisperse. (*See* Reference 7 for a discussion of molecular weight distribution in synthetic polymers.)

Native cellulose is partially crystalline. X-Ray diffraction experiments indicate crystalline cellulose (*Valonia ventricosa*) has space group symmetry P2$_1$ with a = 16.34, b = 15.72, c = 10.38 Å, and γ = 97.0° (8). The unit cell contains eight cellobiose moieties. The molecular chains pack in layers that are held together by weak van der Waals' forces (Figure 2a). The layers consist of parallel chains of anhydroglucose units, and the chains are held together by intermolecular hydrogen bonds. There are also intramolecular hydrogen bonds between the atoms of adjacent glucose residues (Figure 2b). This structure is called cellulose I.

There are at least three other structures reported for modified crystalline cellulose. The most important is cellulose II, obtained by mercerization or regeneration of native cellulose. *Mercerization* is treatment of cellulose with strong alkali. *Regeneration* is treatment of cellulose with strong alkali and carbon disulfide to form a soluble xanthate derivative. The derivative is converted back to cellulose and reprecipitated as regenerated cellulose. The structure of cellulose II (regenerated) has space group symmetry P2$_1$ with a = 8.01, b = 9.04, c = 10.36 Å, and γ = 117.1°, and two cellobiose moieties per unit cell (9). The packing arrangement is modified in cellulose II, and permits a more intricate hydrogen-bonded network that extends between layers as well as within layers (Figure 3). The result is a

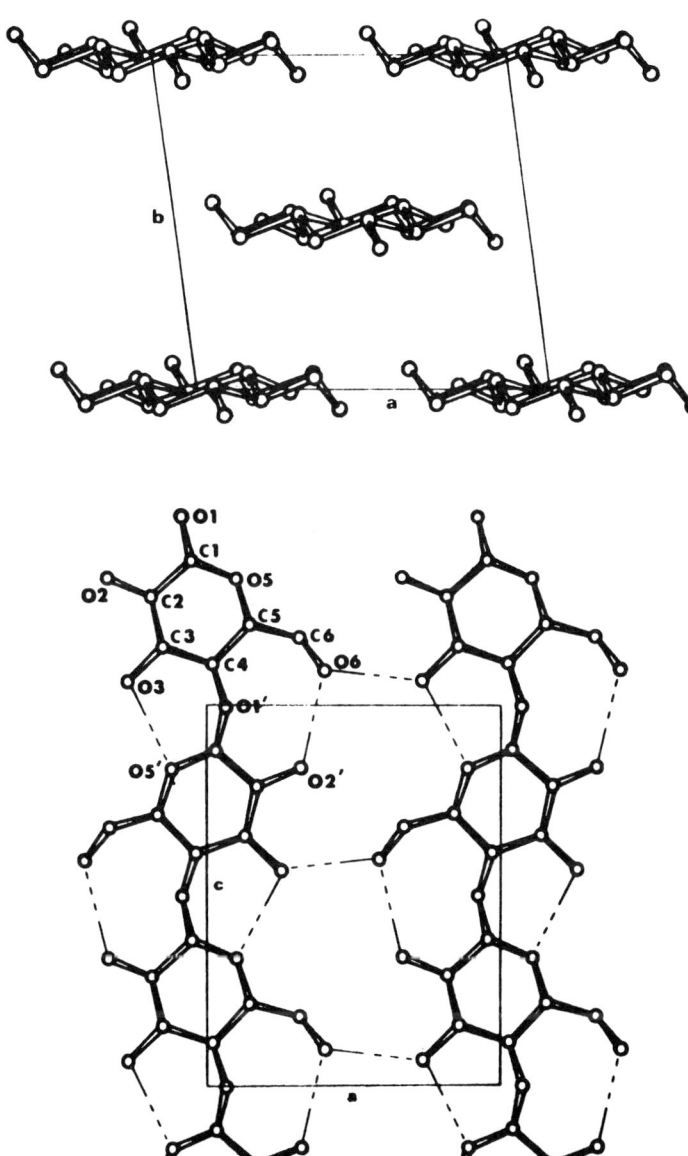

Figure 2. Axial projection (top) and planar projection (bottom) of the crystal structure of cellulose I. The planar projection shows the hydrogen-bonding network within the layers. (Reproduced with permission from Ref. 8. Copyright 1974, Elsevier Scientific Publishing Company, Amsterdam.)

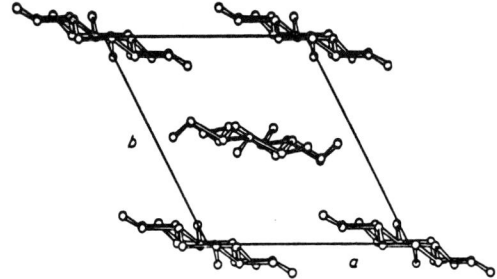

Figure 3. Axial projection of the crystal structure of cellulose II. (Reproduced with permission from Ref. 10. Copyright 1978, Butterworth & Co. (Publishers) Ltd.)

more thermodynamically stable substance. Evidently, all native celluloses have the structure of cellulose I.

Cellulose is insoluble in most solvents including strong alkali. It is difficult to isolate from wood in pure form because it is intimately associated with the lignin and hemicelluloses. Analytical methods of cellulose preparation are discussed in the section on "Analytical Procedures."

HEMICELLULOSES. Hemicelluloses are mixtures of polysaccharides synthesized in wood almost entirely from glucose, mannose, galactose, xylose, arabinose, 4-O-methylglucuronic acid, and galacturonic acid residues. Some hardwoods contain trace amounts of rhamnose. Generally, hemicelluloses are of much lower molecular weight than cellulose and some are branched. They are intimately associated with cellulose and appear to contribute as a structural component in the plant. Some hemicelluloses are present in abnormally large amounts when the plant is under stress; e.g., compression wood has a higher than normal galactose content as well as a higher lignin content (11). Hemicelluloses are soluble in alkali and easily hydrolyzed by acids.

The structure of hemicelluloses can be understood by first considering the conformation of the monomer units (Figure 4). There are three entries under each monomer in Figure 4. In each entry, the letter designations D and L refer to a standard configuration for the two optical isomers of glyceraldehyde, the simplest carbohydrate. The Greek letters α and β refer to the configuration of the hydroxyl group at carbon atom 1. The two configurations are called *anomers*. The first entry is a shortened form of the sugar name. The second entry indicates the ring structure. Pyranose refers to a six-membered ring in the chair or boat form and furanose refers to a five-membered ring. The third entry is an abbreviation commonly used for the sugar residue in polysaccharides.

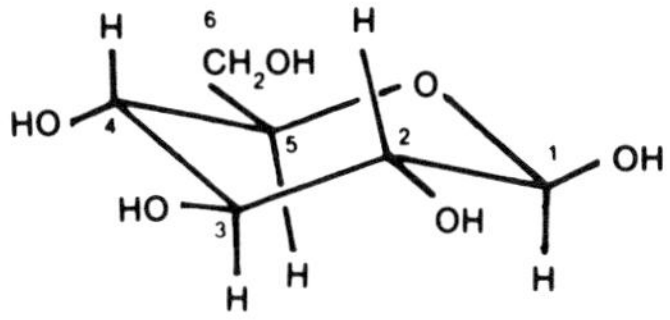

β-D-Glucose
β-D-Glucopyranose
β-D-Glup

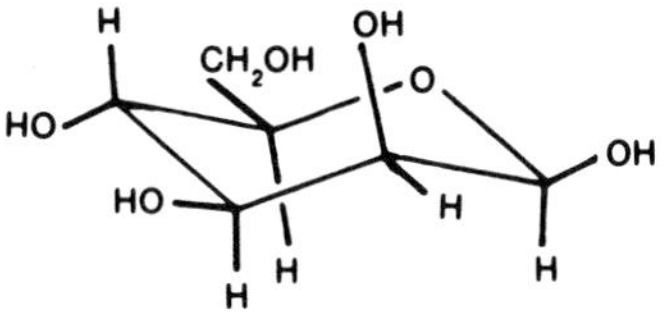

β-D-Mannose
β-D-Mannopyranose
β-D-Manp

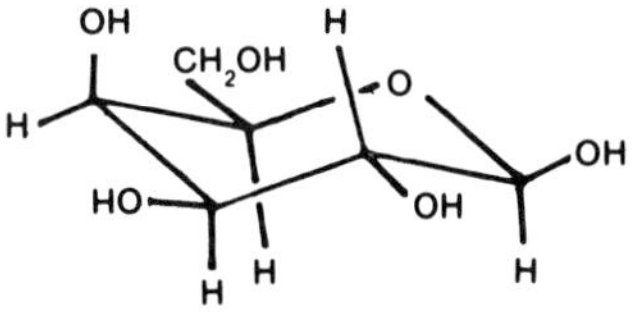

β-D-Galactose
β-D-Galactopyranose
β-D-Galp

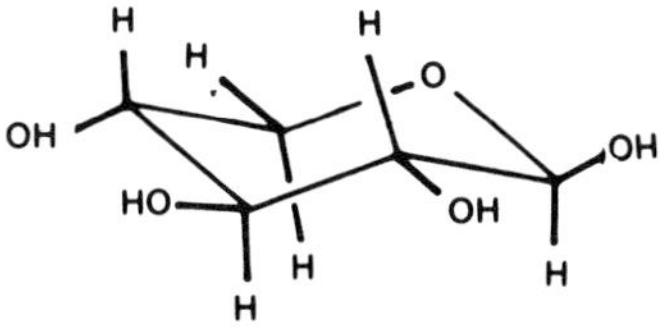

β-D-Xylose
β-D-Xylopyranose
β-D-Xylp

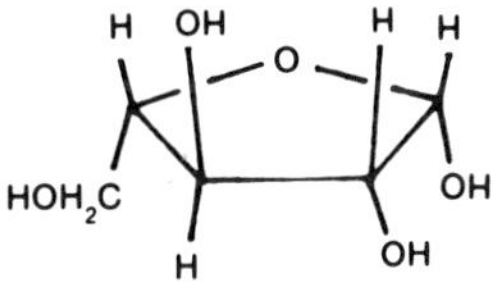

α-L-Arabinose
α-L-Arabinofuranose
α-L-Araf

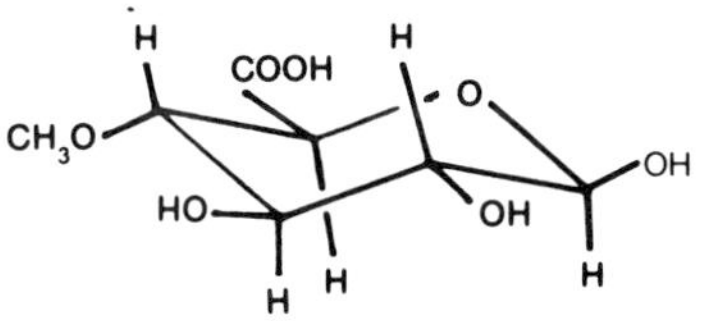

4-O-Methylgucuronic acid
4-O-Methylglucopyranosyluronic acid
4-O-Me-α-D-GlupA

Figure 4. Monomer components of wood hemicelluloses.

Figure 5 shows a partial structure of a common hardwood hemicellulose, O-acetyl-4-O-methylglucuronoxylan. The entire molecule consists of about 200 β-D-xylopyranose residues linked in a linear chain by (1 → 4) glycosidic bonds. Approximately 1 of 10 of the xylose residues has a 4-O-methylglucuronic acid residue bonded to it through the hydroxyl at the 2 ring position. Approximately 7 of 10 of the xylose residues have acetate groups bonded to either the 2 or 3 ring position. This composition is summarized in Figure 5 in an abbreviated structure diagram. Hardwood xylans contain an average of two xylan branching chains per macromolecule. The branches are probably quite short (12).

Table I lists the most abundant of the wood hemicelluloses. The

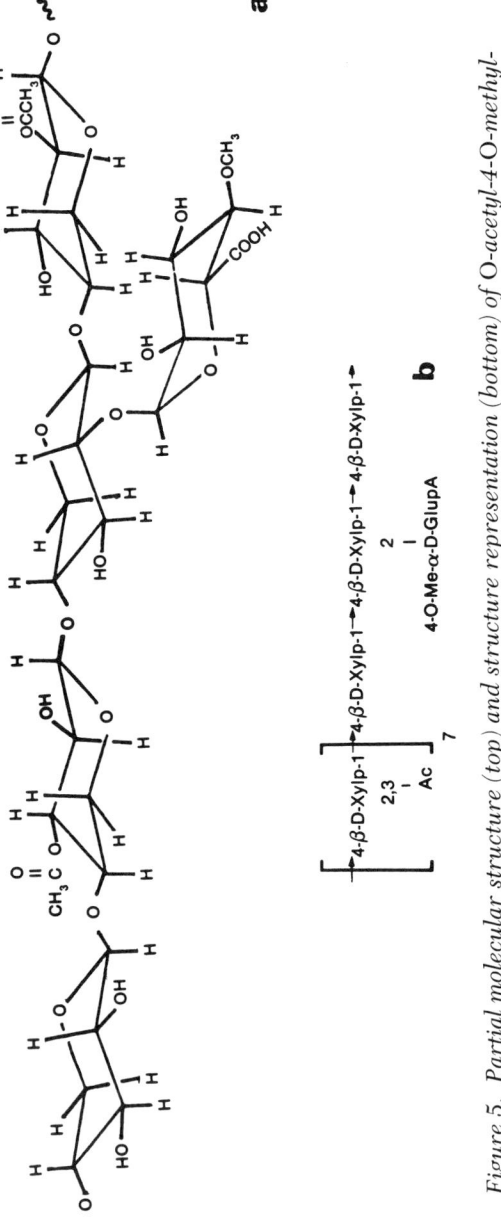

Figure 5. Partial molecular structure (top) and structure representation (bottom) of O-acetyl-4-O-methyl-glucuronoxylan.

Table I. The Major Hemicellulose Components

Hemicellulose Type	Occurrence	Amount (% of wood)	Composition Units	Molar Ratios	Linkage	Solubility[a]	$\overline{DP}_n$[b]
Galactoglucomannan	Softwood	5–8	β-D-Manp β-D-Glup α-D-Galp Acetyl	3 1 1	1 → 4 1 → 4 1 → 6	Alkali, water*	100
(Galacto)Glucomannan	Softwood	10–15	β-D-Manp β-D-Glup α-D-Galp Acetyl	4 1 0.1	1 → 4 1 → 4 1 → 6	Alkaline borate	100
Arabinoglucuronoxylan	Softwood	7–10	β-D-Xylp 4-O-Me-α-D-GlupA α-L-Araf	10 2 1.3	1 → 4 1 → 2 1 → 3	Alkali, dimethyl sulfoxide,* water*	100
Arabinogalactan	Larch wood	5–35	β-D-Galp α-L-Araf β-L-Arap β-D-GlupA	6 2/3 1/3 Little	1 → 3, 1 → 6 1 → 6 1 → 3 1 → 6	Water	200
Glucuronoxylan	Hardwood	15–30	β-D-Xylp 4-O-Me-α-D-GlupA Acetyl	10 1 7	1 → 4 1 → 2	Alkali, dimethyl sulfoxide*	200
Glucomannan	Hardwood	2–5	β-D-Manp β-D-Glup	1–2 1	1 → 4 1 → 4	Alkaline borate	200

[a] The asterisk represents a partial solubility.
[b] $\overline{DP}_n$ is the number average degree of polymerization, usually obtained by osmometry.
(Reproduced with permission from Ref. 6. Copyright 1981, Academic Press.)

methods used for the isolation and structural characterization of each of these materials are beyond the scope of this chapter (13–15).

Lignin. Lignin is a phenolic substance consisting of an irregular array of variously bonded hydroxy- and methoxy-substituted phenylpropane units. The precursors of lignin biosynthesis are *p*-coumaryl alcohol (I), coniferyl alcohol (II), and sinapyl alcohol (III). I is

[Structures of I, II, III showing p-coumaryl alcohol, coniferyl alcohol, and sinapyl alcohol]

a minor precursor of softwood and hardwood lignins; II is the predominant precursor of softwood lignin; and II and III are both precursors of hardwood lignin (15). These alcohols are linked in lignin by ether and carbon–carbon bonds. Figure 6 (15) is a schematic structure of a softwood lignin meant to illustrate the variety of structural components. The 3,5-dimethoxy-substituted aromatic ring number 13 originates from sinapyl alcohol, III, and is present only in trace amounts (<1%) (16). Figure 6 does not show a lignin–carbohydrate covalent bond. There has been much controversy concerning the existence of this bond, but evidence has been accumulating in its support (15, 17).

A structure proposed for hardwood lignin (*Fagus silvatica* L.) is similar to that of Figure 6, except that there are three times as many syringylpropane units as guaiacylpropane units (18). These moieties are derived from III and II, respectively. The ratio of syringyl to guaiacyl moieties is often obtained by measuring the relative amounts of syringaldehyde (3,5-dimethoxy-4-hydroxybenzaldehyde) and vanillin (4-hydroxy-3-methoxybenzaldehyde) generated as products of nitrobenzene oxidation of lignin (19). A better method is to determine the products formed from the two types of moieties on permanganate oxidation of methylated lignins (20).

Lignin can be isolated by one of several methods. Acid hydrolysis of wood isolates Klason lignin, which can be quantified (*see* "Analytical Procedures"), but is too severely degraded for use in structural studies. Björkman's (21) milled wood lignin procedure yields a lignin that is much less degraded and is, thus, more useful

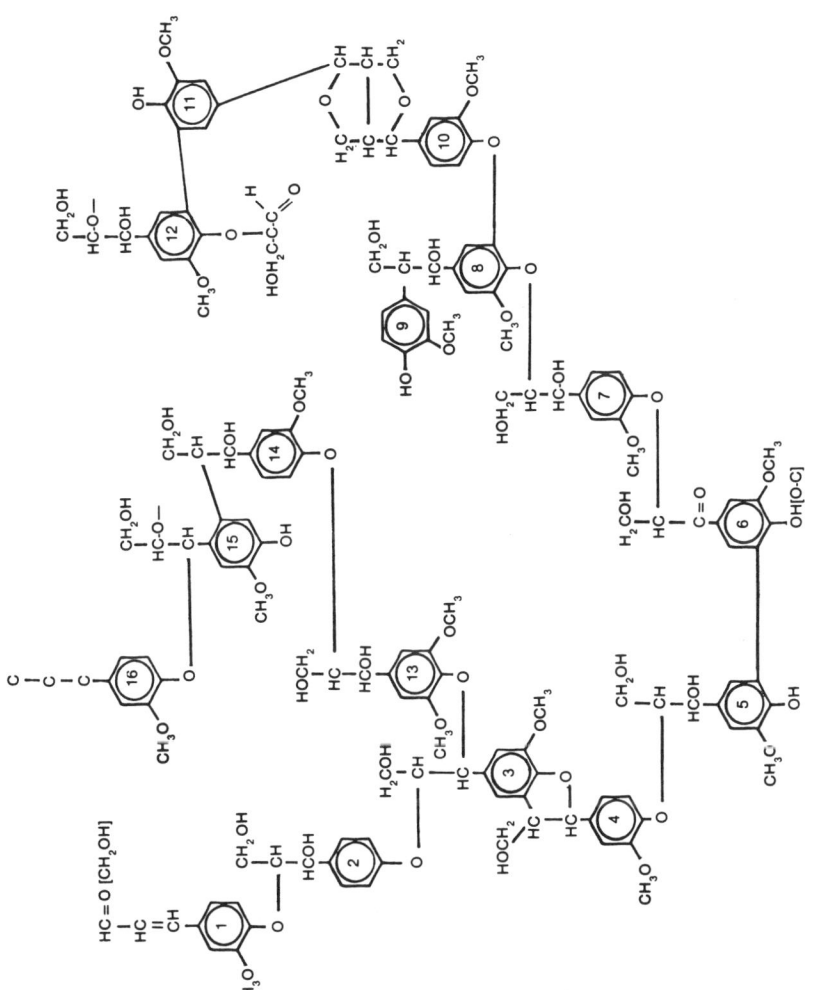

Figure 6. A partial structure of softwood lignin.

for structural studies. The following are examples of the weight average molecular weight of lignins isolated by using the milled wood lignin process: spruce [*Picea abies* (L.) Karst.], 15,000; and sweetgum (*Liquidambar styraciflua* L.), 16,000 (*22*). These values are lower than the molecular weight of the original lignin because fragmentation of the lignin molecules results from the ball milling procedure. Lignin for structural studies can also be obtained by enzymatic hydrolysis of the carbohydrate (*23*). Wood is ground in a vibratory ball mill and then treated with cellulytic enzymes. The isolated lignin contains 12–14% carbohydrate.

Methoxyl content is used to characterize lignins. Elemental and methoxyl analysis of spruce (*Picea abies* (L.) Karst.) milled wood lignin indicates a composition $C_9H_{7.92}O_{2.40}(OCH_3)_{0.92}$ (*15, 24*). Beech (*Fagus silvatica* L.) milled wood lignin has a composition $C_9H_{7.49}O_{2.53}(OCH_3)_{1.39}$ (*24*). This information helps lignin chemists understand what precursors were used for the biosynthesis of lignin. An excellent, comprehensive book on lignin is edited by Sarkanen and Ludwig (*25*).

Extraneous Components. The extraneous components (extractives and ash) in wood are the substances other than cellulose, hemicelluloses, and lignin. They do not contribute to the cell wall structure, and most are soluble in neutral solvents. The detailed chemistry of wood extractives can be found elsewhere (*26*). A review of extractives in eastern U.S. hardwoods is available (*27*).

Extractives—the extraneous material soluble in neutral solvents—constitute 4–10% of the dry weight of normal wood of species that grow in temperate climates. They may be as much as 20% of the wood of tropical species. Extractives are a variety of organic compounds including fats, waxes, alkaloids, proteins, simple and complex phenolics, simple sugars, pectins, mucilages, gums, resins, terpenes, starches, glycosides, saponins, and essential oils. Many of these function as intermediates in tree metabolism, as energy reserves, or as part of the tree's defense mechanism against microbial attack. They contribute to wood properties such as color, odor, and decay resistance.

Ash is the inorganic residue remaining after ignition at a high temperature. It is usually less than 1% of wood from temperate zones. It is slightly higher in wood from tropical climates.

Carbohydrate and Lignin Distribution

Carbohydrates. The morphological parts of the cell wall of a conifer are shown in Chapter 1, Figure 1b. Most of wood carbohydrate is in the massive secondary wall, particularly in S_2. Young tracheids have been isolated (*28*) at various stages of cell wall development, and then the separated fractions were analyzed for the five wood sugars. Table II lists the results obtained by using this method on birch (*Betula verrucosa* Ehrh.) and Scots pine (*Pinus sylvestris* L.) (*29*) fibers. The values are relative and sum to 100% for a given morphological part. This method has difficulty in distinguishing the presence of the very thin S_3. A tentative volume ratio was determined for the lignin-free layers of the pine and birch fibers by using photomicrographs of transverse sections. Taking the proportion to be middle lamella + primary cell wall (ML + P):S_1:S_2:S_3, the values are 2:10:78:10 for pine fibers (*28*) and 3:15:76:6 for birch (*29*). Assuming the density of the cell wall to be constant, the volume ratios become a comparison of amounts of polysaccharide in each layer.

Lignin. The distribution of lignin in the different morphological regions of wood microstructure has been studied using UV microscopy (*30*). In spruce (*Picea mariana* Mill.) tracheids, it was determined that 72% and 82% of the lignin was in the secondary cell walls of earlywood and latewood, respectively (*31*). The remainder was located in the middle lamella and cell corners. In birchwood (*Betula papyrifera* Marsh.), 71.3% of the lignin was of the syringyl type and was found in the secondary walls of the fibers (59.9%) and ray cells (11.4%). An additional 10.9% of the lignin was of the guaiacyl type and was found in the secondary walls of the vessels (9.4%) and the vessel middle lamella (1.5%). The remainder (17.7%) was mixed syringyl- and guaiacyl-type and was in the fiber middle lamella (*32*). Caution is needed in interpreting the syringyl/guaiacyl distribution in hardwood lignins; methoxyl analyses of isolated morphological parts of oak fibers and vessels indicates a rather uniform syringyl/guaiacyl content (*33*).

Analytical Procedures

Carbohydrates. There are a number of analytical determinations associated with the carbohydrate portion of wood.

HOLOCELLULOSE. Holocellulose is the total polysaccharide (cellulose and hemicelluloses) content of wood, and methods for its determination seek to remove all of the lignin from wood without disturbing the carbohydrates. The procedure generally used (*34*) was adopted as Tappi Standard T9m[1] (now useful method 249), and as ASTM Standard D 1104.[2] Extracted wood meal is treated alternately with chlorine gas and 2-aminoethanol until a white residue (holocellulose) remains. The acid chlorite method is also used (*3*). The

---

[1] Tappi standards are maintained by the Technical Association of Pulp and Paper Industry, Atlanta, Ga.
[2] ASTM standards are maintained by the American Society for Testing Materials, Philadelphia, Pa.

Table II. Percentages of Polysaccharides in the Different Layers of the Fiber Wall

Polysaccharide	$Ml + P^a$	S_1	S_2 (outer part)	S_2 (inner part) + S_3
Birch (*Betula verrucosa* Ehrh.)				
Galactan	16.9	1.2	0.7	0.0
Cellulose	41.4	49.8	48.0	60.0
Glucomannan	3.1	2.8	2.1	5.1
Arabinan	13.4	1.9	1.5	0.0
Glucuronoxylan	25.2	44.1	47.7	35.1
Pine (*Pinus sylvestris* L.)				
Galactan	20.1	5.2	1.6	3.2
Cellulose	35.5	61.5	66.5	47.5
Glucomannan	7.7	16.9	24.6	27.2
Arabinan	29.4	0.6	0.0	2.4
Glucuronoarabinoxylan	7.3	15.7	7.4	19.4

[a] Also contains a high percentage of pectic acid.
(Reproduced with permission from Ref. 29. Copyright 1961, John Wiley & Sons.)

product, called chlorite holocellulose, is similar to chlorine holocellulose. The chlorite method removes a fraction more of the hemicelluloses than the chlorine method.

ALPHA CELLULOSE. Alpha cellulose is obtained after treatment of the holocellulose with 17.5% NaOH (see ASTM Standard D 1103). This procedure removes most, but not all, of the hemicelluloses.

CROSS AND BEVAN CELLULOSE. Cross and Bevan cellulose consists largely of pure cellulose, but also contains some hemicelluloses. It is obtained by chlorination of wood meal, followed by washing with 3% SO_2 and 2% sodium sulfite (Na_2SO_3) water solutions. The final step is treatment in boiling Na_2SO_3 solution. The absence of a characteristic red (angiosperm) or brown (gymnosperm) color developed in the presence of chlorinated lignin signals complete lignin removal. For a discussion of the method and its modifications, see Reference 35.

KÜRSCHNER CELLULOSE. Kürschner cellulose is obtained by refluxing the wood sample three times for 1 h with a 1:4 volume mixture of concentrated nitric acid and ethyl alcohol (37). The washed and dried residue is weighed as Kürschner cellulose. The product contains a small amount of hemicelluloses. [The cellulose determined for the Ghanan and Russian woods (see in Tables VI and XI) is Kürschner cellulose]. The method is not widely used because it destroys some of the cellulose and the nitric acid/alcohol mixture is potentially explosive.

PENTOSAN. Pentosan analysis measures the amount of five-carbon sugars present in wood (xylose and arabinose residues). Although the hemicelluloses consist of a mixture of five- and six-carbon sugars (see discussion of hemicelluloses), the pentosan analysis reports the xylan and arabinan content as if the five-carbon sugars were present as pure pentans. Pentoses are more abundant in hardwoods than softwoods; the difference is due to a higher xylose content in hardwoods (see Table XIII for examples).

Tappi standard T 223 outlines the procedure for pentosan analysis. Briefly, wood meal is boiled in 3.85 N HCl with some NaCl added. Furfural is generated and distilled into a collection flask. The furfural is determined colorimetrically with orcinol–iron(III) chloride reagent. Another method also generates furfural, and the furfural is determined gravimetrically by precipitation with 1,3,5-benzenetriol. These and other methods of pentosan analysis are described and discussed in Browning's book (36).

CHROMATOGRAPHIC ANALYSIS OF WOOD SUGARS. This analysis requires acid hydrolysis of the polysaccharide to yield a solution mixture of the five wood sugar monomers, i.e., glucose, xylose, galactose, arabinose, and mannose. The solution is neutralized, filtered, and the sugars chromatographically separated and quantified. Generally this method is accepted as the standard of hydrolysis (37). In this procedure, wood meal is treated with 72% H_2SO_4 at 30 °C for 1 h to depolymerize the carbohydrates. Reversion products (recombined sugar monomers) are further hydrolyzed in 3% H_2SO_4 at 120 °C for 1 h. The solution is then filtered, and the solid residue is washed, dried, and weighed as Klason lignin (see "Lignin" later). The filtrate is neutralized with barium(II) hydroxide or ion exchange resin. The individual sugars are separated by paper, liquid, or gas chromatography (GC). Paper chromatography has been the standard method for many years and all the individual sugar data and hemicellulose data reported in the tables of this chapter were obtained by this method [adopted as Tappi Provisional Test Method T 250 (37)]. This method uses a modified form of the Somogyi colorimetric assay for reducing sugars (38). Timell (39) reports a colorimetric method in which the reducing sugars are reacted with 2-aminobiphenyl hydrochloride. There are many other assay methods for reducing sugars.

Sugar separation by GC requires the preparation of volatile derivatives. Tappi Test Method T 249 pm–75 uses the alditol acetate derivitization (40). Peracetylated aldonitrile (41) or trimethylsilane (42, 43) derivatives can also be prepared and separated by GC. Wood sugar analysis by GC may be useful for specialized problems, but the derivitization steps make it a time-consuming method for routine work.

High performance liquid chromatography (HPLC) is currently the most efficient means for routine separation and quantification of the five wood sugars (44). In this case, no derivitization is necessary, and separation is achieved using water as an eluent. Detection is by a differential refractometer.

URONIC ACID. Uronic acid is determined by measuring carbon dioxide (CO_2) generation when wood is boiled with 12% HCl (45). Results from this method may be somewhat high because of CO_2 evolution from material containing carboxyl groups other than uronic acid. A method developed by Scott (46) is rapid and selective. The sample is treated with 96% H_2SO_4 at 70 °C, and a product, 5-formyl-2-furancarboxylic acid, is derived from uronic acids. This compound reacts selectively with 3,5-dimethylphenol to yield a chromophore absorbing at 450 nm.

ACETYL CONTENT. The acetyl content of wood is determined by saponification of the sample in 1 N NaOH, followed by acidification, quantitative distillation of the acetic acid, and titration of the distillate with standard NaOH (47). A modification here (Forest Products Laboratory) enables acetic acid determination by using GC with propanoic acid as an internal standard. This modification eliminates the tedious, time-consuming distillation step.

WOOD SOLUBILITY IN 1% NaOH. Wood extraction procedures in 1% NaOH (Tappi Standard T 212) extract most extraneous components, some lignin, and low molecular weight hemicelluloses and degraded cellulose. The percent of alkali-soluble material increases as the wood decays (48). The extraction is done in a water bath maintained at 100 °C.

Lignin. The lignin contents of woods presented in the tables of this chapter are Klason lignin, the residue remaining after solubilizing the carbohydrate with strong mineral acid. The usual procedure, as in Tappi Standard T 222 or ASTM Standard D 1106, is to treat finely ground wood with 72% H_2SO_4 for 2 h at 20 °C, followed by dilution to 3% H_2SO_4 and boiling or refluxing for 4 h. An equivalent but shorter method treats the sample with 72% H_2SO_4 at 30 °C for 1 h, followed by 1 h at 120 °C in 3% H_2SO_4 (50). In both cases the determination is gravimetric.

Softwood lignins are insoluble in 72% H_2SO_4 and Klason lignin provides an accurate measure of total lignin content. Hardwood lignins are somewhat soluble in 72% H_2SO_4, and the acid-soluble portion may amount to 10–20% of the total lignin content (51). The *acid-soluble lignin* can be determined spectrophotometrically at 205 nm (51, 52). (Table XIV contains lignin values that add the acid-soluble component measured at 205 nm to the Klason lignin. Lignin contents of hardwoods in all the other tables are low).

METHOXYL. Methoxyl groups are determined by a modified method (53). Methyl iodide is formed by hydrolysis of the methoxyl groups of wood lignin in hydriodic acid and is distilled under CO_2 into a solution of bromine and potassium acetate in glacial acetic acid. Bromine oxidizes iodide to iodate which is then titrated with standard thiosulfate. The method is difficult and time-consuming, and some experience is necessary before satisfactory results can be obtained. Details are in ASTM Standard D 1166 and Tappi Standard T 209 (withdrawn in November 1979). Additional discussion can be found in Reference 54.

Extraneous Components

Wood Solubility. The solubility of wood in various solvents is a measure of the extraneous components content. No single solvent is able to remove all of the extraneous materials. Ether is relatively nonpolar and extracts fats, resins, oils, sterols, and terpenes. Ethanol/benzene is more polar and extracts most of the ether-solubles plus most of the organic materials insoluble in water. Hot water extracts some inorganic salts and low molecular weight polysaccharides including gums and starches. Water also removes certain hemicelluloses such as the arabinogalactan gum present in larch wood (*see* Table I).

ETHANOL/BENZENE. The solubility of wood in EtOH/benzene (benzene is a known carcinogen; toluene can be substituted) in a 1:2 volume ratio will give a measure of the extractives content. This procedure is Tappi Standard T 204 and ASTM Standard D 1107. The wood meal is refluxed 6–8 h in a Soxhlet flask, and the weight loss of the extracted, dried wood is measured. Sometimes the lignin, carbohydrate, and other components are determined on wood that has been extracted previously with EtOH/benzene (see Table XIII).

DIETHYL ETHER. The solubility of wood in diethyl ether is determined in the same way as EtOH/benzene solubility.

Ash Analysis. Ash analysis is performed according to Tappi Standard T 15 and ASTM Standard D 1102. In these standards ash is defined as the residue remaining after dry ignition of the wood at 575 °C. Elemental composition of the ash is determined by dissolving the residue in strong HNO_3 and analyzing the solution by atomic absorption or atomic emission. The inorganic elemental composition of wood can be determined directly by neutron activation analysis. (Table XV contains elemental data using both methods).

Silica (SiO_2) content in wood can be determined by treating the ash with hydrofluoric acid (HF) to form the volatile compound silicon tetrafluoride (SiF_4). The weight loss is the amount of silica in the ash. Silica is rarely present in more than trace amounts in temperate climate woods, but can vary in tropical woods from a mere trace to as much as 9%. More than 0.5% silica in wood is harmful to cutting tools (55).

Moisture Content. The moisture content of wood is determined by measuring the weight loss after drying the sample at 105 °C. Unless specified otherwise, the percent of all other chemical components in wood is calculated on the basis of moisture-free wood. Moisture content is determined on a separate portion of the sample not used for the other analyses.

Recent Improvements in Techniques

The data reported in this chapter were obtained using standard methods. The methods are routine but require much care and time. Some methods have been replaced by better, more efficient methods. For example, the holocellulose, cellulose, and pentosan tests have been replaced by the single five-sugar chromatographic test. The five-sugar test procedure gives more detailed information in a shorter time. The recent change from paper chromatography to HPLC has improved the efficiency of this test. The test for Klason lignin remains in use, as do the acetyl, methoxyl, and uronic acid tests.

Analytical instruments and data processors have helped to remove some of the tedium and to shorten analysis time. The result has been an increase in the number of analyses performed. More significant is the detail possible with advanced instruments. For example, HPLC can separate and quantitate individual uronic acids. This provides more detail of hemicellulose composition. The structure of lignin can be probed further by mass spectrometry and high-resolution NMR spectrometry. Wood extractives can be isolated and characterized by capillary GC/mass spectrometry. A new mass spectrometer has two or more mass analyzers and eliminates the often limiting chromatographic separation step.

More systematic wood composition studies are needed in the future. It would be useful to study the composition of a select number of prominent species and note the content variability with tree parts, climate, soil conditions, and age.

Tables of Composition Data

Tables III–XIV are organized geographically and list chemical composition data for woods from various countries. The data as published originally were of interest to the local pulp and paper industries. This compilation provides a worldwide view of wood composition. Most of the data were obtained using similar test methods (Tappi Standards). When it is known that other test methods were used, the method is footnoted in the tables. Most of the values reported from all sources had one or two figures beyond the decimal point. Except for the ether solubility and ash values (usually less than 1%), values have been rounded off to the nearest percent because this reflects the precision of the sampling and assay methods.

The data in Table III have not been published previously. The same test methods were used for all tree species in Table III. Most of these methods were developed at the laboratory and were later adopted as Tappi standards. Tables IV–XII contain similar data obtained in many test laboratories. The three Taiwanese sources contain data for more than 400 trees. The trees selected for inclusion in Table X were those described in a book published by the Chinese Forestry Association (56). Table XII contains data on trees of unrecorded origin. Except for *Tectonia grandia*, the species reported do not appear in the other tables.

Tables XIII and XIV present more detailed analyses of woods: Table XIII contains data on 30 North American species, and Table XIV contains data on 32 species from the southeastern United States. The lignin values in Table XIV are the sum of Klason and acid-soluble lignins. Pectin (Table XIV) is mainly galacturonic acid. It is the measured total uronic acid value minus the estimated glucuronic acid value. Glucuronic acid content can be estimated from the xylan content by assuming a ratio of xylose to 4-O-methylglucuronic acid of 10:1 (*see* Table I *and* Figure 5). The reported values of the carbo-

Table III. Chemical Composition of U.S. Woods as Determined at U.S. Forest Products Laboratory from 1927 to 1968

	Carbohydrate						Solubility			
Scientific Name/Common Name	Holo-cellulose[a]	Cross and Bevan Cellulose[b]	Alpha Cellulose[c]	Pentosans[d]	Klason Lignin	1% NaOH	Hot Water	EtOH/ Benzene	Ether	Ash
Hardwoods										
Acer macrophyllum Pursh/ Bigleaf maple	—	—	46	22	25	18	2	3	0.7	0.5
Acer negundo L./Boxelder	—	—	45	20	30	10	—	—	0.4	—
Acer rubrum L./Red maple	77 (3)	61 (2)	47 (3)	18 (3)	21 (3)	16 (3)	3 (3)	2 (3)	0.7 (3)	0.4 (3)
Acer saccharinum L./Silver maple	—	56	42	19	21	21	4	3	0.6	—
Acer saccharum Marsh./Sugar maple	—	60	45	17	22	15	3	3	0.5	0.2
Alnus rubra Bong./Red alder	74 (2)	—	44 (3)	20 (3)	24 (3)	16 (3)	3 (3)	2 (3)	0.5 (3)	0.3 (3)
Arbutus menziesii Pursh/ Pacific madrone	—	—	44	23	21	23	5	7	0.4	0.7
Betula alleghaniensis Britton/ Yellow birch	73	64 (2)	47 (2)	23 (2)	21 (2)	16 (2)	2 (2)	2 (2)	1.2 (2)	0.7 (2)
Betula nigra L./River birch	—	57	41	23	21	21	4	2	0.5	—
Betula papyrifera Marsh./ Paper birch	78 (2)	63 (3)	45 (5)	23 (5)	18 (5)	17 (4)	2 (4)	3 (4)	1.4 (4)	0.3 (2)
Carya cordiformis (Wangenh.) K. Koch/ Bitternut hickory	—	56	44	19	25	16	5	4	0.5	—
Carya glabra (Mill.) Sweet/ Pignut hickory	71 (2)	—	49 (2)	17 (2)	24 (2)	17 (2)	5 (2)	4 (2)	0.4 (2)	0.8 (2)
Carya ovata (Mill.) K. Koch/ Shagbark hickory	71	—	48	18	21	18	5	3	0.4	0.6

2. PETTERSEN The Chemical Composition of Wood

Carya pallida (Ashe) Engl. & Graebn./Sand hickory	69	—	—	—	—	—	—	4	0.4	1.0
Carya tomentosa (Poir.) Nutt./Mockernut hickory	71 (2)	—	48 (2)	18 (2)	21 (2)	17 (2)	5 (2)	4 (2)	0.4 (2)	0.6
Celtis laevigata Willd./Sugarberry	—	54	40	22	21	23	6	3	0.3	—
Eucalyptus gigantea Hook. f./—	72	—	49	14	22	16	7	4	0.3	0.2
Fagus grandifolia Ehrh./American beech	77 (2)	61 (2)	49 (2)	20 (2)	22 (2)	14 (2)	2 (2)	2 (2)	0.8 (2)	0.4 (2)
Fraxinus americana L./White ash	51	51	41	15	26	16	7	5	0.5	—
Fraxinus pennsylvanica Marsh./Green ash	—	53 (4)	40 (4)	18 (4)	26 (4)	19 (4)	7 (4)	5 (4)	0.4 (4)	—
Gleditsia triacanthos L./Honey locust	—	—	52	22	21	19	—	—	0.4	—
Laguncularia racemosa (L.) Gaertn./White mangrove	—	52	40	19	23	29	15	6	2.1	—
Liquidambar styraciflua L./Sweetgum	—	60 (3)	46 (4)	20 (4)	21 (4)	15 (4)	3 (3)	2 (4)	0.7 (3)	0.3 (3)
Liriodendron tulipifera L./Yellow-poplar	—	62	45	19	20	17	2	1	0.2	1.0
Lithocarpus densiflorus (Hook. & Arn.) Rehd./Tanoak	71 (2)	—	46 (3)	20 (2)	19 (3)	20 (3)	5 (2)	3 (2)	0.4 (2)	0.7 (2)
Milalenca quinquenervia (Cav.) S. T. Blake/Cajeput	—	56	43	19	27	21	4	2	0.5	—
Nyssa aquatica L./Water tupelo	—	59 (2)	45 (2)	16 (2)	24 (2)	16 (2)	4 (2)	3 (2)	0.6 (2)	0.6
Nyssa sylvatica Marsh./Black tupelo	72	57 (4) / 67	45 (5) / 52	17 (4) / 23	27 (5) / 16	15 (5) / 20	3 (5) / 4	2 (5) / 5	0.4 (5) / 0.9	0.5 (2) / —
Populus alba L./White poplar	—	—	—	—	—	—	—	—	—	—
Populus deltoides Bartr. ex Marsh./Eastern cottonwood	—	64 (3)	47 (3)	18 (3)	23 (3)	15 (3)	2 (3)	2 (3)	0.8 (2)	0.4

Continued on next page

Table III. Continued

Scientific Name/Common Name	Carbohydrate				Klason Lignin	Solubility				Ash
	Holo-cellulose[a]	Cross and Bevan Cellulose[b]	Alpha Cellulose[c]	Pento-sans[d]		1% NaOH	Hot Water	EtOH/Benzene	Ether	
Populus tremoides Michx./Quaking aspen	78 (9)	65 (13)	49 (20)	19 (19)	19 (22)	18 (15)	3 (15)	3 (14)	1.2 (15)	0.4 (11)
Populus trichocarpa Torr. & Gray/Black cottonwood	—	—	49	19	21	18	3	3	0.7	0.5
Prunus serotina Ehrh./Black cherry	85	60	45	20	21	18	4	5	0.9	0.1
Quercus alba L./White oak	67 (2)	—	47 (2)	20 (2)	27 (2)	19 (2)	6 (3)	3 (2)	0.5 (2)	0.4
Quercus coccinea Muenchh./Scarlet oak	63	—	46	18	28	20	6	3	0.4	—
Quercus douglasii Hook & Arn./Blue oak	59	—	40	22[e]	27	23	11	5	1.4	1.4
Quercus falcata Michx./Southern red oak	69	—	42	20	25	17	6	4	0.3	0.4
Quercus kelloggii Newb./California black oak	60	—	37	23[e]	26	26	10	5	1.5	0.4
Quercus lobata Nee/Valley oak	70	—	43	19[e]	19	23	5	7	1.0	0.9
Quercus lyrata Walt./Overcup oak	—	—	40	18	28	24	9	5	1.2	0.3
Quercus marylandica Muenchh./Blackjack oak	—	57	44	20	26	15	5	4	0.6	—
Quercus prinus L./Chestnut oak	76	—	47	19	24	21	7	5	0.6	0.4
Quercus rubra L./Northern red oak	69	—	46	22	24	22	6	5	1.2	0.4
Quercus stellata Wangenh./Post oak	—	55	41	18	24	21	8	4	0.5	1.2

Quercus velutina Lam./Black oak	71	—	48	20	24	18	6	5	0.2	0.2
Salix nigra Marsh./Black willow	—	61 (2)	46 (2)	19 (2)	21 (2)	19 (2)	4 (2)	2 (2)	0.6 (2)	—
Tilia heterophylla Vent./Basswood	77	65	48	17	20	20	2	4	2.1	0.7
Ulmus americana L./American elm	73	61 (3)	50 (3)	17 (3)	22 (3)	16 (3)	3 (3)	2 (3)	0.5 (3)	0.4
Ulmus crassifolia Nutt./Cedar elm	—	—	50	19	27	14	—	—	0.3	—
Softwoods										
Abies amabilis Dougl. ex Forbes/Pacific silver fir	—	61 (3)	44 (3)	10 (3)	29 (3)	11 (3)	3 (3)	3 (3)	0.7 (3)	0.4
Abies balsamea (L.) Mill./Balsam fir	—	58 (16)	42 (16)	11 (16)	29 (16)	11 (16)	4 (16)	3 (16)	1.0 (16)	0.4 (15)
Abies concolor (Gord. & Glend.) Lindl. ex Hildebr./White fir	66	—	49	6	28	13	5	2	0.3	0.4
Abies lasiocarpa (Hook.) Nutt./Subalpine fir	67 (4)	—	46 (4)	9 (4)	29 (4)	12 (4)	3 (4)	3 (4)	0.6 (4)	0.5 (4)
Abies procera Rehd./Noble fir	61	—	43	9	29	10	2	3	0.6	0.4
Chamaecyparis thyoides (L.) B.S.P./Atlantic white cedar	—	53	41	9	33	16	3	6	2.4	—
Juniperus deppeana Steud./Alligator juniper	57	—	40	5	34	16	3	7	2.4	0.3
Larix larcina (Du Roi) K. Koch/Tamarack	64 (3)	—	44 (3)	8 (3)	26 (3)	14 (3)	7	3 (3)	0.9 (3)	0.3 (2)
Larix occidentalis Nutt./Western larch	65 (3)	56 (2)	48 (3)	9 (3)	27 (3)	16 (3)	6 (3)	2 (3)	0.8 (3)	0.4 (2)
Libocedrus decurrens Torr./Incense cedar	56	—	37	12	34	9	3	3	0.8	0.3
Picea engelmanni Parry ex Engelm./Engelman spruce	69 (4)	60 (2)	45 (6)	10 (6)	28 (6)	11 (6)	2 (6)	2 (6)	1.1 (6)	0.2 (2)
Picea glauca (Moench) Voss/White spruce	—	61 (8)	43 (8)	13 (7)	29 (8)	12 (8)	3 (8)	2 (8)	1.1 (8)	0.3 (2)

Continued on next page

Table III. Continued

Scientific Name/Common Name	Carbohydrate				Klason Lignin	Solubility				Ash
	Holo-cellu-lose[a]	Cross and Bevan Cellu-lose[b]	Alpha Cellu-lose[c]	Pento-sans[d]		1% NaOH	Hot Water	EtOH/ Benzene	Ether	
Picea mariana (Mill.) B.S.P./ Black spruce	—	60 (19)	43 (20)	12 (19)	27 (20)	11 (20)	3 (20)	2 (20)	1.0 (20)	0.3 (19)
Picea sitchensis (Bong.) Carr./ Sitka spruce	—	62	45	7	27	12	4	4	0.7	—
Pinus attenuata Lemm./ Knobcone pine	—	—	47	14	27	11	3	1	—	0.2
Pinus banksiana Lamb./Jack pine	66 (6)	58 (25)	43 (27)	13 (27)	27 (27)	13 (27)	3 (26)	5 (27)	3.0 (26)	0.3 (7)
Pinus clausa (Chapm. ex Engelm.) Vasey ex Sarg./ Sand pine	—	57 (3)	44 (4)	11 (4)	27 (4)	12 (2)	2 (2)	3 (2)	1.0	0.4
Pinus contorta Dougl. ex Loud./Lodgepole pine	68 (11)	59 (7)	45 (11)	10 (11)	26 (11)	13 (11)	4 (11)	3 (11)	1.6 (11)	0.3 (11)
Pinus echinata Mill./Shortleaf pine	69	60 (8)	45 (9)	12 (9)	28 (9)	12 (9)	2 (9)	4 (9)	2.9 (9)	0.4 (2)
Pinus elliottii Engelm./Slash pine	64 (3)	59 (13)	46 (15)	11 (15)	27 (15)	13 (15)	3 (15)	4 (15)	3.3 (15)	0.2 (3)
Pinus monticola Dougl. ex D. Don/Western white pine	69 (3)	61 (4)	43 (7)	9 (7)	25 (7)	13 (6)	4 (6)	4 (6)	2.3 (6)	0.2 (3)
Pinus palustris Mill./Longleaf pine	—	59 (7)	44 (5)	12 (7)	30 (6)	12 (7)	3 (5)	4 (7)	1.4 (7)	—
Pinus ponderosa Dougl. ex Laws./Ponderosa pine	68	58	41 (2)	9 (2)	26 (2)	16 (2)	4 (2)	5 (2)	5.5 (2)	0.5

Pinus resinosa Ait./Red pine	71	—	47	10	26	13	4	4	2.5	—
Pinus sabiniana Dougl./Digger pine	—	—	—	—	—	—	—	1 (2)	—	0.2 (2)
Pinus strobus L./Eastern white pine	68 (4)	6c	46 (2)	11 (2)	27 (2)	12 (2)	3 (2)	6 (5)	3.2 (5)	0.2 (3)
Pinus taeda L./Loblolly pine	68	60 (13)	45 (5) 45 (14)	8 (5) 12 (12)	27 (5) 27 (14)	15 (5) 11 (12)	4 (5) 2 (12)	3 (15)	2.0 (12)	—
Pseudotsuga menziesii (Mirb.) Franco/Douglas-fir	66 (9)	60 (42)	45 (50)	8 (50)	27 (50)	13 (50)	4 (50)	4 (50)	1.3 (50)	0.2 (13)
Sequoia sempervirens (D. Don) Endl./Redwood										
Old growth	55	—	43	7	33	19	9	10	0.8	0.1
Second growth	61	—	46	7	33	14	5	<1	0.1	0.1
Taxodium distichum (L.) Rich./Bald cypress	—	55	41	12	33	13	4	5	1.5	—
Thuja occidentalis L./Northern white cedar	59	—	44	14[e]	30	13	5	6	1.4	0.5
Thuja plicata Donn ex D. Don/Western red cedar	—	49	38	9	32	21	11	14	2.5	0.3
Tsuga canadensis (L.) Carr./Eastern hemlock	—	55 (7)	41 (7)	9 (4)	33 (7)	13 (6)	4 (7)	3 (7)	0.5 (7)	0.5 (5)
Tsuga heterophylla (Raf.) Sarg./Western hemlock	67 (2)	53 (22)	42 (22)	9 (22)	29 (22)	14 (22)	4 (22)	4 (22)	0.5 (22)	0.4 (4)
Tsuga mertensiana (Bong.) Carr./Mountain hemlock	60	—	43	7	27	12	5	5	0.9	0.5

NOTE: Numbers in parentheses are independent determinations of the component and in some cases, the trees are from different locations; values are percent moisture-free wood.

[a] Holocellulose is the total carbohydrate content of wood.
[b] Cross and Bevan cellulose is largely pure cellulose but contains some hemicelluloses.
[c] Alpha cellulose is nearly pure cellulose.
[d] Pentosans are the total anhydroxylose and arabinose residues in wood.
[e] Pentosans determined by gravimetric method.

Table IV. Chemical Composition of Woods from South and Central America, Mexico, and Puerto Rico

Scientific Name/Common Name	Carbohydrate			Klason Lignin	Solubility				Ether	Ash	Reference
	Holo-cellu-lose[a]	Alpha Cellu-lose[b]	Pento-sans[c]		1% NaOH	Hot Water	Benzene EtOH	EtOH			

Brazil

Brosimum parinarioides Ducke/ Amapa roxo	—	51	10	26	21	2	6		—	0.2	57
Cecropia juranyiana A. Rich./ Imbauba[d]	69	48	17	25	14	6	3		0.3	0.7	58[e]
Corythophora alta Knuth./ Ripeiro vermelho	—	47	10	30	19	6	4		—	0.5	57
Couepia leptostachya Benth./ Uchi de cutia	—	39	9	33	12	4	<1		—	0.8	57
Eclinusa ucuquirana branca Aubr. et Pellegr./Ucuquirana brava	—	55	15	30	17	4	1		—	0.6	57
Eperua bijuga Mart. et Benth./ Muirapiranga	—	41	12	38	31	11	9		—	0.2	57
Eschweiler odora Poepp. et Miers/Matamata	—	50	13	32	18	6	<1		—	0.9	57
Eucalyptus camaldulensis Dehnh./Red river gum	—	50[f]	17	29	11	2	2		—	0.8	59
Eucalyptus cloeziana F. Muell./ Gympie messmate	—	54[f]	16	28	12	2	3		—	0.3	59
Eucalyptus grandis W. Hillex Maid./Flooded gum	—	54[f]	19	26	16	3	3		—	0.3	59

Eucalyptus kirtoniana F. Muell./—	74	50	15	28	14	3	2	0.3	0.1	60
Eucalyptus saligna Sm./Sydney blue gum	74	50	15	27	14	3	1	0.3	0.2	60
Eucalyptus tesselaris F. Muell./—	—	50^f	21	24	17	5	2	—	0.6	59
Eucalyptus torelliana F. Muell./Cadaga	—	53^f	23	22	19	3	2	—	1.0	59
Eucalyptus urophylla S. T. Blake/Timor white gum	—	53^f	19	24	17	2	2	—	0.4	59
Holopyxidium latifolium (Ducke) Knuth./Jarana	—	50	10	30	17	1	4	—	0.3	57
Licania oblonifolia Standl./Macuco chiador	—	51	20	33	18	2	1	—	0.5	57
Lacuma dissepala (K. Krause) Ducke/Abiurana	74	48	17	25	14	2	2	0.5	1.0	58^e
Micropholis rosadinha brava Aubr. et Pellegr./Rosada brava	—	53	11	28	13	2	1	—	0.8	57
Pouteria guianensis Aubr./Abiurana Abiu	—	54	7	30	13	3	2	—	0.3	57
Protium heptaphyllum March./Breu branco	70	49	17	27	16	5	2	0.4	0.6	58^e
Qualea dinizii Ducke/Pau mulato	69	48	14	28	15	3	2	0.3	0.8	58^e
Schizolobium amazonicum Huber/Parica	—	54	12	26	16	2	2	—	0.8	57
Vantanea parviflora Lam./Macucu murici	—	51	10	37	14	4	2	—	0.2	57

Continued on next page

Table IV. Continued

Scientific Name/Common Name	Carbohydrate				Solubility						
	Holo-cellu-lose[a]	Alpha Cellu-lose[b]	Pento-sans[c]	Klason Lignin	1% NaOH	Hot Water	Benzene EtOH	Ether	Ash	Reference
Eucryphia cordifolia Cav./Ulmo				Chile						
	77	49	15	26	17	3	2	0.3	0.5	60
Laurelia philippiana Looser/Tepa	71	46	16	28	10	2	2	0.4	1.0	60
Nothofagus dombeyi (Mirb.) Oerst/Coigue	70	48	17	23	19	7	6	1.0	0.3	60
				Colombia						
Anacardium excelsum (Bert. & Balb.) Skeels/Caracoli	61	44	10	30	18	6	6	2.9	1.2	60
Ceiba pentandra (L.) Gaertn./Ceiba bruja	62	41	16	25	25	15	2	0.5	2.9	60
Shizolobium parahybum (Vell.) Blake/Gambombo	73	49	14	26	21	2	2	0.5	0.4	60
Spondias purpurea L./Jobo	72	47	17	24	17	3	3	0.7	1.0	60
				Costa Rica						
Anacardium excelsum (Bert. & Balb.) Skeels/Espavel	72	—	8	27	18	7	3	—	1.6	61
Brosimum utile (HBK) Pittier/Baco	79	—	13	26	16	3	2	—	0.4	61
Carapa slateri Standl./Cedro macho	79	—	11	25	14	4	2	—	0.6	61
Caryocar costaricense Donn. Smith/Ajo	75	—	13	24	16	9	3	—	0.4	61

Ceiba pentandra (L.) Gaertn./Ceiba	77	—	10	26	19	7	1	—	2.7	61
Couratari panamensis Standl./Campano	76	—	11	31	12	5	2	—	0.7	61
Dialyanthera otoba (Humb. & Bonpl.) Warb./Bogamani	81	—	12	26	14	4	1	—	0.4	61
Dussia sp./Sangrillo amarillo	82	—	10	28	10	3	1	—	0.6	61
Peltogyne purpurea Pittier/Nazareno	81	—	12	22	13	6	5	—	0.5	61
Platymiscium pinnatum (Jacq) Dugand/Cristobal	76	—	15	26	15	6	6	—	0.6	61
Poulsenia armata Standl./Calugo	81	—	11	36	20	3	1	—	9.7	61
Qualea paraensis Ducke/Masicaran	79	—	11	25	17	5	1	—	1.3	61
Sacoglottis excelsa Druke/Terciopelo	76	—	11	31	19	6	1	—	0.4	61
Sapotaceae sp./Nispero	82	—	14	25	15	3	1	—	1.9	61
Sapotaceae sp./Zapoton	80	—	15	25	18	5	2	—	0.7	61
Symphonia globulifera L.f./Cerillo	78	—	15	24	15	3	3	—	0.4	61
Terminalia amazonia (J.F. Gmel.) Excell./Escobo amarillo	71	—	12	25	17	10	8	—	0.5	61
Uribea tamarindoides Dugand & Romero/Almendro	73	—	12	33	10	4	5	—	1.1	61
Vantanea barbourii Standl./Caracolillo	78	—	11	31	11	3	1	—	0.4	61
Virola sp./Fruta dorada	80	—	15	24	17	4	1	—	0.6	61
Vochysia sp./Mayo negro	82	—	17	22	21	6	4	—	0.9	61

Continued on next page

Table IV. Continued

Scientific Name/Common Name	Carbohydrate			Klason Lignin	1% NaOH	Solubility			Ash	Reference
	Holo-cellu-lose[a]	Alpha Cellu-lose[b]	Pento-sans[c]			Hot Water	EtOH/Benzene	Ether		
Vochysia allenii Standley & L. O. Williams/Mayo blanco	81	—	11	22 Mexico, Yucatan	18	4	3	—	1.1	61
Allophylus psilospermus Radlk./Kanchunup	60	46	12	34	12	4	4	0.5	1.2	60
Brosimum alicastrum Sw./Ramon	63	44	16	27	17	5	2	0.4	1.6	60
Bursera simaruba (L.) Sarg./Chacha	74	46	17	23	20	5	4	0.8	1.6	60
Calyptranthes millspaughii Urb./Chachi	67	47	12	29	15	5	2	0.7	2.7	60
Cecropia obtusifolia Bertol./Kochle	67	45	15	25	19	5	4	0.7	1.7	60
Ceiba pentandra (L.) Caertn./Ceiba	64	40	18	22	28	14	2	0.5	2.4	60
Coccoloba uvifera (L.) Jacq./Boo	69	48	14	28	17	5	2	0.5	1.6	60

Drypetes lateriflora (Sw.) Krug & Erb./Ekulu	69	48	15	26	17	6	4	0.5	2.5	60
Ficus lapathifolia (Liebm.) Miq./Zacamua	66	44	15	30	17	5	2	0.5	1.7	60
Guazuma tomentosa H.B.K./Pixoy	70	45	16	27	16	2	1	0.5	1.2	60
Pisonia sp./Tatsi	76	58	14	20	11	2	1	0.4	1.5	60
Poincianella guameri (Greenm.) Britt. & Rose/Kitanche	62	47	14	25	19	10	7	2.0	1.3	60
Spondias mombin L./Jobo	74	46	18	19	22	6	3	0.7	1.2	60
Puerto Rico										
Cecropia peltata L./Yagrumo hembra	68	46	14	25	16	2	3	0.6	0.7	60
Eucalyptus robusta Sm./Swamp mahogany	67	48	12	28	12	3	2	0.3	0.5	60
Inga vera Willd./Guama	65	50	13	28	11	2	2	0.3	0.2	60

NOTE: Values are percent moisture-free wood.
[a] Holocellulose is the total carbohydrate content of wood.
[b] Alpha cellulose is nearly pure cellulose.
[c] Pentosans are the total anhydroxylose and arabinose residues in wood.
[d] Average of trees from two locations.
[e] The holocellulose, lignin, and pentosans from Ref. 58 are percent extractive-free wood.
[f] Cross and Bevan cellulose is largely pure cellulose but contains some hemicelluloses.

Table V. Supplementary Chemical Composition Data for South and Central American Hardwoods

Scientific Name/Common Name	Carbohydrate		Klason Lignin	Acetyl	Total extractives[b]	Ash
	Alpha Cellulose[a]	Hemi-cellulose				
Guyana (62)						
Couratari pulchra Sandw./Tauary	47	14	31	1.1	5.3	0.8
Eschweilera sagotiana Miers/Kakeralli	49	13	29	1.4	5.8	0.6
Ocotea rodiaei (Rob. Schomb.) Mez./ Greenheart	45	13	31	1.1	9.5[c]	0.2
Honduras (63)						
Cordia alliodora (R. & P.) Cham./ Jaurel blanco	45	17	30	1.3	6.6	1.0
Hymenaea courbaril L./Courbaril	43	20	20	2.2	13.8	0.9
Pseudosamanea guachapele (H.B.K.) Harms./Frijolillo	45	13	24	1.5	13.1	0.6
Tabebuia guayacan (Seem.) Hemsl./ Guayacan	46	14	29	1.1	8.6	0.3
Surinam (63)						
Dicorynia paraensis Benth./ Angelique (64)	45	15	32	1.1	5.4[d]	0.6
Licaria cayennensis (Meissn.) Kosterm./Kaneelhart	46	11	30	0.8	10.4	0.03
Manilkara bidentata (A.D.C.) Chev./ Bulletwood	46	16	26	1.1	7.5	0.4
Ocotea rubra Mez./Determa	48	13	29	0.8	10.1	0.2

NOTE: Analytical methods used for percent moisture-free wood are found in Ref. 3.
[a] Alpha cellulose is nearly pure cellulose.
[b] Total extractives = sum of solubles in ether, 50% EtOH, EtOH/benzene, and hot water (80 °C).
[c] Total extractives = sum of solubles in chloroform, 50% EtOH, and hot water (80 °C).
[d] Total extractives = sum of solubles in ether, 50% EtOH, and hot water (80 °C).

Table VI. Chemical Composition of Woods from Ghana and Mozambique

Scientific Name/Common Name	Carbohydrate		Klason Lignin	Solubility			Ash
	Cellulose[a]	Pentosans[b]		1% NaOH	Hot Water	EtOH/ Benzene	
Ghana[c]							
Gmelina arborea L./Yemane[d]	47	20	29	13	6	4	0.6
Musanga cecropioides R. Br./Odwuma	50	16	26	14	2	2	0.4
Terminalia ivorensis Chev./Emire	45	15	33	16	5	2	0.3
Triplochiton scleroxylon K. Schum./Wawa	40	17	31	19	10	1	1.8
Mozambique[e]							
Acacia nigrescens Oliv./Chicocolo	42	14	20	17	8	14	1.6
Adina microcephala (Del.) Hiern.) Galangola[f]	42	12	27	16	6	10	0.7
Albizzia gummifera (Gmel.) C. A. Sm./Galinga	43	20	23	17	4	5	0.4
Amblygonocarpus andongensis (Welw. ex Oliv.) Excell et Torrey/Banga-uanga	35	12	29	24	9	10	0.4
Androstachys johnsonii Prain/Cimbirre	29	16	29	13	2	16	1.0
Bombax rhodognaphalon K. Schum./Meguza[g]	42	14	30	20	3	8	1.6
Cedrela odorata L./—	37	18	33	16	3	4	1.0
Chlorophora excelsa (Welw.) Benth. et Hook. f./Mahundo[h]	41	15	25	20	5	7	3.1
Crossopteryx febrifuga Benth./Mucobenga	36	16	28	18	8	6	1.8
Dalbergia melanoxylan Guill. et Perr./Ampivi	38	12	26	13	2	14	3.4
Diospyros mespiliformis Hochst. ex A.DC./ Chitomane	38	17	31	20	8	1	4.1
Erythrophloeum guineense D. Don'Chaia	38	11	26	18	4	16	0.0

Continued on next page

Table VI. Continued

Guibourtia conjugata (Bolle) J. Leonard/Chacate	34	16	30	20	10	1.8
Khaya nyasica Stapf. ex Baker f./Imbáua[i]	41	14	28	27	7	1.6
Kirkia acuminata Oliv./Muyumira	39	15	29	17	8	2.0
Lannea discolor (Sond.) Engl./Chumbo	51	18	21	24	5	2.4
Melaleuca leucadendron L./—[g]	41	14	30	31	5	1.9
Morus lactea (Sim.) Mildbr./Mecobze	34	18	28	18	3	1.1
Newtonia buchananii (Bak.) Gilbert et Boutique/Mafamuti[f]	42	15	24	23	7	1.0
Podocarpus falcatus (Thunb.) R. Br. ex Mirb./Gogogo	44	10	29	18	2	0.7
Pterocarpus antunesii (Taub.) Harms/Muchibire	44	16	27	13	6	0.9
Spirotachys africana Sond./Chilingamache	36	15	21	17	4	2.5
Swartzia madagascariensis Desv./Cimbe[g]	37	15	26	16	4	0.2
Syncarpia laurifolia Ten./—	42	15	31	12	7	1.6
Syringa vulgaris L./—	44	19	28	19	3	0.5
Tectona grandis L.f./—[f]						
Sapwood	43	15	25	18	9	1.3
Heartwood	41	14	23	16	12	1.4
Trichilia emetica Vahl/Curre	39	18	31	27	7	3.9
Vitex doniana Sweet/Mucuvo-sique	40	13	31	18	7	2.7
Xylopia holtzii Engl./Mulalabungo	41	17	31	20	4	0.5

[a] Cellulose determined using alcoholic nitric acid (Kürschner cellulose) for Ghanan woods. A mixture of concentrated nitric acid and glacial acetic acid was used to determine cellulose in Mozambique woods. See Refs. 64 and 65 for details.
[b] Pentosans are the total anhydroxylose and arabinose residues in wood.
[c] Data adapted from Ref. 64.
[d] Common name in Burma.
[e] Data adapted from Ref. 65.
[f] Average of three trees.
[g] Average of two trees.
[h] Average of four trees.
[i] Average of five trees.

Table VII. The Chemical Composition of Japanese Woods (66,67)

Scientific Name/Common Name	Carbohydrate				Klason Lignin	Solubility			Ash
	Holo-cellulose[a]	Cross and Bevan Cellulose[b,c]	Alpha Cellulose[d]	Pento-sans[e]		1% NaOH	Hot Water	EtOH Benzene	
Hardwoods									
Acanthopanax sciadophylloides Franch. & Sav./Koshiabura	80	63	45	21	21	23	5	2	0.6
Acer japonicum Thunb./Meigetsukaede	82	61	47	24	21	4	2		0.4
Acer mayrii Schwerin/Beniitaya	78	53	34	26	23	5	2		0.6
Acer mono Maxim./Ezoitaya	81	62	48	22	19	17	4	2	0.4
Acer mono Maxim./Itayakaede	78	—	49	18	24	—	4	2	0.5
Acer palmatum Thunb./Yamanomiji	77	59	42	23	22	24	7	3	0.5
Aesculus turbinata Blume/Tochinoki	79	59	44	22	21	18	5	2	0.3
Aesculus turbinata Blume/Tochinoki	75	—	46	14	27	—	3	1	0.3
Alnus hirsuta Turcz./Keyamahannoki	79	58	43	20	20	22	5	5	0.3
Alnus hirsuta Turcz./Keyamahannoki	73	—	48	15	23	—	4	2	0.3
Alnus japonica Stend./Hannoki	76	56	40	23	22	22	5	4	0.3
Aralia elata Seem./Taranoki	78	57	47	26	20	23	7	4	0.4
Benzoin umbellatum Kuntze/Kuromoji	77	57	34	27	19	26	7	6	0.8
Betula grossa S. et Z./Mizume	78	—	46	27	24	—	2	2	0.4
Betula ermanii Cham./Dakekanba	79	60	46	25	20	17	2	3	0.3
Betula maximowicziana Regel/Udaikanba	82	57	40	26	20	17	2	1	0.2
Betula maximowicziana Regel/Makarba	77	—	47	18	23	—	2	1	0.4
Betula platiphylla Sukatchev/Shirakanba	83	63	46	23	19	16	3	1	0.4

Continued on next page

Table VII. Continued

Scientific Name/Common Name	Holo-cellulose[a]	Carbohydrate			Klason Lignin	Solubility			
		Cross and Bevan Cellulose[b,c]	Alpha Cellulose[d]	Pentosans[e]		1% NaOH	Hot Water	EtOH/ Benzene	Ash
Betula platiphylla Sukatchev/Shirakanba	77	—	56	22	18	—	2	2	0.2
Carpinus cordata Blume/Sawashiba	79	61	43	20	21	23	4	2	0.5
Carpinus laxiflora Blume/Akashide	80	—	46	27	17	—	3	2	0.6
Castanea crenata S. et Z./Kuri	73	52	40	23	26	23	10	3	0.3
Castanea crenata S. et Z./Kuri	70	—	42	15	21	—	11	2	0.8
Cercidiphyllum japonicum S. et. Z/Katsura	78	58	44	23	24	21	6	<1	0.7
Cercidiphyllum japonicum S. et. Z/Katsura	78	—	51	16	26	—	5	3	0.3
Cinnamomum camphora Sieb./Kusunuki	81	—	50	14	29	—	5	2	0.5
Cornus controversa Hemsley/Miznki	82	61	43	23	23	24	5	1	0.3
Cornus controversa Hemsley/Miznki	73	—	46	17	22	—	4	2	0.4
Cyclobalanopsis acuta Oerst./Akagashi	71	—	47	17	25	—	9	4	0.7
Cyclobalanopsis myrsinaefolia Oerst./ Shirakashi	75	—	48	19	23	—	7	2	1.0
Cyclobalanopsis gilva Oerst./Ichiigashi	77	—	48	15	27	—	6	1	1.1
Distylium racemosum S. et Z./Isunoki	73	—	47	17	30	—	5	2	0.5
Euonymus macropterus Rupt./Hirobat-suribana	71	49	33	26	27	21	7	4	0.9
Euonymus oxyphyllus Miq./Tsuribana	76	55	44	24	26	18	5	2	0.6
Fagus crenata Blume/Buna	81	60	45	21	21	17	4	1	0.7
Fagus crenata Blume/Buna[f]	81	—	50	18	24	—	2	1	0.5

2. PETTERSEN   The Chemical Composition of Wood

Fagus japonica Maxim./Inubuna	79	—	47	17	25	—	4	1	0.8
Fraxinum commemoralis Koidzumi/Shioji	78	—	57	14	26	—	3	2	0.5
Fraxinum mandshurica Rupt./Yachidamo	82	59	47	21	20	19	5	1	0.9
Fraxinus mandshurica Rupt./Yachidamo	80	—	51	16	22	—	4	2	1.0
Fraxinus sieboldiana Blume/Aodame	76	55	44	20	23	19	7	4	0.7
Fraxinus sieboldiana Blume/Aodame	75	—	45	17	24	—	6	4	0.9
Ilex macropoda Miq./Aohada	81	49	34	18	16	32	7	5	0.7
Juglans ailanthifolia Carr./Onigurumi	80	61	43	24	21	25	6	4	0.4
Juglans sieboldiana Maxim./Onigurumi	78	—	50	13	22	—	7	4	0.4
Kalopanax pictus Nakae/Harigiri	79	60	48	23	22	18	4	1	0.3
Kalopanax ricinifolium Miq./Harigiri	79	—	51	17	23	—	4	2	0.6
Maackia amurensis Rupt. et Maxim./Inuenju	78	57	45	22	22	24	5	6	0.6
Maackia amurensis Rupt. et Maxim./Inuenju	77	—	53	17	19	—	5	6	0.3
Machilus thunbergii S. et Z./Tabunoki	73	58	49	15	25	—	7	5	0.3
Magnolia kobus Dc./Kobushi	79	61	43	20	26	20	4	1	0.4
Magnolia obovata Thung./Honoki	81	—	44	20	24	17	3	2	0.2
Magnolia obovata Thunb./Honoki	77	—	47	15	30	—	3	2	0.4
Morus bombycis Koidzumi/Yamaguwa	72	50	35	26	21	28	10	9	0.8
Morus bombycis Koidzumi/Yamaguwa	67	—	42	15	21	—	7	8	0.4
Ostrya japonica Sargent/Asada	78	62	44	21	21	19	5	2	0.7
Ostrya japonica Sargent/Asada	80	—	48	19	23	—	4	2	0.5
Paulownia tomentosa Steud./Kiri	72	—	45	16	20	—	9	8	0.2
Phellodendron amurense Rupt./Kihada	80	62	49	21	19	20	5	1	0.6
Phellodendron sachalinense Sargent'/Kihada	80	—	51	14	23	—	4	1	0.1
Picrasma quassiodes Benn./Nigaki	80	62	49	21	19	20	5	1	0.6

Continued on next page

Table VII. Continued

Scientific Name/Common Name	Holo-cellulose[a]	Carbohydrate			Klason Lignin	Solubility			Ash
		Cross and Bevan Cellulose[b,c]	Alpha-Cellulose[d]	Pentosans[e]		1% NaOH	Hot Water	EtOH/Benzene	
Populus maximowiczii A. Henry/Doronoki	81	64	47	22	22	20	3	2	0.6
Populus maximowiczii A. Henry/Doronoki	82	—	53	14	22	—	2	2	0.7
Populus sieboldii Miq./Yamanarashi	81	—	49	19	18	—	3	3	0.5
Pourthiaea villosa Dcne./Vshikoroshi	82	59	45	24	20	19	5	3	0.3
Prunus donarium Sieb./Yamazakura	73	—	48	21	18	—	6	5	0.3
Prunus grayana Maxim./Uwamizuzakura	78	54	39	23	20	21	5	4	0.7
Prunus maximowiczii Komarov/Shirozakura	82	62	46	24	18	24	5	2	0.2
Prunus padus L./Ezonouwamizuzakura	81	49	36	22	21	28	5	2	0.6
Prunus sargentii Rehd./Ezoyamazakura	80	57	44	23	18	28	9	5	0.3
Prunus ssiori Fr. Schmidt/Shurizakura	74	55	40	24	21	27	6	5	0.4
Pterocarya rhoifolia S. et Z./Sawagurumi	83	61	44	21	18	25	4	4	0.3
Pterocarya rhoifolia S. et Z./Sawagurumi (average of 4)	78	—	48	14	24	—	3	2	0.4
Quercus acutissima Carr./Kunugi	78	—	50	18	19	—	4	<1	0.6
Quercus crispula Blume/Mizunara	79	57	45	22	22	22	9	2	0.3
Quercus crispula Blume/Mizunara[g]	75	—	48	20	26	—	6	1	0.2
Quercus dentata Thunb./Kashiwa	73	47	31	24	25	23	9	5	0.6
Quercus serrata Thunb./Konara	78	—	50	17	22	—	6	1	0.6

Rhamnus japonica Maxim./ Ezokuromemodoki	84	59	42	26	21	20	6	2	0.4
Robinia pseudo-acacia L./Harienju	82	61	50	24	21	18	5	3	0.3
Salix bakko Kimura/Bakkoyanagi	82	62	43	22	20	23	3	2	0.4
Salix pet-susu Kimura/ Ezonokinuyanagi	80	59	41	23	22	23	4	3	0.3
Salix sachalinensis Fr. Schmidt/ Nagabayanagi	84	59	38	19	20	25	4	3	0.3
Sambucus sieboldiana Blume/Niwatcko	79	57	46	23	26	18	3	2	0.6
Shiia cuspidata Makino/Kojii	79	—	48	16	23	—	3	2	0.4
Shiia sieboldii Makino/Shiinoki	65	—	37	15	28	—	13	3	0.2
Sorbus alnifolia K. Koch/Azukinashi	80	60	44	22	20	22	3	1	0.4
Sorbus commixta Hedlund/Nanakamado	80	57	46	21	20	24	7	3	0.6
Stewartia monadelpha S. et Z./Himeshara	69	—	44	15	25	—	3	1	0.6
Styrax obassia S. et Z./Hakuunboku	83	59	45	24	21	30	4	2	0.6
Syringa reticulata (Blume) Hara/Hashidoi	78	60	44	22	20	24	6	4	0.4
Tilia japonica Simonkai/Shinanuki	80	59	43	20	17	26	6	7	0.8
Tilia japonica Simonkai/Shinanuki	79	—	46	18	20	—	3	4	0.2
Tilia maximowicziana Shirasawa/ Obabodaiju	82	61	44	23	17	25	5	6	0.6
Tilia maximowicziana Shirasawa/ Obabodaiju	82	—	46	18	21	—	3	3	0.6
Toisusu urbaniana Kimura/Obayanagi	80	—	50	15	21	—	2	2	0.9
Ulmus davidiana Planch./Harunire	80	62	51	20	21	15	3	1	0.9
Ulmus laciniata Mayr./Ohyo	79	56	36	24	23	23	4	2	1.4
Ulmus propinqua Koidzumi/Harunire	79	—	47	15	27	—	2	<1	0.8
Zelkova serrata Makino/Keyaki	75	—	44	16	27	—	8	1	0.8

Continued on next page

Table VII. Continued

Scientific Name/Common Name	Carbohydrate				Klason Lignin	1% NaOH	Solubility			Ash
	Holo-cellu-lose[a]	Cross and Bevan Cellu-lose[b,c]	Alpha Cellu-lose[d]	Pento-sans[e]			Hot Water	EtOH/ Benzene	EtOH	

Softwoods

Scientific Name/Common Name	Holo-cellulose[a]	Cross and Bevan Cellulose[b,c]	Alpha Cellulose[d]	Pentosans[e]	Klason Lignin	1% NaOH	Hot Water	EtOH/Benzene	EtOH	Ash
Abies firma S. et Z./Momi	70	—	49	5	34	—	4	2	2	1.0
Abies homolepis S. et Z./Urajiromomi	77	—	53	6	29	—	2	2	2	0.2
Abies mariesii Masters/Aomoritodomatsu	72	—	50	8	30	—	2	2	2	2.3
Abies mayriana Miyabe & Kudo/Aotodomatsu	74	59	44	13	30	13	3	3	1	0.2
Abies sachalininensis Fr. Schmidt/Todomatsu	70	57	41	13	29	12	5	5	3	0.5
Abies sachalininensis Fr. Schmidt/Todomatsu	74	—	49	5	30	—	3	3	3	0.3
Abies veitchii Lindley/Shirabe	73	—	47	6	29	—	2	2	2	0.2
Chamaecyparis obtusa Endlicher/Hinoki	69	—	39	5	33	—	4	4	5	0.5
Chamaecyparis pisifera S. et Z./Momi	60	—	47	5	29	—	7	7	9	0.4
Criptomeria japonica D. Don/Sugi[h]	71	—	47	7	33	—	3	3	3	0.7
Larix leptolepis Gordon/Karamatsu	67	52	40	12	31	19	7	7	1	0.4
Larix leptolepis Gordon/Karamatsu	69	—	48	6	28	—	10	10	3	0.3
Picea abies (L.) Karst./Doitsutohi	73	54	38	12	29	12	2	2	1	0.4
Picea glehnii Masters/Akaezomatsu	75	60	45	14	27	14	2	2	<1	0.4
Picea glehnii Masters/Akaezomatsu	74	—	50	7	28	—	2	2	2	0.2
Picea hondoensis Mayr./Tohi	64	—	42	5	29	—	3	3	2	0.2
Picea jezoensis Carr./Ezomatsu	75	59	44	14	29	13	3	3	1	0.1
Picea jezoensis Carr./Ezomatsu	71	—	47	6	28	—	4	4	1	0.2

Pinus banksiana Lamb./Bankusumatsu	71	55	40	14	28	13	2	1	0.1
Pinus densiflora S. et Z./ Akamatsu[g]	67	—	45	8	27	—	4	3	0.2
Pinus pentaphylla Mayr./Goyomutsu	71	58	32	12	26	19	6	8	0.1
Pinus pentaphylla Mayr./Himekomatsu	68	—	45	5	27	—	3	8	0.3
Pinus pumila (Pallas) Regel/Haimatsu	63	44	30	12	26	23	9	12	0.2
Pinus strobus L./Sutorobumatsu	71	57	41	13	28	19	4	7	0.5
Pinus thunbergii Parlatore/Kuromatsu	63	—	44	7	26	—	3	3	0.2
Podocarpus macrophyllus D. Don/Inumaki	65	—	49	11	36	—	3	2	0.4
Pseudotsuga japonica Beissner/ Toyasawara	68	—	47	5	33	—	4	4	0.1
Sciadopitys verticillata S. et Z./ Koyamaki	61	—	39	5	29	—	7	11	0.2
Taxus cuspidata S. et Z./Onko	63	58	33	12	29	26	14	14	0.2
Taxus cuspidata S. et Z./Ichii	59	—	38	6	28	—	11	12	0.2
Thuja standishii Carr./Nezuko	70	—	48	6	27	—	11	9	0.3
Thujopsis dolabrata S. et Z./Asunaro	62	—	41	6	32	—	4	4	0.4
Thujopsis dolabrata var, Hondai Makino/Hinokiasunaro	71	56	39	13	29	16	5	4	0.3
Thujopsis dolabrata var, Hondai Makino/Hinokiasunaro	75	—	48	6	33	—	5	4	0.7
Torreya nucifera S. et Z./Kaya	64	—	45	5	35	—	7	7	0.7
Tsuga sieboldii Carr./Tsuga	71	—	51	4	31	—	4	3	0.2

NOTE: Data adapted from Ref. 67 are percent moisture-free wood. Data adapted from Ref. 66 are not defined in the English abstract and table.

[a] Holocellulose is the total carbohydrate content of wood.
[b] Cross and Bevan cellulose is largely pure cellulose but contains some hemicelluloses.
[c] Species with a value for Cross and Bevan cellulose from Ref. 66. All others from Ref. 67.
[d] Alpha cellulose is nearly pure cellulose.
[e] Pentosans are the total anhydroxylose and arabinose residues in wood.
[f] Average of five trees.
[g] Average of four trees.
[h] Average of five trees.

Table VIII. Chemical Composition of Woods from Cambodia, Kalimantan (Borneo), and Papua New Guinea

Scientific Name/Common Name	Carbohydrate		Klason Lignin	Solubility			Ash
	Holo-cellu-lose[a]	Alpha Cellu-lose[b]		1% NaOH	Hot Water	EtOH/ Benzene	
Cambodia (68)							
Anisoptera glabra Kurz/Phdiek	75	50	29	21	5	5	0.9
Dacrydium elatum (Boxb.) Wall/Srol kraham	70	51	35	15	3	3	0.4
Dipterocarpus alatus Boxb./Chhoeuteal sar	73	49	33	24	3	3	0.9
Dipterocarpus insularis Hance/Chhoeuteal bangkuoi	64	44	36	28	5	5	0.4
Hopea pierrei Hance/Koki khsach	69	49	27	30	11	12	0.2
Parkia streptocarpa Hance/Ro yong	78	51	30	15	3	1	0.9
Shorea hypochra Hance/Komnhan	69	47	32	21	6	6	1.3
Tristania sp./Rong leang	72	48	36	20	3	1	0.5
Kalimantan (Borneo) (69)							
Aquilaria sp./Karas	74	50	26	—	6	2	1.5
Artocarpus sp./Keledang	72	51	31	—	4	1	1.6
Cotylelobium sp./Giam	62	46	26	—	11	14	0.8
Dipterocarpus sp./Keruing[c]	74	55	29	—	2	3	0.9
Dryobalanops sp./Kapur	72	50	34	—	7	2	0.7
Dyera sp./Jelutong	72	44	27	—	9	5	1.5
Eugenia sp./Kelat	64	47	35	—	5	6	0.8
Michelia sp./Champaka	73	51	29	—	4	2	4.6
Quercus sp./Borneo oak	74	50	28	—	7	4	0.5
Shorea sp./Balau[d]	65	47	29	—	9	10	0.5
Shorea sp./Bangkirai[c]	70	49	34	—	5	7	0.1

Species / Common name							
Shorea sp./Light red meranti	67	47	35	—	9	5	1.6
Shorea sp./White meranti	69	50	30	—	3	4	0.5
Tarrietia/Teraling	64	45	28	—	4	3	1.4
Vatica sp./Pesak	65	42	27	—	13	12	0.7
Papua New Guinea (70,71)[e]							
Aglai litoralis Talbot/—	74	50	34	17	5	4	1.1
Ailanthus intergrifolia Lam./White siris	74	51	31	11	2	1	0.8
Alstonia scholaris (L.) R.Br./White cheesewood	67	44	34	12	4	1	1.3
Amoora cucullata Roxb./Amoora	68	47	37	20	6	1	0.4
Anthocephalus cadamba (Roxb.) Miq./Labula	74	46	26	16	4	3	0.7
Antiaris toxicaria Lesch./—	73	48	31	12	3	1	1.9
Artocarpus incisa L.f./Kapiak	70	48	31	15	3	3	2.3
Burckella macropoda (Krause) Lam./Burckella	67	50	35	15	4	1	1.9
Calophyllum vexans P. F. Stevens/Calophyllum	71	49	33	16	2	2	0.6
Canarium indicum L./Galip	70	46	28	17	4	1	0.9
Castanospermum australe A. Cunn./—	72	40	28	27	12	12	0.3
Celtis kajewskii Merr. et Perry/Light celtis	73	48	26	17	5	2	1.8
Celtis luzonica Warb./Hard celtis	73	46	23	18	3	1	1.2
Cryptocarya massoy (Oken.) Kosterm/Crytocarya	75	48	25	13	3	2	1.1
Dracontomelum puberulum Miq./P.N.G. walnut	65	46	34	18	8	3	2.2
Dysoxylum arnoldianum K. Schum./—	69	47	32	13	4	2	2.3
Dysoxylum gaudichaudianum (Juss.) Miq./—	69	46	27	12	2	1	1.3
Elaeocarpus sphaericus (Gaertn.) K. Schum./P.N.G. quandong	75	49	27	13	3	2	0.9
Eucalyptus deglupta Blume/Kamarere[d]	73	51	32	10	2	1	0.6
Euodia elleryana F. Muell./—	75	49	29	10	2	1	1.2
Homalium foetidum (Roxb.) Benth./Malas	67	46	32	17	4	2	1.2
Intsia bijuga (Colebr.) O. Kuntze/Kwila	64	41	29	24	10	7	1.0
Neonauclea maluensis S. Moore/Yellow hardwood	69	50	37	10	4	2	0.4

Continued on next page

Table VIII. Continued

Scientific Name/Common Name	Carbohydrate		Klason Lignin	Solubility			Ash
	Holo-cellulose[a]	Alpha-Cellulose[b]		1% NaOH	Hot Water	EtOH/Benzene	
Octomeles sumatrana Miq./Erima	70	48	34	8	2	2	1.0
Palaquium erythrospermum H. J. Lam/Pencil cedar	72	50	30	13	3	1	0.8
Pimelodendron amboinicum Hassk./—	74	48	26	17	4	1	1.7
Planchonella thyrosoidea C. T. White/Planchonella	79	47	21	15	1	2	1.3
Pometia pinnata Forst./Taun[d]	67	46	30	19	6	4	0.6
Pterocymbium beccarii K. Schum./Amberoi	77	47	25	13	4	1	1.6
Sloanea insularis A. C. Smith/Sloanea	77	51	30	13	4	2	1.0
Spondias dulcis Forst./Spondias[c]	74	48	27	16	3	2	1.1
Sterculia parkinsonii F. Muell./Sterculia	78	48	26	18	4	1	1.7
Syzygium sp./Water gum	66	44	29	21	5	7	1.0
Terminalia calamansanai (Blco.) Rolfe/Yellow-brown terminalia	71	49	30	15	5	2	0.9
Terminalia solomonensis Excell./Pale brown terminalia[d]	72	47	33	12	3	1	0.5

NOTE: Values are for percent oven-dry wood.
[a] Holocellulose is the total carbohydrate content of wood.
[b] Alpha cellulose is nearly pure cellulose.
[c] Average of two trees.
[d] Average of three trees.
[e] Common names obtained from Ref. 72.

Table IX. The Chemical Composition of Philippine Woods

Scientific Name/Common Name	Carbohydrate			Klason Lignin	Solubility				Ether	Ash	Reference
	Holo-cellu-lose[a]	Alpha Cellu-lose[b]	Pento-sans[c]		1% NaOH	Hot Water	Benzene[d]	EtOH			
Hardwoods											
Adenanthera intermedia Merr./—	76	40	—	35	17	7	6		2.0	0.8	73
Aegiceras corniculatum (L.) Blanco/Saging-saging	72	—	23	20	23	2	5		—	0.9	74
Aegiceras floridum Roem. & Schult./Tinduk-tindukan	68	—	21	24	24	2	6		—	0.6	74
Aglaia llanosiana C.DC./—	75	37	—	32	10	4	2		0.7	1.3	73
Alangium chinense (Lour.) Rehder/—	81	42	—	29	23	13	10		0.8	0.8	73
Albizzia acle (Blanco) Merr./—	70	32	—	33	17	12	7		0.9	1.1	73
Albizzia falcataria (L.) Fosb./Moluccan sau	72	—	18	24	14	1	2		—	0.6	75
Albizzia lebbeck (Linn.) Benth/—	71	35	—	28	21	11	6		0.5	0.5	73
Albizzia lebbekoides (DC.) Benth/—	79	43	—	29	14	6	5		1.1	0.2	73
Aleurites moluccana Willd./—	78	46	—	20	21	10	1		0.1	2.1	73
Aleurites trisperma Blanco/—	74	38	—	32	22	6	2		0.6	1.7	73
Alphonsea arborea (Blanco) Merr./—	79	41	—	30	13	5	3		0.9	0.7	73

Continued on next page

Table IX. Continued

Scientific Name/Common Name	Carbohydrate			Klason Lignin	Solubility					Ash	Reference
	Holo- cellu- lose[a]	Alpha Cellu- lose[b]	Pento- sans[c]		1% NaOH	Hot Water	Benzene[d]	EtOH	Ether		
Alphanamixis cumingiana (C.DC.) Harms./—	79	40	—	33	18	8	3	3	0.5	2.7	73
Artocarpus cumingiana Trec/—	76	45	—	29	20	7	6	6	0.7	2.3	73
Avicennia marina (Forsk.) Vierh./Bungalon	70	—	25	21	25	4	5	5	—	1.3	74
Avicennia officinalis L./Api-api	69	—	21	17	26	5	7	7	—	2.3	74
Beilschmiedia glomerata Merr./ *Bischofia javanica* Blume/—	73	33	—	25	16	6	3	3	0.7	1.1	73
Bombycidendron vidalianum (Naves) Merr. & Rolfe./—	73	30	—	48	29	3	<1	<1	0.5	1.5	73
Bruguiera gymnorrhiza (L.) Lam./Busaing	66	38	—	29	14	3	2	2	0.4	0.5	73
Bruguiera parviflora (Roxb.) W. A. ex Griff./Langarai	69	—	19	25	19	2	3	3	—	1.1	74
Bruguiera sexangula (Lour.) Poir/Pototan	77	—	22	18	15	2	2	2	—	0.9	74
Caesalpin a sappan Linn./—	69	—	21	24	16	1	4	4	—	1.1	74
Calophyllum blancoi Pl & Tr./Bitanghol	63	29	—	32	24	9	7	7	0.4	0.8	73
	70	—	15	27	14	1	1	1	—	0.3	75
Calophyllum inophyllum Linn./	70	34	—	38	16	4	4	4	0.4	0.5	73

Campostemon philippinense (Vid.) Becc./Gapas-gapas	74	—	20	20	15	1	3	—	1.9	74
Cananga odoratum (Lam.) Hook.f. & Thomas/Ilang-ilang	71	48	13	29	11	2	1	0.3	0.8	76
Canarium aspersum Benth/—	70	32	—	26	29	15	2	0.2	2.1	73
Canarium hirsutum Willd./—	77	45	—	24	20	8	1	0.3	1.6	73
Casuarina rumphiana Miq./Mountain agoho	76	—	21	22	14	1	1	—	0.3	75
Celistocalyx operculatus (Roxb.) Merr. & Perry/Malaruhat	70	—	17	22	21	5	3	—	0.6	75
Celtis philippensis Blanco/—	75	43	—	27	13	7	3	0.5	1.8	73
Ceriops tagal (Perr.) C. B. Rob/Tangal	68	—	20	17	26	6	8	—	1.5	74
Delonix regia (Boj.) Raf/—	78	46	—	25	17	8	4	0.2	1.8	73
Diospyros discolor Willd./—	71	35	—	34	21	8	6	1.4	1.3	73
Diospyros pilosanthera Blanco/—	82	44	—	28	15	7	4	0.5	1.5	73
Diplodiscus paniculatus Turcz/—	80	39	—	33	11	5	2	0.5	3.4	73
Dipterocarpus basilanicus Foxw./Basilan apitong[e]	70	—	13	25	15	1	3	—	0.4	77
Dipterocarpus caudatus Foxw./Leaf-tailed panau	65	—	17	30	23	3	1	—	0.5	77
Dipterocarpus graclis Blume/Panau[f]	65	—	15	27	16	2	4	—	0.6	77
Dipterocarpus grandiflorus Blanco/Apitong[g]	64	—	15	27	22	2	6	—	0.9	77
Dipterocarpus hasseltii Blume/Hasselt panau	63	—	17	29	17	3	4	—	1.2	77

Continued on next page

Table IX. Continued

Scientific Name/Common Name	Carbohydrate			Klason Lignin	Solubility				Ash	Reference
	Holo-cellu-lose[a]	Alpha-Cellu-lose[b]	Pento-sans[c]		1% NaOH	Hot Water	EtOH/Benzene[d]	Ether		
Dipterocarpus kerrii King/Malapanau[f]	65	—	16	28	15	4	3	—	0.8	77
Dipterocarpus orbicularis Foxw./Round-leaf apitong[g]	65	—	16	30	16	2	3	—	0.8	77
Dipterocarpus speciosus Brandis/Broad-winged apitong[e]	65	—	15	29	16	2	3	—	0.7	77
Dipterocarpus warburgii Brandis/Hagakhak[e]	63	—	16	31	14	2	3	—	0.8	77
Drypetes bordenii Pax & K. Hoffm./—	80	42	—	32	16	6	3	0.7	1.7	73
Dysoxylum turczaninowii C.DC./—	77	41	—	35	6	5	1	0.7	1.6	73
Endospermum peltatum Merr./—	81	44	—	31	18	8	3	0.4	0.8	73
Eucalyptus deglupta Blume/Bagras	71	—	16	26	14	1	2	—	0.7	75
Euphoria didyma Blanco/—	69	34	—	36	14	3	2	0.2	1.4	73
Excoecaria aggallocha L./Buta-buta	75	—	22	18	18	3	3	—	1.3	74
Ficus conora King/—	74	35	—	34	18	9	3	0.1	2.6	73
Ficus malunuensis Warb./—	77	43	—	30	13	5	3	0.6	3.0	73

Ficus nota (Blanco) Merr./—	73	33	—	34	18	8	3	0.5	4.0	73
Garciana venulosa (Blanco) Choisy/—	74	38	—	35	22	8	7	4.8	1.5	73
Heritiera littoralis Ait./ Dungon-late	69	—	18	21	22	4	5	—	1.9	74
Hopea plagata (Blanco) Vidal/—	75	31	—	34	24	9	7	6.2	2.0	73
Intsia bijuga (Colebr.) O. Ktze./—	71	41	—	33	22	11	7	1.2	1.3	73
Koordersiodendron pinnatum (Blanco) Merr./—	77	40	—	34	18	2	2	1.0	1.1	73
Lagerstroemia speciosa (Linn.) Pers./—	75	34	—	35	18	9	2	0.2	2.3	73
Lithocarpus lianosii (A.D.C.) Rehd./Ulaian	71	—	17	22	17	5	2	—	0.6	75
Lumnitzera littorea (Jack.) Voigt./Tabau	58	—	15	29	17	3	9	—	1.6	74
Macaranga tanarius (Linn.) Muell-Arg./—	80	40	—	32	15	6	3	0.2	0.9	73
Mangifera altissima Blanco/—	71	38	—	31	14	5	5	0.3	0.7	73
Melanolepsis multiglandulosa (Reinw.) Reichb.f. & Zoll./—	75	38	—	29	25	13	2	0.5	1.3	73
Myristica elliptica Hook.f. & Thomas. Var. Simiarum (A.D.C.) J. Sinal./Tanghas	67	—	15	24	23	6	2	—	0.8	75
Ochroma lagopus Schwartz/—	74	40	—	29	22	4	3	1.2	0.9	73
Osbornia octodonta F. Muell./Tualis	65	—	16	24	20	7	3	—	0.9	74
Pahudia rhamboidea (Blco.) Prain/—	73	33	—	26	26	3	3	0.5	0.9	73

Continued on next page

Table IX. Continued

Scientific Name/Common Name	Carbohydrate			Klason Lignin	1% NaOH	Hot Water	EtOH/ Benzene[d]	Ether	Ash	Reference
	Holo-cellu-lose[a]	Alpha Cellu-lose[b]	Pento-sans[c]		Solubility					
Parashorea malaanonan (Blanco) Merr./—	77	42	—	32	14	7	2	1.3	1.0	73
Parashorea plicata Brandis/ Bagtikan[h]	65	—	15	30	13	2	3	—	1.2	78
Parinarium corymbosum (Blume) Miq./—	74	37	—	36	13	5	3	1.0	3.7	73
Pentacme contorta (Vidal) Merr./White lauan	67	51	9	31	11	2	3	1.0	—	76
Pentacme contorta (Vidal) Merr./White lauan[i]	65	—	14	29	14	2	3	—	0.8	78
Planchonia spectabilis Merr./—	75	37	—	40	20	9	6	1.5	0.4	73
Polyalthia rumphii (Blume) Merr./—	74	34	—	28	20	11	5	0.5	1.9	73
Polyscias nodosa (Blume) Seem/ —	73	36	—	30	25	10	5	0.9	0.9	73
Pometia pinnata Forst./Malugai	68	—	14	27	18	3	2	—	0.7	75
Pterocarpus indicus Willd./—	80	41	—	32	17	10	4	0.7	1.1	73
Pterospermum diversifolium Blume/—	76	38	—	37	15	6	7	0.7	1.2	73
Pterospermum niveum Vidal/—	79	44	—	33	12	2	2	1.0	0.9	73
Pterospermum obliquam Blanco/—	80	45	—	35	13	4	4	0.9	0.6	73

Species										
Pygeum vulgare (Koehne) Merr./—	78	41	—	33	16	3	2	2.4	0.2	73
Quercus bennettii Miq./—	71	41	—	35	16	7	4	0.3	0.3	73
Radermachera pinnata (Blanco) Seem/—	75	34	—	38	14	7	5	0.9	0.8	73
Rhizaphora mucronata Lam./ Bakanan-babae	72	—	18	22	17	1	3	—	0.9	74
Samanea saman (Jacq.) Merr./—	75	38	—	30	20	9	5	0.9	0.3	73
Sandoricum koetjape (Burm.f.) Merr./—	73	40	—	29	18	6	4	2.5	0.6	73
Sapium luzonicum (Vidal) Merr./—	73	44	—	31	16	7	8	0.2	1.6	73
Scyphophora hydrophyllacea Gaertn./Nilad	67	—	23	17	26	2	13	—	0.7	74
Shorea agsaboensis Stern/Tiaong	66	—	12	31	15	1	2	—	0.2	78
Shorea almon Foxw./Almon[f]	67	—	14	26	16	2	5	—	0.3	78
Shorea negrosensis Foxw./Red lauan	62	50	7	34	14	3	2	0.6	—	76
Shorea negrosensis Foxw./Red lauan[j]	58	—	12	35	20	2	5	—	0.3	78
Shorea philippinensis Brandis/ Manggasihoro	64	52	8	34	14	2	2	0.6	—	76
Shorea polysperma (Blanco) Merr./Tangile	61	45	8	37	15	3	2	0.7	—	76
Shorea polysperma (Blanco) Merr./Tangile[j]	64	—	13	32	17	1	3	—	0.3	78
Shorea squamata (Turcz.) Dyer/ Mayapis[i]	64	—	12	30	19	2	5	—	0.3	78

Continued on next page

Table IX. Continued

Scientific Name/Common Name	Carbohydrate			Klason Lignin	Solubility				Ash	Reference
	Holo-cellu-lose[a]	Alpha Cellu-lose[b]	Pento-sans[c]		1% NaOH	Hot Water	EtOH/ Benzene[d]	Ether		
Sonnertia albe J. Sm./Pagatput	63	—	15	26	22	3	5	—	2.2	74
Strombosia philippinensis (Baill.) Rolfe/—	82	41	—	37	12	3	2	0.8	0.6	73
Swietenia mahagoni Jacq./—	73	36	—	25	20	12	7	3.9	0.8	73
Tectona grandis Linn.f./—	73	33	—	35	22	11	4	2.8	1.7	73
Terminalia catappa Linn./—	67	30	—	33	19	11	5	0.4	0.7	73
Terminalia comintana (Blanco) Merr./—	76	36	—	35	16	7	5	0.2	1.8	73
Terminalia edulis Blanco/—	71	36	—	34	20	8	5	0.4	0.4	73
Trema orientalis (L.) Blume/ Anabiong	71	—	17	24	19	3	2	—	0.9	75
Vatica mangachapoi Blanco/—	74	39	—	30	24	7	7	1.8	0.5	73
Vitex parviflora Juss./—	73	36	—	39	7	2	8	0.7	1.6	73
Wallaceodendron celebicum Koord/—	75	40	—	32	14	4	3	1.4	1.2	73
Xylocarpus granatum Koen./ Tabigi	68	—	20	17	26	6	8	—	1.5	74

Species										
Zizyphus talanai (Blanco) Merr./—	76	40	—	32	11	6	4	0.8	1.7	73
Softwoods										
Agathis philippinensis Warb./Almaciga[e]	64	—	8	32	14	1	2	—	0.6	79
Araucaris bidwilli Hook./Bunya pine	67	—	14	28	14	2	3	—	0.5	79
Pinus insularis Endl./Benguet pine[g]	66	—	11	30	14	2	2	—	0.3	79
Pinus merkusii Jungh. & de Vr./Mindoro pine[k]	65	—	10	28	17	2	4	—	0.3	79
Podocarpus imbricatus R.Br./Igem	70	—	10	29	10	1	<1	—	0.2	79
Podocarpus philippinensis Foxw./Malakauayan[e]	58	—	13	38	10	1	2	—	0.4	79

NOTE: Moisture-free wood specified in Refs. 73 and 76. All others were not specified. Analytical methods from Ref. 73 based on methods developed at U.S. Forest Products Laboratory. The values here are 100 − (the sum of percent ash, EtOH/benzene solubles, hot-water solubles, and lignin). Values from Refs. 73 and 76 were experimentally determined.
[a] Holocellulose is the total carbohydrate content of wood.
[b] Alpha cellulose is nearly pure cellulose.
[c] Pentosans are the total anhydroxylose and arabinose residues in wood.
[d] Woods from Ref. 73 extracted with alcohol (probably ethanol).
[e] Average of two trees.
[f] Average of five trees.
[g] Average of three trees.
[h] Average of six trees.
[i] Average of eight tress.
[j] Average of nine trees.
[k] Average of four trees.

Table X. Chemical Composition of Woods from Taiwan

Scientific Name/Common Name	Carbohydrate				Klason Lignin	Solubility				Ash
	Holo-cellulose[a]	Alpha-Cellulose[b]	Pento-sans[c]			1% NaOH	Hot Water	EtOH/Benzene	Ether	
Hardwoods										
Acacia confusa Merr./Taiwan acacia	87	54	19		19	21	7	6	1.5	0.4
Actinodaphne nantoensis Hay./Nantou actinodaphne	87	51	17		26	21	3	3	1.5	0.7
Aleurites montana Wils./Wood oil tree	86	46	23		25	19	3	3	1.5	1.1
Alnus formosana Makino/Formosan alder	86	45	24		24	17	3	2	1.8	0.6
Bischoffia trifoliata Hook./Bishop wood	—	—	15		33	17	—	4	—	0.9
Cassia siamea Lam./Kassod tree	87	51	19		25	16	4	5	1.7	1.6
Castanopsis carlesii Hay. var. Carlessi Li./Candate-leaved chinkapin	78	48	14		23	22	11	3	1.5	0.6
Castanopsis kawakamii Hay./Kawakami chinkapin	84	46	19		26	20	3	4	0.8	0.3
Cinnamomum camphora Sieb./Camphor tree	80	48	17		29	19	5	8	1.6	1.2
Cinnamomum micranthum Hay./Stout camphor tree	86	56	18		20	12	5	3	1.5	0.9
Cinnamomum randaiense Hay./Fragrant cinnamon	86	53	18		22	18	3	5	1.1	0.7
Cryptocarya chinensis Hemsl./Chinese cryptocarya	80	43	16		26	16	7	4	0.4	0.9
Cyclobalanopsis gilva Oerst./Red bark oak	83	46	21		23	21	4	5	1.6	1.7
Cyclobalamopsis longinux Schot./Narrow-leaved oak	84	53	16		22	23	5	3	1.6	0.5
Cyclobalamopsis morii Hay./Mori oak (81)	88	48	17		32	15	2	2	0.2	0.8
Engelhardtia chrsolepis Hance/Taiwan engelhardtia	86	50	16		24	19	2	3	1.6	1.4
Euphoria longana Lam./Dragon's eye lungan	78	53	16		30	28	5	4	0.8	1.7
Lagerstroemia subcostata Koehne/Subcostata crape myrtle	73	37	17		27	18	7	4	1.5	1.4
Lithocarpus amygdalifolius Hay./Almond-leaved tanoak	87	52	23		21	29	8	3	1.5	1.1
Machilus kusanoi Hay./Large-leaved machilus	88	49	17		22	13	4	2	0.5	0.8
Machilus thunbergii S. et Z./Red machilus	81	53	20		19	21	4	5	1.5	1.0
Machilus zuihoensis Hay./Incense machilus	86	49	15		24	23	5	4	1.5	1.9
Michelia formosana Masamune/Formosan michelia	80	43	18		29	15	2	4	1.6	0.5

2. PETTERSEN The Chemical Composition of Wood

Species	Holocellulose[a]	Alpha cellulose[b]	Pentosans[c]						
Pasania brevicaudata Schot./Short-tailed leaf tanoak	82	55	17	26	18	2	3	1.6	0.6
Pasania ternaticupula Schot./Nanban tanoak	80	44	20	26	26	6	3	0.6	0.8
Pasania uraiana Schot./Urai tanoak	82	54	18	23	19	9	3	1.4	0.5
Paulownia kawakamii Ito/Kawakami paulownia	82	54	17	26	15	6	2	0.9	0.7
Sassafras randaiense Rhed./Taiwan sassafras	80	42	19	22	25	5	6	2.4	0.4
Schefflera octophylla Harms./Schefflera tree	84	45	20	22	21	4	3	0.7	0.6
Schima superba G. et Ch./Chinese guger tree	86	47	14	29	19	3	2	1.5	0.5
Ternstroemia gymnanthera Sprague/Japanese ternstroemia	76	42	18	30	21	6	6	1.4	0.5
Trema orientalis Bl./India-charcoal trema	84	50	16	28	24	4	2	1.6	1.6
Trochodendron aralioides S. et Z./Bird-lime tree	86	46	17	29	27	6	6	1.5	0.8
Zelkova formosana Hay./Taiwan zelkova	86	56	17	18	21	7	6	1.4	0.7
Softwoods[d]									
Abies kawakamii Ito/Taiwan white fir	51	35	9	31	16	4	2	—	—
Calocedrus formosana Florin/Taiwan incense cedar	51	33	10	34	14	4	3	—	0.4
Chamaecyparis formosensis Matsam./Taiwan red cypress	50	38	11	33	13	5	4	—	—
Chamaecyparis taiwanensis Matsam. et Suzuki/Taiwan yellow cypress (82)	51	37	10	30	14	5	4	—	—
Cryptomeria japonica D. Don/Japanese fir	47	38	14	33	16	4	4	—	1.4
Cunninghamia lanceolata Hook./China fir	51	39	11	33	13	3	4	—	0.9
Picea morrisonicola Hay./Taiwan spruce	52	38	10	31	15	4	2	—	—
Pinus armandi Franch/Armand pine	54	40	9	33	19	5	7	—	0.8
Pinus luchuensis Mayr./Luchu pine	49	38	10	28	17	6	3	—	—
Taiwania cryptomerioides Hay./Taiwania	45	37	10	32	15	6	7	—	1.2
Tsuga chinensis Pritz./Chinese hemlock	53	42	38[e]	36	13	3	3	—	0.2

[a] Holocellulose is the total carbohydrate content of wood.
[b] Alpha cellulose is nearly pure cellulose.
[c] Pentosans are the total anhydroxylose and arabinose residues in wood.
[d] Values for softwoods are total cellulose obtained by method of Sieber and Walter (83). This method requires successive chlorinations, extractions with 1% aqueous $NaHSO_3$, and bleaching with 0.1% $KMnO_4$ solution.
[e] Probably a typing error in original report.
(Reproduced with permission from Ref. 80. Copyright 1971, Taiwan Foreign Research Institute.)

Table XI. Chemical Composition of Woods from the U.S.S.R.

Scientific Name/Common Name	Carbohydrate		Klason Lignin	Solubility				Ash	Region
	Kürschner Cellulose[a]	Pento-sans[b]		Ether	Alcohol	Water			

Scientific Name/Common Name	Cellulose	Pentosans	Lignin	Ether	Alcohol	Water	Ash	Region
Hardwoods								
Ailanthus glandulosus Desf./Tree of heaven	46	18	14	6.0	3	3	0.9	Caucasus
Alnus glutinosa Medic./European alder	48	24	22	0.9	3	<1	0.3	Leningrad
Ammodendron conollyi Bel./Sandy acacia	43	21	26	4.2	3	2	0.4	Central Asia
Arbutus andrache L./Strawberry tree	38	26	24	0.7	11	1	0.8	Crimea
Betula dahurica Pall./Dahurian birch	50	27	19	1.6	2	<1	0.2	Far Eastern
Betula mandshurika Nakai/Manchurian white birch	43	—	20	1.5	1	3	0.3	Maritime Territory
Betula pubescens Ehrh./White birch	46	29	20	—	—	—	—	Karelia
Betula schmidtii Bgt./Schmidt's birch	47	25	18	1.2	9	<1	0.2	Far Eastern
Betula tianschanica Rupr./Tien shan birch	43	32	19	2.2	2	1	0.3	Central Asia
Buxus sempervirens L./Box tree	40	26	30	0.8	3	1	0.5	Caucasus
Carpinus betulus L./Common hornbeam	47	26	19	0.9	1	1	0.5	Caucasus
Castanea sativa Mill./Sweet chestnut	43	20	22	1.4	8	3	0.4	Caucasus
Celtis austriaca australis L./Hackberry	42	29	21	0.8	7	2	1.3	Crimea
Corylus avellana L./European filbert	47	29	22	0.6	3	2	0.4	Central Chernozem
Cotoneaster vulgaris/Juneberry	44	31	22	0.5	1	<1	0.4	Leningrad
Disopyros lotus L./Date-plum persimmon	45	24	19	2.3	4	2	0.8	Caucasus
Fraxinus excelsior L./Common ash	44	25	25	1.2	3	1	0.5	Central Chernozem
Haloxylon aphyllum Bunge/Black haloxylon	32	21	28	0.7	1.3	3	2.9	Central Asia
Juglans manschurica Max/Manchurian walnut	51	16	20	2.2	4	2	0.4	Far Eastern
Juglans regia L./Persian walnut	49	20	22	2.2	5	1	0.5	Caucasus
Laurus nobilis L./True bay	43	29	21	0.7	5	3	0.7	Crimea
Maclura aurantiaca Nutt./Osage orange	40	21	19	3.0	9	2	0.6	Caucasus
Olea europaea L./Common olive	43	24	20	2.4	14	1	1.0	Crimea
Ostrya carpinifolia Scop./Hop hornbeam	49	24	21	0.8	2	1	0.6	Caucasus
Paulownia tomentosa (Thunb) Steud./Royal pavlownia	46	24	20	1.2	6	2	0.3	Caucasus
Parrotia persica D.A. Med./Persian ironwood	46	26	20	1.4	2	1	0.5	Caucasus
Phellodendron amurense Rupr./Amur cork tree	48	20	22	0.8	2	2	0.4	Far Eastern
Pirus communis L./Common pear	44	26	24	0.7	2	1	0.4	Caucasus
Pirus malus L./Apple tree	45	24	25	0.8	1	1	0.5	Caucasus
Pistacia mutica F./Turkish terebinth	34	23	22	3.3	9	4	0.2	Caucasus

2. PETTERSEN The Chemical Composition of Wood

Species								Location
Platanus orientalis L./Oriental plane	44	21	21	1.2	3	1	1.3	Caucasus
Populus nigra L./Black poplar	48	23	19	1.8	5	1	0.4	Central Eastern
Prunus avium L./Gean tree	45	24	18	2.8	7	1	0.3	Caucasus
Prunus laurocerasus L./Cherry laurel	45	26	27	0.5	1	1	0.5	Caucasus
Prunus padus L./Bird cherry	47	28	20	0.5	1	1	0.2	Leningrad
Punica granatum L./Pomegranate	39	25	21	0.8	4	3	1.2	Crimea
Quercus mongolica Fisch./Mongolian oak	47	24	22	0.9	2	2	0.2	Far Eastern
Quercus sessiliflora Salisb./Sessile oak	44	23	24	0.9	3	2	0.3	Central Chernozem
Salix alba L./White willow	46	25	28	1.2	2	2	0.5	Central Chernozem
Sambucus nigra L./Common alder	48	25	30	0.4	2	1	0.6	Caucasus
Sorbus aucuparia L./Mountain ash	46	30	22	0.9	3	1	0.6	Leningrad
Sorbus torminalis Crtz./Birch	42	27	26	0.4	1	<1	0.7	Caucasus
Tamarix gallica L./Tamarisk	35	21	18	0.7	8	9	5.4	Crimea
Tilia amurensis L./Amur linden	43	23	18	7.7	4	2	0.7	Far Eastern
Tilia cordata Mill/Small-leaved linden	50	23	18	5.7	2	1	0.6	Central Chernozem
Ulmus laevis Pall./Russian elm	52	20	22	1.0	2	2	0.7	Central Chernozem
Zelcova carpinifolia Dipp./Zelkova elm	33	21	20	1.7	15	1	0.8	Caucasus
Softwoods								
Abies holophylla Max./Manchurian fir	43	—	30	1.4	2	3	0.6	Maritime Territory
Abies nephrolepis (Traut.) Maxim./Khingan fir	56	5	28	0.7	—	3	0.4	Far Eastern
Abies nordmannana (Stev.) Spach/Nordmann fir	46	10	29	2.5	4	<1	0.4	Caucasus
Abies sachalinensis Masters/Sakhalin fir	55	6	29	3.7	—	2	0.2	Sakhalin
Abies sibirica Ledeb./Siberian fir	51	5	30	0.9	2	1	0.7	Siberia
Larix dahurica Turcz./Dahurian larch	52	12	27	1.3	1	2	0.2	Far Eastern
Larix sibirica Ledeb./Siberian larch	46	9	30	1.8	2	5	1.0	Siberia
Picea fennica Regel/Finnish Siberian spruce	48	10	29	1.4	—	1	0.3	Karelian ASSR
Picea jesoensis (S. et Z.) Carr./Jeddo spruce	47	7	29	3.1	—	4	0.2	Sakhalin
Picea obovata Led./Siberian spruce	46	10	28	1.5	2	1	0.3	Karelian ASSR
Picea schrenkiana Fish & Meyer/Schrenk spruce	41	13	33	0.6	3	1	0.6	Central Asia
Pinus koraiensis Sieb. & Zuss/Korean pine	44	—	26	6.7	3	8	0.2	Maritime Territory
Pinus sibirica Rupr./Siberian stone pine	53	9	30	2.4	3	2	0.1	Siberia
Pinus sylvestris L./Scotch pine	54	11	28	1.6	1	1	0.2	Leningrad
Taxus baccata L./English yew	43	12	29	2.3	3	1	0.4	Caucasus

[a] Kürschner cellulose is nearly pure cellulose.
[b] Pentosans are the total anhydroxylose and arabinose residues in wood.
(Reproduced with permission from Ref. 84. Copyright 1966, Israel Program for Scientific Translations, Ltd.)

Table XII. Chemical Composition of Woods of Unrecorded Origin

Scientific Name/Common Name	Carbohydrate			Klason Lignin	Solubility		
	Cross and Bevan Cellulose[a]	Alpha Cellulose[b]	Pentosans[c]		1% NaOH	Hot Water	EtOH/ Benzene
Eucalyptus marginata Sm./Jarrah	41	36	11	43	26	7	1
Juniperus procera Hochst./African pencil cedar	42	33	13	37	25	6	7
Mitragyna stipulosa Kuntze/Abura	50	44	17	33	12	5	2
Pinus palustris Mill./Pitch pine							
Highly resinous	45	33	7	21	36	3	24
Slightly resinous	53	41	11	30	15	4	2
Quercus spp./English oak	53	38	23	22	24	10	3
Tectonia grandis L.f./Teak	45	37	13	31	21	7	11
Triplochiton nigericum Sprague/Obeche	49	—	19	33	16	6	3

NOTE: Values are for percent oven-dry wood.
[a] Cross and Bevan cellulose is largely pure cellulose but contains some hemicelluloses.
[b] Alpha cellulose is nearly pure cellulose.
[c] Pentosans are the total anhydroxylose and arabinose residues in wood.
(Reproduced with permission from Ref. 86. Copyright 1939, the Royal Society of Chemistry.)

Table XIII. Chemical Composition of Some North American Woods

Scientific Name/ Common Name	Glucan	Xylan	Galactan	Arabinan	Mannan	Uronic Anhydride	Acetyl	Lignin	Ash	Reference
Hardwoods (Angiosperms)										
Acer rubrum L./Red maple	46	19	0.6	0.5	2.4	3.5	3.8	24	0.2	11
Acer saccharum Marsh./Sugar maple	52	15	<0.1	0.8	2.3	4.4	2.9	23	0.3	86
Betula alleghaniensis Britton/Yellow birch	47	20	0.9	0.6	3.6	4.2	3.3	21	0.3	86
Betula papyrifera Marsh./White birch	43	26	0.6	0.5	1.8	4.6	4.4	19	0.2	11
Fagus grandifolia Ehrh./Beech	46	19	1.2	0.5	2.1	4.8	3.9	22	0.4	11
Liquidambar styraciflua L./Sweetgum	39	18	0.8	0.3	3.1	—	—	24	0.2	87
Platanus occidentalis L./American sycamore										
Fast growth	44	18	2.0	0.7	2.2	5.6	5.3	20	0.8	88
Slow growth	43	15	2.2	0.6	2.0	5.1	5.5	23	0.7	88
Populus deltoides Bartr. ex Marsh./Eastern cottonwood										
Fast growth	42	19	1.3	0.5	2.9	5.5	4.0	24	0.7	88
Slow growth	47	15	1.4	0.6	2.9	4.8	3.1	24	0.8	88
Populus tremuloides Michax./Quaking aspen	49	17	2.0	0.5	2.1	4.3	3.7	21	0.4	11
Quercus falcata Michx./Southern red oak	41	19	1.2	0.4	2.0	4.5	3.3	24	0.8	87
Ulmus americana L./White elm	52	12	0.9	0.6	2.4	3.6	3.9	24	0.3	11
Softwoods (Gymnosperms)										
Abies balsamea (L.) Mill/Balsam fir	46	6.4	1.0	0.5	12	3.4	1.5	29	0.2	11

Continued on next page

Table XIII. Continued

Gingo biloba L./Ginko	40	4.9	3.5	1.6	10	4.6	1.3	33	1.1	89
Juniperus communis L./Common juniper	41	6.9	3.0	1.0	9.1	5.4	2.2	31	0.3	89
Larix decidua Mill./Common larch (sapwood)	46	6.3	2.0	2.5	11	4.8	1.4	26	0.2	89
Larix laricina (Du Roi) K. Koch/Tamarack	46	4.3	2.3	1.0	13	2.9	1.5	29	0.2	90
Picea abies (L.) Karst./Norway spruce	43	7.4	2.3	1.4	9.5	5.3	1.2	29	0.5	89
Picea glauca (Moench) Voss/ White spruce	45	9.1	1.2	1.5	11	3.6	1.3	27	0.3	11
Picea mariana (Mill.)B.S.P./ Black spruce	44	6.0	2.0	1.5	9.4	5.1	1.3	30	0.3	89
Picea rubens Sarg./Red spruce	44	6.2	2.2	1.4	12	4.7	1.4	28	0.3	89
Pinus banksiana Lamb./Jack pine	46	7.1	1.4	1.4	10	3.9	1.2	29	0.2	90
Pinus radiata D. Don/ Australian radiata[a]	42	6.5	2.8	2.7	12	2.5	1.9	27	0.2	91,92
Pinus resinosa Ait./Red pine	42	9.3	1.8	2.4	7.4	6.0	1.2	29	0.4	89
Pinus rigida Mill./Pitch pine	47	6.6	1.4	1.3	9.8	4.0	1.2	28	0.4	89
Pinus strobus L./Eastern white pine	45	6.0	1.4	2.0	11	4.0	1.2	29	0.2	11
Pinus sylvestris L./Scots pine	44	7.6	3.1	1.6	10	5.6	1.3	27	0.4	89
Pinus taeda L./Loblolly pine	45	6.8	2.3	1.7	11	3.8	1.1	28	0.3	87
Pseudotsuga menziesii (Mirb.) Franco/Douglas-fir	44	2.8	4.7	2.7	11	2.8	0.8	32	0.4	87
Thuja occidentalis L./Northern white cedar	43	10.0	1.4	1.2	8.0	4.2	1.1	31	0.2	11
Tsuga canadensis (L.) Carr./ Eastern hemlock	44	5.3	1.2	0.6	11	3.3	1.7	33	0.2	11

NOTE: The values expressed are for percent oven-dry wood and extractive-free wood.
[a] Australian-grown wood. Percent oven-dry wood.

Table XIV. Chemical Composition of Selected Hardwoods from the Southeastern United States (Percent Oven-Dry Wood)

	Carbohydrate		Components of Hemicellulose							
Scientific Name/Common Name	Cellu-lose	Total Hemi-cellu-lose	Gluco-man-nan	Acetyl-glucurono-xylan	Arabino-galactan	Pectin	Lig-nin[a]	Total Extrac-tives[b]	Ash	Loca-tion[c]
Acer rubrum L./Red maple	39.9	28.2	3.5	21.0	1.8	1.9	23.0	8.6	0.3	G
Acer rubrum L./Red maple	40.7	30.4	3.5	23.5	1.6	1.9	23.3	5.3	0.3	T
Aesculus octandra Marsh./Yellow buckeye	40.6	25.8	3.6	18.6	1.0	2.6	30.0	3.1	0.5	T
Carya glabra (Mill.) Sweet/ Pignut hickory	46.2	26.7	1.1	22.1	1.2	2.3	23.2	3.4	0.6	T
Carya illinoensis (Wangenh.) K. Koch/ Pecan	38.7	30.2	1.6	24.7	1.6	2.3	23.3	7.4	0.4	G
Carya sp. Nutt./Hickory	37.7	29.2	0.8	24.9	1.8	1.7	23.0	9.0	1.1	G
Carya tomentosa (Poir.) Nutt./Mockernut	43.5	27.7	1.5	21.5	1.3	3.5	23.6	5.0	0.4	T
Cornus florida L./Flowering dogwood	36.8	35.4	3.4	27.2	1.0	5.0	21.8	4.6	0.3	T
Fagus grandifolia Ehrh./American beech	36.0	29.4	2.7	23.5	1.3	1.8	30.9	3.4	0.4	T
Fraxinus americana L./White ash	48.7	22.4	1.9	16.4	1.7	2.4	23.3	5.4	0.3	G
Fraxinus americana L./White ash	39.5	29.1	3.8	22.1	1.4	1.9	24.8	6.3	0.3	T
Gordonia lasianthus (L.) Ellis/ Loblolly-bay	43.8	29.1	4.1	22.1	1.1	1.8	21.5	5.2	—	G
Liquidambar styraciflua L./Sweetgum[d]	42.8	30.1	3.6	23.6	1.0	1.9	25.7	1.1	0.3	G
Liquidambar styraciflua L./Sweetgum	40.8	30.7	3.2	21.4	1.3	4.9	22.4	5.9	0.2	T
Liriodendron tulipifera L./Yellow-poplar	39.1	28.0	4.9	20.1	0.7	2.4	30.3	2.4	0.3	T
Magnolia virginiana L./Sweetbay	44.2	37.7	4.3	20.2	1.6	1.6	24.1	3.9	0.2	T
Nyssa aquatica L./Water tupelo	45.9	24.0	3.5	18.6	0.8	1.1	25.1	4.7	0.4	G
Nyssa sylvatica Marsh./Black tupelo	44.9	23.2	3.8	17.3	1.2	0.9	28.9	2.6	0.4	G
Nyssa sylvatica Marsh./Black tupelo	42.6	27.3	3.6	18.0	1.0	4.8	26.6	2.9	0.6	T

Continued on next page

Table XIV. Continued

Oxydendron arboreum (L.) DC./Sourwood	40.7	34.6	1.3	31.9	1.0	0.4	20.8	3.6	0.3	T
Persea borbonia (L.) Spreng./Redbay	45.6	25.6	1.0	23.2	0.9	0.5	23.6	5.0	0.2	G
Platanus occidentalis L./Sycamore	43.0	27.2	2.3	22.3	1.4	1.2	25.3	4.4	0.1	G
Populus deltoides Bartr. ex Marsh./ Eastern cottonwood[e]	46.5	24.6	4.4	16.8	1.6	1.8	25.9	2.4	0.6	G
Populus deltoides Bartr. ex Marsh./ Eastern cottonwood[f]	47.0	25.0	5.0	18.4	0.8	0.8	26.0	1.6	0.4	G
Quercus alba L./White oak	43.7	24.2	1.4	18.0	2.2	2.6	24.3	5.4	1.0	G
Quercus alba L./White oak	41.7	28.4	3.1	21.0	1.6	2.7	24.6	5.3	0.2	T
Quercus coccinea Muenchh./Scarlet oak	43.2	29.2	2.3	23.3	1.4	2.2	20.9	6.6	0.1	T
Quercus falcata Michx./Southern red oak	40.5	24.2	1.7	18.6	1.7	2.2	23.6	9.6	0.5	G
Quercus ilicifolia Wangenh./Scrub oak	37.6	27.5	1.0	22.3	1.8	2.4	26.4	8.0	0.5	G
Quercus marylandica Muenchh./Blackjack oak	33.8	28.2	2.0	21.0	2.3	2.9	30.1	6.6	1.3	T
Quercus nigra L./Water oak	41.6	34.8	3.0	28.9	2.2	0.7	19.1	4.3	0.3	G
Quercus prinus L./Chestnut oak	40.8	29.9	2.9	23.8	1.8	1.4	22.3	6.6	0.4	T
Quercus rubra L./Northern red oak	42.2	33.1	3.3	26.6	1.6	1.6	20.2	4.4	0.2	T
Quercus stellata Wangenh./Post oak	37.7	29.9	2.6	23.0	2.0	2.3	26.1	5.8	0.5	G
Quercus velutina Lam./Black oak	39.6	28.4	1.9	23.2	1.1	1.9	25.3	6.3	0.5	T
Quercus virginiana Mill./Live oak	38.1	22.9	1.0	18.3	1.7	1.9	25.3	13.2	0.6	G
Sassafras albidum (Nutt.) Nees/Sassafras	45.0	35.1	4.0	30.4	0.9	<0.1	17.4	2.4	0.2	T
Ulmus americana L./American elm[d]	42.6	26.9	4.6	19.9	0.8	1.6	27.8	1.9	0.8	G
Ulmus americana L./American elm	41.9	29.7	3.2	20.6	1.4	4.3	25.6	2.4	0.5	T

NOTE: The data are for percent oven-dry wood.
[a] Klason lignin + acid soluble lignin.
[b] Total extractives = sum of solubles in petroleum ether, diethyl ether or chloroform, 95% EtOH, and hot water.
[c] G = southeast Georgia (swampy); T = eastern Tennessee (dry, upland).
[d] Average of 20 trees.
[e] Average of 2 trees, age 32 y ears.
[f] Average of 2 trees, age 46 years.
(Data adapted from a private communication with H. L. Hergert and others.)

Table XV. Elemental Composition of Some Woods

Wood	Parts Per Thousand					Parts Per Million				
	Ca	K	Mg	P	Mn	Fe	Cu	Zn	Na	Cl	Reference
Temperate Woods											
Abies balsamea (L.) Mill/Balsam fir[a]	0.8	0.8	0.27	—	0.13	13	17	11	—	—	93
	0.9	0.5	—	—	0.09	—	—	—	18	—	93
Acer rubrum L./Red maple[a]	0.8	0.7	0.12	0.03	0.07	11	5	29	—	—	93
	0.7	0.5	—	—	0.07	—	—	—	5	18	93
Betula papyrifera Marsh./White birch[a]	0.7	0.3	0.18	0.15	0.03	10	4	28	—	—	93
	0.9	0.2	—	—	0.03	—	—	—	9	10	93
Fraxinus americana L./White ash[b]	0.3	2.6	1.8	0.01	—	—	—	—	31	—	94
Liquidambar styraciflua L./Sweetgum[c]											
Bottomland	0.65	0.4	0.37	0.26	0.06	—	—	22	88	—	95
Upland	0.55	0.3	0.34	0.15	0.08	—	—	19	81	—	95
Picea rubens Sarg./Red spruce[a]	0.8	0.2	0.07	0.05	0.14	14	4	8	—	—	93
	0.7	0.1	—	—	0.11	—	—	—	8	0.3	93
Pinus strobus L./Eastern white pine[a]	0.2	0.3	0.07	—	0.03	10	5	11	—	—	93
	0.3	0.1	—	—	0.02	—	—	—	9	19	93

Continued on next page

Table XV. Continued

Populus deltoides Bartr./Eastern cottonwood[a,d]	0.9	2.3	0.29	—	0.02	1 × 10²	—	—	30	9.4 × 10²	—	94
	1.2	2.5	e	—	<0.01	—	—	—	—	1.1 × 10²	30	94
Populus tremuloides Michx./Quaking aspen[a]	1.1	1.2	0.27	0.10	0.03	12	7	17	—	—	—	93
	0.8	0.9	—	—	0.04	—	—	—	—	5	—	93
Quercus alba L./White oak[b]	0.5	1.2	0.31	—	<0.01	—	—	—	—	21	15	94
Quercus falcata Michx./Southern red oak[c]	0.3	0.6	0.03	0.02	0.01	30	73	38	—	44	—	76
	0.1	2.8	0.35	—	—	—	—	—	—	63	38	94
Tilia americana L./Basswood[b]	0.8	0.4	0.11	0.12	0.15	6	5	2	—	—	—	93
Tsuga canadensis (L.) Carr./Eastern hemlock[a]	1.1	0.3	—	—	0.12	—	—	—	—	6	—	93
Tropical Woods[b]												
Eriotheca sp.	0.1	8.7	4.0	—	<0.01	—	—	—	—	1.5 × 10²	2.5 × 10²	93
Peltogyne prophyrocardia Griseb.	0.2	9.8	8.6	—	0.06	—	—	—	—	48	97	93
Stryphnodendron polystachum (Miq.) Kleinh.	0.5	26.1	1.0	—	0.01	—	—	—	—	6.8 × 10²	1.1 × 10³	93

NOTE: Values of parts per thousand or parts per million are for oven-dry wood. Values in the first row obtained by neutron activation methods. Values in second row for same tree species obtained by atomic spectrometric methods.
[a] Values obtained by neutron activation method.
[b] Values obtained by atomic spectrometric methods.
[c] Sawdust.
[d]
[e] Observed, but not measured.

Table XVI. Summary of Carbohydrate, Lignin, and Ash Compositions for Woods of 13 Nations

Country	Holocellulose[a]	Alpha Cellulose[b]	Other Cellulose	Pentosans[c]	Klason Lignin	Ash
Brazil (Table IV)	71.7 ± 26.6(6)	49.4 ± 4.1(18)	52.3 ± 1.9(6)[d]	14.5 ± 4.2(24)	28.6 ± 3.9(24)	0.5 ± 0.3(24)
Cambodia (Table VIII)	71.3 ± 4.3(8)	43.6 ± 2.3(8)	—	—	32.3 ± 3.4(8)	0.7 ± 0.4(8)
Costa Rica (Table IV)	78.1 ± 3.3(22)	—	—	12.3 ± 2.1(22)	26.5 ± 3.7(22)	1.3 ± 2.0(22)
Ghana (Table VI)	—	—	45.5 ± 4.2(4)[e]	17.0 ± 2.2(4)	29.8 ± 3.9(4)	0.8 ± 0.7(4)
Japan (Table VII)						
Hardwoods	78.0 ± 3.7(100)	45.0 ± 4.9(100)	58.0 ± 3.9(56)[d]	20.1 ± 3.7(100)	22.1 ± 3.0(100)	0.5 ± 0.2(100)
Softwoods	68.9 ± 4.8(36)	43.8 ± 5.5(36)	55.8 ± 4.4(12)[d]	8.3 ± 3.5(36)	29.6 ± 2.6(36)	0.4 ± 0.4(36)
Kalimantan (Table VIII)	69.0 ± 4.2(15)	48.3 ± 3.3(15)	—	—	29.9 ± 3.2(15)	0.9 ± 0.5(14)[f]
Mexico (Table IX)	67.8 ± 4.9(13)	46.5 ± 4.1(13)	—	15.1 ± 1.9(13)	25.8 ± 4.1(13)	1.7 ± 0.5(13)
Mozambique (Table VI)	—	—	39.8 ± 4.1(29)[g]	15.1 ± 2.4(29)	27.3 ± 3.4(29)	1.6 ± 1.1(29)
Papua New Guinea (Table VIII)	71.4 ± 3.7(35)	47.4 ± 2.5(35)	—	—	29.8 ± 3.8(35)	1.1 ± 0.6(37)
Philippine Islands (Table IX)						
Hardwoods	71.8 ± 5.5(112)	39.1 ± 5.1(70)	—	16.3 ± 4.1(47)	29.4 ± 5.6(112)	1.2 ± 0.7(108)
Softwoods	65.0 ± 4.0(6)	—	—	11.0 ± 2.2(6)	30.8 ± 3.8(6)	0.4 ± 0.1(6)
Taiwan (Table X)						
Hardwoods	83.3 ± 3.7(33)	48.8 ± 4.7(33)	—	17.9 ± 2.4(34)	25.0 ± 3.8(34)	0.9 ± 0.4(34)
Softwoods	—	—	50.4 ± 2.6(11)[h]	10.4 ± 1.4(11)	32.2 ± 2.1(11)	0.8 ± 0.5(6)

Continued on next page

Table XVI. Continued

U.S.A. (Table III)						
Hardwoods	71.7 ± 5.7(25)	45.4 ± 3.5(39)	59.1 ± 4.3(26)[d]	19.3 ± 2.2(49)	23.0 ± 3.0(40)	0.5 ± 0.3(34)
Softwoods	64.5 ± 4.6(22)	43.7 ± 2.6(35)	58.2 ± 3.0(23)[d]	9.8 ± 2.2(35)	28.8 ± 2.6(35)	0.3 ± 0.1(30)
U.S.A. and Canada (Table XIII)						
Hardwoods	—	—	44.6 ± 4.1(11)[i]	31.7 ± 3.8(10)[j]	22.5 ± 1.8(11)	0.4 ± 0.2(11)
Softwoods	—	—	41.9 ± 1.8(19)[i]	28.5 ± 1.7(19)[j]	29.2 ± 2.0(19)	0.3 ± 0.2(19)
U.S.A. (Table XIV) Softwoods	—	—	41.7 ± 3.3(39)[i]	28.6 ± 3.6(39)[j]	24.5 ± 3.0(39)[k]	0.4 ± 0.3(39)
U.S.S.R. (Table XI)						
Hardwoods	—	—	44.3 ± 5.1(47)[e]	24.2 ± 3.4(46)	21.9 ± 3.2(47)	0.6 ± 0.4(45)[l]
Softwoods	—	—	48.3 ± 4.8(15)	8.8 ± 2.5(12)	29.0 ± 1.6(15)	0.5 ± 0.4(16)

NOTE: Values are mean ± standard deviation (number of data).
[a] Holocellulose is the total carbohydrate content of wood.
[b] Alpha cellulose is nearly pure cellulose.
[c] Pentosans are the total anhydroxylose and arabinose residues in wood.
[d] Cross and Bevan cellulose is largely pure cellulose but contains some hemicelluloses.
[e] Kürschner cellulose is nearly pure cellulose.
[f] One value of 4.6% not included.
[g] Modified Kürschner cellulose.
[h] Modified Cross and Bevan cellulose.
[i] Pure glucan calculated from glucose and mannose content.
[j] Hemicelluloses calculated from five-sugar, acetyl, and uronic acid content.
[k] Klason lignin + acid-soluble lignin.
[l] One value of 5.4% not included.

hydrate components in Table XIV have been adjusted by a hydrolysis-loss factor. This factor was calculated for each species, such that the sum of total extractives, lignin, cellulose, hemicellulose, and ash equals 100%. The hemicellulose components were calculated using the adjusted value of the five individual sugars and the measured values for acetyl and uronic acid.

Table VII reports the trace element composition of some woods. Calcium, potassium, magnesium, and phosphorus are the principal trace elements in temperate woods. The three tropical woods have a higher potassium and magnesium content and a lower calcium content than the temperate woods.

Table XVI is a summary of average wood composition in 13 countries. The mean, standard deviation, and number of data are tabulated for carbohydrate, lignin, and ash compositions. Hardwoods and softwoods are separated when both are available. All other values are only for hardwoods. Be careful comparing values between countries because techniques and methods vary. For example, the mean holocellulose content of Costa Rican hardwoods is 78.1%, higher than that of woods from Brazil (71.7%) and Mexico (67.8%). The holocellulose determined for the Costa Rican hardwoods probably contained some lignin. The mean value of Taiwanese hardwood holocellulose is obviously high (83.3%) because the means for holocellulose and lignin sum to 108%.

Literature Cited

1. Hegnauer, R. "Chemotaxonomie der Pflanzen," Volumes I–VI; Birkhäuser Verlag: Basel and Stuttgart, 1962–1973.
2. Gibbs, R. Darnley "Chemotaxonomy of Flowering Plants," Volumes 1–4; McGill-Queens University Press: Montreal and London, 1974.
3. Wise, L. E.; Murphy, M.; D'Addieco, A. A. *Pap. Trade J.* 1946, 122(2), 35–43.
4. Fengel, Dietrich; Grosser, Dietger *Holz Roh- Werkst.* 1975, 33(1), 32–34.
5. Goring, D. A. I.; Timell, T. E. *Tappi* 1962, 45(6), 454–60.
6. Sjöström, Eero "Wood Chemistry. Fundamentals and Applications"; Academic Press: New York, 1981; p. 56, 65.
7. Billmeyer, F. W., Jr. *J. Polym. Sci., Part C* 1965, No. 8, 161–78.
8. Gardner, K. H.; Blackwell, J. *Biochim. Biophys. Acta* 1974, 343, 232–37.
9. Kolpak, F. J.; Blackwell, J. *Macromolecules* 1976, 9(2), 273–78.
10. Kolpak, F. J.; Weik, M.; Blackwell, J. *Polymer* 1978, 19, 123–31.
11. Timell, T. E. *Wood Sci. Technol.* 1982, 16, 83–122.
12. Timell, T. E. *Wood Sci. Technol.* 1967, 1, 45–70.
13. Timell, T. E. *Adv. Carbohydr. Chem. Biochem.* 1964, 19, 247–302.
14. Timell, T. E. *Adv. Carbohydr. Chem. Biochem.* 1965, 20, 409–83.
15. Adler, Erich *Wood Sci. Technol.* 1977, 11, 169–218.
16. Leopold, B.; Malmström, I. L. *Acta Chem. Scand.* 1952, 6, 49–54.
17. Obst, John R. *Tappi* 1982, 65(4), 109–12.
18. Nimz, H. *Angew. Chem., Int. Ed. Engl.* 1974, 13, 313–21.
19. Creighton, R. N. J.; Hibbert, H. *J. Am. Chem. Soc.* 1944, 66, 37–38.
20. Larsson, Sam; Miksche, Gerhard E. *Acta Chem. Scand.* 1967, 21(7), 1970–71.

21. Björkman, Anders *Sven. Papperstidn.* **1956,** *59*(13), 477–85.
22. Chang, H.-M.; Cowling, E. B.; Brown, W.; Adler, E.; Miksche, G. *Holzforschung* **1975,** *29*(5), 153–59.
23. Pew, John C. *Tappi* **1957,** *40*(7), 553–58.
24. Freudenberg, K.; Neish, A. C. "Constitution and Biosynthesis of Lignins"; Springer-Verlag: Berlin and New York, 1968; p. 113.
25. "Lignins: Occurrence, Formation, Structure and Reactions"; Sarkanen, K. V.; Ludwig, C. H., Eds.; Wiley-Interscience: New York, 1971.
26. "Wood Extractives"; Hillis, W. E., Ed.; Academic Press: New York, 1962.
27. Rowe, John W.; Conner, Anthony H. "Extractives in Eastern Hardwoods— A Review"; General Technical Report FPL 18, Forest Products Laboratory, Madison, WI, 1979.
28. Meier, H.; Wilkie, K. C. B. *Holzforschung* **1959,** *13*(6), 177–82.
29. Meier, Hans *J. Polym. Sci.* **1961,** *51*, 11–18.
30. Scott, J. A. N.; Procter, A. R.; Fergus, B. J.; Goring, D. A. I. *Wood Sci. Technol.* **1969,** *3*, 73–92.
31. Fergus, B. J.; Procter, A. R.; Scott, J. A. N.; Goring, D. A. I. *Wood Sci. Technol.* **1969,** *3*, 117–38.
32. Fergus, B. J.; Goring, D. A. I. *Holzforschung* **1970,** *24*(4), 118–24.
33. Obst, John R. *Holzforschung* **1982,** *36*(3), 143–53.
34. Van Beckum, W. G.; Ritter, G. J. *Pap. Trade J.* **1937,** *105*(18), 127–30.
35. Wise, L. E.; Jahn, E. C. "Wood Chemistry," 2d ed.; Reinhold: New York, 1952; pp. 1148–52.
36. Browning, B. L. "Methods of Wood Chemistry," Volume II; Interscience: New York, 1967; pp. 406, 615–24.
37. Saeman, J. F.; Moore, W. E.; Mitchell, R. L.; Millett, M. A. *Tappi* **1954,** *37*(8), 336–43.
38. Nelson, N. *J. Biol. Chem.* **1944,** *153*, 375–80.
39. Simson, B. W.; Timell, T. E. *Tappi* **1967,** *50*(10), 473–77.
40. Borchardt, L. G.; Piper, C. V. *Tappi* **1970,** *53*(2), 257–60.
41. Seymour, F. R.; Chen, E. C. M.; Bishop, S. H. *Carbohydr. Res.* **1979,** *73*, 19–45.
42. Sweeley, C. C.; Bentley, Ronald; Makita, M.; Wells, W. W. *J. Am. Chem. Soc.* **1963,** *85*, 2497–507.
43. Brower, H. E.; Jeffery, J. E.; Folsom, M. W. *Anal. Chem.* **1966,** *38*(2), 362–64.
44. Wentz, F. E.; Marcy, A. D.; Gray, M. J. *J. Chromatogr. Sci.* **1982,** *20*, 349–52.
45. Bylund, M.; Donetzhuber, A. *Sven. Papperstidn.* **1968,** *71*(15), 505–8.
46. Scott, R. W. *Anal. Chem.* **1979,** *51*(7), 936–41.
47. Wisenberger, E. *Makrochem.* **1947,** *33*, 51–69.
48. Procter, A. R.; Chow, W. M. *Pulp Pap. Mag. Can.* **1973,** *74*(7), 97–100.
49. Effland, Marilyn J. *Tappi* **1977,** *60*(10), 143–44.
50. Musha, Y.; Goring, D. A. I. *Wood Sci.* **1974,** *7*(2), 133–34.
51. Schöning, Arnt G.; Johansson, Gösta *Sven. Papperstidn.* **1965,** *68*(18), 607–13.
52. Swan, Brita *Sven. Papperstidn.* **1965,** *68*(22), 791–95.
53. Zeisel, S. *Monatsh. Chem.* **1885,** *6*, 989–96.
54. Steyermark, A. "Quantitative Organic Microanalysis"; 2d ed.; Academic Press: New York, 1961; p. 422–43.
55. Koeppen, Robert C. in "Papers for Conference on Improved Utilization of Tropical Woods"; Forest Products Laboratory: Madison, WI, 1978; p. 33.
56. Chinese Forestry Association "Important Wood Species of Taiwan"; Taipei, Taiwan, 1967.
57. Corrêa, A. A.; Ribeiro, E. B. P.; Lobato, R. de F. *Assoc. Téc. Bras. Cellul. Pap. Bol.* **1970,** *31*, 95–138.
58. Lauer, Karl *Tappi* **1958,** *41*(7), 334–35.
59. Barrichelo, L. E. G.; Brito, J. O. *IPEF* **1976,** *13*, 9–38.

60. U.S. Forest Products Laboratory, unpublished data for 1948–1957 on file at Madison, WI.
61. Inoue, Hideo; Masao, Moriya; Akiyama, Takeshi *J. Jpn. Tappi* **1972**, *26*(6), 256–62.
62. Wise, Louis E.; Rittenhouse, Ruth C.; Dickey, Edgar E.; Olson, Harry O.; Garcia, Consuelo *J. For. Prod. Res. Soc.* **1952**, *2*(5), 237–49.
63. Wise, Louis E.; Rittenhouse, Ruth C.; Garcia, Consuelo *Tappi* **1951**, *34*(4), 185–88.
64. Smith, J. B.; Primakov, S. F. *Appita* **1977**, *30*(5), 405–6.
65. Oliveira, J. Santos *Rev. Cien. Agron.* **1971**, Ser. B4, No. 3, 3–32.
66. Satonaka, Seiichi *J. Jpn. Wood Res. Soc.* **1963**, *9*, 26–34.
67. Yonezawa, Yasumasa; Kayama, Tsutomu; Kikuchi, Fumihiko; Usami, Kuninori; Takano, Isao; Ogino, Takehiko; Honda, Osamu *Bull. Gov. For. Exp. Stn. (Jpn.)* **1973**, *253*, 55–99.
68. Kayama, Tsutomu; Kikuchi, Fumihiko; Takano, Isao; Usami, Kuninori *Bull. Gov. For. Exp. Stn. (Jpn.)* **1967**, *197*, 155–62.
69. Kayama, Tsutomu *J. Jpn. Tappi* **1968**, *22*(11), 581–90.
70. Working Group on Utilization of Tropical Woods *Bull. Gov. For. Exp. Stn. (Jpn.)* **1977**, *294*, 1–49.
71. Working Group on Utilization of Tropical Woods *Bull. Gov. For. Exp. Stn. (Jpn.)* **1978**, *299*, 85–104.
72. Eddowes, Peter J. "Commercial Timbers of Papua New Guinea"; Hebamo Press: Port Moresby, Papua New Guinea, 1977.
73. Reyes, Luis J. "Philippine Woods"; Bureau of Printing: Manila, 1938; p. 479–82.
74. Francia, P. C.; Escolano, E. U.; Bautista, C. S.; Semana, J. A. *Philippine Lumberman* **1971**, *17*(5), 21–24.
75. Escolano, Eugenia U.; Semana, Jose A. *Philippine Lumberman* **1978**, *24*(12), 16–20.
76. Escolano, E. U.; Francia, P. C.; Semana, J. A.; Bawagan, B. O. *Philippine Lumberman* **1970**, *16*(12), 26–28.
77. U.S. Forest Products Laboratory, unpublished data on file at Madison, WI.
78. Semana, J. A.; Escolano, E.; Francia, P. C.; Bautista, C. S. *Philippine Lumberman* **1968**, *14*(2), 20–25.
79. Francia, P. C.; Escolano, E. U.; Semana, J. A.; Bawagan, B. O. *Philippine Lumberman* **1970**, *16*(11), 26–30.
80. Chao, S. C.; Ku, Y. C.; Lin, S. J.; Pan, T. T. *Bull. Taiwan For. Res. Inst. (Co-Op. Taiwan For. Bur.)* No. 14, **1971**, Taipei, Taiwan.
81. Ku, Y. C. *Bull. Taiwan For. Res. Inst. (Co-Op. Taiwan For. Bur.)* No. 142, **1966**, Taipei, Taiwan.
82. Ku, Y. C.; Haung, L. Y.; Lin, T. N.; Pan, T. T. *Co-Op. Bull. Taiwan For. Res. Inst. Co-Op. Natl. Sci. Counc., Jt. Comm. Rural Reconstr., Taiwan For. Bur.* No. 8, **1965**, Taipei, Taiwan.
83. Sieber, R.; Walter, L. E. *Papier-Falst.* **1913**, *11*, 1179–83.
84. Nikitin, N. I. "The Chemistry of Cellulose and Wood"; trans. from Russian by J. Schmorak. Israel Program for Scientific Translations, Jerusalem, 1966. First published in Russian in 1962.
85. Campbell, W. G.; Bamford, K. F. *J. Soc. Chem. Ind., London, Trans. Commun.* **1939**, *58*, 180–85.
86. Timell, T. E. *Pulp Pap. Mag. Can.* **1958**, *59*(8), 139–40.
87. "The Chemistry of Wood"; Browning, B. L., Ed.; Krieger: Huntington, NY, 1975; p. 70–71.
88. Moore, W. E.; Effland, M. J. *Tappi* **1974**, *57*(8), 96–98.
89. Côté, W. A., Jr.; Simon, B. W.; Timell, T. E. *Sven. Papperstidn.* **1966**, *69*(17), 547–58.
90. Timell, T. E. *Tappi* **1957**, *40*(7), 568–72.
91. Smelstorius, J. A. *Holzforschung* **1971**, *25*(2), 33–39.
92. Smelstorius, J. A. *Holzforschung* **1974**, *28*(2), 67–73.

93. Young, H. E.; Guinn, V. P. *Tappi* **1966,** *49*(5), 190–97.
94. Osterhaus, C. A.; Langwig, J. E.; Meyer, J. A. *Wood Sci.* **1975,** *8*(1), 370–74.
95. Choong, Elvin T.; Chang, Bao-Yih; Kowalczuk, Joseph "LSU Wood Utilization Notes"; No. 27, December 1974.

RECEIVED for review May 6, 1983. ACCEPTED July 22, 1983.

3
Wood–Water Relationships

C. SKAAR
Department of Forest Products, School of Forestry and Wildlife, College of Agriculture and Life Sciences, Virginia Polytechnic Institute and State University, Blacksburg, VA 24061

Wood is a hygroscopic material, and its mass, dimensions, and density, as well as its mechanical, elastic, electrical, thermal, and transport properties are affected by its moisture content. Wood is formed in a water-saturated environment in the living tree, but most of the water is removed prior to use. In use its moisture content and dependent properties change with changes in ambient conditions, particularly relative humidity. Wood is anisotropic with respect to most of its physical properties. The thermodynamics of moisture sorption, including enthalpy, free energy and entropy changes, are moisture dependent. Water sorption by wood is treated in terms of both surface and solution theories. Moisture transport in wood is also treated, particularly in relation to drying.

Wood Moisture and the Environment

Wood differs from most materials used for construction and other purposes in that it is continually exchanging moisture with its surroundings. This is true in both the living tree as well as under conditions of final use.

The moisture content of wood is usually calculated in terms of its dry weight. The *fractional moisture content* m is defined as the ratio of the mass W_w of removable water to the dry mass W_o of the wood (Equation 1).

$$m = W_w/W_o \qquad (1)$$

Moisture content is often expressed in terms of *percent* of dry weight, or

$$M = 100 \times m = 100\,(W_w/W_o) \qquad (2)$$

The definition of M as given above is equivalent to the term regain as used for certain other hygroscopic materials such as textiles (1). The term moisture content is defined on a wet rather than dry weight basis. The *wet basis moisture content* M_w is then related to M by

$$M_w = M/(1 + M/100) \qquad (3)$$

It may be noted that M can be greater than 100% but that M_w is always less than 100%. The dry weight basis, either m or M, will be used throughout this chapter.

Water in the Living Tree. Wood in the living tree is formed and functions in an essentially water-saturated environment. The functioning sapwood cells are a part of the vascular system that conducts water and solutes from the roots to the leaves through a continuous water-saturated network of wood cells (2). When the tree is felled the water in the wood is cut off from the soil water and the wood commences to lose most of its moisture.

Moisture Content of Green Wood. The moisture content of wood in a freshly felled tree is designated as the *green moisture content*. The green moisture content may vary considerably among different kinds of trees and between heartwood and sapwood within a tree. It may also vary with height in the tree and with the season of the year in which the tree is felled.

The green moisture content of the heartwood of 27 different softwood species grown in the United States, based on percent of oven-dry weight, is reported to range from 30 to 121% with a mean of 55% (3). For sapwood of the same softwoods the mean was 149% with a range from 98 to 249%. In contrast, for 34 hardwoods, no consistent difference was found in the green moisture contents of heartwood and sapwood. The mean heartwood value was 81% (range from 44 to 162%), close to the mean of 83% (range from 44 to 146%) for the sapwood of the same trees.

Studies on *Pinus taeda* (4) indicate a strong increase in green moisture content with increasing height in the tree. Similar trends were observed among a number of Appalachian hardwoods and softwoods (5).

Logs cut from trees felled during late winter and early spring in temperate climates generally exhibit higher green moisture contents than those harvested during summer and fall.

Water in green wood is found in three basic forms: *bound* water in the cell walls, *free* or *capillary* water in the cell cavities, and water *vapor*, also in the cell cavities. The total amount of water in vapor form is normally only a small fraction of the total and is negligible at normal temperatures and moisture contents. When green wood dries the water leaves the cell cavities first because it is held with smaller forces than the bound water. Furthermore, most physical properties, such as strength properties and shrinkage, are unaffected by removal of free water (*See* Chapter 5).

The *fiber-saturation point* is defined as the moisture content at which the cell cavities are empty of liquid water but the cell walls are still saturated with bound water (6). The fiber-saturation point is designated as m_f (fraction of dry mass) or M_f (percent of dry mass).

Measuring Water Content of Wood. There are as many as fifteen methods that have been used to measure wood moisture content (7). Some of the more common or useful methods are discussed here.

GRAVIMETRIC METHOD. The moist sample is weighed, W_m, and then dried until a reference weight, W_o, is attained. The difference is taken as the weight of water, W_w, in the moist wood. Ordinarily wood is dried in a convection oven maintained at 103 ± 2 °C. In this case, the atmosphere is at a sufficiently low relative vapor pressure h ($h = p/p_o$; p is the ambient water vapor pressure and p_o is the vapor pressure of pure water at the oven temperature) that h is assumed to be zero.

There are several errors involved in gravimetric moisture measurements. One error is the assumption that h is zero in an ordinary oven. This effect can be minimized by using a vacuum oven or a strong desiccant such as phosphorus pentoxide. Another problem is the evaporation of volatile wood constituents, if present, to give a higher apparent moisture content in the wood. A third problem in accurate moisture measurement is the effect of sample moisture history (8).

A variation of the gravimetric method is to heat the wood in a distillation apparatus containing a water-immiscible liquid such as toluene or xylene. This liquid dissolves the organic volatiles and the water condenses in a separate calibrated trap where it is collected and measured.

KARL FISCHER TITRATION METHOD. In this method the moisture content is measured by titration, using the Karl Fischer reagent, which consists of a solution of pyridine (C_5H_5N), sulfur dioxide, and iodine in methanol (MeOH). This solution reacts with water as follows:

$$3\,\text{(pyridine)} + SO_2 + I_2 + H_2O \rightleftarrows 2\,\text{(pyridine)} \cdot HI + \text{(pyridine)} \cdot SO_3$$

(pyridine)

The end point of the titration may be determined either colorimetrically (free iodine present) or electrically (free water increases the conductivity of the solution).

The Karl Fischer method can be used to measure the moisture contents of many materials besides wood, including solids, liquids, and gases. It gives the best results of any of the standard methods used for measuring wood moisture content (7), but is not practical for large wood samples, particularly those with high moisture contents.

ELECTRICAL RESISTANCE MOISTURE METERS. The electrical resistance of wood is extremely sensitive to its moisture content, approximately doubling for each 1% decrease in moisture content over the hygroscopic range of moisture contents. The development of a successful resistance moisture meter may be attributed primarily to the pioneering work of Stamm (9) who first measured this relationship quantitatively. Because of the nature of electrical conduction in wood there is also a strong increase in resistivity with a decrease in wood temperature. Figure 1 illustrates how the electrical resistivity of wood varies with both moisture content and temperature.

Most resistance moisture meters are essentially megohmeters that measure the resistance between pairs of pin electrodes driven into the wood to various depths. Because the pin electrodes taper along their lengths, the relationship between a resistance reading and the resistivity (resistance of a unit cube) is complex. Therefore the meters are calibrated empirically by using data obtained on a given species at room temperature (10, 11).

Resistance moisture meter scales manufactured for use in North America read directly in moisture content, based on calibration data for Douglas-fir at 27 °C. Figure 2 shows the range in electrical resis-

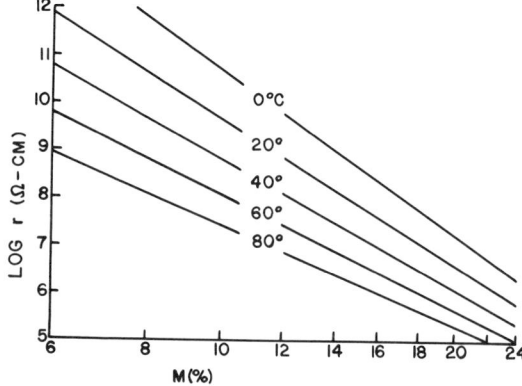

Figure 1. Logarithm of DC resistivity of wood as a function of moisture content (10).

tance among domestic U.S. woods as a function of wood moisture content between the limits of 7 and 25% at (27 °C). The calibration data for Douglas-fir fall approximately midway between the upper and lower curves (*12*).

Note that the curves shown in Figures 1 and 2 are confined to the moisture content limits between 6–7 and 24–25%. Measurements below 6 or 7% are not reliable with ordinary moisture meters because the resistance is too high (above 10^{11} Ω).

At moisture contents above 24 or 25%, readings are less reliable than readings below 24 or 25% for two reasons. First, the rate of change of resistance with moisture content decreases markedly, so the sensitivity is reduced. Second, the moisture content reading decreases substantially with time because of polarization effects. The latter effect can be minimized by the use of alternating current (AC) rather than the direct current (DC) instruments traditionally used for resistance meters.

Another method proposed for minimizing polarization and related effects is to use short repetitive current pulses rather than continuous voltage on the sample (*13*). This method also reduces the ohmic heating effect at higher moisture contents. Some contemporary resistance meters have provisions for switching to the pulsed current mode for wood moisture contents greater than 12% and retain the DC mode at lower moisture contents.

A resistance meter reads moisture contents higher than the true values when used on hot wood, and vice versa for cold wood. Therefore, the readings must be adjusted for this temperature factor. A family of curves used to adjust measurements made on wood at temperatures from −40 °F (−40 °C) to 160 °F (71 °C) is reproduced in Figure 3 (*14*). It is probable that individual species, in addition to showing variations from the standard curve of resistance against moisture content, also show variation with respect to the temperature adjustment factors (*10*). Some modern meters are provided with adjustable meter calibration for direct temperature compensation (*11*).

Resistance moisture meters are useful for determining the magnitude of moisture gradients in wood, particularly during drying. This is accomplished by measuring the moisture content at different depths from the surface because the meter readings are most affected by the wettest point of penetration. For the same reason, if the wood surface has been wetted by rain or high humidity conditions the surface rather than interior moisture content is measured. This effect can be minimized by use of probes that are insulated along their lengths, except for the penetrating tips that serve as the electrodes.

DIELECTRIC MOISTURE METERS. These moisture meters use AC, usually at radio frequencies. There are two general types: the *capac-*

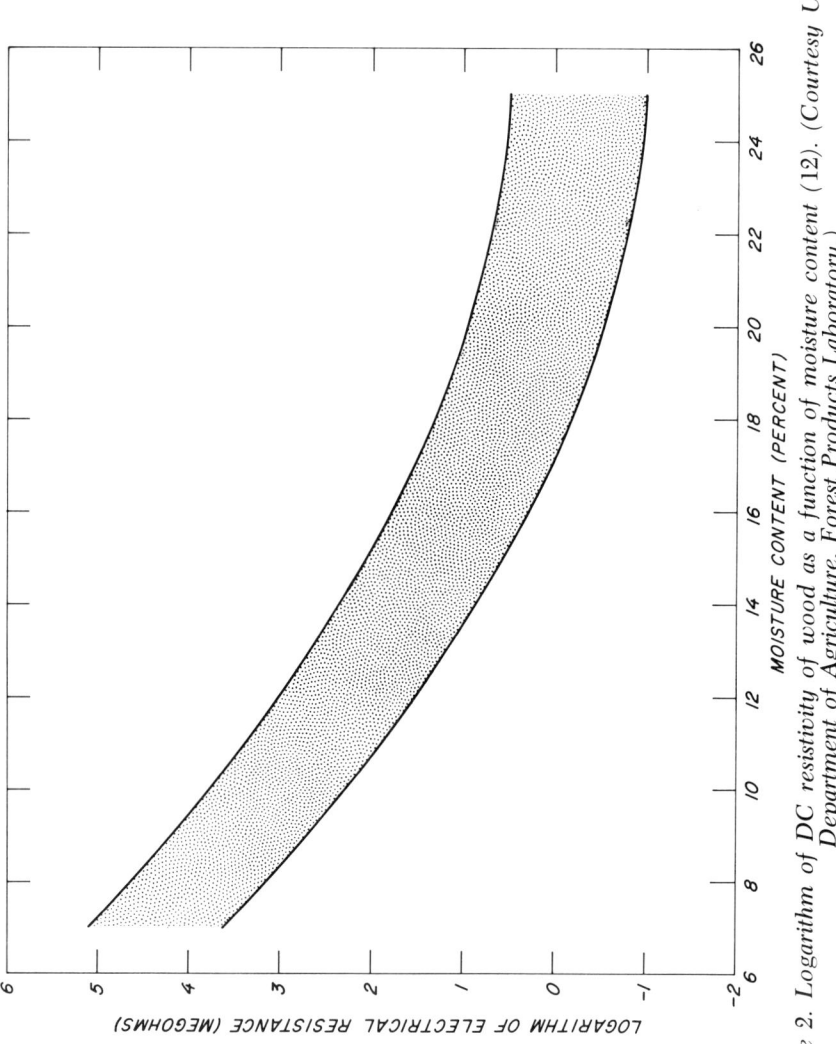

Figure 2. Logarithm of DC resistivity of wood as a function of moisture content (12). (Courtesy U.S. Department of Agriculture, Forest Products Laboratory.)

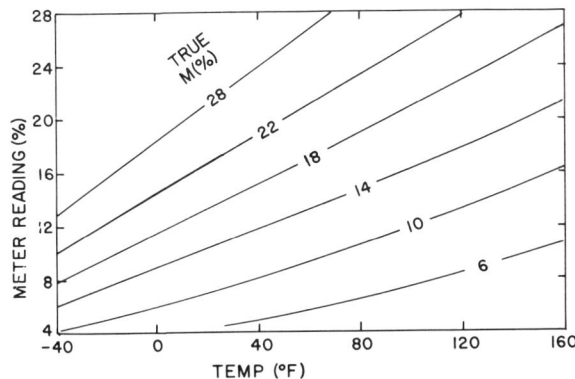

Figure 3. Temperature calibration curves for a DC resistance moisture meter (14). (Courtesy U.S. Department of Agriculture, Forest Products Laboratory.)

itance type which measures primarily the dielectric constant of the wood, and the *power-loss* type which measures the rate of energy absorption by wood from an oscillating electric field.

The capacitance type essentially measures the dielectric constant of wood. At a given frequency, the dielectric constant increases with wood density, moisture content (Figure 4), and increasing temperature (10). The most effective electrode configuration for a capacitance meter appears to be a pair of flat parallel electrodes contacting each of two opposite faces of the wood to be measured. There is then

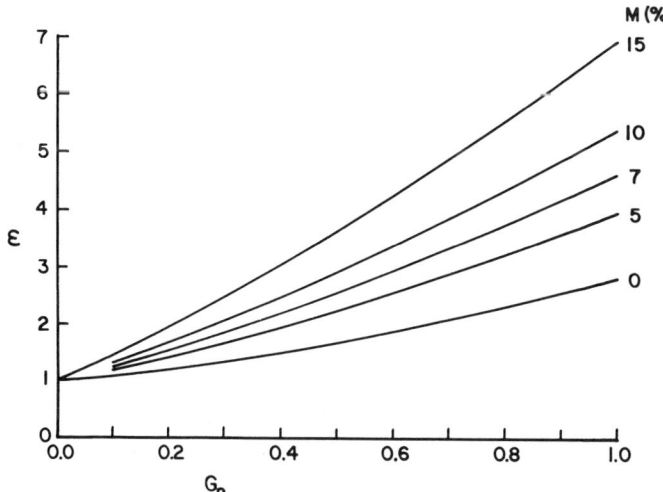

Figure 4. Dielectric constant ϵ vs. dry wood specific gravity G_o for several different moisture contents. (Reproduced with permission from Ref. 10. Copyright 1972, Syracuse University Press.)

a simple geometrical relationship between the measured capacitance and the dielectric constant of the wood. However, most meters of the capacitance type, as well as of the power-loss type use concentric electrodes placed on one wood surface (14). This type of electrode is more practical for use but tends to read the moisture content near the wood surface rather than in the interior.

The power-loss meter is the most common type of dielectric moisture meter. It senses the product of the dielectric constant and loss factor. Generally, the loss factor increases with wood moisture content but may exhibit variations from this behavior depending on the frequency of measurement (10, 11, 14). An increase in temperature produces effects similar to increasing moisture content, with interaction between these two parameters. Therefore, temperature adjustments of meter readings are complex, sometimes increasing and sometimes decreasing the scale reading as temperature increases (14).

MISCELLANEOUS METHODS. Several other methods have been explored for measuring wood moisture content, some of which are discussed briefly.

Nuclear Magnetic Resonance (NMR). NMR techniques have been applied to wood moisture measurements in the laboratory (15). This technique is based on the fact that the hydrogen nucleus is a nuclear magnetic dipole due to its characteristic spin. When it is subjected to a static magnetic field of strength, H_o, the magnetic dipole precesses about the direction of H_o with a frequency γ_o which is directly proportional to H_o. For the basic hydrogen nucleus (proton) $\gamma_o = 4.257\ H_o$ where γ_o is in kHz when H_o is measured in Gauss (15).

Two different techniques of NMR have been applied to measure wood moisture content based on the presence of the hydrogen nuclei in water. In one of these, designated as a steady-state method, the wood is subjected to an alternating magnetic field of constant frequency, with H_o varied slowly so as to resonate γ_o with respect to the applied frequency. At resonance a strong absorption of energy occurs, and the width and intensity of this absorption curve give information on the moisture content of the wood (16).

The second general NMR technique applied to wood (15) is the pulsed NMR method. In this case "a short intense burst of a magnetic field oscillating in resonance with the spin precession frequency is applied at right angles to H_o," (15). A voltage is induced by the pulse in a coil surrounding the sample. This voltage decays exponentially, and an analysis of this *free induction decay* gives information on the nature of the molecules containing the hydrogen nuclei, as well as to their number. Figure 5 shows a plot of the amplitude of the free

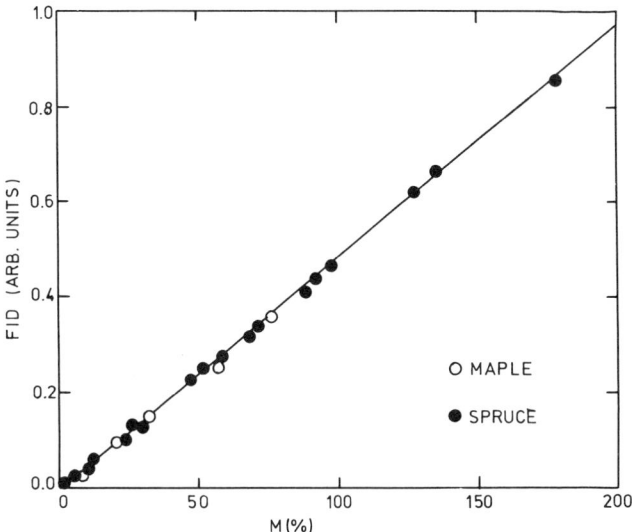

Figure 5. Free induction decay (FID) voltage vs. moisture content. (Reproduced with permission from Ref. 15. Copyright 1978, Wood Fiber.)

induction decay voltage 50 μs after pulsing as a function of wood moisture content for spruce and maple wood (15).

Neutron Moisture Meter. A neutron moisture meter can also be used to measure wood moisture content (10). This consists of a fast neutron generator which is a source of high-energy neutrons. These are directed into the wood (Figure 6) where some are moderated into slow neutrons by the hydrogen atoms and scattered back toward a slow-neutron detector. The number moderated and detected is proportional to the amount of water in wood because of the high content of hydrogen in water. Such neutron meters have been developed for field use in measuring soil moisture content (17).

The neutron moisture measurement technique gives information on the amount of water per unit volume of the wood. To reduce this to a weight basis the density of the wood must also be known. This

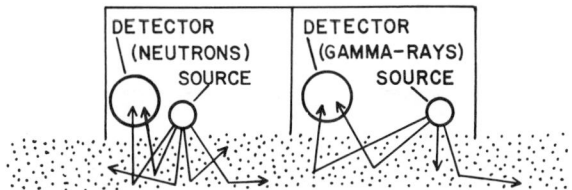

Figure 6. Schematic diagram of a nuclear gauge for moisture measurement of bulk materials. (Adapted from Nuclear-Chicago Corporation.)

may be accomplished by using a supplemental system such as a γ-radiation and detection system (Figure 6). A beam of γ rays directed into the wood is absorbed in proportion to the wood density. The γ rays not absorbed are detected and are inversely proportional to the density of the wood. The output data from the neutron and γ detectors can be combined to obtain the moisture content on a weight basis.

Moisture Sorption Isotherms. Green wood loses moisture to the atmosphere and approaches a moisture content designated as the *equilibrium moisture content* (EMC) for the particular atmospheric conditions. The EMC is a function of relative humidity, temperature, previous exposure history (hysteresis), species, and other miscellaneous factors.

EFFECT OF RELATIVE HUMIDITY AND SORPTION HISTORY. An indirect method for estimating wood moisture content is to measure its equilibrium relative vapor pressure h. This is related to wood moisture content by a sorption isotherm. The percent relative humidity (H) or relative vapor pressure (h) ($H = 100\ h$) is the most important factor in determining the EMC for wood. A curve showing EMC as a function of percent relative humidity or relative vapor pressure at constant temperature is called a *moisture sorption isotherm*.

Figure 7 shows three typical sorption isotherms for Douglas-fir at 90 °F (32 °C) (*18*). The general sigmoid shapes for all three curves is apparent, but each curve represents the isotherm for a different

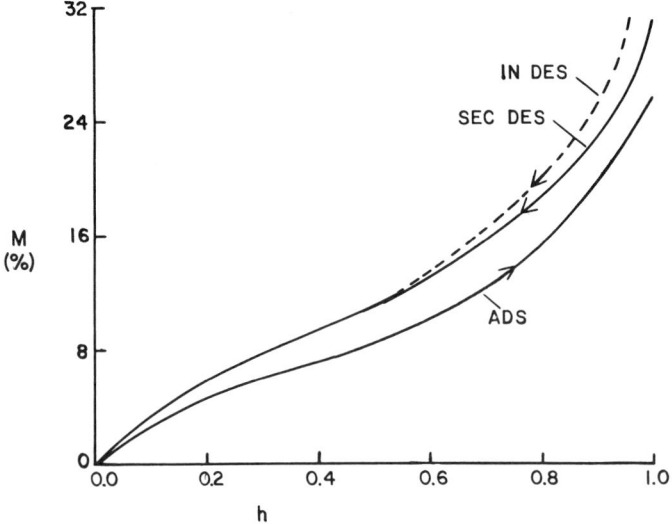

Figure 7. Initial desorption (IN DES), adsorption (ADS), and secondary desorption (SEC DES) isotherms for Douglas-fir. (Adapted from Ref. 18.)

sorption exposure history. The uppermost curve is that for the initial desorption or drying from the green condition. The lowest curve is the adsorption isotherm obtained by exposing the wood, after vacuum drying, to successively higher relative humidities. The intermediate curve is the secondary desorption isotherm obtained by re-exposing the sample to successively lower humidities after first equilibrating it to essentially 100% relative humidity.

A sample taken through repetitive cycles of relative humidity exposure between 0 and 100% tends to follow the adsorption and secondary desorption curves repetitively. The adsorption isotherm (A) is always lower than the corresponding desorption isotherm (D) and their ratio, designated as the A/D ratio, cannot exceed unity.

The A/D ratio varies with relative humidity and different kinds of wood (19) (Figure 8). At room temperature it generally ranges between 0.8 and 0.9, and tends to decrease with increasing temperature (20).

Sorption hysteresis in wood is beneficial from the viewpoint of wood utilization. This is because wood exposed to cyclic humidity conditions shows smaller changes in moisture content for given humidity changes than would be the case if there were no hysteresis (21). Sorption hysteresis reduces the effective slope dM/dH of the sorption isotherm and the dimensional changes associated with humidity changes.

EFFECT OF TEMPERATURE. The sorption isotherms for wood generally decrease with increasing temperature (Figure 9) above 0 °C.

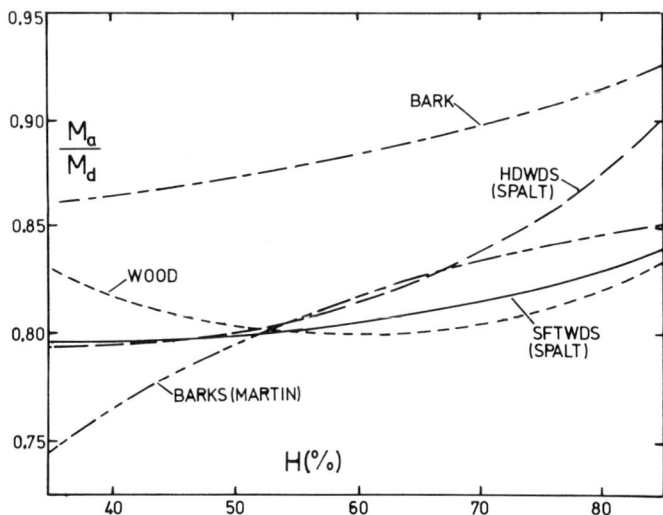

Figure 8. Representative A/D (M_a/M_d) ratios as functions of relative humidity H for different woods and bark (19).

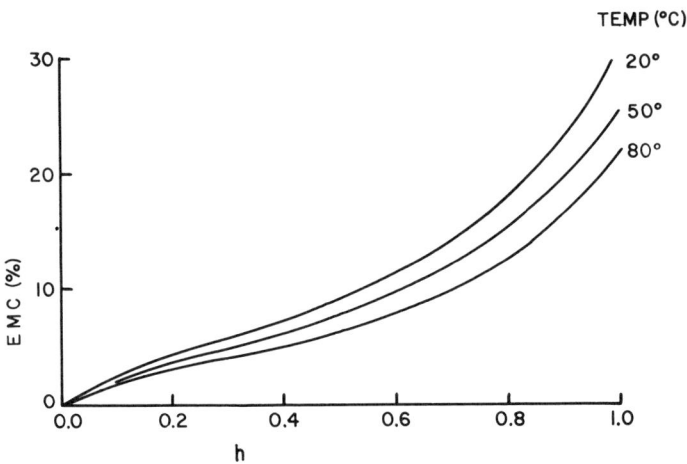

Figure 9. Sorption isotherms as affected by temperature.

This result is as expected based on thermodynamic considerations and is discussed later in this chapter. The apparent fiber-saturation point M_f, which is obtained by extrapolating the sorption isotherm to 100% relative humidity, decreases approximately 0.1%/°C rise in temperature (22).

Above the boiling point of water the sorption isotherms apparently continue to decrease with increasing temperature (23). It is difficult to measure isotherms above 100 °C because the vapor pressure of water is greater than atmospheric pressure. Therefore, to attain relative humidities near 100% it is necessary to carry out the measurements in a pressurized system.

If measurements are made at atmospheric pressure the maximum relative humidities that can be attained decrease with increasing temperature (Figure 10). The maximum relative humidity possible at any temperature is equivalent to the ratio of the prevailing atmospheric pressure to the vapor pressure of water at that temperature, expressed in percent. The practice of drying lumber at high temperatures (above 100 °C) has created a renewed interest in the sorption isotherms of wood at these temperatures (23).

Below 0 °C the hygroscopicity of wood decreases with decreasing temperature, the opposite of the trend above 0 °C (10).

EFFECT OF WOOD SPECIES AND EXTRACTIVES. The sorption isotherms of all woods are generally similar in shape. However, there may be considerable variations among them with respect to the absolute values of hygroscopicity. This variation may be because of differences in the proportion of the primary wood constituents, such as cellulose, hemicellulose, and lignin in different woods; or more importantly, because of differences in the kind and quantity of ex-

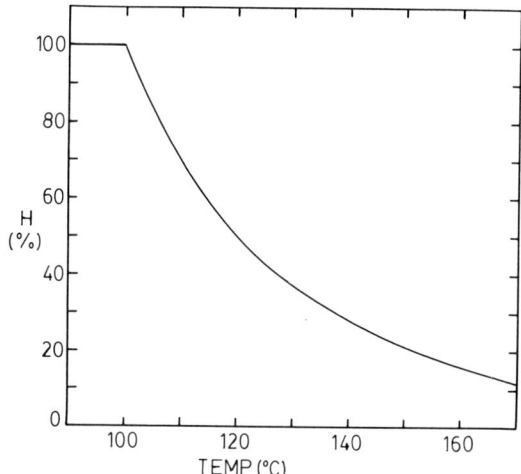

Figure 10. Maximum possible relative humidities at atmospheric pressure and temperatures above 100 °C (23).

tractives. The adsorption isotherms shown in Figure 11 indicate that hemicelluloses are the most hygroscopic, and lignin the least hygroscopic, of the primary chemical constituents of wood (24).

The hygroscopicities of woods with high extractive contents are generally lower than those without extractives. For example, the heartwood of nine tropical woods showed an increase in apparent

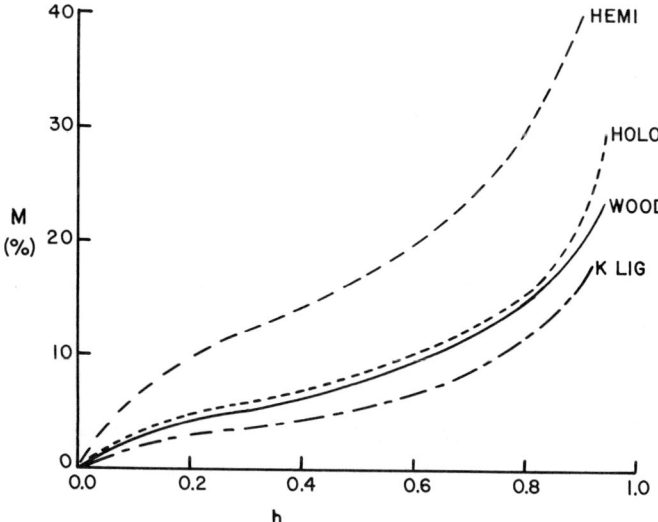

Figure 11. Adsorption isotherms for wood hemicellulose (HEMI), holocellulose (HOLO), Klason lignin (KLIG), and wood at 25 °C (24).

fiber saturation, based on the adsorption isotherm, from a mean of 21.9% for unextracted wood to 27.6% following successive extractions with benzene–alcohol, 95% alcohol and water, for 10–20 d, using a Soxhlet apparatus (25). The corresponding mean desorption fiber-saturation point increased from 28.3 to 33.7%.

OTHER FACTORS AFFECTING HYGROSCOPICITY. Several other factors affect the hygroscopicity of wood. One of these factors is the effect of mechanical stress (26). Compressive stresses decrease the moisture content of wood, and tensile stresses increase it. This effect is related to the swelling pressure of wood.

The hygroscopicity of wood may be reduced appreciably by heating (22), the effect increasing with increasing temperature and time of heating (27).

Moisture Content of Wood in Use. Wood retains its hygroscopic characteristics after it is put into use. It is then subjected to fluctuating humidity, the dominant factor in determining its EMC. These fluctuations may be more or less cyclical such as the 24-h diurnal changes or the annual seasonal changes.

In order to minimize the changes in wood moisture content in service, wood is usually dried to a moisture content that approximates the average EMC conditions to which it will be exposed. These conditions vary with respect to wood intended for interior compared with exterior use in a given geographic location. They also vary with geographical location. For example, the target moisture contents of 8% for wood intended for interior use and 11% for wood intended for exterior use are recommended (28) in most of the continental United States. Corresponding figures for the dry southwestern states are 6 and 9%, respectively, and those for the damp coastal areas of the southeast are 11 and 12%, respectively.

The primary reason for drying wood to a moisture content equivalent to its mean EMC under use conditions is to minimize dimensional changes in the final product.

Shrinking and Swelling of Wood

The moisture content of wood in the living tree is always above the fiber-saturation point. Therefore, the changes in wood moisture content that occur during the life of the tree are essentially limited to changes in the levels of water in the cell cavities, that is, to the so-called *free* water. The cell walls in green wood are, therefore, in the fully saturated condition and no hygroscopic shrinking or swelling occurs, except that resulting from changes in fiber-saturation points already referred to, which are a function of temperature.

However, when trees are felled and the cell walls lose moisture, shrinkage occurs in proportion to the extent of loss of this *bound* water. Because wood in use is generally exposed to cycling relative humidity, swelling also occurs during the adsorption of water by the cell wall of wood.

The Cell Wall. Before considering the dimensional changes in the cell wall of wood associated with gain or loss of moisture it is desirable to first consider the density ρ' of the cell wall and how it varies with moisture content.

DENSITY OF THE DRY CELL WALL. The dry cell wall of wood has a density of approximately 1.5 g/cm³ when measured by pycnometric or volume-displacement methods. Somewhat higher values are obtained when using water as opposed to nonswelling displacement media such as toluene or benzene (22).

The apparent density ρ_o' of the cell wall of wood has also been measured by optical methods. In this case the relative fractions of void and cell-wall volumes are determined optically by using thin microtomed sections of wood (29). These data are then combined with measurements of the dry wood density ρ_o to give ρ_o', based on Equation 4.

$$\rho_o' = \rho_o(V_o' + V_o'')/V_o' \qquad (4)$$

where V_o' and V_o'' are the cell wall and void volumes, measured optically on the microtomed wood sections.

Measurements of the dry cell wall density based on microscopic observations generally give lower values (1.42 g/cm³) than those obtained using pycnometrically (1.47 g/cm³) with toluene as a displacement medium (29). This discrepancy is attributed to various uncontrollable factors such as cell-wall ruptures produced during preparation of the microtomed sections.

For the purpose of the discussion that follows the density of the dry cell wall will be taken as 1.5 g/cm³, and its specific gravity G_o' as 1.5.

MAXIMUM SHRINKING AND SWELLING OF THE CELL WALL. When dry wood is immersed in water the cell wall swells in proportion to the volume of water adsorbed. If it is assumed that the sorbed water has the same density as free liquid water, the percent swelling Sw_m' of the cell wall can be approximated by Equation 5.

$$Sw_m' = MG_o' \qquad (5)$$

Thus, with G_o' taken as 1.5, the percent volumetric swelling of the cell wall from the dry condition is 1.5 times the percent moisture content M.

The *maximum possible swelling* Sw_{max}' of the cell wall is obtained when the cell wall is saturated, that is when $M = M_f$. The fiber-saturation point can be measured in a number of different ways

(Stamm (30) has listed nine such methods). Somewhat different values are obtained using different methods. There also appear to be variations among woods. A mean value of approximately 35% for Sw'_{max} was calculated (29) based on measurements of 18 woods native to the continental United States. Lower values have also been found (10, 18, 22, 30) and 30% will be taken here to be the nominal value of M_f at room temperature for the purpose of calculating the maximum possible swelling of the cell wall of wood.

The swelling of the cell wall at fiber saturation Sw'_f is equal to Sw'_{max}. Therefore, from Equation 5 it can be written that:

$$Sw'_f = M_f G_o' \qquad (6)$$

Taking M_f' as 30% and G_o' as 1.5, the maximum volumetric swelling of the cell wall is 45%, based on the assumptions given above.

Conversely it can be shown that the percent shrinkage Sh'_m of the cell wall is given by Equation 7

$$Sh'_m = (M_f - M)G_f' \qquad (7)$$

for a percent moisture content change from M_f to the lower moisture content M where G_f' is the specific gravity of the cell wall based on oven-dry weight W_o and a fully swollen volume V_f. The maximum shrinkage Sh_f' from M_f to $M = 0$ is therefore given by Equation 8.

$$Sh'_f = M_f G_f' \qquad (8)$$

The ratio Sw'_f/Sh'_f therefore is equal to the ratio G_f'/G_o', based on Equations 6 and 8.

The specific gravity G_m' of the cell wall at any moisture content M is given by

$$G_m' = G_o'/(1 + G_o'm) \qquad (9)$$

where $m = M/100$. At $M = M_f$ the specific gravity G_f' is given by $G_f' = G_o'/(1 + G_o'm_f)$. Taking G_o' as 1.5 and m_f as 0.30, $G_f' = 1.5/[1 + 1.5(0.3)] = 1.035$.

The Gross Wood. The dimensional changes in the gross wood are not generally the same as those for the cell wall material for several reasons. First, the cell cavities affect the shrinkage of the gross wood. Second, the cell wall structure is anisotropic, resulting in differences in swelling and shrinkage in different directions in the cell wall. Third, the cell structure varies among different kinds of woody tissue, such as ray tissue compared with longitudinal tissue. Finally, mechanical stresses affect the extent and direction of dimensional changes. These factors all contribute to the overall dimensional instability of wood associated with moisture changes.

In the discussion that follows, the volumetric shrinking and swelling of the gross wood will be treated first, followed by discussion of anisotropy, and finally the effect of stress.

VOLUMETRIC SHRINKING AND SWELLING. The volumetric swelling of the cell wall of wood is proportional to the volume of water absorbed. The gross wood however contains air spaces; therefore, its volumetric swelling depends on what happens to the air spaces during water sorption by the cell wall.

Tiemann (31) has indicated that there are three possibilities for these air spaces during water sorption, shown schematically in Figure 12. First, all or part of the swelling may take place into the cell cavities (Figure 12b) with reduction in lumen volume. If all of the swelling takes place into the cell cavities there would be no external swelling in the gross wood. Second, the cell cavities may be unaffected by the cell wall swelling and remain the same size (Figure 12c). Third, the cell cavity may swell to a lesser or greater extent than the cell wall itself (Figure 12d).

If it is hypothesized that the cell cavity remains constant in size as wood changes moisture content it can be shown (10) that the volumetric shrinkage Sh_f of a wood of swollen volume specific gravity G_f can be predicted, based on a modification of Equation 8, as in Equation 10.

$$Sh_f = M_f G_f \qquad (10)$$

Stamm and Loughborough (32) first reported that this relationship has been reported (32) to be approximately valid for woods of the continental United States. The mean value of the ratio Sh_f/G_g was 27 for 107 hardwood species and 26 for 52 softwood species of the United States. These ratios should be equivalent to the fiber-saturation point M_f if the green volume specific gravity G_g is taken to be

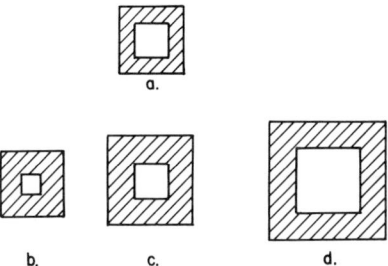

Figure 12. Volumetric swelling of a single cell showing the cell. Key: a, before swelling; b, all swelling into cell cavity; c, all swelling external; d, both cavity and external swelling (10).

equal to G_f at the fiber-saturation point. This is a valid assumption if there is no shrinkage above M_f with a change of wood moisture content. This is generally true unless collapse of cell cavities occurs during removal of free water.

Volumetric shrinkage data on other woods have also indicated that the ratio Sh_f/G_g tends to approximate the fiber-saturation moisture content M_f. For example, a mean ratio was found for Sh_f/G_g of 27 for 170 Australian woods (33). Data on tropical woods suggest somewhat lower values for this same ratio. The mean value for 140 Indian woods was approximately 20, considerably lower than the values for U.S. woods. This may indicate that tropical woods are less hygroscopic than temperate-zone woods, possibly because of their higher mean extractive contents.

The reason the cell cavity tends to change only a small extent if at all during moisture changes is probably resident in the microfibril orientation in the typical cell wall of wood (32). Figure 13 is a simplified diagram of the woody cell wall. The central or S_2 layer is the thickest layer. Its microfibrils are nearly parallel to the cell axis and tend to swell transversely as moisture content increases. The microfibrils in the S_1 and S_3 layers however are oriented nearly perpendicular to the cell axis. Therefore, although they are thin, they tend to restrain swelling of the cell wall because of the high strength of microfibrils along their length. Transverse swelling and shrinking of individual cells and, therefore, of the gross wood are also reduced.

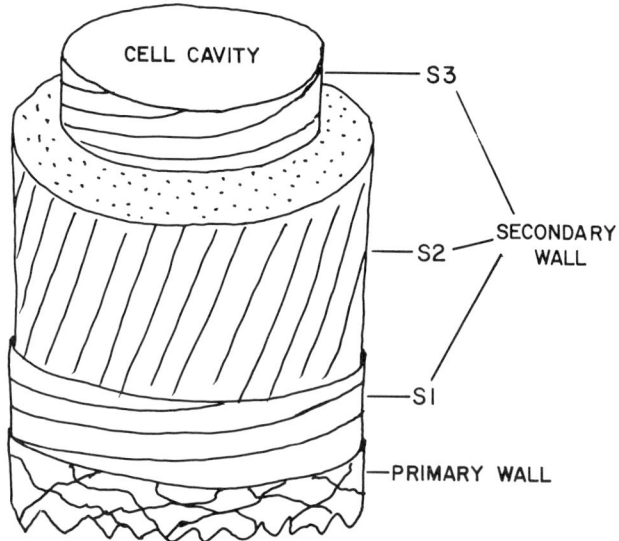

Figure 13. Cell wall schematic diagram showing S_1, S_2, and S_3 of secondary wall, primary wall, and their fibril orientations θ with respect to the cell axis (10).

Thus the cell cavity tends to remain nearly constant as the cell wall shrinks or swells.

Fortunately, from the utilization standpoint, wood does not shrink and swell to the same extent as does the cell wall. If this were not so, all woods would shrink and swell volumetrically, for a given moisture change, as much as the cell wall, rather than in proportion to their specific gravities. Therefore, they would shrink or swell more than they actually do. It should also be noted that the magnitude of the fiber-saturation point of a given wood directly affects its dimensional changes. The fiber-saturation point may be reduced by the restraining effects of the cell wall layers because of hygroelastic effects, as is discussed later.

Moisture-induced dimensional changes in wood have been described traditionally in terms of shrinkage Sh (based on green dimensions) or of swelling Sw (based on dry dimensions), as given above. However, it is sometimes more appropriate to describe these changes in terms of the dimensions at some intermediate moisture content. For volume changes a hygroexpansion coefficient X_v may be defined as follows,

$$X_v = (1/v)(dv/dm) \qquad (11)$$

where v is the wood volume at moisture content m and dv/dm is the change of volume per unit moisture content change.

Figure 14 shows the linear idealized increase in volume v of wood as its moisture content m increases from zero to a moisture content greater than fiber saturation m_f. In the idealized case shown here the volume increases linearly with m from zero to m_f, with a constant slope dv/dm. The magnitude of X_v however decreases as m increases, because the volume v in Equation 11 increases with m.

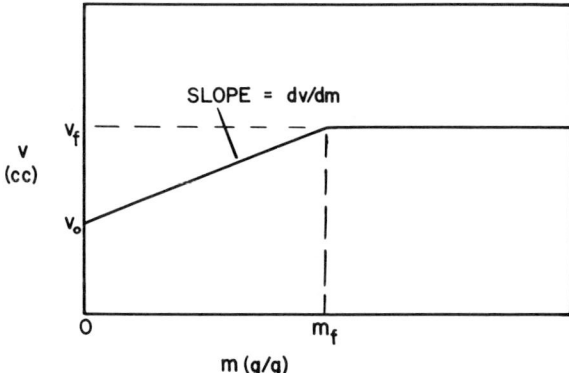

Figure 14. Idealized linear curve of wood volume V vs. moisture content.

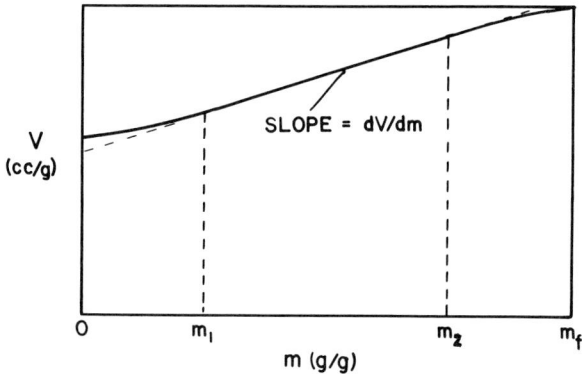

Figure 15. Actual form of curve of wood volume V vs. moisture content.

The value X_{vo} of X_v at $m = 0$ is related to Sw_f by

$$X_{vo} = Sw_f/M_f \qquad (12)$$

Similarly X_{vf}, the value of X_v at m_f, is related to Sh_f by

$$X_{vf} = Sh_f/M_f \qquad (13)$$

Because the percent swelling Sw_f is greater than the percent shrinkage Sh_f it is evident that X_{vo} is greater than X_{vf}, and that these values define the limits of X_v for a given wood.

Comparison of Equations 13 and 10 reveals that, for the case of a wood whose cell cavities remain constant in size, $X_{vf} = G_f$ and, therefore, G_g. Similarly, $X_{vo} = G_o$ under the same conditions. The ratio X_{vo}/G_o was measured (34) for a number of woods and was plotted as a function of moisture content in each case. The slope dv/dm is not constant over the entire hygroscopic range of moisture contents. As indicated in Figure 15, the slope is lower at the lower and upper moisture contents but is essentially constant over most of the hygroscopic moisture range. The mean value of the ratio X_{vo}/G_o should be unity if the cell cavity remains constant with swelling of the cell wall. In some woods the mean value of the ratio X_{vo}/G_o is less, and in others greater than unity, indicating deviations from the hypothetical assumption that the cell cavity remains constant in size.

ANISOTROPY IN SHRINKING AND SWELLING. Wood is anisotropic— that is, different in different directions—with respect to dimensional changes. The least shrinkage occurs along the grain and the most shrinkage in the tangential direction; radial shrinkage is about half that of tangential shrinkage.

Directional dimensional changes can be expressed in terms of hygroexpansion coefficients, one for each of the three principal axes. These may be written as follows:

$$X_l = (1/l)(dl/dm) \quad (14)$$

$$X_r = (1/r)(dr/dm) \quad (15)$$

$$X_t = (1/t)(dt/dm) \quad (16)$$

where X_l, X_r, and X_t are the longitudinal, radial, and tangential hygroexpansion coefficients, respectively and l, r, and t are the corresponding dimensions in the respective directions.

Longitudinal or axial shrinkage is almost negligible in normal mature wood, ranging from 0.1 to 0.3% when such wood dries from the green to oven-dry condition. This shrinkage is so small that it causes no problems in ordinary use. However, reaction wood (compression wood in softwoods and tension wood in hardwoods) and also juvenile wood (wood from near the pith) usually show much higher axial shrinkages, which may cause excessive crooking, bowing, or twisting when wood dries. There has been an increasing trend toward harvesting younger trees which contain a larger proportion of juvenile wood than do mature trees. It is anticipated therefore that excessive longitudinal shrinkage and the related warping problems will become increasingly common.

Excessive longitudinal shrinkage usually is associated with high microfibril angles in the S_2 layer, the thickest layer of the secondary wall. This angle Θ, which refers to the long axis of the cell, is small in normal wood (Figure 13). However, in juvenile wood and in reaction wood the angle Θ may increase from the normal values of less than 30° to values in excess of 45°. Because dimensional changes occur primarily at right angles to the microfibrils the component of shrinkage along the cell axis (and grain direction) increases as Θ increases. Figure 16 shows that the observed shrinkage Sh_f of *Pinus jeffreyi* increases with fibril angle Θ for angles greater than 30° (35). The figure also shows how the tangential shrinkage decreases with increasing Θ, as expected.

Several quantitative models have been proposed for explaining the effect of fibril angle θ on both axial and transverse wood shrinkage (10). Only the model of Barber (36) will be discussed here. In this model the typical wood cell is assumed to be circular in cross section and the cell wall is considered to consist of two layers. One of these is the thick S_2 layer whose microfibril angle Θ is the principal independent variable (Figure 13). The second layer is the S_1 layer,

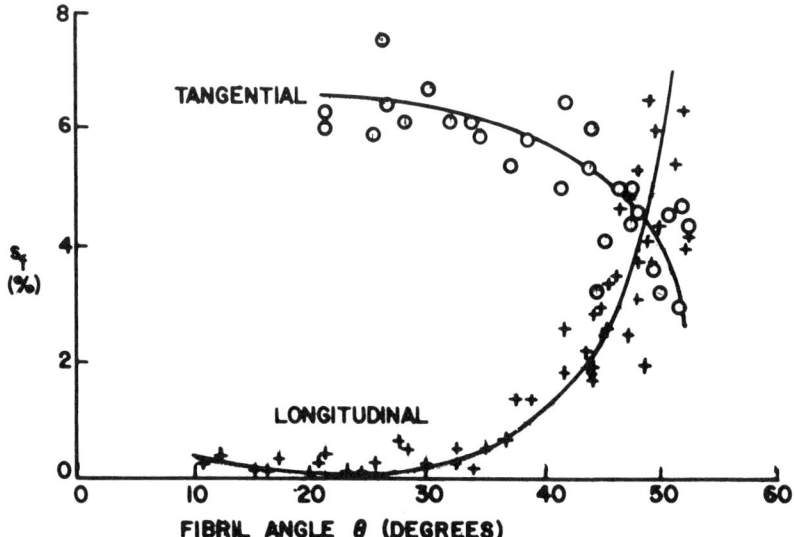

Figure 16. Experimental points and fitted curves of longitudinal and tangential shrinkages of Pinus jeffreyi as functions of fibril angle θ. (Adapted from Ref. 37.)

which is considered to be a thin constraining sheath, surrounding the S_2 layer and having its microfibrils oriented perpendicular to the cell axis, thus resisting the hygroexpansion of the S_2 layer. Half of the microfibrils in the S_2 layer are assumed to be oriented in a spiral which makes an angle Θ with the cell axis. The other half spiral, in the opposite direction, has the same angle Θ with respect to the cell axis but in the opposite sense. This is to overcome the twisting tendency associated with a single direction of orientation. In actual wood the S_2 layers of adjacent cell walls are oriented in opposite directions to prevent twisting of individual cells.

The equations derived by Barber (36) are given in his paper and are not reproduced here. However, the curves shown in Figure 17 illustrate the predicted ratios of longitudinal (ϵ_x), transverse (ϵ_1), and cell cavity (ϵ_2) swellings of the model cell to the unrestrained isotropic swelling (ϵ_o) as functions of the angle Θ. Figure 17 shows two curves each of ϵ_x/ϵ_o, ϵ_1/ϵ_o, and ϵ_2/ϵ_o, each curve differing by a factor of 10 in the relative stiffness of the restraining sheath and the swelling cell wall material. As anticipated, when this stiffness ratio is high there is less external cell swelling both longitudinally (ϵ_x/ϵ_o) and transversely (ϵ_1/ϵ_o) and also for the cell cavity (ϵ_2/ϵ_o). In fact for the higher fibril angles Θ, there is some negative swelling or shrinkage of the cell cavity (ϵ_2/ϵ_o) for the higher stiffness ratio of 50.

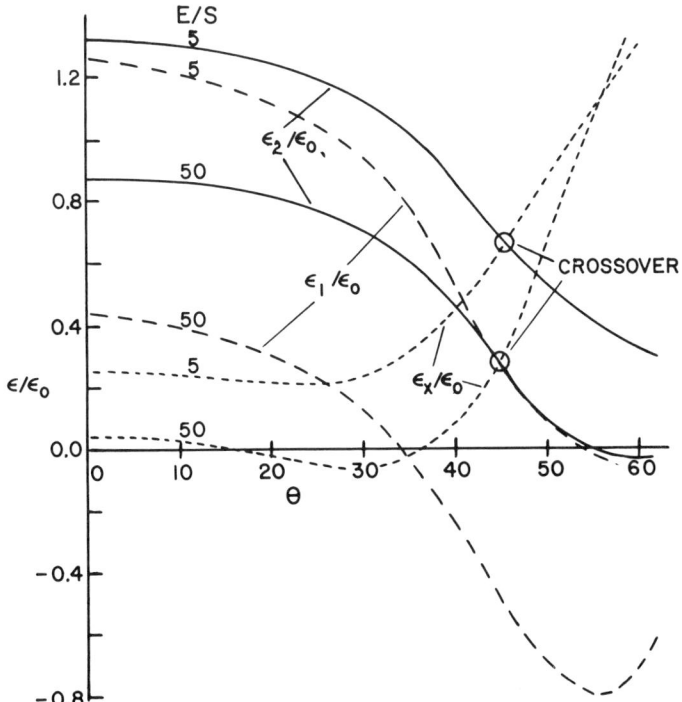

Figure 17. Calculated strain ratio curves L (ϵ_x/ϵ_o), T (ϵ_1/ϵ_o) and C (ϵ_2/ϵ_o) for two stiffness of restraining sheath/stiffness of swelling cell wall material (E/S) ratios as given by Barber (36).

Looking at the individual curves for a stiffness ratio of 50, it may be noted that both the longitudinal (ϵ_x/ϵ_o) and transverse (ϵ_1/ϵ_o) swelling curves strongly resemble the experimental curves shown in Figure 16. In both figures the longitudinal and transverse curves show the same shrinkage (or swelling) at a fibril angle Θ of near 45°. The longitudinal swelling curve of Figure 17 predicts a slight negative swelling for Θ between 20 and 35° for a stiffness ratio of 50. This negative swelling has been observed in some cases (37).

Hygroexpansion is usually greater transversely than longitudinally. However, there is considerable anisotropy in the transverse direction because the tangential hygroexpansion coefficient X_t is about twice the radial coefficient X_r. Traditionally the ratio of these two coefficients X_t/X_r was called the tangential/radial (T/R) shrinkage ratio, with a mean value among woods of about two.

The high T/R ratio is the primary reason for the warping in a cross section of wood which occurs in boards when they are first dried or when they are subjected to moisture changes in use. Figure

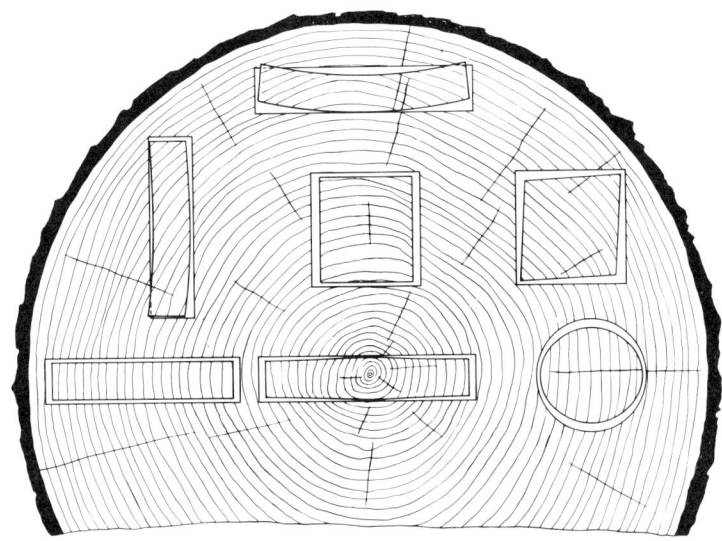

Figure 18. Cross-sectional distortion after drying of flats, squares, and rounds from representative locations log. (Courtesy of U.S. Department of Agriculture, Forest Products Laboratory.)

18 illustrates some of the more common kinds of cross-sectional warping that are caused by tangential/radial anisotropy when wood dries from the green condition. The pronounced cupping in flat-sawn boards is probably the most troublesome kind of cross-sectional distortion.

Several theories have been proposed to explain transverse shrinkage anisotropy. These fall into two general categories (10, 22, 38). One category is based on differences in the structure of the radial and tangential walls of wood cells. The other category is based on differences in the shrinkages of different wood tissues such as ray vs. longitudinal tissues or of earlywood vs. latewood tissues.

The theories based on cell wall structure are of two kinds: those that relate to fibril angle differences in the radial and tangential walls and those that relate to the thickness and behavior of the middle lamella between cells. The mean fibril angle in the radial walls may be greater than in the tangential walls, which results in less radial shrinkage than tangential shrinkage in the individual cells because of the strong effect of fibril angle on transverse shrinkage (Figure 17).

Boyd (39) has reviewed the published data and theories of anisotropic transverse shrinkage. He has concluded, in agreement with Bosshard's (40) contention, that the dominant factor is the greater degree of lignification in the radial walls. This characteristic reduces sorption of water (Figure 11). Boyd also attributes a significant effect to the preponderance of radially flattened thick-walled cells in the latewood of some woods, particularly conifers.

There are at least two theories based on shrinkage differences among wood tissues. The most general of these is the *ray restraint theory*. This theory proposes that the rays shrink less than the longitudinal tissues in the radial direction and therefore reduce the extent of radial shrinkage. This theory applies to at least some cases. The plotted points and calculated curve shown in Figure 19, for example, show how the radial shrinkage of beech wood decreases as the volume of ray tissues increases (*41*). It is probable that additional tangential shrinkage is induced by the radial restraint because of Poisson's ratio effect.

The second theory, based on tissue shrinkage differences, is the *earlywood–latewood interaction theory*. This theory attributes the anisotropy to the alternation of earlywood and latewood layers in many woods. Latewood is often two or three times more dense than earlywood. It shrinks more, and is stronger. Therefore, the tangential shrinkage is high because it is dominated by the high latewood shrinkage which forces the weak earlywood to shrink tangentially more than it would if it were allowed to shrink independently of the latewood. The radial shrinkage is equal to the effective weighted average of the two tissue shrinkages. However, the effective radial shrinkage of earlywood is reduced because of Poisson's ratio effect and excessive tangential shrinkage (*10*).

Moisture Effects on the Physical Properties of Wood

All of the physical and mechanical properties of wood are affected by its moisture content. The effect on mechanical properties will be discussed first, followed by consideration of some other important physical properties. In all cases the discussion is limited to clear wood, free from defects such as knots.

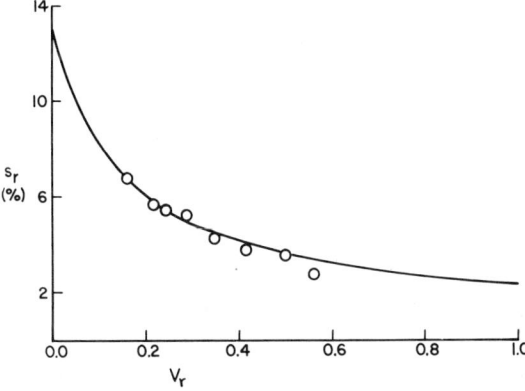

Figure 19. Experimental points and fitted curve of radial shrinkage s_r against fraction of ray tissue V_r. (Adapted from Ref. 41.)

Mechanical Properties. Many of the important mechanical properties of wood increase exponentially as the moisture content decreases below the fiber-saturation point (28) (*see* Chapter 5). This relationship can be expressed as

$$S_2/S_1 = \exp\left[-(r/100)(M_2 - M_1)\right] \quad (17)$$

where S_1 and S_2 are the magnitudes of a particular strength property at moisture contents M_1 and M_2, respectively, and r is a coefficient that represents the percent increase in a particular strength property S for a 1% decrease in wood moisture content M.

The value of the coefficient r is about two for the modulus of elasticity, and four for the modulus of rupture in static bending. The maximum value of about six applies to compressive strength parallel to the grain (42).

Some strength properties, such as toughness and shock resistance, may decrease with decreasing wood moisture content because these properties are proportional to the deformation of a wood member under load and the stress sustained. Moist wood deforms more than dry wood and the product of stress and deformation, which is a measure of toughness, may actually be greater for moist wood.

The strength of wood is generally not affected by changes in moisture content above fiber saturation because the excess water accumulates in the cell cavities. Therefore, it does not affect the strength of the cell wall itself, which determines the overall wood strength.

Other Physical Properties. In addition to its important effect on the strength of wood, moisture also affects wood's other physical properties. Moisture's effect on electrical properties was described in the section on "Electrical Resistance Moisture Meters" (p. 130). Other properties such as specific gravity and thermal properties are discussed here.

SPECIFIC GRAVITY AND DENSITY. The decrease that occurs in specific gravity with an increase in moisture content was discussed in connection with wood swelling. The specific gravity G_m decreases with increasing moisture content up to fiber saturation but above this there is no change because the volume remains constant and the weight is based on the oven-dry condition.

The density ρ_m of wood, however, always increases with increasing moisture content. This increase is because density, in contrast to specific gravity, is always based on the wet weight of the wood. Figure 20 illustrates how both specific gravity and density change with increasing moisture content based on the assumption of constant cell cavity volume. Because the cell cavity or pore volume

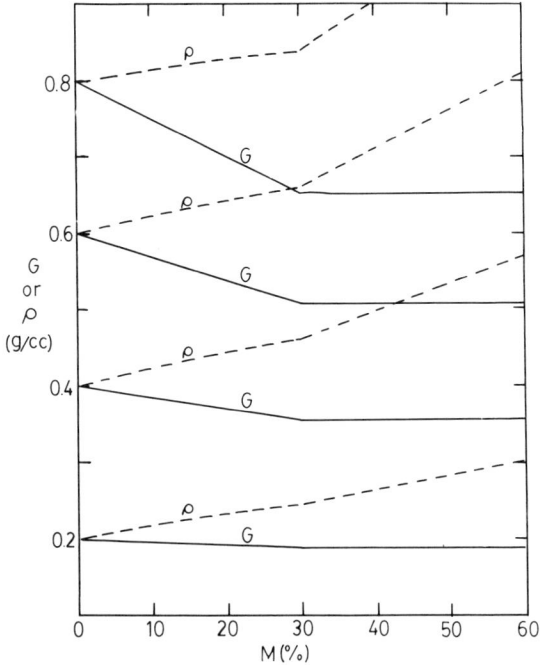

Figure 20. Calculated curves of specific gravity G and of density ρ (g/cm³) of wood vs. moisture content which assume a constant cell cavity volume and a 30% fiber-saturation point. (Adapted from Ref. 45.)

remains approximately constant, the fraction of void volume decreases with increasing moisture content due to swelling of the cell wall.

THERMAL PROPERTIES. Some of the important thermal properties of wood are affected by its moisture content. These include specific heat, thermal conductivity, and thermal diffusivity.

The specific heat c_o of dry wood, which is about 0.295 at 25 °C, increases linearly with temperature (10) from about 0.27 (cal/g °C) at 0 °C to 0.38 at 100 °C. The specific heat c_w of liquid water is 1.0. Therefore, by using the method of mixtures with the assumption that bound water has the same specific heat as liquid water, the specific heat c for moist wood can be calculated as

$$c = (c_o + mc_w)/(1 + m) \qquad (18)$$

Experimental measurements (43) and also theoretical considerations (10, 44) have yielded higher values for c than are predicted based on Equation 18. However, for many practical purposes Equation 18 is adequate as a first approximation.

The thermal conductivity of wood also increases greatly with its moisture content, as well as with other factors such as temperature and specific gravity. A number of proposed empirical equations indicate that the thermal conductivity of wood increases with increasing moisture content. These equations have been summarized by Siau (45) who proposed that the primary effect of water was to swell the cell wall and, thus, to provide a larger heat conduction area for a given wood.

An empirical equation that relates the thermal conductivity (K_h) to wood density ρ (g/cm^3) and moisture content m is Equation 19.

$$K_h = [0.60 + \rho(4.1 + 5.1m)] \times 10^{-4} \text{ cal/cm s } °C \quad (19)$$

The thermal diffusivity (D_h) is equal to the ratio $K_h/(\rho c)$. Its variation with moisture content can be determined by the effect of m on K_h, ρ, and c, where c is the specific heat. Because ρc increases more rapidly with moisture content than does K_h, there is a slight decrease in D_h with increasing moisture content, on the order of 0.5% per percent increase in moisture content M.

Thermodynamics of Moisture Sorption

When wood below the fiber-saturation point interacts with water, heat is evolved, and there are changes in the free energy and entropy of the sorbed water. Furthermore, the wood exerts swelling forces that can be measured. These effects can be treated by classical thermodynamic methods although moisture sorption by wood is not a perfectly reversible process because sorption hysteresis is involved, as was pointed out in the section on "Moisture Sorption Isotherms" (p. 136).

Enthalpy Changes. The three forms of water found in wood have different energy or enthalpy levels, as shown in Figure 21. Water vapor in the cell cavities has the highest enthalpy. The enthalpy of liquid water in the cell cavities of green wood is considerably lower, essentially equal to that of free liquid water, if the effects of capillary forces and dissolved materials are neglected. The difference in enthalpy between liquid water and water vapor is the *heat of vaporization* [Q_o (cal/g water)] of free water. The bound water in the cell wall of wood is at still lower energy level, Q_L (cal/g water) below that of liquid water. The sum of the *heat of sorption*, Q_L and Q_o, is equal to Q_v (cal/g water), which is the heat required to evaporate bound water from the cell wall.

The curve for the energy level of bound water shown in Figure 21 indicates that Q_L and Q_v increase with decreasing wood moisture content below fiber saturation M_f. This increase means that more

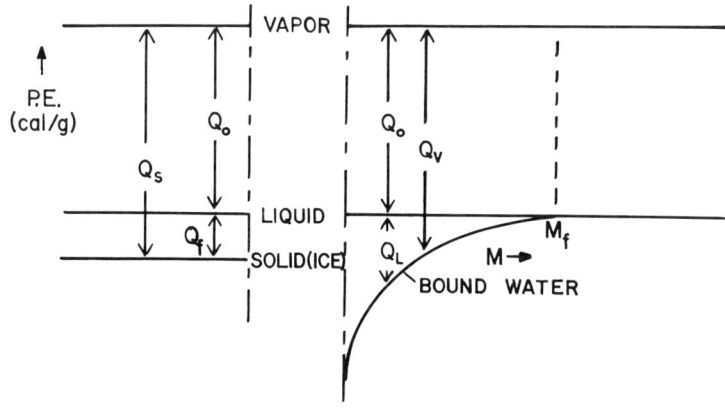

Figure 21. Diagram showing the relative energy levels, in terms of Q, for water vapor, liquid water, ice, and bound water in wood at different moisture contents. (Reproduced with permission from Ref. 10. Copyright 1972, Syracuse University Press.)

energy is required to evaporate 1 g of water from wood as M decreases (below M_f).

Stamm and Loughborough (46) first calculated Q_v and Q_L as functions of wood moisture content M by applying the Clausius–Clapeyron equation to the moisture sorption isotherms for wood at several temperatures. For example, Q_L can be obtained by replotting sorption isotherms, such as are shown in Figure 9, into the form of isosteres of constant moisture content, of $\ln h$ against the reciprocal of Kelvin temperature. These plots yield a family of essentially straight lines, each at a different moisture content. The magnitude of Q_L at any moisture content is calculated from

$$Q_L = -(R/18)d(\ln h)/d(1/T) \qquad (20)$$

where R is the gas constant and $d(\ln h)/d(1/T)$ is the slope of the curve for the specified moisture content.

The enthalpy changes associated with moisture sorption also can be measured calorimetrically. In this case the heat of wetting W (cal/g of wood) is usually measured. The heat of wetting is defined as the heat generated when wood at some initial moisture content m is thoroughly wetted by liquid water. When measuring the heat of wetting calorimetrically, the wood is ground into small particles in order to expedite thorough wetting by water. The wood particles are then conditioned to a desired initial moisture content prior to insertion into the calorimeter.

The heat of wetting W for wood at any initial moisture content can be calculated if Q_L is known as a function of moisture content.

The heat of wetting is equivalent to the integral of $Q_L dm$ between the limits of initial moisture content m and complete saturation. Thus

$$W = \int_m^\infty Q_L dm \qquad (21)$$

Conversely, if W is known as a function of moisture content m, Q_L can be calculated as in Equation 22.

$$Q_L = -dW/dm \qquad (22)$$

The earliest recorded measurements of W vs. m were given, according to Stamm (22), by Volbehr in 1896. The results of these measurements can be represented (47) by

$$W = W_o \exp(-Bm) \qquad (23)$$

where W_o is the value of W for initially dry wood ($m = 0$), and B is an empirical constant. More recent measurements are summarized in Reference 48.

Figure 22 contains plots of Q_L (log scale) vs. wood moisture content. The logarithm of Q_L appears to decrease linearly with in-

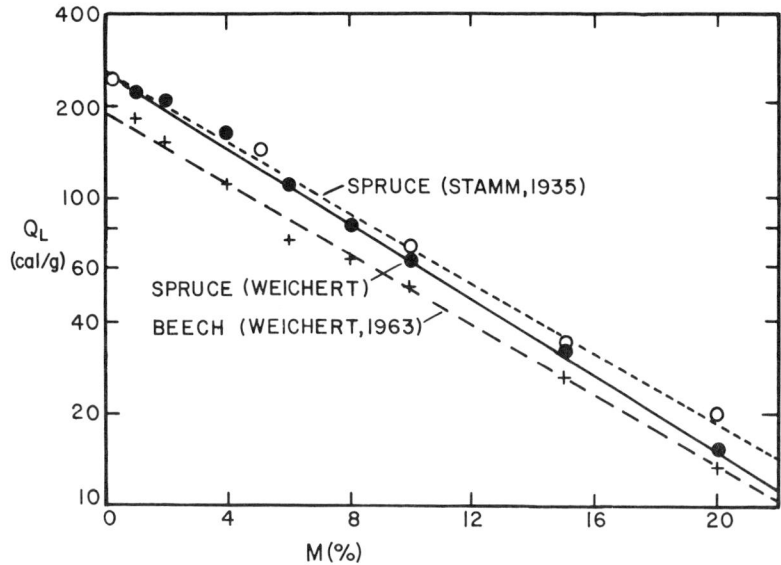

Figure 22. Curves of Q_L (log scale) vs. wood moisture content for European spruce and beech and for Sitka spruce. (Reproduced with permission from Ref. 10. Copyright 1972, Syracuse University Press.)

creasing moisture content, as is anticipated based on Equations 22 and 23. Equations 22 and 23 can be combined to give

$$Q_L = Q_{Lo} \exp(-Bm) \qquad (24)$$

where $Q_{Lo} = BW_o$, the value of Q_L at $m = 0$.

The heats of wetting W and sorption Q_L are interrelated, but they have different interpretations in terms of moisture sorption. For example, the heat of sorption is presumed to be a measure of the excess energy required to break the bond between bound water and the sorption sites, and the heat of wetting is a measure of the total number of sorption sites accessible to water (10).

The total heat of wetting W_o generally ranges between 15 and 20 cal/g of wood, but may be lower for woods with high extractive content. It increases with decreasing particle size and generally with removal of extractives (48). As shown in Figure 23, the heat of wetting at a given moisture content is higher for partial desorption than for partial adsorption, possibly because there are more sorption sites available during desorption.

Free Energy and Entropy Changes. The heat of sorption Q_L consists of two parts, the free energy change ΔG and the entropy change ΔS. The loss in free energy ΔG (cal/g of water) can be calculated at any moisture content by

$$\Delta G = (RT/18) \ln(1/h) \qquad (25)$$

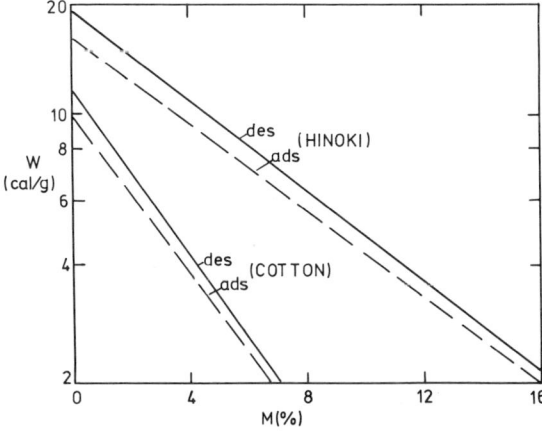

Figure 23. Curves of heat of wetting W (log scale) for Hinoki cypress and for cotton against the moisture content (48).

where h is the relative vapor pressure (assumed to be equivalent to the activity) of the wood at equilibrium with the moisture content m, R is the gas constant, and T is the Kelvin temperature. The excess energy associated with the entropy change ΔS is defined as the difference in Q_L and ΔG. Thus

$$T\Delta S = Q_L - \Delta G \qquad (26)$$

Figure 24 shows curves of Q_L, ΔG, and $T\Delta S$ plotted against wood moisture content. All energy terms are negative (heat is given off) when wood takes up water from the liquid state. The decrease in entropy indicates that bound water is more ordered than liquid water, in analogy to the greater order of water in ice compared with the liquid state. As the moisture content approaches fiber saturation the distinction between liquid water and water in wood decreases toward zero. However, even above fiber saturation the water in cell cavities may be different from ordinary liquid water because of capillary forces and/or dissolved materials.

Swelling Pressure of Wood. Wood swells when its moisture content increases. If it is restrained from swelling it will exert a

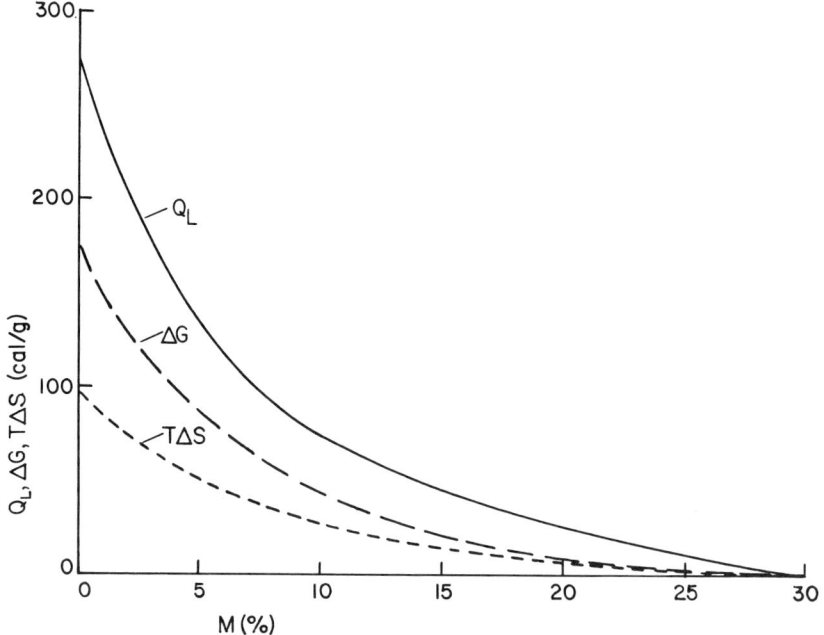

Figure 24. Curves of Q_L, ΔG, and $T\Delta S$ when liquid water is taken up by wood at various moisture contents. (Reproduced with permission from Ref. 10. Copyright 1972, Syracuse University Press.)

swelling pressure against the restraining medium. The order of magnitude of the swelling pressure Π of the cell wall, when it is initially at equilibrium with relative vapor pressure h and then immersed in liquid water, can be calculated from the osmotic pressure equation for solutions written as

$$\Pi = \rho \Delta G = (\rho RT/18) \ln (1/h) \qquad (27)$$

where ρ is the density of water in the cell wall and the other terms are as defined in Equation 25.

When wood is restrained from swelling while soaking in water, the pressures measured are much smaller than those predicted from Equation 27 because swelling takes place into the cell cavities. Therefore, the maximum swelling pressure developed is a function of the strength of the wood (49).

The maximum transverse swelling pressures P have been measured (49) in wood dowels which had previously been densified to different specific gravities. After densification, each sample was conditioned to equilibrium with $h = 0.3$ and then inserted into a steel restraining ring equipped with strain gages to measure the transverse swelling stress. The wood dowels were then exposed to liquid water. The results showed that the maximum observed swelling pressure increased exponentially with increasing specific gravity of the test samples (Figure 25). The extrapolation of their curves to the specific gravity of the dry cell wall, 1.5, yields a value of 13,200 psi (91 MPa). The theoretical value based on Equation 27 is 158 MPa at room temperature. The discrepancy between these two values may be be-

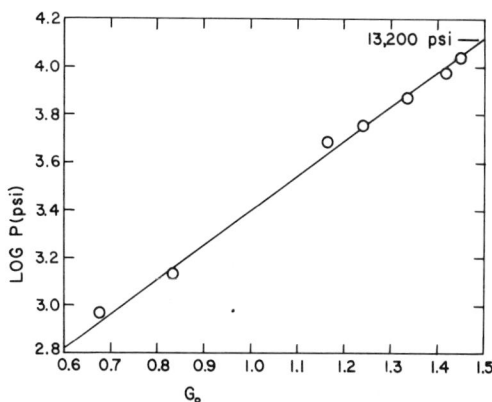

Figure 25. Curve of swelling pressure (log scale) vs. dry-volume specific gravity G_0 (49). (Reproduced with permission from Ref. 10. Copyright 1972, Syracuse University Press.)

cause the swelling along the grain was neglected and because of the compressibility of the cell wall itself. This discrepancy is treated in the Barkas theory of hygroelasticity discussed next.

Hygroelastic Effects. When hygroscopic materials such as wood are restrained from swelling freely they not only exert a swelling pressure but also come to a lower moisture content than if unrestrained. The reverse effect also holds true: wood restrained from shrinking exhibits a higher equilibrium moisture content than if unrestrained. This is the *hygroelastic effect,* sometimes called the *Barkas effect* because Barkas was the first to treat it quantitatively (26).

Barkas proposed a generalized osmotic pressure theory for hygroscopic gels such as wood, based on the generalized Porter equation in the form

$$(dV/dm)_p dP_m = vdp \qquad (28)$$

where $(dV/dm)_p$ is the apparent specific volume of the sorbed water at constant vapor pressure p (because V is the swollen volume of the gel per unit dry mass, and m is the fractional moisture content); dp_m is the increase in hydrostatic pressure required to raise the vapor pressure of the gel by the increment dp at constant moisture content m; and v is the specific volume of water vapor.

Equation 28 reduces to the osmotic pressure equation (Equation 27) if it is integrated between the limits of p and p_o (where $p = hp_o$), assuming that $(dV/dm)_p = 1/\rho$ and that the ideal gas law applies to water vapor, where $P_m = \Pi$. Barkas objected to applying the osmotic pressure equation to hygroscopic gels such as wood because this equation neglects the properties of the gel itself, such as rigidity, when compared with solutions. Furthermore, it does not distinguish between the swelling pressure at constant m and at constant V.

Barkas, therefore, proposed a more generalized swelling pressure theory for gels, which uses the Porter equation, in which the terms P, V, m, and p are interrelated based on empirical data obtained from wood. Figure 26 shows a pressure–volume diagram for the cell wall of Sitka spruce which was calculated by applying the Porter equation to the data of Stamm and Seborg (50) on Sitka spruce. The diagram indicates that the bulk modulus (VdP/dV) of the cell wall is greater at constant moisture content m (proportional to the slopes of the solid lines of constant m) than at constant relative vapor pressure h (proportional to the slopes of the broken lines of constant h).

By using the simple osmotic pressure equation (Equation 27) we assume that no moisture change occurs in the wood when it is completely restrained from swelling during exposure to water. According

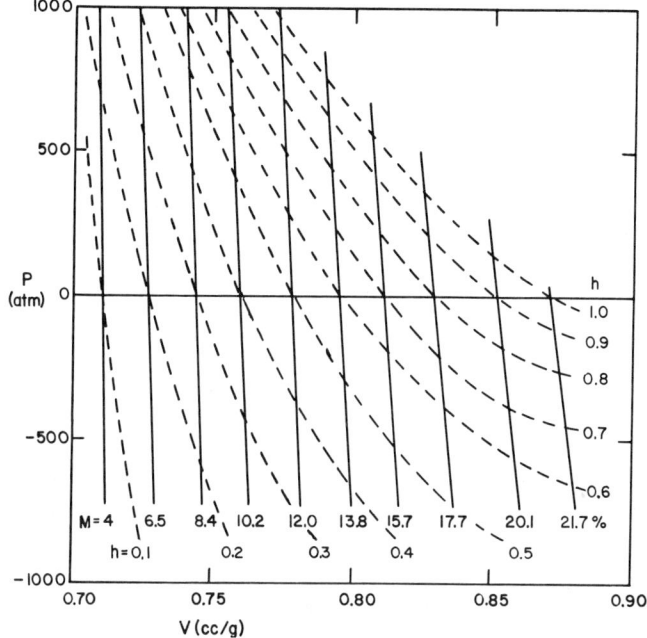

Figure 26. Curves of hydrostatic pressure P vs. specific volume V at constant moisture content m or relative vapor pressure h, from Barka (26). (Reproduced with permission from Ref. 10. Copyright 1972, Syracuse University Press.)

to the constant m curves of Figure 26, if the volume V is to remain constant the moisture content m must increase when the wood cell wall is at an equilibrium moisture content m and the relative vapor pressure is increased. This increase in moisture content effectively increases the initial value of h so that the swelling pressure is lower than that calculated from the simple osmotic pressure equation. For example, in the experiment of Tarkow and Turner (49) previously cited, the effective value of h would be somewhat higher than 0.3 due to a slight increase in m, and the calculated value of swelling pressure by using Equation 27 would be lower than the 158 MPa when calculated by assuming that $h = 0.3$.

Theories of Water Sorption

Many theories have been proposed to account for the sorption of water by hygroscopic materials such as wood. One of the earliest theories (Peirce, Reference 51) suggests that water is sorbed by textiles in two forms, one strongly attached to primary sorption sites and the other bound more weakly.

Most subsequent sorption theories, including those discussed here, have followed this general approach and postulate two forms of sorbed water. These theories may be classified into at least two general types based on the sorption mechanism assumed. One type assumes sorption on internal surfaces and is represented by the Dent theory (52), which is a modification of the classic Brunauer, Emmett, and Teller (BET) theory (53). The second type assumes that the wood–water system forms a solution, exemplified by the Hailwood–Horrobin theory. There have been other theories, not discussed here, that have also been applied to explain water sorption by hygroscopic materials (10, 54, 55).

Dent's Surface Sorption Theory. The Dent sorption theory or model (52), in the simple form, is a modification of the BET model, which is itself an extension of the earlier Langmuir model (56). The Langmuir model assumes that a gas (water vapor in the case of wood) is sorbed onto sorption sites on the substrate or sorbent in a monolayer only. The fraction of sorption sites occupied by the vapor or sorbate is a function of the vapor pressure of the sorbate and approaches unity as the vapor pressure increases.

The BET model (53) extends the Langmuir model to permit more than one layer of condensate on any particular sorption site. Furthermore, it postulates that the gas condensed in layers above the first layer has the same thermodynamic properties as ordinary condensate (liquid water in the wood–water system). The basic sorption isotherm predicted by the BET model fits the isotherm for wood reasonably well at relative vapor pressures less than ~0.3 but not at higher values. A modification that limits the maximum number of layers to a finite number, say five or six gives a better fit at high relative vapor pressure (Figure 27). However, the Dent model gives a more satisfactory fit over most of the hygroscopic moisture range.

Figure 28 shows the wood substrate containing primary sorption sites (vertical lines), some occupied by primary water molecules (dark circles), and some containing both primary and secondary (open circles) water molecules. In both the BET and Dent models the primary sorption sites are assumed to be high energy binding sites, such as accessible hydroxyl groups, and the secondary sites are of lower binding energy. The BET model assumes that the thermodynamic properties of the secondary water molecules are the same as those of ordinary liquid water, whereas the Dent model assumes that they are different.

If the three fundamental Dent constants, m_o, k_1, and k_2 are known the sorption isotherm can be written as

$$m = m_o k_1 h / [(1 - k_2 h)(1 + k_1 h - k_2 h)] \qquad (29)$$

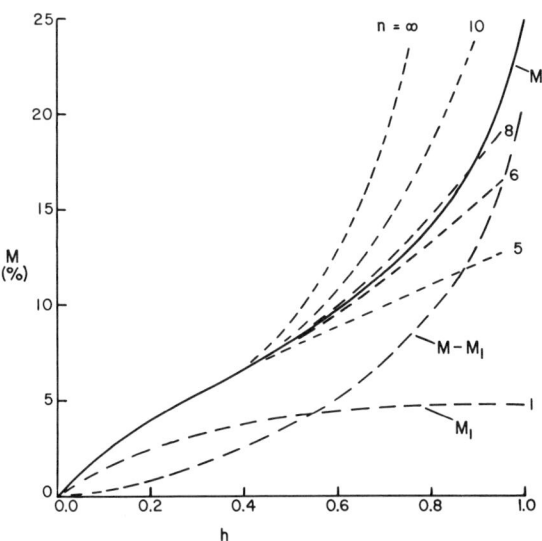

Figure 27. Sorption isotherms predicted by BET sorption theory for various values of n *(broken lines), compared with the experimental isotherm (solid line). Also shown is the monolayer moisture content* M_1 *(10). (Reproduced with permission from Ref. 10. Copyright 1972, Syracuse University Press.)*

where m_o is the moisture content corresponding to complete monolayer coverage (all primary sites occupied by a single molecule on each site); k_1 and k_2 are equilibrium constants related to the binding energies of the primary and secondary water layers, respectively; and h is the relative vapor pressure. The total moisture content m consists of two components; the primary water m_1, and the secondary water m_2. These can be calculated as

$$m_1 = m(1 - k_2 h) = m_o k_1 h/(1 + k_1 h - k_2 h) \tag{30}$$

$$m_2 = mk_2 h = m_o k_1 k_2 h^2/[(1 - k_2 h)(1 + k_1 h - k_2 h)] \tag{31}$$

The free energy change ΔG_1 (cal/g of water) associated with sorp-

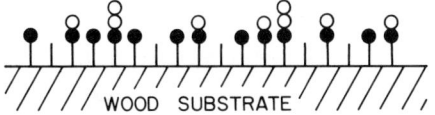

Figure 28. Schematic diagram showing sorption sites (vertical lines), some occupied by primary water molecules (dark circles) and some by secondary water molecules (open circles) (59).

tion of primary water, and ΔG_2 (sorption of secondary water) can be calculated from the equilibrium constants k_1 and k_2 (Equation 32).

$$\Delta G_1 = -(RT/18)\ln k_1; \Delta G_2 = -(RT/18)\ln k_2 \qquad (32)$$

The coefficients m_o, k_1, and k_2 can be calculated from actual sorption isotherms by modifying Equation 29 to Equation 33 (where A, B, and C are constants determined empirically from sorption isotherm data).

$$h/m = A + Bh - Ch^2 \qquad (33)$$

When A, B, and C are known, the fundamental constants k_1, k_2, and m_o can be evaluated as

$$k_2 = [-B + (B^2 + 4\,AC)^{1/2}]/(2\,A) \qquad (34)$$
$$k_1 = 1/[1 - k_2(B/C)] = (B/A) + 2\,k_2 \qquad (35)$$
$$m_o = 1/[Ak_2(k_1 + 1)] = 1/(Ak_1) \qquad (36)$$

The first step in evaluating the coefficients for a particular sorption isotherm is to calculate the ratio h/m from experimental data of h and m at each data point. The h/m values are then fitted to a parabola, using least squares regression procedures, with h as the independent variable (Equation 33). The values of m_o, k_1, and k_2 are then evaluated from the regression coefficients A, B, and C.

Figure 29 shows curves of H/M ($=h/m$) vs. H ($=100h$) calculated from experimental sorption data (also plotted) on wood and bark at 25 °C, for both adsorption and desorption. Figure 30 shows the curves of the total moisture content M, and of M_1 and M_2, all expressed in percent ($M = 100\,m$). These curves were obtained using values of m_o, k_1, and k_2 calculated from the curve in Figure 29 for the adsorption isotherm of wood. The curves labelled M_h and M_s are derived from the Hailwood–Horrobin sorption isotherm model.

Hailwood–Horrobin Solution Sorption Theory. The Hailwood–Horrobin (57) model treats moisture sorption as hydration of the polymer, taken here to be dry wood, by some of the sorbed water called water of hydration, m_h. The hydrate forms a partial solution with the remaining sorbed water, called water of solution, m_s. An equilibrium is assumed to exist between the dry wood and water and the hydrated wood with an equilibrium constant K_1. Equilibrium is also assumed to exist between the hydrated wood and water vapor at relative vapor pressure h, with equilibrium constant K_2. A third constant m_o is defined as the moisture content corresponding to com-

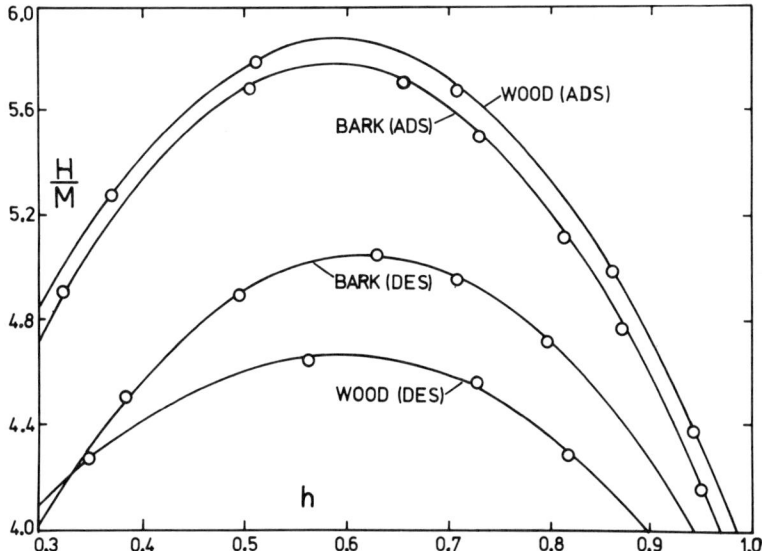

Figure 29. Plotted points and fitted curves of the ratios H/M (=h/m) vs. relative vapor pressure h for mean adsorption and desorption data on 10 woods and barks (19).

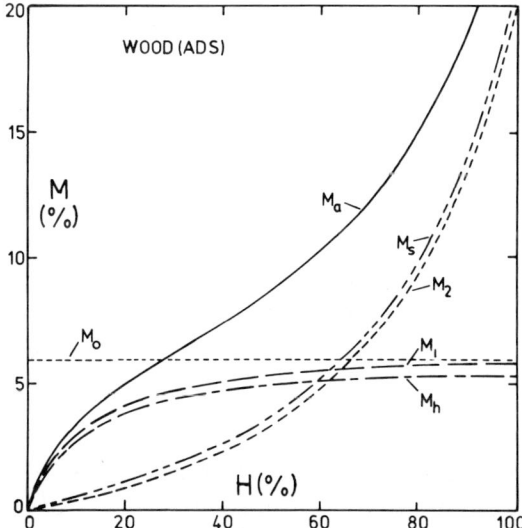

Figure 30. Mean adsorption isotherms calculated from uppermost curve of Figure 29 and curves of M_1, M_2, M_h, and M_s vs. H. Also shown is M_0 (19).

plete hydration of the wood. The model also assumes that the solution of hydrated wood and dissolved water behaves ideally, an assumption that has been criticized (58).

The Hailwood–Horrobin single hydrate model predicts a sorption isotherm of the same form as the Dent model, that is, a parabolic relationship between h/m and h as given in Equation 33. Furthermore, two of the fundamental constants, m_o and K_2 are identical with the Dent constants m_o and k_2. The third constant K_1 is analogous to k_1 of the Dent model but not identical. They are related by

$$K_1 = K_2(k_1 + 1) \tag{37}$$

The water of hydration m_h and of solution m_s are analogous and almost equal to the primary m_1 and secondary m_2 moisture contents, respectively (Figure 30). Equations for the free energy changes associated with the water of hydration and of solution, in analogy with Equation 32 are given by

$$\Delta G_h = -(RT/18)\ln K_1; \Delta G_s = -(RT/18)\ln K_2 \tag{38}$$

Moisture Transport

Water in wood is rarely in static equilibrium. It is continually adjusting to changes in its environment. The most dramatic change occurs when green wood is first dried. However, even in use wood is exposed to cycles of changing humidity, both daily and seasonally.

The rate of change of wood moisture content is determined by several factors. These factors include the current moisture content and gradients, specific gravity, dimensions and grain orientation of the wood, and the temperature, relative humidity, and air velocity surrounding the wood. It is convenient to discuss these parameters in terms of their effects on two phenomenological coefficients that have been customarily used to express moisture transport in wood and other materials. These coefficients are the moisture diffusion coefficient which determines the rate of movement internally through the wood and the surface emission coefficient which determines the rate of transport between the wood surface and its surroundings. The internal transport coefficient will be discussed first, followed by consideration of the surface coefficient.

The Moisture Diffusion Coefficient. The one-dimensional moisture flux (J) (g/cm^2 s) of water through wood customarily is given as the product of the moisture diffusion coefficient D (cm^2/s) and the gradient dc_m/dx of moisture concentration c_m (g/cm^3) in the direction of flow, or

$$D = -J/(dc_m/dx) \quad (39)$$

The driving potential assumed for moisture movement based on Equation 39 is the moisture concentration c_m. Other driving potentials may also be assumed. Table I lists the potentials that have been proposed, the resulting transport coefficients, and their relationships to D in each case (59). Although one or more of these other potentials may be more descriptive of the driving force for moisture movement, the discussion that follows will be restricted to the diffusion coefficient because it is so well established in the literature, and can be related to any of the others. Furthermore, it appears unchanged in the unsteady-state diffusion equation (Fick's second law), unlike any of the other coefficients. Thus Fick's second law may be written, for one dimension, as

$$dc_m/dt = d(Ddc_m/dx)/dx \quad (40)$$

The coefficient D is affected by many factors, the most important of which are temperature, moisture content, specific gravity, and grain orientation with respect to direction of flow as first calculated quantitatively by Stamm (60). For example, Figure 31 shows the strong increase in D with temperature as well as the much higher value along (D_l) than across (D_t) the grain for wood of 0.5 specific gravity (45). Figure 31 also indicates the complex effect of wood mois-

Table I. Some Moisture Transport Coefficients Used for Wood, Their Assumed Potentials, and Their Relationships to the Diffusion Coefficient D

Assumed Potential	Symbol (cgs units)	Transport Coefficient (cgs units)	Relation to Diffusion Coefficient
Moisture Concentration	c_m (g/cm^3)	$D = -J/(\partial c_m/\partial x)$ (cm^2/s)	$D = D$
Fractional Moisture Content	m (g/g)	$K_m = -J/(\partial m/\partial x)$ (g/cm s)	$K_m = D(\partial c_m/\partial m)$
Percent Moisture Content	M (g/100g)	$K_M = -J/(\partial M/\partial x)$ (g/100 cm s)	$K_M = D(\partial c_m/\partial M)$
Vapor Pressure	P (dyne/cm^2)	$K_p = -J/(\partial p/\partial x)$ (g cm/dyne s)	$K_p = D(\partial c_m/\partial p)$
Relative Vapor Pressure	h (ratio)	$K_h = -J/(\partial h/\partial x)$ (g/cm s)	$K_h = D(\partial c_m/\partial h)$
Osmotic Pressure	Π (dyne/cm^2)	$K_\Pi = -J/(\partial \Pi/\partial x)$ (g cm/dyne s)	$K_\Pi = D(\partial c_m/\partial \Pi)$
Spreading Pressure	ϕ (dyne/cm)	$K_\phi = -J/(\partial \phi/\partial x)$ (g/dyne s)	$K_\phi = D(\partial c_m/\partial \phi)$

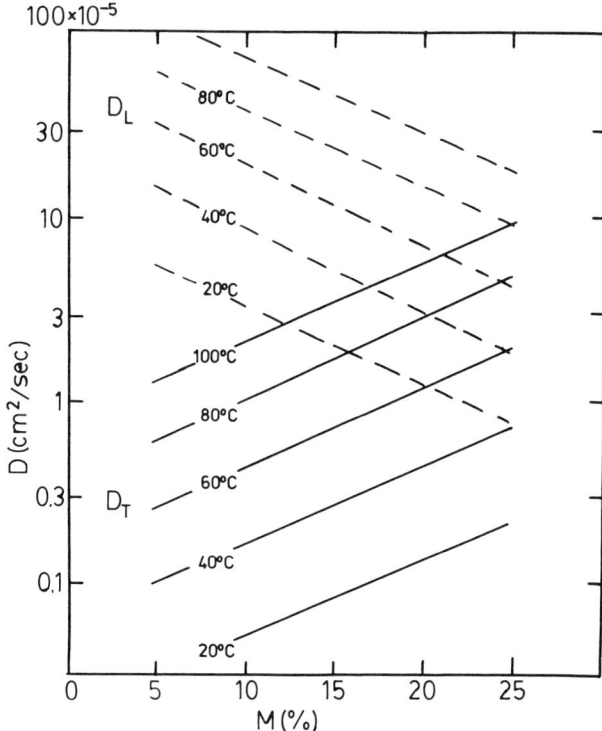

Figure 31. Curves of D_l *and* D_t *vs. wood moisture content for various temperatures. (Reproduced with permission from Ref. 45. Copyright 1971, Syracuse University Press.)*

ture content, with D_l generally decreasing and D_t increasing, as moisture content increases over the hygroscopic range from 5 to 25%. This difference is because the rate of vapor flow through the elongated cell cavities limits longitudinal diffusion and the rate of bound water flow through the cell walls determines the rate of transverse diffusion (60).

The strong increase in the diffusion coefficient D with increasing moisture content may be related to the decrease in the activation energy E_B for moisture diffusion in the cell wall with increasing moisture content, as is shown in Figure 32. According to the diagram the energy E_B is less than the energy E_V required to vaporize the water from the bound water level to the vapor state (61). This diagram is similar to Figure 21 except that the energy levels for the activated molecules are also shown.

Above fiber saturation, the effect of moisture content on D is even more complex because of the great variability in capillary flow

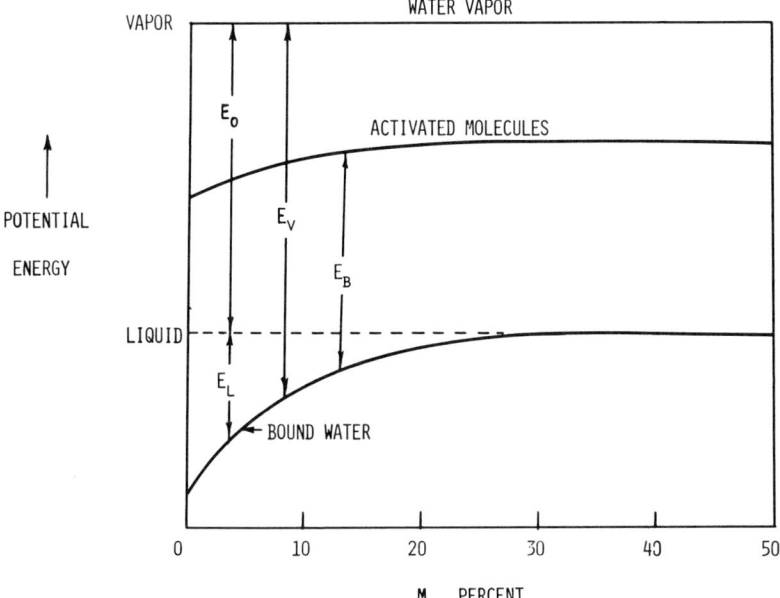

Figure 32. Curves showing relative energy levels of water vapor, activated molecules, and liquid and bound water (61).

through and particularly between wood cells. These cells are connected by pores whose dimensions and numbers vary by several orders of magnitude between and even within woods (62). This extreme variability makes quantitative estimates of D virtually impossible for moisture movement above fiber saturation (63).

Hawley (64) first demonstrated the complex nature of moisture flow through wood above the fiber-saturation point resulting from capillary forces associated with air bubbles and pores of variable radii interconnecting cells. Using Comstock's (65) simplified structural model for softwoods, Spolek and Plumb (66), however, were able to predict the capillary pressures in southern yellow pine as a function of percent of water saturation of the cell cavities. Such a quantitative analysis would be more difficult to implement in the case of woods other than southern yellow pine because their structures and permeabilities are more variable in most cases. However, computer modeling techniques are developing to the point where more general models may become feasible.

The Surface Emission Coefficient. During wood drying, particularly of thin wood such as veneers, flakes, and chips, the limiting rate factor may be the rate at which moisture can be removed from the wood surface. This is proportional to the surface emission coefficient S_e, defined as

$$S_e = J/(c_{m_s} - c_{m_e}) \qquad (41)$$

where c_{m_s} is the moisture content (g/cm^3) at the wood surface, and c_{m_e} is the value for the wood at equilibrium with the drying air.

As is the case with the diffusion coefficient D, S can be related to similar coefficients based on assumed potentials other than c_m. These coefficients are related to S as defined above in the same way that the alternate coefficients shown in Table I are related to D. In the case of S, however, the fundamental potential is probably the vapor pressure difference $(p_s - p_e)$ because vapor moves essentially in response to vapor pressure differences.

Rosen (67) has given solutions of the diffusion equation for wood from which the surface moisture content can be predicted at various stages of wood drying as a function of the transport ratio L, defined as Sa/D, where a is half the thickness of the wood. Based on these solutions and on experimental drying data, Rosen showed that the surface moisture contents calculated at various stages of drying were essentially equivalent to the values obtained by using the psychrometric approach given by Hart (68).

Rosen (69) showed that the coefficient S increases with increasing air velocity over the range from 1 to 12 m/s. The rate of increase became less pronounced at the higher air velocities.

Literature Cited

1. Morton, W. E.; Hearle, J. W. S. "Physical Properties of Textile Fibers"; 2nd ed.; William Heinemann Ltd.: London, 1975; p. 161.
2. Zimmerman, M. H.; Brown, C. L. "Trees—Structure and Function"; Springer-Verlag: New York, 1971.
3. Peck, E. C. "The Sap or Moisture in Wood"; U.S., For. Prod. Lab., Rep. No. 768, Madison, WI, 1953.
4. Zobel, B.; Matthias, M.; Roberts, J. H.; Kellison, R. C. "Moisture Content of Southern Pine Trees"; N.C. Sch. For. Tech. Rep. 37, Raleigh, NC, 1968.
5. Lawrence, W. E., Jr. "Field-Drying Logging Residue as an Industrial Fuel"; MS thesis, Virginia Polytechnic Inst. and State Univ., Blacksburg, VA, 1981.
6. Tiemann, H. D. "Effect of Moisture on the Strength and Stiffness of Wood", U.S. For. Serv. Bull. No. 70, Washington, DC, 1906.
7. Kollmann, F.; Hockele, G. *Holz Roh- Werkst.* 1962, 20(12), 461–73.
8. Christensen, G. N.; Hergt, H. F. A. *J. Polym. Sci., Part A-1* 1969, 7(8), 2427–30.
9. Stamm, A. J. *Ind. Eng. Chem.* 1927, 19, 1021–25.
10. Skaar, C. "Water in Wood"; Syracuse Univ. Press: Syracuse, 1972.
11. Beall, F. C. "Relative Humidity and Moisture Content Instrumentation";

Proc., Symposium on Wood Moisture Content, Temperature and Humidity Relationships, U.S., For. Prod. Lab. Rep., Madison, WI, 1979.
12. U.S. Forest Products Laboratory, "Wood Handbook"; U.S.D.A. Handbook No. 72, rev., 1974.
13. Couture, R. F.; Hill, J. L. *For. Prod. J.* **1974**, *24*(4), 17–23.
14. James, W. L. "Electric Moisture Meters for Wood"; U.S., For. Prod. Lab. Rep. FPL-6, Madison, WI, 1975.
15. Sharp, A. R.; Riggin, M. T.; Kaiser, K.; Schneider, M. H. *Wood Fiber* **1978**, *10*(2), 74–81.
16. Nanassy, A. J. *Wood Sci.* **1973**, *5*, 187–93.
17. Hillel, D. "Fundamentals of Soil Physics"; Academic Press: New York, 1980.
18. Spalt, H. A. *For. Prod. J.* **1958**, *8*(10), 288–95.
19. Okoh, K. A. I.; Skaar, C. *Wood Fiber* **1980**, *12*(2), 98–111.
20. Weichert, L. *Holz Roh- Werkst.* **1963**, *21*(8), 290–300.
21. Skaar, C. "Moisture Sorption Hysteresis in Wood"; Proc., Symposium on Wood Moisture Content, Temperature and Humidity Relationships, U.S. For. Serv. For. Prod. Lab., Madison, WI, 1979.
22. Stamm, A. J. "Wood and Cellulose Science"; Ronald Press, 1964.
23. Simpson, W. T.; Rosen, H. N. *Wood Fiber* **1981**, *13*(3), 150–58.
24. Christensen, G. N.; Kelsey, K. E. *Holz Roh- Werkst.* **1959**, *17*(5), 178–88.
25. Wangaard, F. F.; Granados, L. A. *Wood Sci. Technol.* **1967**, *1*(4), 253–77.
26. Barkas, W. W. "The Swelling of Wood Under Stress"; *For. Prod. Res. Bull. (G.B.)* London, 1946.
27. Mitchell, R. H. Ph.D. dissertation, Virginia Polytechnic Inst. and State Univ., Blacksburg, VA, 1981.
28. U.S. Forest Products Laboratory. "Wood Handbook", U.S.D.A. Handbook No. 72 (rev.), U.S. GPO: Washington, DC, 1974.
29. Kellogg, R. W.; Wangaard, F. F. *Wood Fiber* **1969**, *1*(3), 180–204.
30. Stamm, A. J. *Wood Sci.* **1971**, *4*(2), 114–28.
31. Tiemann, H. D. "Wood Technology"; 2nd ed., Pitman Publ. Co.: New York, 1944.
32. Stamm, A. J.; Loughborough, W. K. "Variation in Shrinking and Swelling of Wood"; Trans. ASME, Louisville, KY, 1941.
33. Greenhill, W. L. "The Shrinkage of Australian Timbers, I"; Austr., CSIRO, Div. For. Prod. Technol. Pap. No. 21, Melbourne, 1936.
34. Keylwerth, R. *Holz Roh- Werkst.* **1964**, *22*(7), 255–58.
35. Meylan, B. A. *For. Prod. J.* **1968**, *18*(4), 75–78.
36. Barber, N. F.; Meylan, B. A. *Holzforsch.* **1968**, *22*(4), 97–103.
37. Meylan, B. A. *Wood Sci. Technol.* **1972**, *6*(4), 293–301.
38. Kollmann, F.; Cote, W. A., Jr. "Principles of Wood Science and Technology", Vol. I; Springer-Verlag: New York, 1968.
39. Boyd, J. D. *Mokuzai Gakkaishi* **1974**, *20*(10), 473–82.
40. Bosshard, H. H. *Holz Roh- Werkst.* **1956**, *14*(8), 285–94.
41. McIntosh, D. C. *For. Prod. J.* **1955**, *5*(5), 355–59.
42. Panshin, A. J.; de Zeeuw, C. H. "Textbook of Wood Technology", 4th ed.; McGraw-Hill: New York, 1980.
43. Hearman, R. F. S.; Burcham, J. N. *Nature (London)* **1955**, *176*, 978.
44. Kelsey, K. E.; Clarke, L. N. *Austr. J. Appl. Sci.* **1956**, *7*, 160–75.
45. Siau, J. F. "Flow in Wood"; Syracuse Univ. Press: Syracuse, 1971.
46. Stamm, A. J.; Loughborough, W. K. *J. Phys. Chem.* **1935**, *39*, 121–32.
47. Cooper, D. N. E.; Ashpole, D. K. *J. Text. Inst.* **1959**, *50 T*, 223–32.
48. Kajita, H. "The Heat of Wetting of Wood in Water (I)", Bull. No. 20 of the Kyoto Prefectural Univ. Forests, Kyoto, 1976; pp. 49–61.
49. Tarkow, H.; Turner, H. D. *For. Prod. J.* **1958**, *8*(7), 193–97.
50. Stamm, A. J.; Seborg, R. M. *J. Phys. Chem.* **1935**, *39*, 133–42.
51. Peirce, F. T. *J. Text. Inst.* **1929**, *20*, T133–50.
52. Dent, R. W. *Text. Res. J.* **1977**, *47*(2), 145–52.

53. Brunauer, S.; Emmett, P. H.; Teller, E. *J. Am. Chem. Soc.* **1938**, *60*, 309–19.
54. Simpson, W. T. *Wood Fiber* **1973**, *5*(1), 41–49.
55. Chirife, J.; Iglesias, H. A. *J. Food Technol.* **1978**, *13*, 159–74.
56. Langmuir, I. *J. Am. Chem. Soc.* **1918**, *40*, 1361.
57. Hailwood, A. J.; Horrobin, S. *Trans. Faraday Soc.* **1946**, *42B*, 84–92.
58. Barrie, J. A. in "Diffusion in Polymers"; Crank, J.; Park, G. S., Eds.; Academic Press: London, 1968.
59. Skaar, C.; Babiak, M. *Wood Sci. Technol.* **1982**, *16*, 123–38.
60. Stamm, A. J. "Passage of Liquids, Vapors and Dissolved Materials Through Softwoods"; U.S.D.A. Tech. Bull. No. 929, U.S. Dept. Agric., Washington, DC, 1946.
61. Skaar, C.; Siau, J. F. *Wood Sci. Technol.* **1981**, *15*, 105–12.
62. Smith, D.; Lee, E. "The Longitudinal Permeability of Some Hardwoods and Softwoods"; Spec. Rep. For. Prod. Res. No. 13, London, 1958.
63. Banks, W. B. *Wood Sci. Technol.* **1981**, *15*, 171–77.
64. Hawley, L. F. "Wood–Liquid Relations"; U.S.D.A. Tech. Bull. No. 248, U.S. Dept. Agric.: Washington, DC, 1931.
65. Comstock, G. L. *Wood Fiber* **1970**, *1*, 283–89.
66. Spolek, G. A.; Plumb, O. A. *Wood Sci. Technol.* **1981**, *15*, 189–99.
67. Rosen, H. N. *Wood Sci.* **1982**, *14*(3), 134–37.
68. Hart, C. A. *Wood Sci.* **1977**, *9*(4), 194–201.
69. Rosen, H. N. *Wood Fiber* **1978**, *10*(3), 218–28.

RECEIVED for review May 9, 1983. ACCEPTED July 7, 1983.

PROPERTIES AND REACTIVITY

4

Penetration and Reactivity of Cell Wall Components

ROGER M. ROWELL

U.S. Department of Agriculture, Forest Service, Forest Products Laboratory, Madison, WI 53705

Chemical modification of wood to increase its resistance to biodegradation and photodegradation, to improve its dimensional stability, and to decrease its flammability depends on adequate distribution of reacted chemicals in the water-accessible regions of the cell wall. The chemicals used for modifying wood must be capable of swelling the wood to facilitate penetration and must react with the cell wall polymer hydroxyl groups under neutral or mild alkaline conditions at temperatures at or below 120 °C. The chemicals should react quickly with the hydroxyl groups to yield stable chemical bonds with no by-products. The modified wood must retain the desired properties of the untreated wood. Chemicals used to modify wood include anhydrides, acid chlorides, carboxylic acids, isocyanates, aldehydes, alkyl chlorides, lactones, nitriles, and epoxides. Reaction of these chemicals with wood yields a modified wood with good biological resistance and greatly improved dimensional stability. The reaction takes place in the cell wall and is evident when the increases in the wood volume approach the volume of chemical added, when the leach resistance of the modified wood is high, and by IR data. Studies on the distribution of the bonded chemical show good penetration into the cell wall structure. The lignin component is highly substituted although the carbohydrate components are less substituted.

Physical Properties and Chemical Modification

Wood is a three-dimensional, polymeric composite made up primarily of cellulose, hemicellulose, and lignin. These polymers make up the cell wall and are responsible for most of the physical and chemical properties exhibited by wood. Wood is a preferred building and engineering material because it is economical, low in processing energy, renewable, strong, and aesthetically pleasing. It has, however, several undesirable properties, such as biodegradability, flammability, dimensional instability with varying moisture contents, and degradability by UV light, acids, and bases. These properties are all the result of chemical reactions involving degradative environmental agents. Wood is degraded biologically because organisms recognize the polysaccharide polymers in the cell wall (e.g., the cellulose and hemicelluloses) and have very specific enzyme systems capable of hydrolyzing these polymers into digestible units. Biodegradation of the high molecular weight cellulose weakens the wood because cellulose primarily is responsible for the strength in wood (*see* Chapter 5). Strength is lost as the cellulose polymer undergoes degradation through oxidation, hydrolysis, and dehydration reactions. The same types of reactions take place in the presence of acids and bases (*see* Chapter 15).

Wood changes dimension with changing moisture content because the cell wall polymers contain hydroxyl and other oxygen-containing groups that attract moisture through hydrogen bonding (*see* Chapter 3). This moisture swells the cell wall, and the wood expands until the cell wall is saturated with water. Water beyond this saturation point is free water in the void structure and does not contribute to further expansion. This process is reversible, and the wood shrinks as it loses moisture.

Wood burns because the cell wall polymers undergo hydrolysis, oxidation, dehydration, and pyrolysis reactions with increasing temperature to give off volatile, flammable gases. The lignin component contributes more to char formation than do the cellulose components, and the charred layer helps insulate the wood from further thermal degradation (*see* Chapter 13).

Wood exposed to the outdoors undergoes photochemical degradation caused by UV light. This degradation takes place primarily in the lignin component and causes characteristic color changes. The lignin acts as an adhesive in wood, holding cellulose fibers together. Consequently, the wood surface becomes richer in cellulose content as the lignin degrades. In comparison to lignin, cellulose is much less susceptible to UV degradation. These poorly bonded fibers are washed off the surface during a rain, which exposes new lignin to the degradative reactions. In time, this "weathering" process can account for a significant loss in surface fibers (*see* Chapter 11).

Because these types of degradation are chemical in nature, it should be possible to eliminate them or decrease their rate by modifying the basic chemistry of the wood cell wall polymers. Chemical modification of wood is any chemical reaction between some reactive part of a wood component and a simple single chemical reagent, with or without catalyst, that forms a covalent bond between the two components. The most abundant reactive chemical sites in wood are the hydroxyl groups on cellulose, hemicellulose, and lignin.

Most of the research done in the area of chemical modification involves the reaction of hydroxyl groups. For example, biodegradation can be prevented by reacting chemicals with the hydroxyls on the cellulose component. When this is done, the highly specific biological enzymatic reactions cannot take place because the chemical configuration and molecular conformation of the substrate have been altered. More research has centered around improving dimensional stability. Changes in dimension can be reduced by bulking the cell wall with bonding chemicals. This method works because the treatment puts the wood in a partially, if not completely, swollen state.

These techniques demonstrate that it is possible to change the basic chemistry and, therefore, the properties of wood cell wall polymers through chemical reactions. These chemical modifications can greatly enhance the properties of wood products.

Chemical modification of wood for biological resistance is based on the theory that the enzymes (cellulases) must directly contact the substrate (wood cellulose), and the substrate must have a specific configuration. If the substrate is chemically changed, this highly selective reaction cannot take place. One way to chemically modify the substrate is to change the hydrophilic nature of the wood. In some cases water, a necessity for decay organisms, is excluded from biological sites. The chemicals used for modification need not be toxic to the organism because their action renders the substrate unrecognizable as a food source to support microbial growth. In other words, the organisms starve in the presence of plenty.

Research involving cell wall bulking treatments has shown that the increase in wood volume is directly proportional to the theoretical volume of chemical added (1). The wood volume increases with increasing chemical addition to about a 25% gain in weight, at which point the treated volume is approximately equal to the green wood volume (2). When this bulked wood comes into contact with water, very little additional swelling can take place. This is how bulking treatments are effective for dimensional stability.

Several terms are used to describe the degree of dimensional stability given to wood by various treatments: antishrink efficiency (ASE), swelling percent, dimensional stabilization efficiency, antiswelling efficiency, and percent reduction in swelling. Generally the volumetric swelling coefficient is calculated by

$$S = \frac{V_2 - V_1}{V_1} \times 100$$

where V_2 is the wood volume after humidity conditioning or wetting with water, and V_1 is the wood volume of oven-dried sample before conditioning or wetting. Then ASE, which is the reduction in swelling or antishrink efficiency resulting from a treatment, can be calculated from

$$\text{ASE} = \frac{S_1 - S_2}{S_1} \times 100$$

where S_2 is the treated volumetric swelling coefficient, and S_1 is the untreated volumetric swelling coefficient.

Reaction Requirements

Penetration. In whole wood, accessibility of the treating reagent to the reactive chemical sites is a major consideration. To increase accessibility to the reaction site, the chemical must penetrate the wood structure. Penetration can be achieved by causing the wood structure to swell. If a reagent potentially capable of modifying wood does not cause the wood substance to swell, then catalyst may be necessary. If both the reagent and catalyst are unable to cause the wood to swell, a workable cosolvent could be added to the reaction system.

The swelling of wood by various organic liquids has been studied (3–9). For the most part, these studies consisted of soaking oven-dried blocks of wood for prolonged periods in anhydrous organic liquids at room temperature. The degree of swelling, or swelling coefficient, represents an unadjusted average swelling coefficient and is usually expressed as a three-dimensional function, i.e., volumetric swelling coefficient. For comparative purposes, volumetric swelling coefficients are usually standardized to a volumetric swelling coefficient compared to water, setting water at 10. If, for example, the unadjusted average volumetric swelling coefficient for water was experimentally determined to be 11.7, this could be standardized to 10 by dividing 1.17 into 11.7. All other volumetric swelling coefficient values obtained would then be divided by 1.17 to standardize them to a water value of 10. In other words, adjustment is made by dividing experimental volumetric swelling coefficient values by one-tenth the average volumetric swelling coefficient value for wood blocks treated with water.

Tables I, II, and III give volumetric swelling coefficients for southern pine sapwood for various potential reagents, catalysts, and solvents (10). These coefficients were determined under two sets of conditions. Specimens from oven-dried southern pine sapwood blocks were measured and their volumes were determined. Ten specimens from this set were submerged in a solution and either treated at 120 °C and a pressure of 150 lb/in.^{2} for 1 h, or they were soaked at 25 °C for 48 h.

Table I. Volumetric Swelling Coefficients (S) for Southern Pine Sapwood in Various Reagents

Reagent	120 °C; 150 lb/in.^{2}; 1 h	25 °C; Soaking
Methyl isocyanate	52.6	5.1
Acetic anhydride	12.3	1.5
Formaldehyde solution	12.3	12.3
Water	10.0	10.0
Epichlorohydrin	6.9	5.9
Acrolein	6.7	7.0
Propylene oxide	5.2	5.0
Acrylonitrile	4.6	4.5
Methyl isothiocyanate	4.5	4.1
Butylene oxide	4.1	0.7

Table I shows that the volumetric swelling coefficients for the potential reagents under the two conditions are nearly the same except in the cases of methyl isocyanate and acetic anhydride. The amount of swelling for these two reagents is much greater at 120 °C than at 25 °C because at 120 °C a reaction with wood has occurred using both methyl isocyanate and acetic anhydride. The large increase in volume at 120 °C is caused by reacted chemicals bulking the cell wall. Much less consistency between the two treating conditions is seen in Table II with catalysts. Piperidine and aniline have high swelling coefficients at 120 °C but very low swelling coefficients

Table II. Volumetric Swelling Coefficients (S) for Southern Pine Sapwood in Various Catalysts

Reagent	120 °C; 150 lb/in.^{2}; 1 h	25 °C; Soaking
n-Butylamine	15.5	15.2
Piperidine	13.3	0.0
Dimethylformamide	12.8	12.5
Pyridine	11.3	13.1
Acetic acid	11.1	8.8
Aniline	11.0	0.5
Water	10.0	10.0
Diethylamine	5.0	11.0
N-Methylaniline	2.6	0.8
N-Methylpiperidine	2.2	1.6
N,N-Dimethylaniline	0.3	0.5
Triethylamine	-0.1	2.1

at 25 °C. If the soaking at 25 °C is continued, piperidine reaches an equilibrium swelling coefficient of 13.1 after 90–100 d (3, 4) and aniline reaches an equilibrium swelling coefficient of 10 after 90 d (4). Triethylamine actually causes the wood structure to shrink slightly at 120 °C. This is also observed with hexane (Table III).

There are no striking differences between the high and low temperature soaking values for the solvents listed in Table III. In fact, there seems to be a correlation between hydrophilic nature and degree of swelling. Stamm (7) has suggested that some solvents actually swell the carbohydrate-type polymers in the cell wall and others swell lignin. Attempts have been made (4–7) to correlate observed swelling behavior with certain physiochemical properties of the liquids. Trends were noted between the degree of swelling and the dielectric constant, or between the surface tension, dipole moment, molecular size, or the tendency to hydrogen bond with methanol. Every trend, however, had its exceptions.

It is known that swelling is directly related to the density of the wood (7, 11, 12). Because latewood of most species has a density more than twice that of earlywood, latewood is a major contributor to swelling.

Many physical differences exist between latewood and earlywood (see Chapter 1). In softwoods, earlywood tracheids have thin

Table III. Volumetric Swelling Coefficients (S) for Southern Pine Sapwood in Various Solvents

Reagent	120 °C; 150 lb/in.^{2}; 1 h	25 °C; Soaking
Dimethyl sulfoxide	13.3	11.7
Dimethylformamide	12.8	12.5
Cellosolve	10.6	10.2
Methyl cellosolve	10.3	10.0
Water	10.0	10.0
Methanol	9.0	9.3
1,4-Dioxane	6.5	0.6
Tetrahydrofuran	5.4	7.2
Acetone	5.1	5.6
Dichloromethane	3.8	3.3
Methyl ethyl ketone	3.6	5.0
Ethyl acetate	2.4	4.2
Cyclohexanone	2.3	0.5
4-Methyl-2-pentanone	0.4	1.5
Xylenes	0.1	0.2
Cyclohexane	0.1	0.1
Hexanes	−0.2	0.2

cell walls, large lumens with ends overlapping those of other tracheids, and large and numerous pit-pairs distributed along the radial face. However, pit-pairs are most abundant on the other ends where tracheids overlap each other. Latewood tracheids have thick cell walls, narrow lumens, and fewer and smaller pit-pairs on the radial wall. In addition, the latewood tracheids are predominantly pitted on their tangential walls (13, 14).

The thick cell walls of latewood [mainly caused by a thicker S_2 layer in the cell wall (15)] result in less pit aspiration on drying (16–18). The main flow of liquids in softwoods is through the lumens of tracheids by way of bordered pit-pairs.

Several studies have been concerned with the penetration of liquids into latewood and earlywood (11, 16–23). Under atmospheric pressure, the penetration of nonpolar liquids into softwood latewood may be caused, in part, by capillary action in the very small lumens and passage through unaspirated pit membranes. In aspirated earlywood this penetration would not occur. Penetration of nonpolar liquids may also be through drying checks in the thick latewood cell walls. As the temperature and pressure of the liquid are raised, penetration of polar liquids in earlywood would be expected to increase because of softening of the pit structure and displacement of the pit membrane. Because the cell wall of earlywood is thinner than that of latewood, penetration into earlywood walls would be quicker and facilitated by swelling. Incrustation occurs in the pit membranes of southern pine latewood (24); this would retard liquid penetration.

Reactants. Cellulose, hemicelluloses, and lignin are distributed throughout the wood cell wall. These three hydroxyl-containing polymers make up the solid phase of wood. The void structure or lumens in wood can be viewed as a bulk storage reservoir for chemical reactants, which could be used to modify the cell wall polymers. For example, the void volume of southern pine earlywood with a density of 0.33 g/cm^3 is 0.77 cm^3 voids/cm^3 wood or 2.3 cm^3/g. For latewood with a density of 0.70 g/cm^3, the void volume is 0.52 cm^3/cm^3 or 0.74 cm^3/g. The cell wall can also swell and act as a chemical storage reservoir. For southern pine, the change in cell wall storage volume from oven-dry to water-swollen is 0.077 cm^3/cm^3. These data show that there is more than enough volume in the voids in wood to house sufficient chemical reactants for a reaction to take place with the cell wall polymers.

Potential reactants must contain functional groups that will react with hydroxyl groups of the wood components. There are many literature reports of chemicals that failed to react with wood components when, in fact, the chemicals did not contain functional groups that could react.

The chemical bond desired between the reagent and the wood component is of major consideration. For permanence, this bond should have enough stability to withstand environmental stresses. In such cases, the ether linkage may be the most desirable covalent C–O bond. Ether linkages are more stable than the glycosidic acetal bonds between sugar units in the wood polysaccharides; therefore, the polysaccharides would degrade before the bonded ethers. Less stable bonds can also be formed which would be useful for the release of a bonded chemical under environmental stresses. Acetals and esters are less stable than ether bonds and could be used to bond biological agents or fire retardants to the wood in such a way that they would be released under certain conditions. Unless all of the reagent skeleton becomes bonded to the wood, i.e., no by-products are generated, economics may dictate that a recovery system be implemented.

Gas reactants are difficult to handle because they require high pressure equipment. Also, the level of chemical substitution is usually lower in gas than in liquid systems, and gas penetration can be very difficult. The best success, to date, of chemical systems is with low boiling liquids that swell wood easily. If the boiling point is too high, it is difficult to remove excess reagent after treatment. Generally, the lowest member of a homologous series is the most reactive and has the lowest boiling point.

Some chemicals react completely with a single hydroxyl group. Such is the case, for example, with methylation using methyl iodide. Other chemicals, such as epoxides, in the course of reacting form a new hydroxyl group that reacts further. In other words, cases such as methylation involve single-site substitution, whereas cases such as epoxidation involve polymer formation from a single graft point. This will be discussed in detail later in this chapter.

From the standpoint of industrial application of reagents for wood, toxicity, corrosivity, and cost are important factors in selecting a chemical. The reacted chemicals should not be toxic or carcinogenic in the finished product, and the reactant itself should be as nontoxic as possible in the treating stage. This is somewhat difficult to achieve because chemicals that react easily with wood hydroxyl groups will also react easily with blood and tissue hydroxyl-containing polymers. The reactants should be as noncorrosive as possible to eliminate the need for special treatment of equipment. In the laboratory experimental stage, the high cost of chemicals is not a major consideration. Chemical cost is important, however, in commercialization of a process.

Conditions. There are certain experimental conditions that must be considered before a reaction system is selected. The temperature required for complete reaction must be low enough that it causes little or no wood degradation. However, the rate of reaction must be relatively fast. A safe upper limit is about 120 °C, because little wood degradation occurs at this temperature over a short period of time.

It is impractical to dry wood to less than 1% moisture, but the water content of the wood during reaction is, in most cases, critical. The hydroxyl in water is more reactive than the hydroxyl groups available in wood components, i.e., hydrolysis is faster than substitution. The most favorable condition is a reaction system in which the rate of reagent hydrolysis is relatively slow.

It is also important to keep the reaction system simple. It is best to avoid multicomponent systems that require complex separation procedures to recover the chemicals after the reaction. The optimum system would be when the reacting chemical swells the wood structure and acts as the solvent.

Almost all chemical reactions require a catalyst. Strong acid catalysts cannot be used with wood because they cause extensive degradation. The most favorable catalyst from the standpoint of wood degradation is a weakly alkaline one. Alkaline catalysts are also favored because many of them swell the wood structure and give better penetration (see Table II). The catalyst used should be effective at low reaction temperatures, easily removed after reaction, nontoxic, and noncorrosive. In most cases, the organic tertiary amines are best suited for this purpose.

The reaction conditions must be mild enough that the reacted wood still possesses the desirable properties of wood. The wood should remain strong, retain its natural color (unless a color change is desirable), still be a good electrical insulator, not become dangerous to handle, and be gluable and paintable.

Reactions with Wood

Esters. Acetylation. The most studied of all the chemical modification treatments for wood has been acetylation. The early work was done with acetic anhydride catalyzed with pyridine (25) or zinc chloride (26). In the reaction with acetic anhydride, acetylation occurs, and acetic acid is split out as a by-product:

$$\text{Wood-OH} + \text{CH}_3-\overset{\overset{\text{O}}{\|}}{\text{C}}-\text{O}-\overset{\overset{\text{O}}{\|}}{\text{C}}-\text{CH}_3 \rightarrow$$

$$\text{Wood}-\text{O}-\overset{\overset{\text{O}}{\|}}{\text{C}}-\text{CH}_3 + \text{CH}_3-\overset{\overset{\text{O}}{\|}}{\text{C}}-\text{OH}$$

The reaction is acid or base catalyzed. Many catalysts have been tried, including potassium acetate and sodium acetate (27), dimethylformamide (DMF) (28–30), urea ammonium sulfate (29), magnesium perchlorate (31–33), trifluoroacetic acid (32), boron trifluoride (30), sodium acetate (31), potassium hydrogen phosphate (34), and γ-rays (35). The best acetylation condition, however, is uncatalyzed acetic anhydride in xylene at 100–130 °C (36).

Acetylation is a single-site reaction, that is, one acetyl per reacted hydroxyl group with no polymerization. This means that all the weight gain in acetyl can be converted directly into units of hydroxyl groups blocked. When polymer chains are formed the weight gain cannot be converted into units of blocked hydroxyl groups.

At weight gains above 17%, acetylated wood was found in soil-block tests (90 d) to resist attack by the fungi *Coniophora puterana* (36), *Lentinus lepideus* (36), *Poria incrassata* (36, 37), *Polyporus versicolor* (36–38), *Gloeophyllum trabeum* (36, 38, 39), *Poria monticola* (36), *Poria microsporia* (37), and *Coniophora cerebella* (40–43). Acetylated, laminated veneers of yellow birch in ground contact stake tests at 19.2% weight gain had an average life of 17.5 years compared to 2.7 years for untreated controls (44).

Acetylation to a weight gain of 20–25% showed a 70% reduction in swelling or ASE (37, 38, 45, 46). Southern yellow pine weathered for 12 months decreased slightly in acetyl content. Its ASE dropped from 78 to 64% (38).

DENSITY. Acetylated wood is more dense than untreated wood and has fewer fibers of lignocellulose per unit volume (47). This effect is caused by the bulking of the acetate, which is more dense than water. Wood usually gets slightly darker after acetylation with uncatalyzed acetic anhydride; it also loses much of its natural brilliance (48). The change in color with catalyzed acetylation varies depending on the reaction conditions and catalyst. Color changes from a slight darkening to almost black with pyridine and DMF have been found.

Acetylated wood is less permeable to gases than untreated wood (49). This may be caused by the bulking chemicals which restrict the pore space. Moisture absorption decreases by a factor of two to three (50) as does overall water resistance (51, 52). Acetylation in a N_2O_4–N,N-DMF–pyridine system causes a permanent loss of cellulose crystallinity (53, 54). The loss of crystallinity yields a uniform distribution of acetyl groups in cellulose.

The mechanical properties of acetylated wood are generally equal to those of untreated wood. However, shear strength parallel to the grain decreases in treated wood (47), and the modulus of elasticity decreases slightly (54). Impact strength (38) or modulus of elasticity (or stiffness) are unchanged (47). Wet and dry compressive strength (38, 47), hardness, fiber stress at proportional limit (47), and work to proportional limit (47) are increased. Modulus of rupture is increased for softwoods but decreased for hardwoods (47).

Results of a 2-year paint study indicate acetylated wood is a better painting surface (37) than untreated wood. UV radiation darkens unacetylated wood, but there is no change or a slight bleached effect with acetylated wood (37). In general, acetylation reduces the adhesive strength of wood (48). Adhesive strength is reduced with urea–formaldehyde resins (54, 55) and casein glues (55), but there is very little effect with resorcinol–formaldehyde resins (55).

Many of the properties of acetylated wood depend on the method of acetylation. The temperature of treatment, time of reaction, and type and amount of catalysts play a significant role in the extent that fibers degrade during treatment. The amount of moisture present in the wood also is important. Some moisture (2–5%) seems to be needed to obtain the best reaction, but above this level the water hydrolyzes the acetic anhydride to acetic acid. This loss by hydrolysis accounts for a 5.7% loss of anhydride with each 1% of water in the wood (36). The rate of acetylation decreases as moisture content increases (37).

The anhydride method of acetylation gives an acid by-product that results in an acidic condition in the wood and a loss of 50% of the reaction chemical. These by-products must be removed to prevent degradation. Acetic acid, the by-product of acetylation with acetic anhydride, is virtually impossible to remove completely from wood. This results in a product that smells of acetic acid, acid conditions that catalyze the removal of more acetyl groups, acid hydrolysis of cellulose fibers which results in strength losses over a long term, and acid corrosion of metal fasteners used in the wood product.

Acetylation can also be done by vapor-phase treatments, but the diffusion rate varies inversely as the square of the thickness (37, 56). Because of this effect, vapor-phase treatment has been applied only to thin veneers.

Another method for the acetylation of wood involves reaction with ketene gas dissolved in acetone or toluene (57–61):

$$\text{Wood}-\text{OH} + \text{CH}_2=\text{C}=\text{O} \rightarrow \text{Wood}-\text{O}-\overset{\overset{\text{O}}{\|}}{\text{C}}-\text{CH}_3$$

Reactions carried out at 55–60 °C for 6–8 h produce weight gains of 22% (59). Much of the work with ketene, however, has resulted in much lower weight gains. At the higher level of treatment, the acetylated wood shows a reduction in water absorption by 35%, tangential swelling by 77%, and radial swelling by 69% (58).

Vapor-phase acetylation with ethanethioic acid produces a modified wood with slightly lower weight gains than acetylation with acetic anhydride (62). At weight gains of about 17%, the treated wood has an ASE of 48%. Ethanethioic acid is less corrosive than acetic anhydride, but the treated wood continues to emit hydrogen sulfide because of the entrapment of small amounts of ethanethioic acid.

In spite of the vast amount of research in the acetylation of wood, the process has not been applied commercially. Two attempts, one in the United States (38) and one in Russia (63, 64), came close to commercialization but were discontinued, presumably because they were not cost effective.

PHTHALYLATION. A wood product that has a very high initial ASE can be obtained by using phthalic anhydride. The initial ASE decreases if the wood is soaked repeatedly in water (65). Starting at an ASE of 100% on the first soak cycle, the ASE value drops to about 70% on the second cycle, 60% on the third cycle, and down to 50% on the sixth cycle. There is a corresponding loss of bonded chemical after each soaking, which shows the susceptibility to hydrolysis of the phthalyl group (66). Phthalyl groups have a greater affinity for water than do the hydroxyl groups in wood, so phthalylated wood is more hygroscopic than untreated wood (66, 67). Whereas the mechanism of ASE effectiveness by acetylation is by chemical blocking of the hydroxyl groups, phthalylation operates mainly by mechanical bulking of the submicroscopic pores in the wood cell wall (68). Phthalylation produces very high weight gains (65, 69). Most researchers have found that acetylation weight gains range from 15 to 21%, whereas phthalylation weight gains range from 40 to 130% (65, 69). These high weight gains may result from a polymerization reaction.

OTHER ANHYDRIDES. Other anhydrides have been reacted with wood, including propionic and butyric anhydrides in xylene without catalyst. These compounds react slower than acetic anhydride (36). After a 10-h reaction time (in xylene at 125 °C with ponderosa pine) acetylation produces weight gains of 17%, compared to less than 4% for propionylation and no weight gain for butyrylation. After 30 h of reaction, propionic anhydride produces a weight gain of 10%. Reaction with butyric anhydride produced little or no weight gain (36).

ACID CHLORIDES. Acid chlorides can also be used in esterification reactions (70). The product is the ester of the reacted acid chloride, with hydrochloric acid as a by-product:

$$\text{Wood}-\text{OH} + \text{R}-\overset{\overset{\text{O}}{\|}}{\text{C}}-\text{Cl} \rightarrow \text{Wood}-\text{O}-\overset{\overset{\text{O}}{\|}}{\text{C}}-\text{R} + \text{HCl}$$

Using lead acetate as a catalyst with acetyl chloride, Singh et al. (71) found a lower acetyl content than with the acetic anhydride method. They obtained much higher ASE values, however, with acetyl chloride (60–84% for acetyl chloride vs. 47% for acetic anhydride). By using a 20% lead acetate solution, the amount of free HCl released in the reaction is reduced. This very strong acid causes extensive degradation of the wood, and because of this very little work has been done in this area.

CARBOXYLIC ACIDS. Carboxylic acids have been esterified to wood catalyzed with trifluoroacetic anhydride (72, 73). Several unsaturated carboxylic acids react with wood by the trifluoroacetic anhydride *impelling* method to give an increase in oven-dry volume and ASE, and a decrease in wood crystallinity and moisture content (74). Reactions of wood with β-methylcrotonic acid (Reaction 1) give a degree of substitution high enough to make the esterified wood soluble in acetone and $CHCl_3$ to the extent of 30% (75).

$$H_3C-\overset{H-C-COOH}{\underset{}{C}}-CH_3 + Wood-OH \rightarrow Wood-O-\overset{O}{\underset{H_3C-\overset{\|}{C}-CH_3}{\overset{\|}{C}}}-C-H \quad (1)$$

Further esterification increases the solubility but is accompanied by considerable degradation of wood components. Solubilization seems to be hindered by both lignin and hemicellulose (76, 77).

Isocyanates. A nitrogen-containing ester is formed in the reaction of wood hydroxyls with isocyanates:

$$Wood-OH + R-N=\overset{O}{\overset{\|}{C}} \rightarrow Wood-O-\overset{O}{\overset{\|}{C}}-NHR$$

Wood veneer swollen in DMF was exposed to vapors of phenyl isocyanate at 100–125 °C (29). The wood gained no weight, but the ASE was as high as 77%. The modified veneers showed increased mechanical strength with little or no change in color. Baird (28) reacted DMF-soaked cross sections of white pine and Engelmann spruce with ethyl, allyl, butyl, *tert*-butyl, and phenyl isocyanates. Vapor-phase reactions of butyl isocyanate in DMF gave the best results. The reaction produced ASE values of 47% with a 14% gain in weight and 67% with a 31% gain in weight. Weight gains were as high as 50% with an ASE of 75–80%. The samples treated to 67% ASE had about a 25% reduction in toughness and abrasion resistance.

White cedar was reacted with 2,4-tolylene diisocyanate (78) with and without a pyridine catalyst to a maximum nitrogen content of 3.5 and 1.2%, respectively. This corresponds to weight gains of 21.8 and 7.5%. This high weight gain was accompanied by an ASE of 50%.

Compressive strength and bending modulus increased with increasing nitrogen content. Beech wood modified with a diisocyanate (79) up to 50% weight gain lost 4.5–8.1% weight after 6 weeks of attack by the fungi *Coniophora cerebella* and *Polystictus versicolor*. After fungal attack, the modified wood lost almost 20% of its static bending strength as compared to the modified wood before fungal attack. At chemical add-ons over 18%, wood modified with methyl, ethyl, *n*-propyl, and *n*-butyl isocyanates was resistant to attack by *Gloeophyllum trabeum* (80).

Methyl isocyanate reacts very quickly without catalyst to give weight gains up to approximately 75% (81). Maximum ASE values of 60% are obtained at weight gains of 25–30%. Above this level of bonded weight gain, the ASE values start to decrease. Scanning electron micrographs show that high levels of chemical add-ons to the cell wall polymers cause splitting in the tracheid wall and not in the intercellular spaces (80). In some cases the splits go through the bordered pits. When the tracheid wall splits, the ASE starts to drop and continues to drop as chemical weight gain increases. Splitting exposes new fiber surfaces where water can cause swelling. Swelling beyond the green wood volume takes place because the cell wall is ruptured and no longer acts as a restraint to swelling.

Ethyl, *n*-propyl, *n*-butyl, and phenyl isocyanates also react with wood without the need for a catalyst; but *p*-tolyl isocyanate, 1,6-diisocyanatohexane, and tolylene 2,4-diisocyanate require either DMF or triethylamine as a catalyst (80). High weight gains are observed with these last three isocyanates, but little or no dimensional stability results from the reaction. Therefore, polymerization must be taking place in the void structure.

Isocyanates are sensitive to moisture; therefore, the reaction needs to be done on dry wood (82). As wood moisture content increases before reaction, more nonbonded polymers are formed after reaction. Reacted moist wood shows very high ASE values on the first water-soak test, but leaching causes a significant loss in ASE. This shows that the bulking chemical is not bonded to the cell wall but comes out upon water leaching.

Acetals. FORMALDEHYDE. Wood hydroxyls and formaldehyde react in two steps (Reaction 2). Because the bonding is between two hydroxyl groups, the reaction is called cross-linking.

$$\text{Wood}-\text{OH} + \text{H}-\overset{\text{O}}{\underset{\|}{\text{C}}}-\text{H} \rightarrow \underset{\text{(hemiacetal)}}{\text{Wood}-\text{O}-\overset{\text{OH}}{\underset{|}{\text{C}}}\text{H}_2} \xrightarrow{\text{Wood}-\text{OH}} \underset{\text{(acetal)}}{\text{Wood}-\text{O}-\text{CH}_2-\text{O}-\text{Wood}} \quad (2)$$

The two hydroxyl groups may come from (1) hydroxyls within a single sugar residue; (2) hydroxyls on different sugar residues within a single cellulose chain; (3) hydroxyls between two different cellulose chains; (4) same as in (1), (2), and (3) except reaction occurs on the hemicelluloses; (5) hydroxyls on different lignin residues; and (6) interaction between cellulose, hemicelluloses, and lignin hydroxyls. The possible cross-linking combinations are many, and theoretically all of them are possible. Because the reaction is a two-step mechanism, some of the added formaldehyde will be in the noncross-linked form of hemiacetals. These bonds are very unstable and would not survive long after treatment.

The reaction is best catalyzed by strong acids, such as HCl (83–86), HNO_3 (85), SO_2 (87, 88), p-toluene sulfonic acid, and zinc chloride (84, 89). Weaker acids, such as sulfurous and formic acid, do not work (85). Bases, such as lime water or tertiary amines, can initiate the reaction (90), but attempts with triethylamine were unsuccessful (91).

When its weight is increased by 2%, formaldehyde-treated wood is not attacked by fungi (92). This is far short of the quantity of cross-linking needed to prevent attack on the basis of hydroxyl blocking for enzyme inhibition. Cross-linking, which is effective at these low levels, must be tying together structural units (92). An ASE of 47% is achieved at a weight gain of 3.1%, an ASE of 55% at 4.1%, an ASE of 60% at 5.5%, and an ASE of 90% at 7% (85, 89). Thus, a weight gain of 4% results in 4 times the ASE as would be found by bulking treatments such as acetylation.

The mechanical properties observed in formaldehyde-treated wood are reduced compared to those observed in untreated wood. Toughness and abrasion resistance decrease greatly (85, 89), crushing strength and bending strengths decrease about 20% (93), and impact bending strength decreases up to 50% (93). The measurements done thus far on the last two properties have been done on γ-ray-treated wood; consequently part of the strength reduction may be due to the γ-ray treatment. The loss in toughness is directly proportional to the ASE; i.e., a 60% ASE is equal to a 60% loss in toughness (85).

Formaldehyde treatment causes wood to become brittle. This embrittlement may be caused by the short inflexible cross-linking unit of the $-O-C-O-$ type. If the inner carbon unit were longer, there would be more flexibility in this unit, and the embrittlement should be reduced. Most of the loss in wood strength properties is probably caused by the hydrolysis structural cellulose units with a strong acid catalyst.

OTHER ALDEHYDES. Acetaldehyde (85) and benzaldehyde (85, 94) react with wood by using either HNO_3 or zinc chloride catalysts. Acetaldehyde modification produces a high ASE, but benzaldehyde modification yields an ASE of only 40%. Mechanical properties of these treated woods are the same as those of formaldehyde-treated wood.

Difunctional aldehyde (dialdehydes) reactions have been catalyzed with zinc chloride, magnesium chloride, phenyldimethylammonium chloride, and pyridinium chloride (94). Glyoxal, glutaraldehyde, and α-hydroxyadipaldehyde all show ASE values of 40% with weight gains of 15% and the highest ASE (50%) at 20% weight gain. With these three compounds, cross-linking is possible; however, with the low ASE at high weight percent gain, it is clear that bulking is the mechanism for the ASE achieved.

Chloral (trichloroacetaldehyde) with no catalyst gives a 60% ASE at 30% weight gain (94). After 15 weeks at 70% relative humidity, however, all weight gain was lost as well as the ASE. This shows a very unstable, perhaps reversible, bond formation.

Phthaldehydic acid in acetone catalyzed with p-toluenesulfonic acid gives an ASE of 40% at a weight gain of 34% (94). The ASE reaches 50–70% when phthaldehydic acid or its derivatives are cured at 100 °C uncatalyzed for 16–24 h (95).

Other aldehydes and related compounds have been reacted either alone or catalyzed with sulfuric acid, zinc chloride, magnesium chloride, ammonium chloride, or diammonium phosphate (94). Compounds such as 1,3-bis(hydroxymethyl)-2-imidazolidone, glycol acetate, acrolein, chloroacetaldehyde, heptaldehyde, o- and p-chlorobenzaldehydes, furfural, p-hydroxybenzaldehyde, and m-nitrobenzaldehyde all achieve the ASE by a bulking mechanism and not by low-level cross-linking. At weight gains of 15–25%, the highest ASE reported is 40%.

Ethers. METHYLATION. The simplest ether is the methyl ether. Reaction of wood with dimethyl sulfate and NaOH (54, 55), or methyl iodide and silver oxide (54) are two systems that have been reported. Methylation up to 15% weight gain did not affect the adhesive properties of casein glues. The mechanical properties of methylated wood are greatly reduced because of the severe reaction conditions required.

ALKYL CHLORIDES. In the reaction of alkyl chlorides with wood, HCl is formed as a by-product (Reaction 3).

$$\text{Wood–OH} + \text{R–Cl} \rightarrow \text{Wood–O–R} + \text{HCl} \quad (3)$$

Because of this, the treated wood is not very strong. Reaction of wood with allyl chloride in pyridine (96, 97) or aluminum chlorides gives high initial ASE; but when the wood is dried and resoaked, the effects of allylation are lost (97). In the allyl chloride–pyridine case, the ASE is not caused by the formation of allyl ethers in cellulose or lignin, but by the bulking caused by the formation of allyl pyridinium chloride polymers, which are water soluble and easily leached out (98).

Other alkyl chlorides tested are crotyl chloride (99) and n- and *tert*-butyl chlorides (100) catalyzed with pyridine. Again, the ASE is only temporary and the liberated HCl causes severe degradation.

β-PROPIOLACTONE. The reaction of β-propiolactone with wood produces different products depending on the pH of the reaction. Acid conditions (Reaction 4) result in an ether bond to the hydroxyl group, along with a free-acid end group.

$$\text{Wood-OH} + \begin{array}{c} CH_2-CH_2 \\ | \quad \quad | \\ O \; - \; C=O \end{array} \xrightarrow{H^+} \text{Wood}-O-CH_2-CH_2-COOH \quad (4)$$

$$\xrightarrow{OH^-} \text{Wood}-O-\overset{O}{\overset{\|}{C}}-CH_2-CH_2-OH \quad (5)$$

Under basic conditions (Reaction 5), an ester bond is formed with a primary alcohol end group.

Uncatalyzed β-propiolactone reactions in southern yellow pine (pH = 5) give a carboxyethyl derivative (101). High concentrations of β-propiolactone cause delamination and splitting because of the very high degree of swelling (91).

At a 25% weight gain, treated wood strongly resists (2% weight loss or less) rot (101, 102) caused by *Lentinus lepideus, Lenzites trabea, Poria monticola,* and *Coniophora puteana* in soil-block tests. Increasing the weight gain to 45% does not change the rot resistance in either weathered or unweathered samples. At 30% weight gain, the treated wood has an ASE of 60%.

The major problem in β-propiolactone reactions is that β-propiolactone has been labeled a very active carcinogen. For this reason, little future research can be expected on this chemical. It would be interesting, however, to look at this chemical reaction under the basic conditions that produce ester formation.

ACRYLONITRILE. When acrylonitrile is reacted with wood in the presence of an alkaline catalyst, cyanoethylation occurs (Reaction 6).

$$\text{Wood-OH} + CH_2 = CH-CN \rightarrow \text{Wood}-O-CH_2CH_2CN \quad (6)$$

With NaOH, a weight gain up to 30% has been achieved. At this level, the wood has an ASE of 60%. At a weight gain of 25%, there was no loss in sample weight in soil-block tests with *Poria monticola,* *Coniophora puteana, Lenzites trabea,* or *Lentinus lepideus* (*101*). With a nitrogen content of 8.5%, the treated wood is resistant to *Poria vaporaria* (*103*). With only 1% fixed nitrogen, the wood resists attack by *Lentinus lepideus, Poria monticola, Lenzites trabea,* and *Polyporus versicolor* (*104, 105*). Cyanoethylated stakes in ground contact at 15% weight gain have an average life of 7.8 years, compared to 3.9 years for untreated stakes (*44*).

To show that the decay resistance observed is caused by a bulking mechanism and not the toxicity of acrylonitrile or its reaction products, cyanoethylated wood was extracted with hot water (*104*) to show that the leachate had no toxic effects on *Lenzites trabea*. The leached blocks lost their decay resistance, which may be caused by the reaction of acrylonitrile with the ammonia catalyst that was used to form water-soluble polymers in the cell wall.

Cyanoethylated wood (which was prepared with NaOH as catalyst) had a lower impact strength than untreated wood (*101*). Exposure of 25% acrylonitrile in MeOH to 10^7 rads of ionizing radiation gave an ASE of only 40% at a weight gain of 29% (*106*). This low ASE may be caused by the acrylonitrile reacting with the MeOH and forming polymers in the lumen rather than in the cell wall.

Epoxides. The reaction between epoxides and hydroxyl groups is an acid- or base-catalyzed reaction; however, all work in the wood field has been with base-catalyzed reactions:

$$\text{Wood-OH} + \text{R-CH}\underset{\diagdown\diagup}{\overset{\text{O}}{-}}\text{CH}_2 \rightarrow \text{Wood-O-CH}_2\overset{\text{OH}}{\underset{|}{\text{C}}}\text{H-R}$$

The simplest epoxide, ethylene oxide, catalyzed with trimethylamine, has been used as a vapor-phase treatment. At a weight gain of 20%, there is a 60% ASE (*107*). An ASE of 82% with a weight gain of 10% for the same process or with propylene oxide has been claimed also (*108*). Under similar conditions, a weight gain of 22% gives less than 1% tangential and radial shrinkage (*109*). By using an oscillating pressure rather than a constant pressure system with ethylene oxide and trimethylamine, an ASE of 42% is found for a weight gain of 11% (*110*). More work (*111*) with propylene oxide, butylene oxide, and epichlorohydrin shows an ASE of 70% at weight gains of 22–25%. If NaOH is used with ethylene oxide in a vapor treatment, extensive swelling results, which causes bursting of the wood structure (*112*).

As with the methyl isocyanate system, high weight gains with propylene and butylene oxides cause the ASE to fall (Figure 1) (*2*). For propylene oxide, the maximum ASE (60–70%) is attained at a weight gain between 25 and 33%. For butylene oxide, a wider range

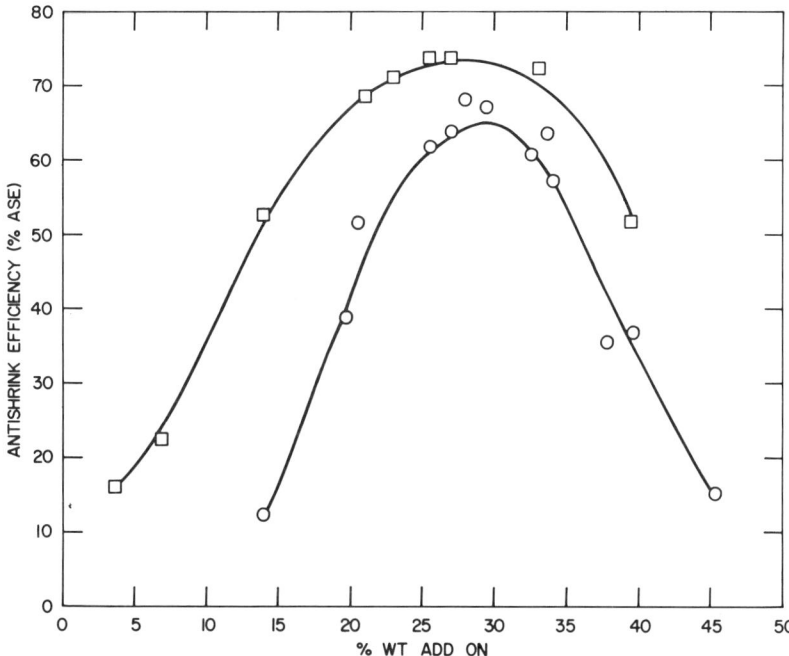

Figure 1. Relationship between antishrink efficiency (ASE) and chemical add-on caused by epoxide modification. Key: □, butylene oxide; and ○, propylene oxide.

of maximum ASE values is observed: 60–73% ASE for weight gains between 21 and 33%. The difference between these two examples may be caused by the greater hydrophobicity of butylene oxide and the difference in molecular weight. Both treatments show a downward trend in ASE above 33% weight gain.

Scanning electron micrographs show the effects of adding large amounts of chemicals. Figure 2 A shows a radial-split sample of untreated southern pine. Figure 2 B shows a radial-split sample of southern pine treated with propylene oxide to a 29.5% weight gain. The tracheid walls are intact, and there are no visible effects of the chemical added. Figure 2 C shows the same type of sample at a 32.6% weight gain. Note that checks are starting to form in the tracheid walls. In Figure 2 D, at a weight gain of 45.3%, the checks in the tracheid wall are very large. The splitting is always in the tracheid wall, not in the intercellular spaces; in some cases, the splits go through the border pits. Most of the checks are in the latewood portion of the treated wood. The less dense earlywood may be able to accommodate more chemical add-on before the cell wall would rupture. It is also possible that there is less chemical add-on in the

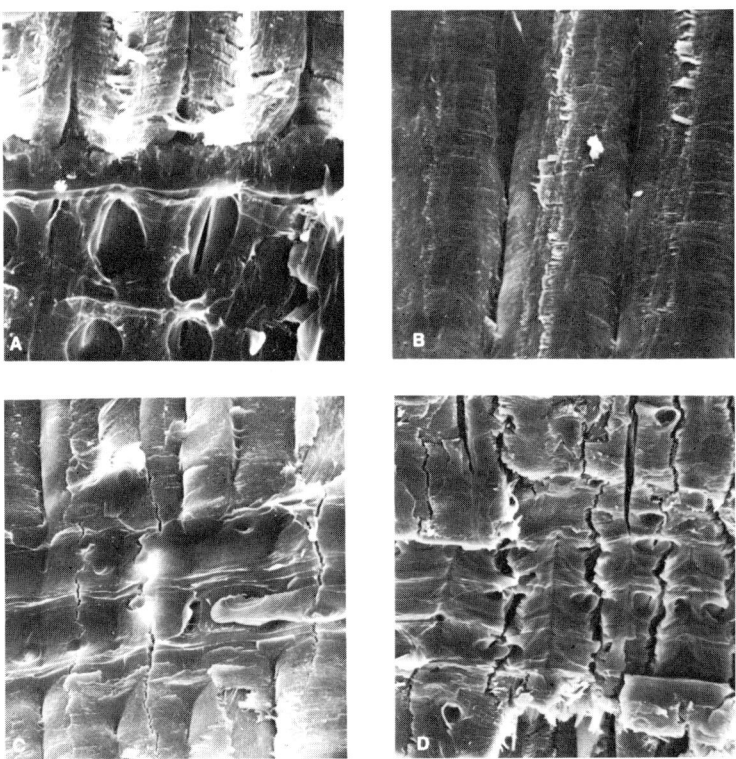

Figure 2. Scanning electron micrographs of radial-split southern pine, showing the swelling of wood when treated with propylene oxide–triethylamine. No swelling is shown in A, the untreated control (990 ×). In B, wood is treated to the green volume (990 ×). In C, wood is superswollen above the green volume at 32.6% weight gain and cell wall rupture is apparent (540 ×). In D, rupture is pronounced at 45.3% weight gain (495 ×).

earlywood. If so, and because the weight percent gain is an average for the whole sample, the weight gain in the latewood would be higher than 33% when the cell walls rupture.

Only the epoxide and isocyanate treatments have been reported to add to wood cell wall components to such a degree that they cause the wood structure itself to break apart (2, 80). Other chemical substitution treatments of wood components add to wood up to about 35% weight gain with no cell wall rupture. The epoxide and isocyanate systems seem to swell the cell wall, react with it, and continue to swell and react to the point of cell wall rupture and beyond.

In the case of the epoxy system, after the initial reaction with a cell wall hydroxyl group, a new hydroxyl group originating from the epoxide is formed. From this new hydroxyl, a polymer begins to form. The ionic nature of the reaction and the availability of alkoxyl ions in the wood components probably produce chain transfer, thereby yielding a short chain length. The formation of a polymer in the cell wall may be the cause of cell wall rupture at high chemical weight gains. At a weight gain of approximately 20% the volume of the treated wood is equal to the original untreated green wood volume (*49*). Where the weight gain is more than about 30%, the volume of the treated wood is greater than that of green wood. This is the level where the ASE starts to drop, which may mean the polymer loadings are now so high they have broken the cell wall and allowed the wood to superswell above the green wood volume.

The simple epoxides are sensitive to moisture levels in the wood during reaction (*82*). The propylene oxide reaction system seems to be the most affected by moisture, as is shown by high weight losses by extraction of nonbonded chemical and by losses in ASE. The butylene oxide system is less sensitive to moisture, but still results in formation of large amounts of nonbonded glycols.

Soil burial tests with epichlorohydrin- or dichlorohydrin-treated specimens show no decay after 2 months (*113*). Longer field tests show that butylene oxide stakes, treated above 20% weight gain, resist attack by ground organisms after 7 years in northern U.S. exposures but show some decay in southern exposures (*44, 114*). Laboratory soil-block tests with the brown-rot fungi *Gloeophyllum trabeum* (*111, 114*) and *Lentinus lepideus* (*111*) and with the white-rot fungus *Coriolus* (*Polyporus*) *versicolor* (*114*) show butylene oxide-modified wood to be resistant to attack above about 17% weight gain.

Butylene oxide-modified blocks were resistant in laboratory tests to attack by subterranean termites (*Reticulitermes flavipes*) (*114, 115*). Resistance seems attributable primarily to the wood's unpalatability. Whereas the wood treated to higher weight percent gains lost little weight under attack, termite mortality paralleled that for a starvation set. Mortality may be attributable to either an enhanced starvation effect or a slow-acting toxic effect. These two options are difficult to assess because pathogenic microbes in groups of starvationally weakened termites confound data interpretation.

Figure 3 shows that after only 2 weeks of termite attack the control specimen is almost completely destroyed. The epoxide-modified block suffered only minor damage, because the termites did some surface grazing but did not attack.

Unmodified control specimens are destroyed by marine borer attack in less than 1 year in a marine environment. Epoxide-modified specimens have been tested for over 5 years with very little marine borer attack (*114*). The mechanism of effectiveness of modified wood in resisting attack by marine organisms is unknown. As with the

Figure 3. Butylene oxide-modified block (left) and control (right) after 2 weeks of termite attack.

laboratory termite tests, unpalatability may be the largest single factor.

Most of the mechanical properties of propylene oxide-modified wood are reduced (116). The modulus of elasticity is reduced 14%, modulus of rupture is reduced 17%, fiber stress at proportional limit is reduced 9%, and maximum crushing strength is reduced 10% (116). Ethylene oxide-modified wood showed no reduction in static bending tests (107).

Radial, tangential, and longitudinal hardness indexes of propylene oxide-modified wood were the same as for untreated controls (116). The diffusion coefficient to water vapor was increased 29%.

The thermal stability of modified wood as shown by both thermogravimetric analysis and evolved gas analysis decomposition temperatures was slightly increased by epoxide bonding, the same with acetyl bonding, and slightly lowered by isocyanate bonding, as compared to controls (117). The amount of char generated during pyrolysis was nearly the same for untreated wood, acetyl- and isocyanate-bonded wood, and less for epoxide-bonded wood. The epoxide bond seems to stabilize the components that degrade at 325 °C—hemicelluloses—which apparently gasifies with the cellulose component at 385 °C. The ether linkage is chemically more stable and apparently thermally more stable than the acetyl linkage that bonds the polysaccharides. Thus, the epoxide may still have been bonded to the carbohydrate at the temperature at which carbohydrate pyrolysis occurred.

Acetyl- and isocyanate-bonded chemicals did not stabilize the components degrading at 325 °C, but showed the same thermogravimetric and evolved gas analysis profiles as did the controls. Because ester and urethane bonds are not as stable toward pyrolysis as ether linkages at high temperatures, there was a partial release of bonded chemical at low temperatures (117).

Evolved gas analysis showed that the epoxide-bonded wood had a higher heat of combustion of volatiles than did the control. The heat of combustion of volatile products from acetyl isocyanate- and methyl isocyanate-bonded wood was almost the same as that of the control. The high heat of combustion of volatiles observed for epoxide-bonded wood is primarily due to the hydrocarbon content attached to the bonding group. The epoxide bonding group, $-CH_2-CH-$, accounts for part of the heat of combustion, but the $-CH_3$ and $-CH_2CH_3$ added by propylene oxides and butylene oxides, respectively, also contribute (117).

Types of Wood. For the most part, chemical modification has been done with relatively few species of wood. Among the softwoods, Douglas-fir, ponderosa pine, and southern pine have been used; among the hardwoods, hard maple and birch have been used. It is easy to generalize on the type of wood used and extrapolate information to an untried species with the rationale that if it worked on one it will work on the other. This is a dangerous assumption and more often than not, it is incorrect.

In a recent study, 13 species of wood were treated with either propylene oxide or butylene oxide and catalyzed with triethylamine (118). Weight gain was determined and, in most cases, so were ASE values (Table IV). Species such as radiata, southern and ponderosa pines, hard maple, walnut, elm, cativo, and eucalyptus all had acceptable weight gains and medium to high ASE values. Red oak and teak gave good weight gains with little or no dimensional stability. The reason for this is not clear, but the extractives in teak seem to interfere with both chemical penetration and reactivity. If additional species were used in further research, even greater variability would be expected.

Proof of Bonding

Three criteria have been used as evidence that a chemical has reacted in the cell wall and that it has bonded with the cell wall polymers: (1) increases in wood volume as a result of reaction, (2) resistance to leaching of added chemical after reaction, and (3) IR data.

Increases in Wood Volume. Oven-drying green southern pine causes a shrinkage of 6–10% from the original green wood volume (Table V). When wood is treated to a weight gain of about 20%, the oven-dry volume of the treated wood is equal to the original untreated green wood volume. Table VI shows that for propylene oxide, methyl isocyanate, and acetic anhydride, volume expansion in the wood is nearly equal to the volume of chemical added (1). Although this is strong evidence that the bulking chemicals are in the cell wall, these results do not indicate whether or not the chemical is bonded.

Table IV. Weight Percent Gains for Various Wood Treatments Applied to Several Wood Species

Species	Treatment[a]	Time (min)	Weight Percent Gain	ASE[b]
Red oak	PO	30	21.8	0
	PO	40	25.6	2.1
Hard maple	PO	35	27.3	41.1
	BO	60	18	52.2
	BO	180	32	61.0
Teak	PO	30	20.5	0
	PO	60	20.7	0
Walnut	PO	3	26.2	46
	BO	240	28.3	53
Elm	PO	40	28.2	46.3
Cativo	PO	40	29.7	42.2
	BO	240	22.8	64.2
Persimmon	BO	180	22	—
	BO	240	33	—
Eucalyptus obligva	BO	240	22	46.4
Radiata pine (sapwood)	PO	40	34.2	67.3
(heartwood)	PO	40	32.1	52.3
Southern pine (sapwood)	PO	40	35.5	68.3
(heartwood)	PO	40	24.6	59.7
Ponderosa pine	PO	40	26.9	36.5
Douglas-fir	BO	300	20.7	—
	BO	360	24.6	—
Spruce	PO	40	32.6	—
	BO	360	30.4	—

[a] Treatments: PO, propylene oxide; and BO, butylene oxide.
[b] Antishrink efficiency after one water soak.
Conditions: temperature, 120 °C; solvent, epoxide/triethylamine, 95/5, v/v; and pressure, 150 lb/in.^{2}.

For acrylonitrile, there is a greater volume of chemical added than there is an increase in wood volume. This means that not all the chemical in the wood is in the cell wall. This is very evident when using methyl methacrylate, which shows a very large addition of chemical volume with very little increase in wood volume. The methacrylate polymer is mainly in wood lumens.

Resistance to Leaching. If the chemical that caused the cell wall to swell is bonded to the cell wall polymers, then solvent extraction cannot leach it out. Nonbonded chemicals will leach out resulting in weight loss. Table VII shows that methyl isocyanate-, butylene oxide-, and acetic anhydride-modified wood are very resis-

Table V. Changes in Volume of Southern Pine upon Drying and Chemical Treatment

Green Volume (in.³)	Oven-Dry Volume (in.³)	ΔV (%)	Treatment	Oven-Dry Volume After Treatment (in.³)	Weight Percent Gain
3.48	3.24	6.9	Propylene oxide	3.42	15.9
3.60	3.24	10.0	Propylene oxide	3.60	21.1
3.66	3.42	6.6	Propylene oxide	3.66	26.1
3.60	3.30	8.3	Propylene oxide	3.66	34.1
3.60	3.36	6.7	Propylene oxide	3.72	41.0
2.33	2.11	9.4	Acetic anhydride	2.30	13.9
2.39	2.15	10.0	Acetic anhydride	2.33	17.5
2.41	2.17	9.9	Acetic anhydride	2.39	19.5
2.37	2.13	10.1	Acetic anhydride	2.37	22.8

Table VI. Volume Changes in Southern Pine upon Chemical Treatment

Treatment	Weight Percent Gain	Increase in Wood Volume with Treatment[a] (cm^3)	Calculated Volume of Chemical Added[b] (cm^3)
Propylene oxide	26.5	7.1	7.5
	28.8	6.4	7.2
	34.3	8.4	8.0
	36.2	8.9	9.0
Methyl isocyanate	12.4	0.16	0.14
	25.7	0.21	0.27
	47.7	0.46	0.54
	51.2	0.54	0.58
Acetic anhydride	17.5	3.0	2.9
	19.5	3.6	3.3
	22.8	3.9	4.0
Acrylonitrile	25.7	0.46	0.77
	28.7	0.26	0.39
	36.0	0.74	1.2
Methyl methacrylate	58.0	0.6	7.6
	91.4	0.9	10.1

[a] Difference in volume between treatments is due to different sample size.
[b] Density used in volume calculations: propylene oxide, 1.01; methyl isocyanate, 0.967; acetic anhydride, 1.049; acrylonitrile, 0.806; and methyl methacrylate, 0.94.

tant to the leaching of added chemical(s). The starting chemicals and nonbonded by-products would be very soluble in benzene or water. Soxhlet extraction of ground-modified wood (20–40 mesh) is a severe environment that exposes a very large internal surface area to the extracting solvent. Propylene oxide-modified wood shows more weight loss than the three aforementioned chemical systems. Propylene oxide is more moisture sensitive than butylene oxide, and thus forms more nonbonded polymers during reaction. Acrylonitrile modification using ammonium hydroxide as catalyst results in almost no permanently bonded chemical even in a mild water-soaking test. Wood treatments using NaOH as catalyst show a lower weight loss in water than do wood treatments using ammonium hydroxide as a catalyst. However, weight loss is still significantly higher than any other chemically bonded system.

Another test for resistance to leaching of bonded chemical can

Table VII. Oven-Dry Weight Loss of Chemically Modified Wood Leached with Various Solvents

Reagent	Weight Percent Gain	Benzene/Ethanol 4 h Soxhlet 20 Mesh (%)	Benzene 24 h Soxhlet 40 Mesh (%)	Water 24 h Soxhlet 40 Mesh (%)	Water 7 d Soaking Blocks (%)
Control	0	2.3	4.7	11.2	0.6
Methyl isocyanate	10.0	2.9	—	—	1.0
	23.5	6.5	—	—	1.0
	47.2	9.6	—	—	9.7
Propylene oxide	29.2	5.2	—	—	4.0
	38.0	6.8	10.8	12.5	—
Butylene oxide	27	3.8	—	11.7	1.6
Acetic anhydride	16.3	2.3	—	9.7	1.0
	22.5	2.8	—	12.2	1.2
Acrylonitrile + NH$_4$OH	26.1	—	—	—	21.7
Acrylonitrile + NaOH	25.7	—	—	—	13.5

be seen in data generated for repeated water-leaching ASE tests (1). Table VIII shows that wood modified with propylene or butylene oxide, methyl isocyanate, and acetic anhydride maintains a 50–60% ASE value even after four soaking–drying cycles. This value shows that the bulking chemical is staying in the cell wall. Acrylonitrile-modified wood catalyzed with both ammonium hydroxide and sodium hydroxide loses bulking chemical even after one soaking cycle. The ASE value on the second soaking cycle is negative, which means the modified wood is less dimensionally stable than the control. This loss in stability may be due to hemicellulose extraction during reaction under strong alkaline catalyst conditions.

IR Data. Evidence that a chemical reaction has taken place with the wood cell wall hydroxyl groups is seen in the IR spectra of methyl isocyanate-modified southern pine (Figure 4). Samples were first milled to pass a 40-mesh screen and extracted with benzene/ethanol (2/1, v/v) followed by water in a Soxhlet extractor. Any unreacted reagent and isocyanate homopolymer formed during the reaction would be removed by this procedure. The spectrum for unreacted wood in the region of 1730 cm^{-1} shows some carbonyl stretching vibrations (Figure 4 A). After the wood is modified to 17.7% weight gain, the carbonyl band is stronger (Figure 4 B). At 47.2% weight gain (Figure 4 C) this band becomes one of the major bands in the IR spectra. The increase in carbonyl is due to the formation of R–O–C–N–R in the urethane bond. There is also an increase in the absorption bands as the weight percent gain increases: at 1550 cm^{-1}, N–H deformation frequencies of secondary amines; at 1270 cm^{-1}, C–N vibration of disubstituted amines; and at 770–780 cm^{-1}, N–H deformation of bonded secondary amines. No unreacted reagent remains in the samples, as shown by the absence of isocyanate absorption at 2275–2240 cm^{-1} (Figures 4 B and 4 C).

The strong absorption at 3400 cm^{-1} and 2950 cm^{-1} in all the IR spectra is caused by hydroxyl absorption. Because substitution is not high enough to eliminate all hydroxyl groups, these bands are always present.

The holocellulose (cellulose and hemicellulose) from a sample modified by methyl isocyanate to a weight gain of 17.7% was isolated by the sodium chlorite procedure (119). The IR spectrum of the holocellulose (Figure 4 D) shows that urethane bonding has taken place in the carbohydrate component of wood. The IR spectrum of lignin isolated from a methyl isocyanate-modified sample at 47.2% weight gain by the H_2SO_4 procedure (120) shows that urethane bonding has occurred in the lignin component of wood (Figure 4 E). The lignin spectrum shows the characteristic aromatic skeletal vibration at 1515 cm^{-1} (121). This band is missing from the modified

Table VIII. Volumetric Swelling Coefficients (S) and ASE as Determined by the Water-Soaking Method

Treatment	Weight Percent Gain[a]	S_1[b]	ASE_1[c]	S_2[d]	ASE_2[e]	S_3[f]	ASE_3[g]	S_4[h]	ASE_4[i]
Propylene oxide	0	15.8	—	15.8	—	15.9	—	15.9	—
	29.2	6.0	62.0	9.0	43.8	7.8	50.9	7.9	50.3
Butylene oxide	0	13.6	—	12.4	—	12.4	—	12.9	—
	27.0	3.6	73.5	5.7	54.0	5.2	58.1	5.6	56.6
Acetic anhydride	0	13.8	—	13.3	—	13.6	—	13.3	—
	16.3	5.1	63.0	5.1	61.7	5.3	61.0	5.3	60.2
	22.5	4.1	70.3	3.8	71.4	4.0	70.6	4.1	69.2
Acrylonitrile	0	14.1	—	13.9	—	14.0	—	—	—
+ NH$_4$OH	26.1	2.7	80.9	15.3	neg.	14.4	neg.	—	—
Acrylonitrile	0	20.3	—	16.8	—	16.7	—	—	—
+ NaOH	25.7	10.5	48.3	18.8	neg.	17.5	neg.	—	—
Methyl isocyanate	0	14.0	—	13.8	—	13.7	—	13.5	—
	21.6	5.5	60.4	6.6	52.0	6.5	52.6	6.3	53.3
	29.9	4.7	66.4	6.0	56.8	4.8	65.0	5.3	60.7

[a] Samples recorded at 0% are controls.
[b] Volumetric swelling coefficient determined from initial oven-dry volume and first water-swollen volume.
[c] Antishrink efficiency based on S_1.
[d] Determined from first water-swollen volume and reoven-drying.
[e] Based on S_2.
[f] Determined from reoven-dry volume and second water-swollen volume.
[g] Based on S_3.
[h] Determined from second water-swollen volume and second reoven-drying.
[i] Based on S_4.

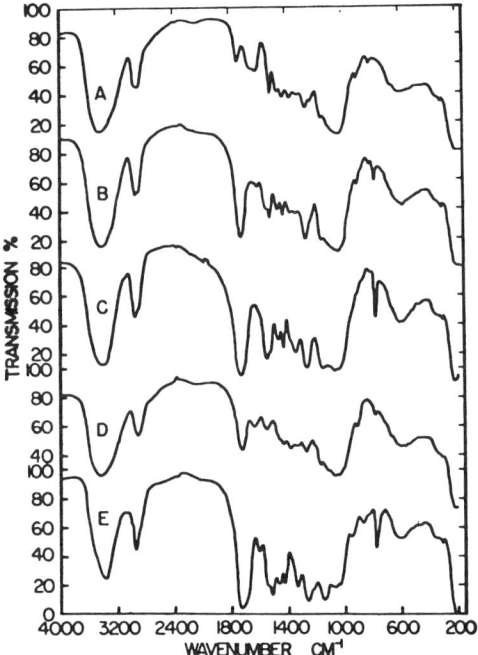

Figure 4. IR spectra of methyl isocyanate-modified southern pine. Key: A, southern pine control; B, methyl isocyanate-modified southern pine to 17.7% weight gain; C, methyl isocyanate-modified southern pine to 47.2% weight gain; D, holocellulose fraction from methyl isocyanate-modified southern pine to 17.7% weight gain (0.07% lignin); and E, lignin fraction from methyl isocyanate-modified southern pine to 47.2% weight gain.

holocellulose curve (Figure 4 D), showing that the chlorite procedure does remove substituted lignins.

Similar results are observed on IR spectra of acetylated wood (122). As the weight percent gain increases upon acetylation, the absorption band at 1730 cm^{-1} increases because of the carbonyl group in the acetyl bond.

Distribution of Bonded Chemical

Chemical modification of wood to impart decay resistance and to provide dimensional stability depends on adequate distribution of reacted chemicals in water-accessible regions of the cell wall. It is important, therefore, to determine the distribution of bonded chemicals. This information may also lead to a better understanding of how chemical modification of wood changes the chemical properties of cell wall polymers.

The distribution of bonded chemical as a function of depth of penetration was determined by tracing the fate of chlorine in epichlorohydrin-reacted wood (123) and acetyl groups in acetylated wood (122). Outside, middle, and inner samples were taken from specimens 1.27 × 1.27 cm to 5.08 × 5.08 cm prepared from epichlorohydrin-modified southern pine. In samples up to 3.81 × 3.81 cm, no significant differences were observed in chlorine content from the three sections analyzed. Beyond 3.81 cm a concentration difference occurred between the outside and the inner part of the wood treated. Results were similar for acetylated wood (122).

The epichlorohydrin reaction system was also used to determine the distribution of chlorine in earlywood, latewood, sapwood, and heartwood of southern pine (124). The very polar epoxy system reacts more quickly, and with greater weight gains, with the earlywood—as opposed to latewood—cell wall components. Although the amount of the epoxides was larger in heartwood, benzene extraction of the reacted wood effected a greater weight loss from the heartwood than from the sapwood. This weight loss may have been caused by reaction of the epoxides with heartwood extractives, which were then removed on benzene extraction.

A study of soft-rot decay patterns showed that the tangential cell wall is reacted to a higher degree than radial cell walls in ponderosa pine reacted with butylene oxide at 8% weight gain (125). The radial wall in latewood is nearly twice as thick as the tangential wall, so the radial wall may not be totally penetrated by the epoxide system.

Energy X-ray analysis of bromine in wood acetylated with tribromoacetyl bromide showed that bromine was distributed throughout the entire secondary wall, suggesting chemical reaction with lignin (126). Using a similar technique, the greatest percentage of chlorine in epichlorohydrin-modified wood reference was found in the S_2 layer of the cell wall. This is the thickest cell wall layer and contains the most cellulose.

By taking apart the cell wall of a modified wood specimen and separating the cell wall components from one another, it is possible to determine the distribution of bonded chemicals in the cell wall polymer. It is more difficult to delignify modified wood than unmodified wood, which means that the lignin has been substituted (122, 127, 128). This is true for wood reacted with both acetic anhydride and methyl isocyanate. Table IX shows that the lignin component is always more substituted than the holocellulose components (128). This would indicate that the lignin is either more accessible for reaction than holocellulose or that it is more reactive than holocellulose. Lignin was found to be more reactive than cellulose toward acetylation (129).

Table IX. Degree of Substitution of Hydroxyl Groups in Methyl Isocyanate-Modified Southern Pine

Weight Percent Gain	Lignin	Holocellulose	Lignin: Holocellulose
5.5	0.17	0.025	7.4
10.0	0.28	0.047	6.0
17.7	0.41	0.084	4.9
23.5	0.59	0.117	5.1
47.2	0.89	0.209	4.3

For acetylated wood at approximately 25% chemical weight gain, all of the lignin hydroxyls were found to be substituted (*122, 130*). More bonded acetyl is found on the cellulose than the hemicelluloses at low (13.5%) chemical weight gain, but this was reversed at higher (24.5%) weight gain (*130*).

Hydroxyl substitution calculations are based on the assumption that all hydroxyl groups are accessible and that reaction with acetic anhydride or methyl isocyanate is a single-site substitution reaction — i.e., only one reagent reacting with one hydroxyl and no polymerization. Only 60% of the total hydroxyl groups in spruce wood are accessible to tritiated water (*131*). About 65% of the cellulose in wood is crystalline and, therefore, probably not accessible for reactions involving these hydroxyl groups (*7*). Based on these estimates and assuming that only 35% of the cellulose hydroxyls are accessible for substitution, the degree of substitution in the holocellulose component is much higher in the accessible regions than shown in Table IX.

The data on the distribution of bonded chemicals suggest that a high rate of lignin substitution does not contribute significantly to the overall protection mechanism of wood from decay or dimensional stabilization. The degree of substitution in lignin was high in samples at lower weight percent gain of bonded chemical where little or no protection from decay or dimensional stabilization was observed. If the degree of substitution in lignin does have an effect on these mechanisms, it is only observed at very high levels. The degree of substitution in the holocellulose components seems to be the most important factor in decay resistance and dimensional stability.

Conclusion

Chemical modification of wood will be important in the future because of its ability to enhance the properties of the end products in use (*132*). If, for example, fire retardancy is important in a wood material, the fire retardant chemical could be bonded permanently to the cell wall of the wood. If the level of chemical addition were high enough, dimensional stability and some degree of resistance to biological attack would also be achieved at no additional cost.

The greatest single application of current research may be in reconstituted products in which standard operating procedures call for dry wood materials, spray chemical addition for maximum distribution, small sample size for good penetration, and high temperature and pressure in product formation. These are exactly the procedures required for successful chemical modification. Permanently bonded chemicals that provide fire retardancy, UV stabilization, color changes, dimensional stability, and resistance to biological attack may be possible through chemical modification.

Literature Cited
1. Rowell, R. M.; Ellis, W. D. *Wood Fiber* **1978**, *10*(2), 104–11.
2. Rowell, R. M.; Gutzmer, D. I.; Sachs, I. B.; Kinney, R. E. *Wood Sci.* **1976**, 9(1),51–54.
3. Stamm, A. J. "Colloid Chemistry of Cellulosic Materials"; USDA Misc. Publ. No. 240, Madison, WI, 1936.
4. Nayer, A. N., Ph.D. thesis, University of Minnesota, Minneapolis, 1948.
5. Kumar, V. B. *Nor. Skogind.* **1957**, *11*(7), 259.
6. Kumar, V. B. *Nor. Skogind.* **1958**, *12*(9), 337.
7. Stamm, A. J. "Wood Science"; Ronald: New York, 1964.
8. Ashton, H. E. *Wood Sci.* **1973**, 6, 159–66.
9. Ashton, H. E. *Wood Sci.* **1974**, 6(4), 368–74.
10. Rowell, R. M.; Hart, S. V. Final Report Project 3212–5–76–3, *U.S., For. Serv., For. Prod. Lab. Rep.,* **1981**.
11. Erickson, H. D. *For. Prod. J.* **1955**, 5(4), 241–50.
12. Haslam, J. H.; Werthan, S. *Ind. Eng. Chem.* **1931**, *23*(2), 226–33.
13. Phillips, E. W. *J. Forestry* **1933**, 7(1), 109–20.
14. Thomas, R. J.; Scheld, J. L. North Carolina State College Tech. Rep. No. 19, Raleigh, NC, 1964.
15. Meier, H. "The Formation of Wood in Forest Trees"; Zimmerman, M. H., Ed.; Academic: New York, 1964; p. 137–51.
16. Griffin, C. J. *J. For.* **1919**, *17*(7), 813–22.
17. Huber, B.; Merz, W. *Planta* **1958**, *51*, 660–72.
18. Nicholas, D. D.; Siau, J. F. in "Wood Deterioration and Its Prevention by Preservative Treatments"; Nicholas, D. D., Ed.; Syracuse Univ.: Syracuse, NY, 1973.
19. Erickson, H. D. *Minn., Agric. Exp. Stn., Tech. Bull.* **1937**, *122*.
20. Gjovik, L. R.; Roth, H. G.; Lorenz, L. F. *Proc.—Annu. Meet. Am. Wood-Preserv. Assoc.* **1970**, *66*, 260–62.
21. MacLean, J. D. *Proc.—Annu. Meet. Am. Wood-Preserv. Assoc.* **1924**, *20*, 44–73.
22. Nicholas, D. D. *For. Prod. J.* **1972**, *22*(5), 31–36.
23. Verrall, A. F. *U.S. For. Serv., South For. Exp. Stn., For. Surv. Release* No. 157, **1957**.
24. Thomas, R. J. *Wood Fiber* **1969**, *1*(2), 110–23.
25. Stamm, A. J.; Tarkow, H. U.S. Patent 2 417 995, 1947.
26. Ridgway, W. B.; Wallington, H. T. British Patent 579 255, 1946.
27. Tarkow, H. *U.S. For. Prod. Lab., Rep.* **1959**.
28. Baird, B. R. *Wood Fiber* **1969**, *1*, 54–63.
29. Clermont, L. P.; Bender, F. *For. Prod. J.* **1957**, 7, 167–70.

30. Risi, J.; Arseneau, D. F. *For. Prod. J.* **1957,** *7,* 210–13.
31. Ozolina, I.; Svalbe, K. *LLA Raksti* **1972,** 47–50.
32. Arni, P. C.; Gray, J. D.; Scougall, R. K. *J. Appl. Chem.* **1961,** *II,* 163–70.
33. Truksne, D.; Svalbe, K. *LLA Raksti* **1977,** *130,* 26–31.
34. Lorenz, L. F.; Kirk, T. K. *U.S., For. Prod. Lab., Rep.* **1972.**
35. Svalbe, K.; Ozolina, I. *Plast. Modif. Drev.* **1970,** 145–46.
36. Goldstein, I. S.; Jeroski, E. B.; Lund, A. E.; Nielson, J. F.; Weater, J. M. *For. Prod. J.* **1961,** *11,* 363–70.
37. Tarkow, H.; Stamm, A. J.; Erickson, E. C. O. *U.S., For. Serv., For. Pest Leafl. Rep.* No. 1593, **1950.**
38. Koppers' Acetylated Wood. New Materials Technical Information (RDW–400) E–106, Pittsburgh, PA, 1961.
39. Peterson, M. D.; Thomas, R. J. *Wood Fiber* **1978,** *10*(3), 149–63.
40. Ozolina, I. O.; Svalbe, K. P. *Latv. PSR Zinat. Akad. Vestis* **1966,** 56–59.
41. Rugevitsa, A. *LLA, Raksti* **1977,** *130,* 42–44.
42. Svalbe, K.; Ozolina, I. O.; Truskne, D.; Vitolins, J. *Biopovrezhdeniya Mater. Zashch. Nikh.* **1978,** 166–74.
43. Bekere, M.; Svalbe, K.; Ozolina, I. *LLA Raksti* **1978,** *163,* 31–35.
44. Gjovik, L. R.; Davidson, H. L. *U.S., For. Serv., Res. Note FPL–02* **1979.**
45. Sidorenko, A. K.; Kuzmin, N. F.; Reshtnikov, E. K. *Lesn. Zh. (Archangel, USSR)* **1973,** *16*(5), 157–58.
46. Krylova, A. N. *LLA Raksti* **1970,** *130,* 59.
47. Dreher, W. A.; Goldstein, I. S.; Cramer, G. R. *For. Prod. J.* **1964,** *14,* 66–68.
48. Goldstein, I. S.; Weaver, J. W. U.S. Patent 3 094 431, 1963.
49. Kumar, S.; Singh, S. P.; Sharma, M. *J. Timber Dev. Assoc. India* **1979,** *25*(3), 5–9.
50. Rugevitsa, A.; Iurevitsa, S.; Svalbe, K. *LLA Raksti* **1977,** *130,* 45–49.
51. Darzin'sh, T. A. *Abstr. Bull. Inst. Pap. Chem.* **1975,** *48,* 3860.
52. Shiraishi, N.; Yokota, T.; Kimura, T.; Sumizawa, K. *J. Jpn. Wood Res. Soc.* **1972,** *18*(5), 215–22.
53. Shiraishi, N.; Okumura, M.; Yokota, T. *J. Jpn. Wood Res. Soc.* **1976,** *22*(4), 232–37.
54. Narayaramurti, D.; Handa, B. K. *Das Papier* **1953,** *7,* 87–92.
55. Rudkin, A. W. *Aust. J. Appl. Sci.* **1950,** *1,* 270–83.
56. Avora, M.; Rajawat, M. S.; Gupta, R. C. *Holzforsch. Holzverwert.* **1979,** *31*(6), 138–41.
57. Tarkow, H. *U.S., For. Prod. Lab., Rep.* **1945.**
58. Karlson, I. M.; Svalbe, K. P. *Uch. Zap. Latv. Gos. Univ. im. Petra Stuchka* **1972,** *166,* 89–94.
59. Karlson, I. M.; Svalbe, K. P. *Uch. Zap. Latv. Gos. Univ. im. Petra Stuchka* **1972,** *166,* 98–104.
60. Svalbe, K. *LLA Raksti* **1977,** *130,* 3–9.
61. Karlson, I.; Svalbe, K. *LLA Raksti* **1977,** *130,* 10–21.
62. Singh, S. P.; Dev, I.; Kumar, S. *Wood Sci.* **1979,** *11*(4), 268–70.
63. Otlesnov, Y.; Nikitina, N. *LLA Raksti* **1977,** *130,* 50–53.
64. Nikitina, N. *LLA Raksti* **1977,** *130,* 54–55.
65. Risi, J.; Arseneau, D. F. *For. Prod. J.* **1958,** *8,* 252–55.
66. Popper, R.; Bariska, M. *Holz Roh- Werkst.* **1972,** *30*(8), 289–94.
67. Popper, R.; Bariska, M. *Holz Roh- Werkst.* **1973,** *31*(2), 65–70.
68. Popper, R.; Bariska, M. *Holz Roh- Werkst.* **1975,** *33*(11), 415–19.
69. Popper, R. *Bull. Schweizer. Arbeitsgemeinsch. Holzforsch.* **1973,** *1* (1), 3–12.
70. Suida, H. Austrian Patent 122 499, 1930.
71. Singh, S. P.; Dev, I.; Kumar, S. *Int. J. Wood Pres.* **1981,** 169–71.
72. Arni, P. C.; Gray, J. D.; Scougall, R. K. *J. Appl. Chem.* **1961,** *11,* 157–63.
73. Nakagami, T.; Amimoto, H.; Yokota, T. *Bull. Kyoto Univ. For.* **1974,** *46,* 217–24.

74. Nakagami, T.; Yokota, T. *Bull. Kyoto Univ. For.* **1975**, *47*, 178–83.
75. Nakagami, T.; Ohta, M.; Yokota, T. *Bull. Kyoto Univ. For.* **1976**, *48*, 198–205.
76. Nakagami, T.; Yokota, T. *J. Jpn. Wood Res. Soc.* **1978**, *24*(5), 311–17.
77. Nakagami, T.; Yokota, T. *J. Jpn. Wood Res. Soc.* **1978**, *24*(5), 318–23.
78. Wakita, H.; Onishi, H.; Jodai, S.; Goto, T. *Zairyo* **1977**, *26*(284), 460–64.
79. Lutomski, K. *Mater. Org.* **1975**, *10*(4), 255–62.
80. Rowell, R. M.; Ellis, W. D. in "Urethane Chemistry and Applications," Edwards, Kenneth N., Ed.; ACS SYMPOSIUM SERIES No. 172, ACS: Washington, D.C., 1981; pp. 263–84.
81. Rowell, R. M.; Ellis, W. D. *Wood Sci.* **1979**, *12*(1), 52–58.
82. Rowell, R. M.; Ellis, W. D. *Wood Fiber* in press.
83. Burmester, A. German Patent 1 812 409, 1970.
84. Ueyama, A.; Araki, M.; Goto, T. *Mokuzai Kenkyu* **1961**, *26*, 67–73.
85. Tarkow, H.; Stamm, A. J. *J. For. Prod. Res. Soc.* **1953**, *3*, 33–37.
86. Minato, K.; Mizukami, F. *Mokuzai Gakkaishi* **1982**, *28*(6), 346–54.
87. Dewispelaere, W.; Raemdonck, J. V.; Stevens, M. *Mater. Org.* **1977**, *12*(3), 211–22.
88. Stevens, M.; Schalck, J.; Raemdonck, J. V. *Int. J. Wood Pres.* **1979**, 57–68.
89. Stamm, A. J. *Tappi* **1959**, *42*, 39–44.
90. Schuerch, C. *For. Prod. J.* **1968**, *18*, 47–53.
91. Rowell, R. M., unpublished results. Forest Products Laboratory, Madison, WI.
92. Stamm, A. J.; Baechler, R. H. *For. Prod. J.* **1960**, *10*, 22–26.
93. Burmester, A. *Holzforschung* **1967**, *21*, 13–20.
94. Weaver, J. W.; Nielson, J. F.; Goldstein, I. S. *For. Prod. J.* **1960**, *10*, 306–10.
95. Kenaga, D. L. U.S. Patent 2 811 470, 1957.
96. Kenaga, D. L.; Sproull, R. C.; Esslinger, J. *South. Lumberman* **1950**, *180*, 45.
97. Kenaga, D. L.; Sproull, R. C. *J. For. Prod. Res. Soc.* **1951**, *1*, 28–32.
98. Risi, J.; Arseneau, D. F. *For. Prod. J.* **1957**, *7*, 293–95.
99. Risi, J.; Arseneau, D. F. *For. Prod. J.* **1957**, *7*, 245–46.
100. Risi, J.; Arseneau, D. F. *For. Prod. J.* **1957**, *7*, 261–65.
101. Goldstein, I. S.; Dreher, W. A.; Jeroski, E. B. *Ind. Eng. Chem.* **1959**, *51*, 1313–17.
102. Goldstein, I. S. U.S. Patent 2 931 741, 1960.
103. Fuse, G.; Nishimoto, K. *Mokuzai Gakkaishi* **1961**, *7*, 157–61.
104. Baechler, R. H. *For. Prod. J.* **1959**, *9*, 166–71.
105. Baechler, R. H. U.S. Patent 2 959 496, 1960.
106. Kenaga, D. L. U.S. Patent 3 077 417, 1963.
107. McMillin, C. W. *For. Prod. J.* **1963**, *13*, 56–61.
108. Liu, C.; McMillin, C. W. U.S. Patent 3 183 114, 1965.
109. Aktiebolag, M. D. French Patent 1 408 170, 1965.
110. Barnes, H. M.; Choong, E. T.; McIlhenny, R. L. *For. Prod. J.* **1969**, *19*, 35–39.
111. Rowell, R. M. *Wood Sci.* **1975**, *7*(3), 240–46.
112. Zimakov, P. V.; Pokrovskil, V. A. *Zh. Prikl. Khim.* **1954**, *27*, 346–48.
113. Pihl, L. O.; Olsson, I. British Patent 1 101 777, 1968.
114. Rowell, R. M. Swedish Forest Products Research Laboratory, STFI Series A, No. 772, 1982; 32–49.
115. Rowell, R. M.; Hart, S. V.; Esenther, G. R. *Wood Sci.* **1979**, *11*(4), 271–74.
116. Rowell, R. M.; Moisuk, R.; Meyer, J. A. *Wood Sci.* **1982**, *15*(2), 90–96.
117. Rowell, R. M.; Susott, R. A.; DeGroot, W. F.; Shafizadeh, F. *Wood Fiber* in press.
118. Rowell, R. M.; Ellis, W. D. *U.S., For. Serv., For. Prod. Lab., Res. Pap.* in press.

119. Green, J. W. in "Methods in Carbohydrate Chemistry," Vol. III; Whistler, R. L.; Wolfrom, M. O., Eds.; Academic: New York, 1963; p. 9–21.
120. Moore, W. E.; Johnson, D. B. USDA Forest Serv. unnumbered report, Forest Prod. Lab., Madison, WI, 1967.
121. Sarkanen, K. V.; Chang, H.-m.; Ericsson, B. *Tappi* **1967**, *50*(11), 572–75.
122. Rowell, R. M. *Wood Sci.* in press.
123. Rowell, R. M. *Wood Sci.* **1977**, *9*(3), 144–48.
124. Rowell, R. M. *Wood Sci.* **1978**, *10*(4), 193–97.
125. Nilsson, T.; Rowell, R. M. International Res. Group on Wood Preservation; Document No. IRG/WP/3211, 1982.
126. Peterson, M. D.; Thomas, R. J. *Wood Fiber* **1978**, *10*(3), 149–63.
127. Callow, H. J. *J. Indian Chem. Soc.* **1951**, *28*, 605–10.
128. Rowell, R. M. *Wood Sci.* **1980**, *13*(2), 102–10.
129. Callow, H. J. *J. Text. Inst.* **1952**, *43*, T247–49.
130. Truksne, D. *LLA Raksti* **1977**, *130*, 32–39.
131. Sumi, Y.; Yale, R. D.; Meyer, J. A.; Leopold, B.; Ranby, B. G. *Tappi* **1964**, *47*(10), 621–24.
132. Rowell, R. M. *Proc.—Annu. Meet. Am. Wood-Preserv. Assoc.* **1975**, *71*, 41–51.

RECEIVED for review May 9, 1983. ACCEPTED July 7, 1983.

5

The Chemistry of Wood Strength

JERROLD E. WINANDY and ROGER M. ROWELL

U.S. Department of Agriculture, Forest Service, Forest Products Laboratory, Madison, WI 53705

The source of strength in solid wood is the wood fiber. Generally, cellulose is responsible for strength in the wood fiber because of its high degree of polymerization and linear orientation. Hemicellulose acts as a matrix for the cellulose and increases the packing density of the cell wall. The actual role of hemicellulose in wood strength is unknown, but hemicellulose and lignin are closely associated. Lignin not only holds fibers together, but also holds cellulose molecules together within the fiber cell wall. The chemical components of wood that are responsible for mechanical properties can be viewed from three levels: macroscopic (cellular), microscopic (cell wall), and molecular (polymeric). Mechanical properties change with changes in the chemical environment. Changes in temperature, pressure, humidity, pH, chemical adsorption from the environment, UV radiation, fire, or biological degradation can have significant effects on the strength of wood.

THE SOURCE OF STRENGTH IN WOOD is the wood fiber. Wood is basically a series of tubular fibers or cells cemented together. Each fiber wall is composed of various quantities of three polymers: cellulose, hemicellulose, and lignin (*see* Chapter 1). Cellulose, primarily, is responsible for strength in the wood fiber because of its high degree of polymerization and linear orientation. Hemicellulose acts as a matrix for the cellulose and increases the packing density of the cell wall. The role of hemicellulose as a contributor to strength is largely unknown; however, hemicellulose may act as a link between the fibrous cellulose and the amorphous lignin. Lignin, a phenolic compound, not only holds the fibers together, but also acts as a stiffening agent for the cellulose molecules within the fiber cell wall. All three cell wall components contribute in different degrees to the strength of wood. Together the tubular structure and the polymeric construction are responsible for most of the physical and chemical properties exhibited by wood.

The strength of wood can be altered by environmental agents. The changes in pH, moisture, and temperature; the influence of decay, fire, and UV radiation; and the adsorption of chemicals from the environment can have a significant effect on strength properties. Environmentally induced changes must be considered in any discussion on the strength of treated or untreated wood.

The strength of wood can also be altered by preservative and fire-retardant compounds used to prevent environmental degradation. In some cases, the loss in mechanical properties caused by these treatments may be large enough that the treated material can no longer be considered the same as the untreated material. The treated wood may now resist environmental degradation but may be structurally inferior to the untreated material. Except for preservative-treated wood used in marine environments, these possible changes are not accounted for in the structural design process because chemical degradation of the polymers responsible for strength is assumed to be marginal or nonexistent. With fire-retardant treated wood, a 10% reduction in strength is incorporated into the structural design process.

This chapter presents a theoretical model to explain the relationship between the mechanical properties and the chemical components of wood. This model is then used to describe the effects of altered composition on those mechanical properties. Many of the theories presented are unproven. They should be considered as a starting point for dialogue between chemists and engineers that will eventually lead to a better understanding of the chemistry of wood strength.

Mechanical Properties

Even wood that has no discernible defects has extremely variable properties as a result of its heterogeneous composition and natural growth patterns. Wood is an anisotropic material in that the mechanical properties vary with respect to the three mutually perpendicular axes of the material (radial, tangential, and longitudinal). These natural characteristics are compounded further by the environmental influences encountered during the growth of the living tree. Yet wood is a viable construction material because workable estimates of the mechanical properties have been developed.

Mechanical properties relate a material's resistance to imposed loads (i.e., forces). Mechanical properties include: (1) measures of resistance to deformations and distortions (elastic properties), (2) measures of failure-related (strength) properties, and (3) measures of other performance properties. To preface any discussion concerning mechanical properties, two concepts need to be explained: stress (σ) and strain (ϵ).

Stress is a measure of the internal forces exerted in a material as a result of an application of an external load. Three types of primary stress exist: tensile stress, which pulls or elongates an object (Figure 1a); compressive stress, which pushes or compresses an object (Figure 1b); and shear stress, which causes two contiguous parts of a body to slide through some plane of the object (Figure 1c). Bending stress (Figure 1d) is a combination of all three of the primary stresses and causes rotational distortion or flexure in an object.

Strain is the measure of a material's ability to deform—that is, elongate or compress—while under stress. Over the elastic range of a material, stress and strain are related to each other in a linear manner. In elastic materials a unit of stress will cause a corresponding unit of strain. This concept, known as Hooke's law ($E = \sigma/\epsilon$), applies to all elastic materials at points below their elastic or proportional limits (1).

Elastic properties relate a material's ability to be deformed by a stress to its ability to regain its original dimensions when the stress is removed. The criterion for elasticity is not the amount of deformation, but the ability of a material to completely regain its original dimensions when the stress is removed. The opposite quality is vis-

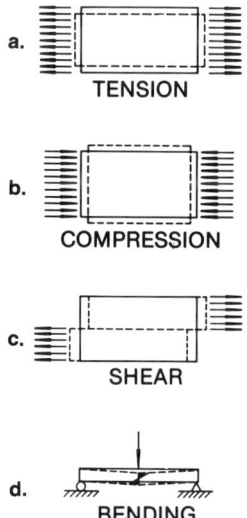

Figure 1. Examples of the three axial and one flexural types of stress.

cosity, which can also be thought of as plasticity. A perfectly plastic body is one that makes no recovery of its original dimensions upon the removal of a stress. Wood is not ideally elastic; it will not completely recover deformation immediately on unloading, but in time, residual deformations tend to be recoverable. Wood is considered a viscoelastic material. For simplicity's sake, however, wood will be considered as an elastic material in this chapter.

The two main elastic properties are modulus of elasticity, which describes the relationship of load (stress) to deformation (strain), and modulus of rigidity or shear modulus, which describes the internal distribution of shearing stress to shear strain or, more precisely, angular displacement within a material.

Strength values are numerical estimates of the material's ability to resist applied forces. The major strength properties are limit values for the stress–strain relationship within a material. Strength, in these terms, is the quality that determines the greatest unit stress a material can withstand without fracture or excessive distortion. In many cases the unqualified term strength is somewhat vague. It is sometimes more useful to think of specific strengths, such as compressive, tensile, shear, or ultimate bending strengths.

The American Society for Testing and Materials (ASTM) is an organization that standardizes testing procedures to provide reliable and universally comparable estimates of strength. The ASTM Standards (2–6) outline procedures for determining basic mechanical properties and deriving allowable design stresses. In performing a test, a load is applied to a specimen in a particular manner and the resulting deformation is monitored. The load information allows the internal forces within the specimen (stress) to be calculated. The deformation information allows the internal distortion (strain) to be calculated. When stress and strain are plotted against each other on a graph, a stress–strain diagram is developed (Figure 2).

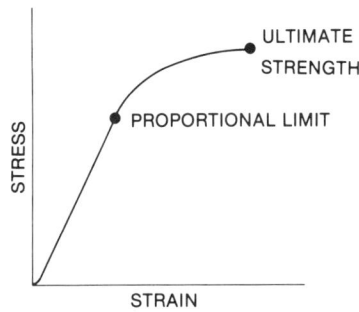

Figure 2. A typical stress–strain diagram for wood.

The unit stress corresponding to the upper limit of the linear segment of the stress–strain diagram is known as the proportional limit (Figure 2, Point A). This proportional or elastic limit measures the boundary of a material's completely recoverable strength. At stress levels below the proportional limit, a perfectly elastic material will regain its original dimensions and form. At stress levels in excess of the proportional limit, an elastic material will not regain its original shape; it will be permanently distorted.

The unit stress represented by the maximum ordinate is the ultimate (maximum) strength (Figure 2, Point B). This point estimates the maximum stress at the time of failure. Many of the mechanical properties of interest to the engineer, such as maximum crushing strength or ultimate bending strength, describe this point of maximum stress.

Factors Affecting Strength. MATERIAL FACTORS. *Specific Gravity.* Specific gravity is the ratio of the weight of a given volume of wood to that of an equal volume of water. As specific gravity increases, strength properties increase (7) because internal stresses are distributed among more molecular material. Mathematical approximations of the relationship between specific gravity and various mechanical properties are shown in Table I.

Growth Characteristics. As a fibrous product from living trees, wood is subjected to many environmental influences as it is formed and during its lifetime. These environmental influences can increase the variability of the wood material and, thus, increase the variability of the mechanical properties. To reduce the effect of this inherent variability, standardized testing procedures using small, clear specimens of wood are often used. Small, clear specimens do not have knots, checks, splits, or reaction wood. However, the wood products used and of economic importance in the real world have these defects. Strength estimates derived from small clear specimens are reported because most chemical treatment data have been generated from small clear specimens.

Because strength is affected by material factors such as specific gravity and growth characteristics, a coefficient of variation is used to approximate the variability associated with each strength property. The estimated coefficient of variation of various strength properties can be found in Table II.

ENVIRONMENTAL FACTORS. *Moisture.* Wood, which is a hygroscopic material, gains or loses moisture to equilibrate with its immediate environment. The equilibrium moisture content (EMC) is the steady-state level that wood achieves when subjected to a particular relative humidity and temperature. The eventual EMC of two similar specimens will differ if one approaches EMC under adsorbing

Table I. Functions Relating Mechanical Properties to Specific Gravity of Clear, Straight-Grained Wood

Property		Specific Gravity–Strength Relation[a]	
		Green Wood	Air-Dried Wood[b]
Static bending			
Fiber stress at proportional limit	lb/in.^{2}	10,200 $G^{1.25}$	16,700 $G^{1.25}$
Modulus of elasticity	million lb/in.^{2}	2.36 G	2.80 G
Modulus of rupture	lb/in.^{2}	17,600 $G^{1.25}$	25,700 $G^{1.25}$
Work to maximum load	in.-lb/in.^{3}	35.6 $G^{1.75}$	32.4 $G^{1.75}$
Total work	in.-lb/in.^{3}	103 G^2	72.7 G^2
Impact bending			
Height of drop causing complete failure	in.	114 $G^{1.75}$	94.6 $G^{1.75}$
Compression parallel to grain			
Fiber stress at proportional limit	lb/in.^{2}	5,250 G	8,750 G
Modulus of elasticity	million lb/in.^{2}	2.91 G	3.38 G
Maximum crushing strength	lb/in.^{2}	6,730 G	12,200 G
Compression perpendicular to grain			
Fiber stress at proportional limit	lb/in.^{2}	3,000 $G^{2.25}$	4,630 $G^{2.25}$
Hardness			
End	lb	3,740 $G^{2.25}$	4,800 $G^{2.25}$
Side	lb	3,420 $G^{2.25}$	3,770 $G^{2.25}$

SOURCE: Ref. 7.
[a] The properties and values should be read as equations. For example: Modulus of rupture for green wood = 17,600 $G^{1.25}$, where G represents the specific gravity of oven-dry wood, based on the volume at the moisture condition indicated.
[b] 12% moisture content.

Table II. The Average Coefficient of Variation for Some Mechanical Properties of Clear Green Wood

Property	Coefficient of Variation[a] (%)
Specific gravity	10
Shrinkage	
Radial	15
Tangential	14
Volumetric	16
Static bending	
Fiber stress at proportional limit	22
Modulus of rupture	16
Modulus of elasticity	22
Work to proportional limit	38
Work to maximum load	34
Impact bending	
Height of drop causing complete failure	25
Compression parallel to grain	
Fiber stress at proportional limit	24
Maximum crushing strength	18
Modulus of elasticity	29
Compression perpendicular to grain	
Fiber stress at proportional limit	28
Shear parallel to grain	
Maximum shearing strength	14
Tension perpendicular to grain	
Maximum tensile strength	25
Hardness	
End	17
Side	20
Toughness	34

SOURCE: Ref. 9.
[a] Values given are based on results of tests of approximately 50 species of green wood. Values for wood in the air-dried condition (12% moisture content) may be assumed to be approximately of the same magnitude.

conditions and the other approaches EMC under desorbing conditions. For example, if the relative vapor pressure of the environment is 0.65, two similar specimens exposed under either adsorbing or desorbing conditions will equilibrate at moisture contents of approximately 11% and 13.8%, respectively (Figure 3).

Wood strength is related to the amount of water in the wood fiber cell wall (7–10). At moisture contents from oven-dry (OD) to the fiber-saturation point, water accumulates in the wood cell wall

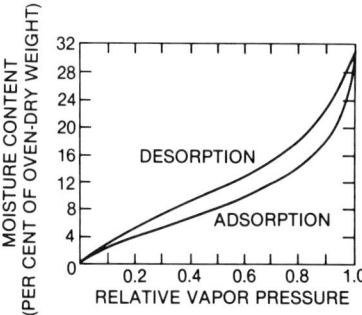

Figure 3. Adsorption–desorption isotherms for water vapor by spruce at 25 °C (17).

(bound water). Above the fiber-saturation point, water accumulates in the wood cell cavity (free water), and there are no tangible strength effects associated with a changing moisture content. However, at moisture contents between OD and the fiber-saturation point, water does affect strength. Increased amounts of bound water interfere with and reduce hydrogen bonding between the organic polymers of the cell wall (11) which decreases the strength of wood. The approximate relationships are shown in Tables III and IV.

Not all mechanical properties change with moisture content. The performance of wood under dynamic loading conditions is a dual function of the strength of the material, which is decreased with increased moisture content, and the pliability of the material, which is increased with increased moisture contents. Changes in strength and pliability offset one another and, therefore, mechanical properties that deal with dynamic loading conditions are not usually affected by a changing moisture content.

Temperature. Strength is related to the temperature of the working environment (7, 9, 10). At constant moisture content, the immediate effect of temperature on strength is linear (Figure 4) and usually recoverable when the temperature returns to normal. In general, the strength of wood is higher in cooler temperatures and lower in warmer temperatures. However, this relationship can be dramatically influenced by increasing moisture content and temperature duration, which can induce permanent (nonrecoverable) effects.

The immediate effects of increased temperature are an increase in the plasticity of the lignin and an increase in spatial size, which reduces intermolecular contact and is, thus, recoverable. Permanent effects manifest themselves as an actual reduction in wood substance or weight loss via degradative mechanisms, and are thereby nonrecoverable.

LOAD FACTORS. *Duration of Load.* The ability of wood to re-

Table III. Approximate Change in the Mechanical Properties of Clear Wood When Subjected to Change in Moisture Content

Property	Change Per 1% Change in Moisture Content[a] (%)
Static bending	
Fiber stress at proportional limit	5
Modulus of rupture	4
Modulus of elasticity	2
Work to proportional limit	8
Work to maximum load	0.5
Impact bending	
Height of drop causing complete failure	0.5
Compression parallel to grain	
Fiber stress at proportional limit	5
Maximum crushing strength	6
Compression perpendicular to grain	
Fiber stress at proportional limit	5.5
Shear parallel to grain	
Maximum shearing strength	3
Hardness	
End	4
Side	2.5

SOURCE: Ref. 9.
[a] These adjustments can be calculated more precisely by the formulas given in Reference 7.

sist load is dependent upon the length of time the load is applied (7, 12, 13). The load required to cause failure over a long period of time is much less than the load required to cause failure over a very short period of time. Wood under impact loading (duration of load ~1 s) can sustain nearly twice as great a load as wood subjected to long-term loading (duration of load >10 years). This time-dependent relationship can be seen graphically in Figure 5.

Fatigue. Cyclic or repeated loadings often induce fatigue failures. Fatigue resistance is a measure of a material's ability to resist repeating, vibrating, or fluctuating loads without failure. Fatigue failures often result from stress levels far lower than those required to cause static failure. Repeated or fatigue-type stresses usually result in a slow thermal buildup within the material and initiate and propagate tiny microchecks that eventually grow to a terminal size. When wood is subjected to repeated stress (e.g., 5.0×10^7 cycles), fatigue-

Table IV. The Relationship Between Some Mechanical Properties and Moisture Content

Property	Moisture Content				
	Green	19%	12%	8%	Oven-Dry
	Douglas-Fir				
Modulus of rupture	62	76	100	117	161
Compression parallel to grain	52	68	100	124	192
Modulus of elasticity	80	88	100	108	125
	Loblolly Pine				
Modulus of rupture	57	72	100	121	175
Compression parallel to grain	49	66	100	127	203
Modulus of elasticity	78	87	100	109	128
	Aspen				
Modulus of rupture	61	75	100	118	165
Compression parallel to grain	50	67	100	126	199
Modulus of elasticity	73	83	100	111	137

NOTE: All values are expressed as a percent of property at 12% moisture content.

related failures may be induced by stress levels as low as 25–30% of the anticipated ultimate stress under static conditions (14).

Mechanical Properties. To design with any material, mechanical property estimates need to be developed. ASTM standard test methods detail the procedures required to determine mechanical properties via stress–strain relationships (2–6).

FLEXURAL LOADING PROPERTIES. Flexural (bending) properties are important in a wood design. Many structural designs recognize either bending strength or some function of bending, such as deflection, as the limiting design criterion. Structural examples in which bending-type stresses are often the limiting consideration are bridges or bookshelves. Five mechanical properties are derived from the stress–strain relationship of a standard bending test: modulus of rupture (MOR), fiber stress at proportional limit (FSPL), modulus of elasticity (MOE), work to proportional limit (WPL), and work to maximum load (WML).

Modulus of Rupture. The MOR is the ultimate bending strength of a material. Thus, MOR describes the load required to cause a wood beam to fail and can be thought of as the ultimate resistance or strength that can be expected (Figure 2, Point B; Figure 6, Point B) from a wood beam exposed to bending-type stress. MOR is derived by the flexure formula

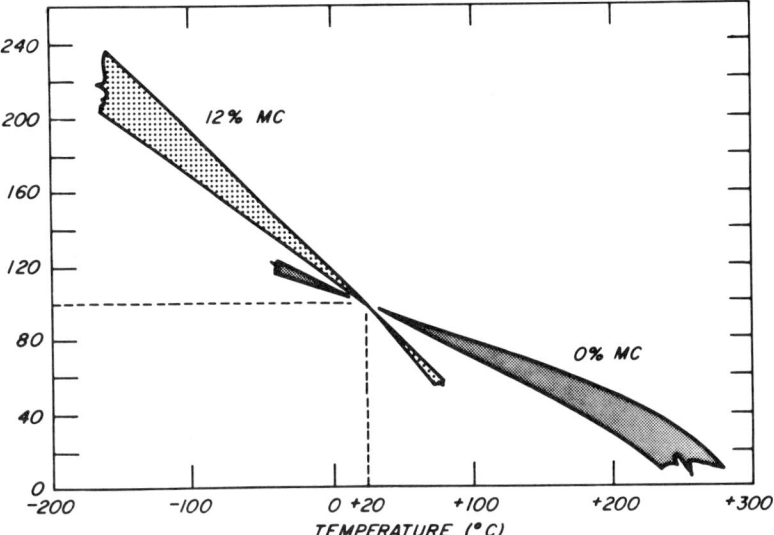

Figure 4. Immediate effect of temperature on strength properties expressed as a percent of value at 20 °C (7). Trends illustrated are composites from several studies on three strength properties (MOR, $T_\perp$, and $C_\parallel$).

$$\text{MOR} = \frac{M}{S}$$

where M is the maximum bending moment and S is the section modulus, which relates the moment to the geometric shape of the beam.

For a rectangular or square beam in bending under centerpoint loading, the flexure formula is varied to reflect loading conditions and beam geometry

$$\text{MOR} = \frac{1.5 \cdot P \cdot l}{bh^2}$$

where P is the ultimate load, l is the span of beam, b is the width of beam, and h is the height of beam.

Fiber Stress at Proportional Limit. The FSPL is the maximum bending stress a material can sustain under static conditions and still exhibit no permanent set or distortion. It is by definition the amount of unit stress on the y-coordinate at the proportional limit of the material (Figure 2, Point A; Figure 6, Point A). FSPL is derived using the flexure formula

$$\sigma_{\text{FSPL}} = \frac{M}{S}$$

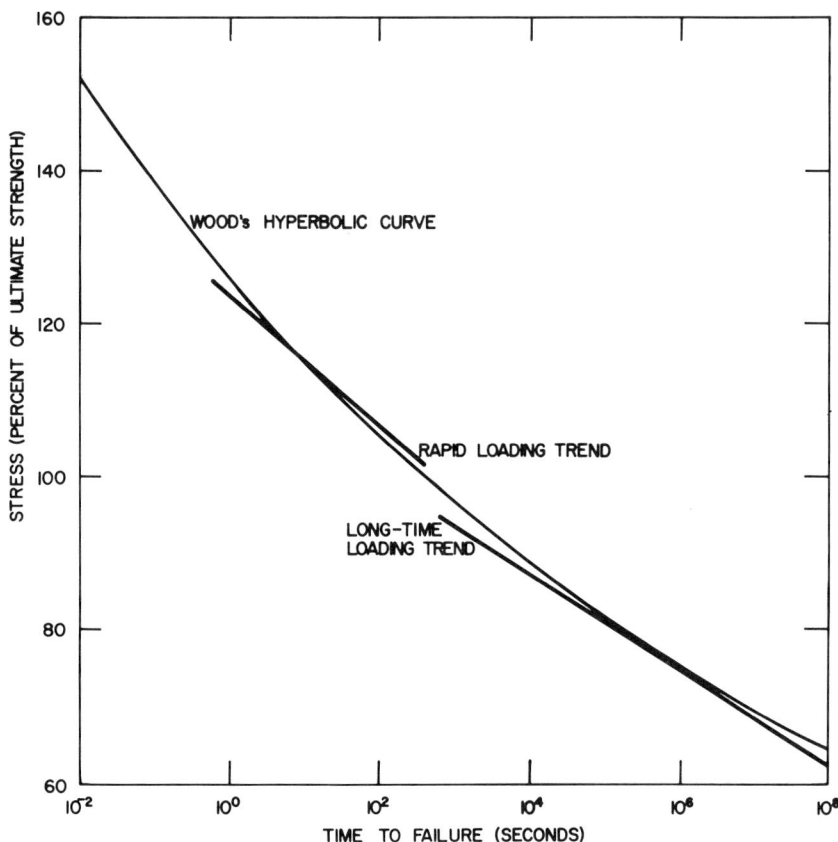

Figure 5. Hyperbolic load-duration curve with rapid loading and long-time loading trends for bending (12,13).

where M is the bending moment at the proportional limit and S is the section modulus.

Modulus of Elasticity. The MOE quantifies a material's resistance to deformation under load. The MOE corresponds to the slope of the linear portion of the stress–strain relationship from zero to the proportional limit (Figure 6). Stiffness is often incorrectly thought to be synonymous with MOE. However, MOE is solely a material property and stiffness depends on the size of the beam. Large and small beams of similar material would have similar MOEs but different stiffnesses. The MOE can be calculated from the stress–strain curve as the change in stress causing a corresponding change in strain.

$$\text{MOE} = \frac{\Delta \sigma}{\Delta \epsilon}$$

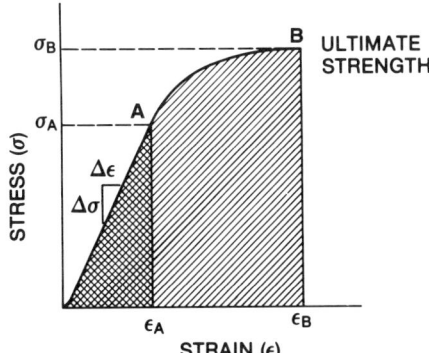

Figure 6. Example of the relationship between a typical stress–strain diagram and some mechanical properties. Key: A, proportional limit; B, ultimate strength; σ_B, MOR; σ_A, FSPL; $\Delta\sigma/\Delta\epsilon$ (from origin to A), MOE; $\int_0^{\epsilon_A} \sigma \, d\epsilon$, WPL; and $\int_0^{\epsilon_B} \sigma \, d\epsilon$, WML.

Work to Proportional Limit. The WPL is the measure of work performed in going from an unloaded state to the elastic or proportional limit of a material (Figure 6). For a beam of rectangular cross section under centerpoint loading, WPL is found by calculating the area under the stress–strain curve from zero to the proportional limit.

$$\text{WPL} = \frac{P\partial}{2bhl}$$

where P is the load, ∂ is the deflection, b is the width of beam, h is the height of beam, and l is the span of beam.

Work to Maximum Load. WML is the amount of work needed to actually fracture or fail a material, and it is the area under the stress–strain curve from zero to the ultimate strength of the material (Figure 6). Because WML is a measure of work both below and beyond the proportional limit it is derived by either graphical approximations or by means of calculus.

AXIAL LOADING PROPERTIES. Because of the anisotropic and heterogeneous nature of wood, there can be profound differences in the strength in various directions. Wood is stronger along the grain (parallel to the longitudinal axis of the log or longitudinal axis of the wood cell) than perpendicular to the grain (at right angles to the longitudinal axis). Axial loads describe forces that have the same line of action and are, thus, both parallel and concurrent. Because there is no eccentricity in the application of these forces, they do not induce flexure or bending moments.

Mechanical properties dealing with axial loading conditions are maximum crushing strength (compression parallel to the grain), compression perpendicular to the grain, tension parallel to the grain, and tension perpendicular to the grain.

Compression Parallel to the Grain. If wood is considered a bundle of straws bound together, then a compression parallel to the grain ($C_\parallel$) can be thought of as a force trying to compress the straws from end to end. The distance through which compressive stress is transmitted does not increase or magnify the stress, but the length over which the stresses are carried is important. If the length of the column is far greater than the width, the specimen may buckle. This stress is analogous to bending-type failure rather than axial-type failure. As long as specimen width is great enough to preclude buckling, $C_\parallel$ is solely an axial property.

Examples of wood in compression parallel to the grain are wooden columns or the top chord of a roof truss. Compression-parallel-to-the-grain strength or the maximum crushing strength is derived at the ultimate limit value of a standard stress–strain curve.

The strength of wood in $C_\parallel$ is derived by

$$\sigma_{C_\parallel} = \frac{P}{A}$$

where $\sigma_{C_\parallel}$ is the stress in compression parallel to grain, P is the maximum load, and A is the area over which load is applied.

Compression Perpendicular to the Grain. Compression perpendicular to the grain ($C_\perp$) can be thought of as stress applied perpendicular to the length of the wood cell. Therefore, in our straw example, the straws (or wood cells) are being crushed at right angles to their length. Until the cell cavities are completely collapsed, wood is not as strong perpendicular to the grain as it is parallel to the grain. However, once the wood cell cavities collapse, wood can sustain a nearly immeasurable load in $C_\perp$. Because a true ultimate stress is nearly impossible to achieve, maximum $C_\perp$ in the sense of ultimate load-carrying capacity is undefined and discussions of $C_\perp$ are usually confined to stress at the proportional limit.

Compression-perpendicular-to-the-grain stresses are found whenever one member is supported upon another member at right angles to the grain. Examples of compression perpendicular to the grain are the bearing areas of a beam, truss, or joist.

The $C_\perp$ strength is derived by

$$\sigma_{C_\perp} = \frac{P}{A}$$

where $\sigma_{C_\perp}$ is the stress in compression perpendicular to the grain, P is the proportional limit load, and A is the area.

Tension Parallel to the Grain. A tension parallel to the grain ($T_\parallel$) stress is a force trying to elongate the wood cells, or straws in our straw example. Wood is extremely strong in $T_{\|}$. The distance through which tensile stress is transmitted does not increase the stress. The $T_{\|}$ is difficult to measure because of the difficulty in securely gripping the tensile specimen in the testing machine. Often $T_{\|}$ is conservatively estimated by the MOR (ultimate strength in bending). This conversion is accepted because failure often occurs on the lower face of a bending specimen where the lower face fibers are under tensile-type stresses. An example of tension parallel to the grain would be the bottom chord of a truss that is under tensile stress.

The $T_{\|}$ strength of wood is derived by the formula

$$\sigma_{T_{\|}} = \frac{P}{A}$$

where $\sigma_{T_{\|}}$ is the stress in tension parallel to the grain, P is the maximum load, and A is the area.

Tension Perpendicular to Grain. Tension perpendicular to the grain $(T_\perp)$ is induced by a tensile force applied perpendicular to the longitudinal axis of the wood cell. In this case, the straws (or wood cells) are being pulled apart at right angles to their length. The $T_\perp$ is extremely variable and is often avoided in discussions on wood mechanics. However, $T_\perp$ stresses often cause cleavage or splitting failures along the grain, which can dramatically reduce the structural integrity of large beams. Failures from $T_\perp$ are sometimes found in large beams that dry while in service. For example, if a beam is secured by a top and a bottom bolt at one end, shrinkage may eventually cause cleavage or splitting failures between the top and bottom boltholes. Wood can be cleaved by $T_\perp$ forces at a relatively light load. It is this weakness that is often exploited in karate and other demonstrations of human strength. The $T_\perp$ strength of wood is derived by

$$\sigma_{T_\perp} = \frac{P}{A}$$

where $\sigma_{T_\perp}$ is the stress in tension perpendicular to the grain, P is the maximum load, and A is the area.

OTHER MECHANICAL PROPERTIES. *Shear.* Shear parallel to the grain (γ) measures the ability of wood to resist the slipping or sliding of one plane past another parallel to the grain. Shear strength is derived in a manner similar to axial properties, by using the equation

$$\gamma = \frac{P}{A}$$

where γ is the shear stress parallel to grain.

Hardness. Hardness is used to represent the resistance to indentation and/or marring. Hardness (2) is measured by the load required to embed a 1.128-cm steel ball one-half its diameter into the wood.

Shock Resistance. Shock resistance or energy absorption is a function of a material's ability to quickly absorb and then dissipate energy via deformation. This is an important property for baseball bats, tool handles, and other articles that are subjected to frequent shock loadings. High shock resistance on energy absorption properties requires both the ability to sustain high ultimate stress and the ability to deform greatly before failing.

Shock resistance can be measured by several methods. With wood, two of the most often used methods are impact bending tests and toughness tests. Both test methods yield measures of strength and pliability. These measures are similar but are not particularly relative to one another. Impact bending is tested by dropping a weight onto a beam from successively increasing heights (2) and is recorded as the height of drop causing complete failure in a beam. Toughness is the ability of a material to resist a single impact-type load from a pendulum device (2). Thus, it is similar to impact bending in that it is a measure of resistance, but it refers to a single load rather than multiple loads.

Chemical Components of Strength

Relationship of Structure to Chemical Composition. The chemical components responsible for the strength properties of wood can be theoretically viewed from three distinct levels: the macroscopic (cellular) level, the microscopic (cell wall) level, and the molecular (polymeric) level.

MACROSCOPIC. Wood with its inherent strength is a product of growing trees. Wood exists as concentric bands of cells oriented for specific functions. Thin-walled earlywood cells act as conductive tissue; thick-walled latewood cells provide support. Each of these cells is a single fiber. Softwood fibers average about 3.5 mm in length and 0.035 mm in diameter. Hardwood fibers are generally shorter (1–1.5 mm) and smaller in diameter (0.015 mm). The fibers comprise a large mat, bonded together by a phenolic adhesive, lignin. The mat is anisotropic in character but is reinforced in two of the three axial directions by longitudinal parenchyma and ray parenchyma cells. These parenchyma cells function as a means of either longitudinal or radial nutrient conduction and as a means of providing lateral support by increased stress distribution (Figure 7). Because wood is a reinforced composite material, its structural performance at the cellular level has been likened to reinforced concrete (*15*, *16*).

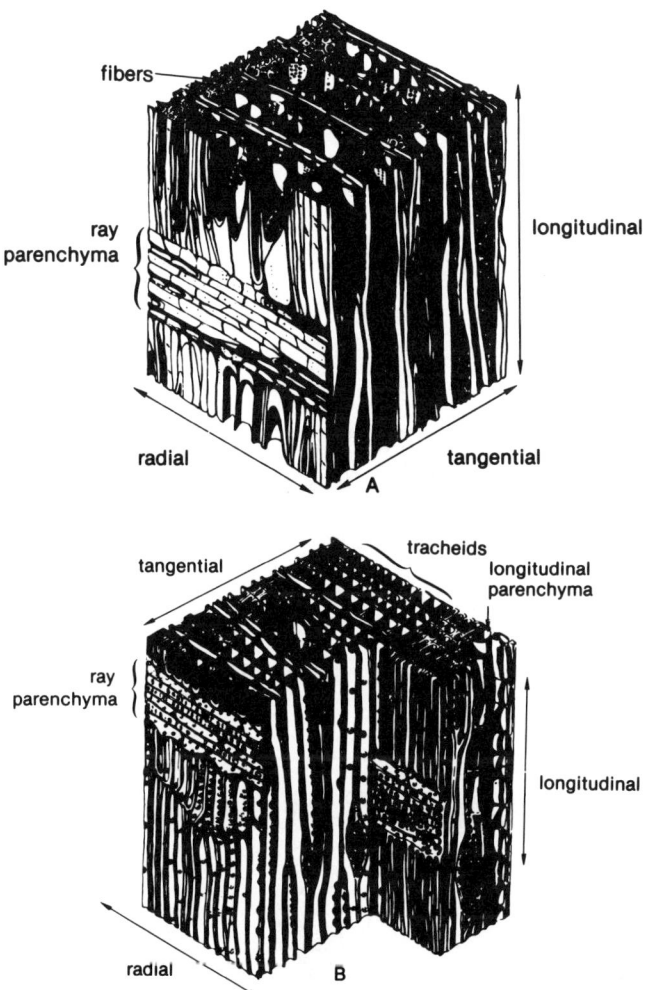

Figure 7. Two representative diagrams of three-dimensional anisotropic wood cubes showing the radial, tangential, and longitudinal direction of each cube. Key: A, a typical hardwood; and B, a typical softwood.

The macroscopic level of consideration takes into account fiber length and differences in cell growth such as earlywood, latewood, reaction wood, sapwood, heartwood, mineral content, resin content, etc. Differences in growth chemistry can cause significant differences in the strength of wood.

MICROSCOPIC. At the microscopic level, wood has been compared to multipart systems such as filament-wound fiber products (16). Each component complements the other in such a manner that, when considering the overall range of physical performance, the components together outperform the components separately.

Composition. Within the cell wall are distinct regions (Figure 8), each of which has distinct composition and attributes. For a typical softwood the middle lamella and primary wall are mostly lignin (8.4% of the total weight) and hemicellulose (1.4%), with very little cellulose (0.7%). The S_1 layer consists of cellulose (6.1%), hemicellulose (3.7%), and lignin (10.5%). The S_2 layer is the thickest layer and has the highest carbohydrate content; it is mostly cellulose (32.7%) with lesser quantities of hemicelluloses (18.4%) and lignin (9.1%). The S_3 layer, the innermost layer, consists of cellulose (0.8%), hemicelluloses (5.2%), and very little lignin.

The large number of hydrogen bonds existing between cellulose molecules results in such strong lateral associations that certain areas of the cellulose chains are considered crystalline. More than 60% of the cellulose (*17*) exists in this crystalline form, which is stiffer and stronger than the less crystalline or amorphous regions. The crystalline areas are approximately 60 nm long (*18*) and are distributed throughout the cell wall.

Microfibril Orientation. Microfibrils are highly ordered groupings of cellulose that may also contain small quantities of hemicellulose and lignin. The exact composition of the microfibril and its relative niche between the polymeric chain and the layered cell wall

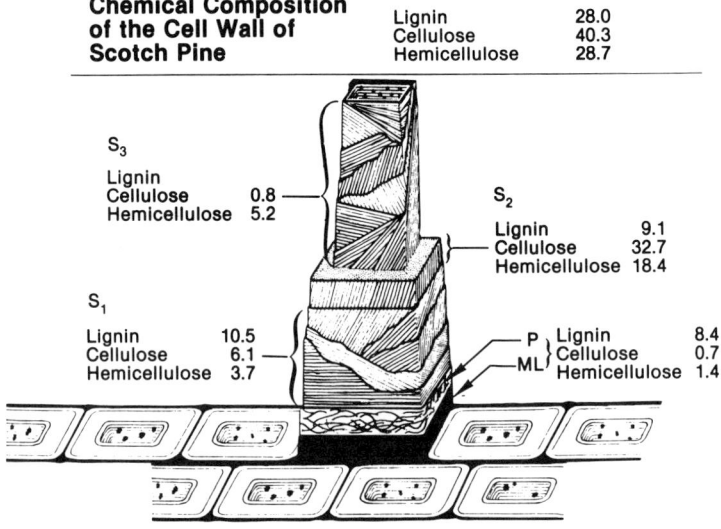

Figure 8. Representation of the microfibril orientation for each cell wall layer of Scotch pine with the chemical composition as a percent of total weight.

are subjects of great discussion (16). The microfibril orientation (fibril angle) is different and distinct for each cell wall layer (Figure 8). The entire microfibril system is a grouping of rigid cellulosic chains analogous to the steel reinforcing bars in reinforced concrete or the glass or graphite fibers in filament-wound reinforced plastics. Most composite materials use an adhesive of some type to bond the entire material into a system. In wood, lignin fulfills the function of a matrix material. Yet, it is not truly or solely an adhesive and by itself adds little to strength (19). Lignin is a hydrophobic phenolic material that surrounds and encrusts the carbohydrate complexes (Figures 9 and 10). It aids in holding the cell components together at the microscopic level. Lignin also seems to be responsible for part of the stiffness of wood. Rubbery wood, a viral disease of certain varieties of apple (*Malus* sp.), is characterized by extremely flexible wood. The affected wood has been shown (20) to have cells rich in cellulose but low in lignin.

MOLECULAR. At the molecular level the relationship of strength and chemical composition deals with the individual polymeric components that make up the cell wall. The physical and chemical properties of cellulose, hemicelluloses, and lignin play a major role in the chemistry of strength. However, our perceptions of wood polymeric properties are based on isolated polymers that have been removed from the wood system and, therefore, possibly altered. The individual polymeric components may be far more closely associated with one another than has heretofore been believed.

Cellulose is an unbranched, rigid chain, linear polymer com-

Figure 9. Photomicrograph of softwood fiber embedded in lignin (×4000).

Figure 10. Photomicrograph of delignified cell-wall softwood fibers (×16,000).

posed of anhydro-D-glucopyranose ring units bonded together by β-1-4-glycosidic linkages. The greater the length of the polymeric chain, the higher the degree of polymerization, the greater the strength of the unit cell (16, 21) and, thus, the greater the strength of the wood. The cellulose chain may be 5000–10,000 units long. Cellulose is extremely resistant to tensile stress because of the covalent bonding within the pyranose ring and between the individual units. Hydrogen bonds within the cellulose provide rigidity to the cellulose molecule via stress transfer and allow the molecule to absorb shock by subsequently breaking and reforming.

The hemicelluloses are carbohydrate molecules that consist of various elementary sugar units, primarily the six-carbon sugars, D-glucose, D-galactose, and D-mannose, and the five-carbon sugars, L-arabinose and D-xylose (see Chapter 2). Hemicelluloses have linear chain backbones that are highly branched and have a lower degree of polymerization than cellulose. The sugars in the hemicellulose structure exhibit hydrogen bonding both within the hemicellulose chain as well as between other hemicellulose and cellulose chains. Most hemicelluloses are found in the amorphous regions of the cellulose chains and in close association with the lignin. Hemicellulose may be the connecting material between cellulose and lignin.

Lignin is often considered nature's adhesive. It is the least understood and most chemically complex polymer of the wood-structure triad. Its composition is based on highly organized three-dimensional phenolic polymers rather than linear or branched carbohydrate chains. Lignin is the most hydrophobic (water-repelling) component of the wood cell. Its ability to act as an encrusting agent on and around the carbohydrate fraction, and thereby limit water's influence on that carbohydrate fraction, is the cornerstone of wood's ability to retain its strength and stiffness as moisture is introduced to the system. Dry delignified wood has nearly the same strength as normal dry wood, but wet delignified wood has only approximately 10% of the strength of wet normal wood (19). Thus, wood strength is due in part to lignin's ability to limit the access of water to the carbohydrate moiety and thereby lessen the influence of water on wood's hydrogen-bonded structure.

Relationship of Chemical Composition to Strength. To relate chemical composition to strength properties, to work and toughness properties, and eventually to elastic parameters, a model or theory must be developed that explains the relationship between strength and wood composition. The theoretical relationship of stress to strain can be graphically represented by a diagram (Figure 11). If wood is assumed to be an elastic material, a linear region, A, represents the constant relationship below the proportional limit, and the nonlinear regions, B and C, represent nonconstant relationship beyond the proportional limit. If each region of the stress–strain relationship is examined at each of the three distinct levels of wood structure (macroscopic, microscopic, and molecular), the relationship between strength and composition can be hypothetically explained.

As loads are applied to a wood system, stresses are immediately introduced and distributed throughout the material. The stresses cause two types of strain or distortion: immediate, under which wood can be described as an elastic material, and time dependent. Even at low stress levels, permanent set or distortion will eventually be induced in a wood member. This phenomenon, known as creep,

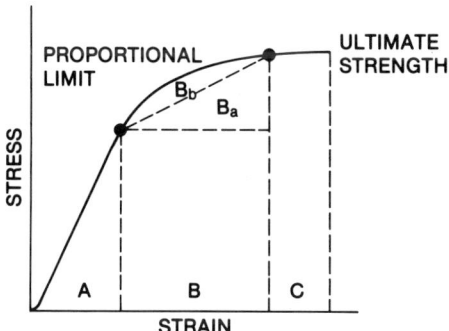

Figure 11. Typical stress–strain curve showing the three theoretically identifiable regions of mechanical behavior. Key: A, elastic region; B, elastic (B_a)–plastic (B_b) region; and C, plastic region.

requires that wood be considered viscoelastic. But for purposes of simplification this discussion will be confined to immediate strain or distortion and will consider wood as an elastic material. Immediate strain or distortion can be conceptualized at each of the three distinct levels of wood structure.

BELOW PROPORTIONAL LIMIT (ELASTIC STRENGTH). When a load is applied to a piece of wood, at the molecular level, hydrogen bonds between and within individual polymer chains are breaking, sliding (uncoiling), and subsequently reforming (Figure 12); C–C and C–O bonds are distorting within the ring structures (Figure 13).

At the microscopic level, hydrogen bonds between adjacent microfibrils are breaking and reforming (Figure 12), to allow the microfibrils to slide by one another with only the disruption of the hydrogen bonds that are subsequently reformed. Additionally, the individual cell wall layers are distorting in relation to each other, but no permanent set or distortion is occurring between these individual cell wall layers.

At the macroscopic level, there is distortion between the individual cells, but it is not permanent because the stresses are being distributed between the individual cells such that no permanent translocation or set is introduced.

Within the limits of the elastic model, all strain or distortion resulting from the accumulation of stress in this material has been

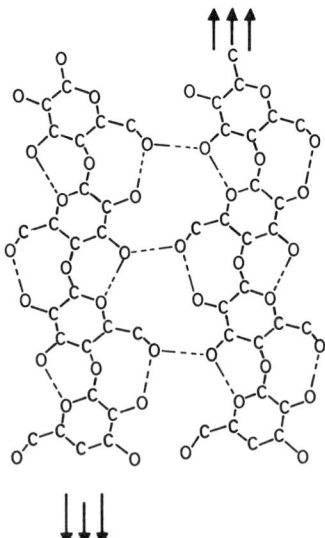

Figure 12a. Hydrogen bonding (bonded) between polysaccharide chains under shear forces.

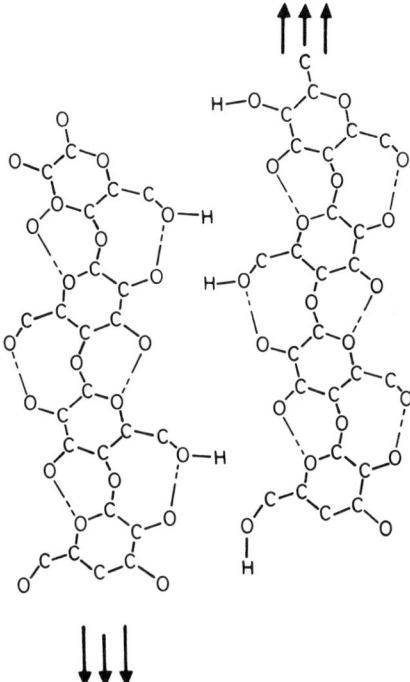

Figure 12b. Hydrogen bonding (sliding, unbonded) between polysaccharide chains under shear forces.

recoverable up to this point. As the proportional limit is approached, the wood material can no longer distribute the stress in a linear manner.

BEYOND PROPORTIONAL LIMIT (PLASTIC STRENGTH). As the proportional limit is exceeded (Figure 11, Region B), the stress–strain relationship is no longer linear. Stresses are now great enough to induce covalent bond rupture and permanent distortion at all three structural levels.

At the molecular level, the limit of reversible or recoverable hydrogen bonding has been exceeded. Covalent C–C and C–O bonds are breaking, thus reducing larger molecules to smaller ones. This reduction in degree of polymerization by covalent bond scission is nonrecoverable.

At the microscopic level, stresses develop within the crystalline region of the carbohydrate microfibrils. Failure of the microfibril from stress overload causes actual covalent bond rupture and excessive microfibril disorientation. Additionally, the cell wall layers distort such that permanent microcracks occur between the various cell wall layers. Separation of the cell wall layers is soon noticeable.

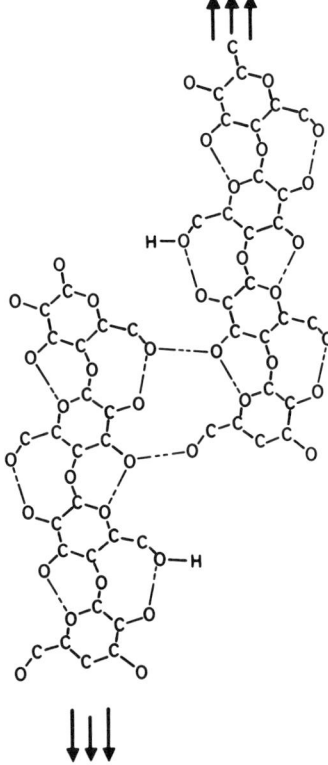

Figure 12c. Hydrogen bonding (rebonded) between polysaccharide chains under shear forces.

At the macroscopic level, entire fibers actually distort in relation to one another, such that recovery of original position is now impossible. The wood cells or wood fibers are actually failing either by scission of the cell, in which the cell actually fails by tearing into two parts to give a brash type of failure, or by cell-to-cell withdrawal (middle lamella failure) where the cells actually pull away from one another to give a splintering type of failure.

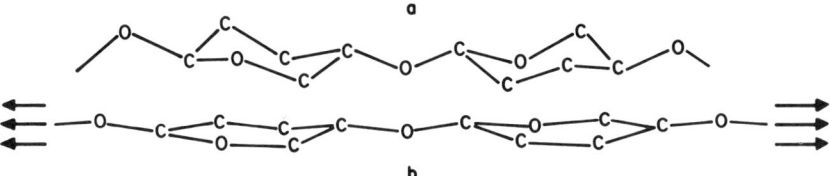

Figure 13. Flexing and elongation of polysaccharide molecule under tensile force. Key: a, no tensile force—no elongation; and b, tensile force— elongation.

Permanent set is now evident at all levels of consideration, and eventual failure is imminent. In approaching the ultimate strength (Figure 11, Region C), molecular level failures occur by C–C and C–O bond cleavage. Stress redistribution within the individual polymers is now impossible. At the microscopic level, the cell walls are distorting without additional stress. These walls are actually deforming at such an exaggerated rate that they can be thought of as being completely viscous or plastic, and they can no longer handle additional stress. The cell wall is being sheared or torn apart. At the macroscopic level, failure is related to cell wall scission or cell-to-cell withdrawal.

Relationship of Structure to Strength. The mechanism of strength, as it relates to wood composition, has been discussed as a theoretical elastic model. To better understand this proposed model, we will look at what may be happening at each of the three structural levels.

MOLECULAR LEVEL. At the molecular level, strength is elastic or recoverable because the polymeric structure can flex and, thus, absorb energy without fracturing the important covalent bonds (Figure 11, Region A). The second region of the stress–strain diagram (Figure 11, Region B) consists of Region B_a, which is indicative of residual elastic strength such as represented in Region A, and Region B_b, which represents plastic strength or, more appropriately, strength associated with initial permanent set or distortion.

Section B_b is representative of C–C and C–O cleavage at the intrapolymer level which cannot be recovered. Examples of C–C bond breakage are lignin–hemicellulose copolymer separation, hemicellulose depolymerization, and amorphous cellulose depolymerization.

In Region C (Figure 11), elastic deformation essentially ends; there is now nearly pure plastic flow in the stress–strain relationship. Strain is continuing with little additional increase in stress and ultimate failure is imminent. This region is characterized by all the same mechanisms as in B_b, but a new and terminal intrapolymeric factor is introduced—the crystalline cellulose failure. As crystalline cellulose failure occurs, the main framework of the wood material at the molecular level is disintegrating.

MICROSCOPIC LEVEL. At the microscopic level, the strength of the phenolic matrix is usually great enough that the cell wall stress reaches failure level in the carbohydrate framework. The S_1-layer microfibrils are oriented in both a right-hand (S helix) and a left-hand (Z helix) arrangement whereas the S_2 and S_3 have only the S-helix arrangement (Figure 8). The S_3 layer can be bihelical or monohelical, but, for the purpose of simplification, it has been assumed to be monohelical in this example. Because of the different linear elongation of the bihelical S_1 layer as compared to the monohelical arrangement of the S_2 and S_3 layers, the cell wall initially fails by S_1–S_2 separation (16). As S_1–S_2 separate, the S_2–S_3 layers assume the transferred stresses, and sustained stress increases, which will eventually cause either a brash-type failure (carbohydrate covalent bond failure) or a slow buildup to ultimate stress yielding a fibrous-type failure (phenolic covalent bond failure).

Below the proportional limit (Figure 11, Region A), there is elastic transfer of stresses between the S_1–S_2–S_3 cell wall layers. As Region B is entered, stress is still transferred between the S_1–S_2–S_3 cell wall layers as characterized by Section B_a. But S_1–S_2 separation is initiating, causing a sizeable transfer of stresses to the S_2–S_3 layers characterized by Section B_b. In Region C, ultimate strength is now dictated by the S_2–S_3 cell wall layer's ability to sustain additional stress until eventual failure of the substantial S_2 layer.

MACROSCOPIC LEVEL. At the macroscopic level, it is necessary to consider wood a viscoelastic material. As stress is applied to a wooden member, minute cracks initiate, propagate, and terminate throughout the collective cellular system in all directions. They develop in all regions of the stress–strain relationship at the macroscopic level, but only in the elastic region (Figure 11, Regions A and B_a) is crack propagation controlled and eventually terminated. In the tangential direction, the concentric ring structure of thin-walled earlywood and thick-walled latewood in softwoods, and porous early-season vessels and dense late-season fibers in hardwoods act as the elements of elastic stress transfer. In the radial direction, the ray structures and the linear arrangement of fibers and vessels are the elements of elastic stress transfer. Every cell in the radial direction is aligned closely with the next cell because each cell in the radial direction has originated from the same cambial mother cell. Thus, the material can transfer stress elastically until an induced crack or a natural growth defect interrupts this orderly cellular arrangement. As stresses are built up within the material, cracks are initiated in the areas where elastic stress transfer is interrupted. These cracks continue to propagate until they are either terminated via dispersion of the energy away from the crack by the structural elements of stress transfer, or by eventual terminal failures as graphically characterized by Regions B_b and C (Figure 11).

Altered Composition Effects

When wood is exposed to environmental agents of deterioration, such as chemical treatments or elevated temperatures, each mechanical property reacts differently. Most commonly, ultimate strength properties are reduced and properties dealing with the proportional limit show little or no effect. However, the strain-to-failure (strain rate) is often considerably reduced, which, due to embrittlement of the fibers, is reflected as a reduction in pliability.

As individual wood components are altered in size, stature, or composition, the strength of the wood material is dramatically affected. Hypothetically, when ultimate stress is reduced 5% (Figure 14, U_1–U_2) and the proportional limit is not affected, the properties dealing with proportional limit (FSPL, MOE, WPL) reflect this. The mechanical properties dealing with the point of ultimate stress (MOR in bending tests, $C_{\|}$ in axial-type compression tests, and $T_{\|}$ and $T_{\perp}$ in axial-type tensile tests) are reduced 5%. Work to maximum load can be reduced 33% because it is a function of both stress and strain. If the stress level at the proportional limit is reduced, and both the ultimate stress and strain levels are significantly reduced (Figure 15), larger decreases will occur in proportional limit properties (MOE, FSPL, WPL), ultimate strength properties (MOR, $C_{\|}$, $T_{\|}$, $T_{\perp}$), and in work to maximum load. The examples in both Figures 14 and 15 show evidence that WML and related properties, such as toughness and impact bending, are usually affected long before the other properties dealing with ultimate strength and the proportional limit are significantly affected.

What causes the phenomenon of stress and strain reduction and why is the reduction in impact and work properties so visible at small or negligible changes in elastic modulus and ultimate strengths? As discussed previously, mechanical properties deal with stress and strain relationships that are simply functions of chemical bond strength. At the molecular level, strength is related to both covalent and hydrogen intrapolymer bonds. At the microscopic level, strength

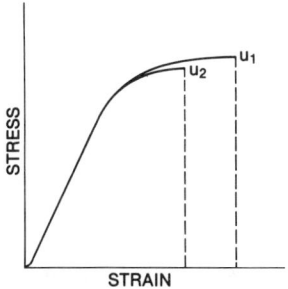

Figure 14. Hypothetical example of the effect of no change in proportional limit and a 5% reduction in ultimate bending strength on a few mechanical properties: MOE is not affected, MOR is reduced 5%, but WML is reduced by 33% because it is a dual function of stress and strain.

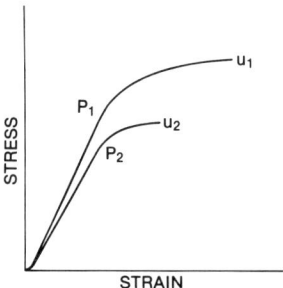

Figure 15. Hypothetical example of the effect of a 20% reduction in proportional limit and a 30% reduction in ultimate bending strength on a few mechanical properties: MOE is reduced 20%, MOR is reduced 30%, and WML is reduced 50%, because it is a dual function of stress and strain.

is related to both covalent and hydrogen interpolymer bonds and cell wall layer bonds (S_1–S_2 and S_2–S_3). At the macroscopic level, strength is related to fiber-to-fiber bonding with the middle lamella acting as the adhesive. Thus, any chemical or environmental agent that affects those bondings also affects strength.

Environmental Effects. In considering the structural performance of the polysaccharide and phenolic polymers in wood fiber, the chemical environment of the fiber is of great importance. Chemicals can swell, hydrolyze, pyrolyze, oxidize, and, in general, depolymerize wood polymers, causing a loss in strength properties due to wood fiber network degradation. Other environmental agents such as UV light, heat, and biological organisms have a similar influence in changing strength properties.

ACIDS AND BASES. The average pH of wood is between 3 and 5.5 (*17*) due to the acetyl content, the presence of acid extractives, and the adsorption of cations that comprise the ash. Even after several hundred years, this naturally mild acidic state does not induce any appreciable strength losses as long as the wood is protected from biological attack (*7*).

If the pH of the environment substantially changes, strength properties can be reduced. These effects are further compounded by time and temperature. In general, the longer the time or the higher the temperature at which wood is exposed to an acid or a base, the greater is the degradative effect on strength (*see* Chapter 14). Heartwood is generally more resistant to acid than is sapwood, probably because of heartwood's lower permeability and higher extractives content. Hardwoods are usually more susceptible to degradation by either acids or alkalies than are softwoods. This may be due to hardwood's lower lignin content and higher proportion of pentosan hemicelluloses. Oxidizing acids, such as HNO_3, have a greater degradative action on wood fiber than do nonoxidizing acids. Alkaline solutions are more destructive to wood fibers than are acidic solutions (7) because wood adsorbs alkaline solutions more readily than acidic solutions. Acids with pH values above 2 and bases with pH values below 10 do not degrade the wood fiber greatly over short periods of time at low temperatures (22). Mild acids, such as acetic acid, have little effect on strength, whereas strong acids, such as H_2SO_4, cause extensive strength losses (23).

ADSORPTION OF ELEMENTS. Chemicals other than acids and bases can also be adsorbed and can cause degradation of the wood fiber. For example, fibers of southern pine exposed to the ocean air can be degraded badly (Figure 16). Salt crystals deposited in the void structure (Figure 17) can cause extensive chemical and physical damage.

Other materials can be adsorbed from the environment if a hydrolytic solvent (e.g., water) is available. When water is available, wood will adsorb iron from oxidized metal (rust) and cause decomposition of the cellulose (24). This is also true for copper, chromium, tin, zinc, and other similar reactive metals.

SWELLING SOLVENTS. Solutions that swell wood tend to plasticize it and reduce its strength properties. Water, for example, swells the intrapolymeric spaces, reduces cross-linking and, thus, reduces strength. In general, the greater the swelling, the greater the strength loss. Nonswelling liquids generally do not decrease strength

Figure 16. Photomicrograph of NaCl deposited on the surface (left) of southern pine (×40).

Figure 17. Photomicrograph of NaCl deposited in southern pine (×400).

properties. For example, oven-dry wood and wood saturated with water-free benzene have virtually the same strength (25, 26).

UV DEGRADATION. Wood exposed to the outdoors undergoes chemical reactions due to UV radiation (see Chapter 11). UV radiation causes photochemical degradation primarily in the lignin component, which gives rise to characteristic color changes. Southern pine, for example, changes from a light-yellow natural color to brown and eventually to gray. As the lignin degrades, the wood surface becomes richer in cellulose content. Although the cellulose is much less susceptible to UV degradation (27) it is washed off the surface with water during rain, which exposes new lignin-rich surfaces that then start to degrade. As this process continues, the wood surface is said to *weather*.

Because UV radiation does not penetrate wood more than a few cells deep, weathering is considered a surface phenomenon. Over time, it can account for a significant loss in surface fiber (Figure 18). As the degradative process continues, the loss in fiber may eventually cause a reduction in the material's load-carrying capacity.

HEAT DEGRADATION. Wood strength is inversely related to temperature. A nearly linear decrease in strength is observed on increasing the temperature from −200 to +160 °C (Figure 4) with a corresponding loss in strength from two- to three-fold (10, 22). Heat has two types of effects on wood: immediate effects that occur only as long as the increased temperature is maintained, and permanent effects that result from thermal degradation of wood's polymer components (see Chapter 13). The immediate effects of heat are recov-

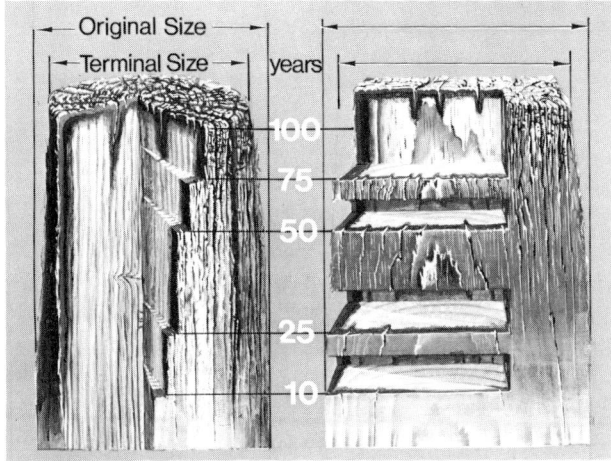

Figure 18. Wood loss in southern pine due to weathering.

erable but permanent effects are not. The combination of immediate and permanent effects is multiplicative rather than additive.

In an environment without adequate humidity, the initial effect of heating wood is dehydration. As temperatures approach 55–65 °C for extended periods (2–3 months), hemicellulose and cellulose depolymerization begins (28). Pyrolysis and volatilization of cell wall polymers occur at about 250 °C followed by char formation in the absence of air and combustion in the presence of air.

Heating Douglas-fir in an oven at 102 °C for 335 d reduced MOE by 17%, MOR by 45%, and fiber stress at proportional limit by 33% (29–31). The same losses might be observed in 1 week at 160 °C. In the absence of air, heating softwood at 210 °C for 10 min reduced MOR by 2%, hardness by 5%, and toughness by 5% (32). Under the same conditions at 280 °C MOR is reduced 17%, hardness is reduced 21%, and toughness is reduced 40%. Both examples illustrate the compound effect of heat and time.

Comparison of photomicrographs of southern pine at 25 °C (Figure 19a) and the same sample after heating from 20 to 295 °C under nitrogen over a period of 15 min (Figure 19b) shows the cell structure still intact, but the cell wall components have been darkened by pyrolysis. Although the cell structure still appears somewhat normal, the strength properties are greatly reduced because of the thermal degradation of the wood material.

MICROBIAL DEGRADATION. When certain organisms come into contact with wood, several types of degradation occur (*see* Chapter 12). The mechanical damage caused by metabolic action can result

Figure 19. Southern pine at 25 °C (×400) (top), and southern pine after heating in a nitrogen environment from 20 to 295 °C over 15 min (×400).

in significant losses in strength. Microbial activity via enzymatic pathways induces wood fiber degradation by chemical reactions such as hydrolysis, dehydration, and oxidation. Brown-rot fungi preferentially metabolize holocellulose, especially the strength-critical cellulose fraction (33). White-rot fungi metabolize nearly all fractions of the wood, but strength is not affected to the same degree as with a brown-rot fungus. An initial 10% weight loss for sweetgum (Liquidambar styraciflua L.) attacked by a brown-rot fungus reduced the degree of polymerization of the holocellulose from 1500 to 300 (Figure 20) (33). As the polymers responsible for strength in the wood fiber are degraded, mechanical properties of the wood decrease.

The large drop in degree of polymerization represents a large drop in wood strength properties without corresponding weight loss (21, 33, 34) and indicates that, in the initial stages of biological

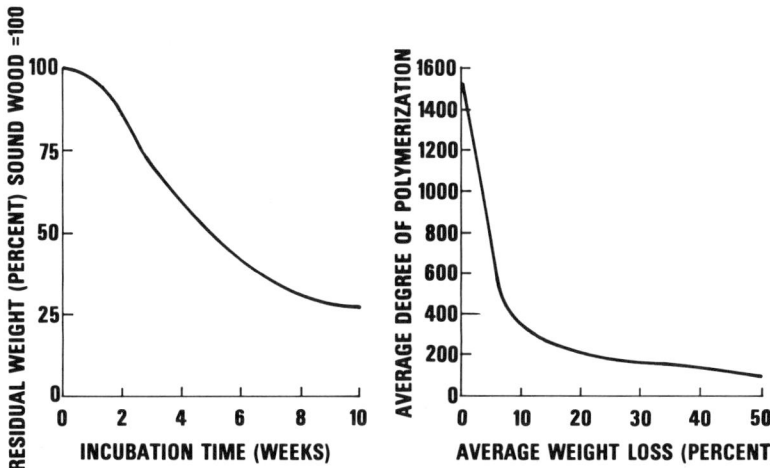

Figure 20. Action of a typical brown-rot fungus (Poria monticola) *on the holocellulose of sweetgum sapwood (33).*

attack, hydrolytic chemical reactions play an important part. In this phase of degradation, large polymers are broken into smaller, more digestible units, but the initial degradation products are not actually consumed by the organisms. Hydrogen peroxide and iron might be involved in this rapid depolymerization (24). The initial enzymatic attack by microorganisms is probably not only hydrolytic but oxidative in nature.

NATURALLY OCCURRING CHEMICALS. Some woods have a higher acidic extractives content that can cause greater strength loss due to hydrolysis. This may be a problem in some of the tropical species coming into the market. These more acidic woods will not only be more susceptible to strength loss but will also increase the corrosion potential of iron fasteners used with the wood. It is not uncommon to find silica and calcium salt crystals in wood fiber (Figure 21), and they can lead to strength losses by abrasion of the fibers during machining. Naturally occurring chemicals may also cause increased hygroscopicity and hydrolysis when salts dissolve in water.

Treatment Effects. BIOLOGICAL RESISTANCE. Salts such as chromated copper arsenate (CCA), ammoniacal copper arsenate (ACA), and other metallic salts of copper, arsenic, zinc, and tin are used to increase the service life of wood in use against biological attack (*see* Chapter 8). One major advantage of many waterborne preservatives is their resistance to leaching by water. These particular metal salts undergo hydrolytic reduction upon contact with the reducing sugars found in wood. In this process, known as fixation, the cell wall is oxidized and subsequently strength is affected. At most

Figure 21. Calcium carbonate crystals deposited in lumens of southern pine (×800).

usable concentrations these preservatives are sufficiently acidic or alkaline to cause some cell wall hydrolysis (35, 36, 37). Most waterborne preservative salts increase the hygroscopicity of the wood. This causes an increased EMC, which further influences strength.

With waterborne treatments such as CCA or ACA, most treated dimension lumber is dried twice. Lumber must be dry to achieve adequate waterborne preservative penetration, so the material is kiln-dried before treatment, and then the treated lumber often is kiln-dried a second time to remove residual water remaining from the waterborne preservative treatment process and to finalize the fixation process. Limited research has been done to investigate the effects of this second drying process. The effects are probably not the same as those due to the initial drying process because the combination of hydrolytic chemicals and high temperatures may be substantially more degrading to strength properties. In general, waterborne preservative treatments have no apparent effect on MOE; they reduce MOR, $T_\|$, and FSPL by about 5%; reduce WML by about 20%; and reduce toughness by about 30% (Table V). The magnitude of these effects can be increased or decreased by controlling the severity of treating and processing parameters.

After fixation, precipitated preservative salts can be seen in the wood structure. Comparison of photomicrographs of untreated southern pine (Figure 22a) and southern pine treated with 2.5 lb CCA/ft^3 (Figure 22b) reveals that the salts in the treated wood form a rough coating on the lumen walls. The exact final location of these fixed salts is still subject to discussion. Undoubtedly, most of these injected hydrolytic preservatives remain in the cell lumen, but the extent to which these preservatives enter the cell wall and react with the cell wall substance dictates the preservative's effect on strength.

Table V. Effects of Preservatives on the Mechanical Properties of Wood

Chemical[a]	Species[b]	Drying	Retention[c] (lb/ft³ or % concentration)	Treatment Process	MOR	MOE	FSPL	WML	Toughness	Impact Bending	Maximum Crushing Strength	$T_\parallel$	Reference
Water	SP	Air	—	Pressure	−2	2		−7					50
		Kiln	—	Pressure	−6	−5		−11					50
ACA	Hdwd veneer	Air	0.1–4	Pressure					−24				51
	SP	Air	0.25	Pressure	1	2		−7					50
		Kiln, 60 °C	0.25	Pressure	−5	−3		−13					50
		Air	0.60	Pressure	−3	−2		−10					50
		Kiln, 60 °C	0.60	Pressure	−2	1		−10					50
		Air	1.00	Pressure	0	2		−7					50
		Kiln, 60 °C	1.00	Pressure	−3	2		−16					50
		Air	2.50	Pressure	−8	−5		−19					50
		Kiln, 60 °C	2.50	Pressure	−7	−2		−21					50
	Combo	—	2.50	Pressure	−32	−21		−67			−26		41
ACC	Aspen CWM	—	—	Pressure	−20								52
	Hdwd veneer	Air	0.1–4	Pressure					−36				51
	Scotch pine	—	—	Pressure	0								53
CCA	Hdwd veneer	Air	0.1–4	Pressure					−17				51
	Slash pine	Air	2.00	Pressure	0	0							54
	Combo	—	2.50	Pressure	−32								41
CCA–A	—	—	—	Pressure	0								55
	SP	Air	0.25	Pressure	−5	−4		−13					50

Continued on next page

Table V. Continued

Chemical[a]	Species[b]	Drying	Retention[c] (lb/ft³ or % concentration)	Treatment Process	MOR	MOE	FSPL	WML	Toughness	Impact Bending	Maximum Crushing Strength	$T_\parallel$	Reference
CCA-A (Continued)		Kiln, 60 °C	0.25	Pressure	−2	1		−10					50
		Air	0.60	Pressure	−4	0		−17					50
		Kiln, 60 °C	0.60	Pressure	−10	−2		−26					50
		Air	1.00	Pressure	−9	−4		−29					50
		Kiln, 60 °C	1.00	Pressure	−12	−5		−33					50
		Air	2.50	Pressure	−19	1		−53					50
		Kiln, 60 °C	2.50	Pressure	−16	7		−53					50
		Air	1.00	Pressure	4	2	−6	23	−12		3		37
		Air	1.20	Pressure	3	−10	−2	5	−8		1		37
		Air	2.50	Pressure	−6	−10	−11	−23	−32		18		37
		Air	1.00	MSU process	1	−1	−14	12	−3		−4		37
		Air	1.20	MSU process	0	−1	−9	22	−10		1		37
		Air	2.50	MSU process	−2	1	−4	−12	−31		11		37
CCA-B	Beech	—	—	Soak 6 mos	12					−16	−8		56
		—	—	Soak 24 mos	5					−6	−8		56
		—	—	Soak 36 mos	9					−3	−8		56
		Combo —	1.0%	Soak 12 mos						−4			57
			1.0%	Soak 60 mos						−8			57
			2.0%	Soak 3 mos						−7			57
			2.0%	Soak 12 mos						−2			57
			5.0%	Soak 3 mos						−5			57
			5.0%	Soak 12 mos						−6			57
			5.0%	Soak 12 mos						−14			57
			5.0%	Soak 60 mos						−10			57
		Combo Air	0.5%	Soak	12					−7	7		56

Species	Conditioning	Retention	Process						Ref
Scotch pine	Air	2.0%	Soak	8					56
	Air	3.0%	Soak	11			−8	7	56
	Air	—	Soak				−13	7	57
	Air	—	Pressure				−13	10	58
SP	—	—	Pressure						59
	Air	0.25	Pressure	−4	0	−2	−4	−4	50
	Kiln, 60 °C	0.25	Pressure	0	−5	−2			50
	Air	0.60	Pressure	−4	−2	−2			50
	Kiln, 60 °C	0.60	Pressure	−5	−3	−15			50
	Air	0.60	Pressure	−9	2	−24			50
	Kiln, 60 °C	1.00	Pressure	−1	−2	−13			50
	Air	1.00	Pressure	−7	−3	−24			50
	Kiln, 60 °C	2.50	Pressure	7	2	−26			50
	Air	2.50	Pressure	11		−39			50
CCA–C DF plywood	—	—	Pressure	−47					52
Aspen	—	—	Pressure	−50					52
Aspen CWM	—	—	Pressure	−50		−13			52
Combo	—	1.00	Pressure	−5		−38			60
	—	3.00	Pressure	−17		−20			60
SP	—	0.60	Pressure			−41			61
	—	2.50	Pressure						61
ACA + Creo	Combo	—	Pressure	−34	−19	−48		−19	40
CCA + Creo	Combo	—	Soak	−51	−33	−53		−30	41
CFA	Combo	—	Pressure	−15		−7			57
	Combo	—	Pressure						62
	Combo	Air	0.40 Pressure					+ −15	62

Continued on next page

Table V. Continued

Chemical[a]	Species[b]	Drying	Retention[c] (lb/ft³ or % concentration)	Treatment Process	MOR	MOE	FSPL	WML	Toughness	Impact Bending	Maximum Crushing Strength	$T_\parallel$	Reference
ACF	Scotch pine	—	0.53	Pressure							+	−36	62
CZC	—	—	—	Soak									63
CKB	Combo	Air	—	Pressure									64
CC	Hdwd veneer	Air	0.1–4	Soak						−11			57
NaPCP	Beech	—	0.5–4%	Pressure	0				−5				51
PCP			0.5–4%	Soak	0						0		65
				Soak	5						0		65
Creo	Combo	Air	—	Pressure	−22	−19		−40			+	−7	62
			29.0	Pressure									41

NOTE: Values are expressed as the percent change found by the respective source.
[a] CCA = chromated copper arsenate (types A, B, C are different formulations); ACA = ammoniacal copper arsenate; ACC = acid copper chromate; CFA = chromated fluorarsenate; ACF = ammoniacal copper fluoride; CZC = chromated zinc chloride, copperized; CKB = chromated copper borate; CC = copper chromate; NaPCP = sodium pentachlorophenol; PCP = pentachlorophenol; and Creo = creosote.
[b] SP = southern pine; Combo = various species used; DF = Douglas-fir; CWM = composite wood material (flakeboard).
[c] In lb/ft³ where 1.0 lb/ft³ = 16.0 kg/m³, or % concentration.

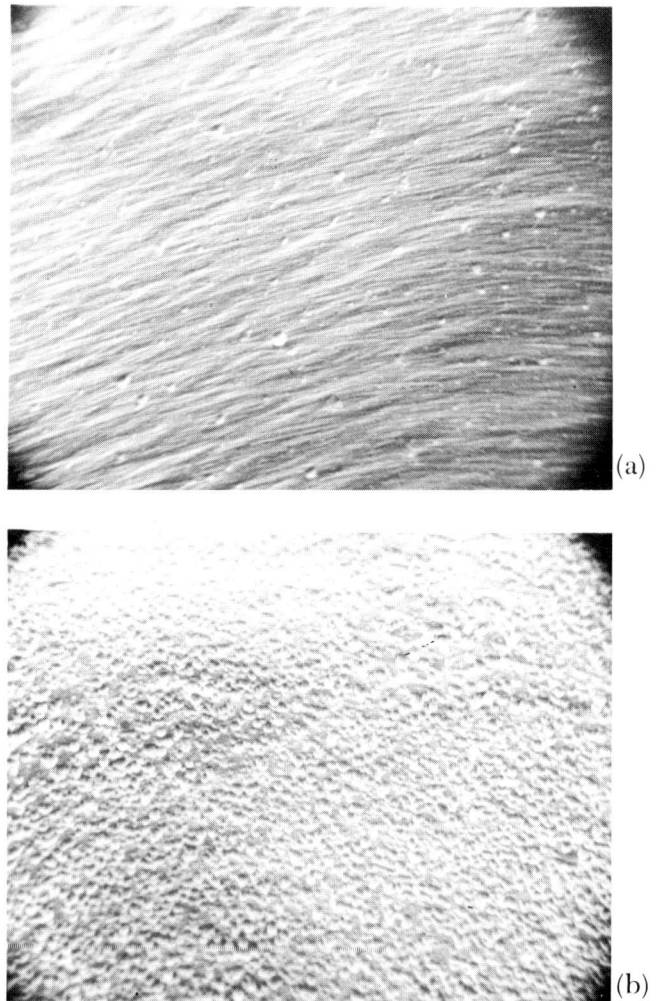

Figure 22. Untreated southern pine shows the absence of crystalline materials. (×4000) (a), and southern pine treated with the wood preservative CCA to a retention of 2.5 lb/ft³ shows the presence of crystalline material in the cell wall (b) (×4000).

Preservative formulations that contain copper and chromium salts reportedly promote afterglow in treated wood subjected to fire. Once the treated wood starts to burn or glow, the wood may continue to glow until the entire member is consumed, even when no flame is present (38, 39). This characteristic can cause serious problems in utility poles, fenceposts, and highway signs, structures that might be subjected to accidental fires or controlled ground fire which is used as an agricultural management tool.

Petroleum-based chemicals such as creosote and pentachlorophenol in oil are also used as wood preservatives. These organic preservatives are inert toward the cell wall substance and do not seem to cause any appreciable strength losses (40-42). Because of the higher treating temperatures commonly used with organic preservatives, strength losses (Table V) have been attributed to the treatment conditions (high temperature and pressure) rather than the organic preservative chemicals (43).

FIRE RETARDATION. Salts such as sodium tetraborate, diammonium phosphate, trisodium phosphate, diammonium sulfate, and salts of boric acid have long been used as fire retardants. (*See* Chapter 14.) Hygroscopicity, corrosion of fasteners, and increased acidity are also problems with these salts.

Like the preservative salts, these salts also precipitate in the cell cavity and the cell wall (Figure 23). They appear on the fiber surface of pine treated with 4.2 lb of ammonium dihydrogen phosphate ($NH_4H_2PO_4$) per cubic foot of wood.

Many fire-retardant salt treatments are very hygroscopic. Accordingly, most formulations are not recommended for use where relative humidity is over 80% (44). Recently, fire-retardant treatments have come on the market that claim to have overcome hygro-

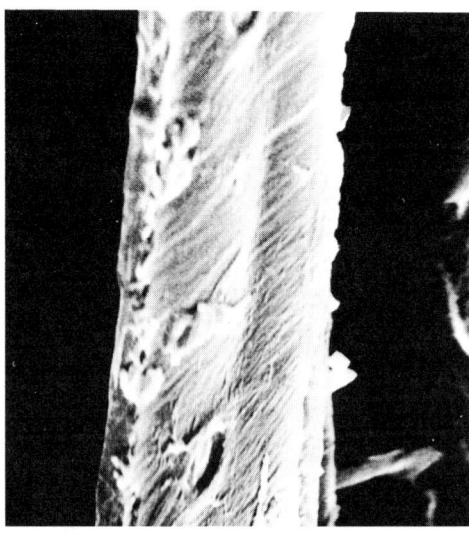

Figure 23. Southern pine fiber treated with the fire-retardant $NH_4H_2PO_4$ (× 1600).

Table VI. Effects of Fire Retardants on the Mechanical Properties of Wood

Chemical[a]	Species[b]	Drying	Retention[c] (lb/ft³)	MOR	MOE	FSPL	WML	WPL	Toughness	Impact Bending	Reference
AP	SP	Kiln, 60 °C	6.0	−19	−12		−21				46
APAS	SP	Kiln	6.0	−17	−3		−44				46
		Kiln	6.0	−16	−3		−45				46
	Red pine	Kiln, 65.5 °C	—	−29	−5	−16	−62		−25		65
	DF	Kiln		−3	−1	−5	−24		−41		65
	DF plywood	Kiln		−15	−9	−22	−6		−19		65
	Red pine	Kiln		−23	−5	−13	−55		−18		65
	DF	Kiln		−13	−1	−17	−29		−36		65
	DF plywood	Kiln		−16	−9	−23	−2		−13		65
FCAP	DF plywood	—	—	−5	−6					−8	66
FRT	Combo	—	—	−13	−5		−34				46
		Air		−10	0				−30		67
		Kiln, 65 °C		−30					−50		67
		Kiln, 90 °C		−50	−13						67
	SP	Kiln		−11	−8		−26				46
		Kiln		−17	−13		−33				46
	DF	Air		−11	−5		−55				46
		Air		−5	0						49
		Air		1	−2						49
		Kiln		−10	−8						46
		Kiln, 60 °C		−15	−3		−40				49
		Kiln, 70 °C		−14	−1						49
Minalith	DF plywood	—	—	−8	−5					−57	68
Noncom X	DF	—	—	−16	−5	−7	−32	−12			46
	SP	—	—	−14	−8	−2	−24	4			46
	SP Glulam	Kiln		−12	−2						46

Continued on next page

Table VI. Continued

Chemical[a]	Species[b]	Drying	Retention[c] (lb/ft³)	MOR	MOE	FSPL	WML	WPL	Toughness	Impact Bending	Reference
Pyrosote	SP	—	—	-10	-8	4	-28	20			46
	DF	—	—	-17	-5	-17	-32	-28			46
	DF	Air	—	-29	-12		-37				46
FRT A	DF plywood	—	5.45	-6	-6		-2			-20	68
FRT B	DF plywood	—	4.94	-11	-24		-31			-41	68
ZAB	Red pine	Kiln, 65.5 °C	—	-10	2	2	-42		-10		67
	DF	Kiln	—	-2	2	-7	-12		-28		67
	DF plywood	Kiln	—	-13	-10	-21	-8		-13		67

NOTE: Values are expressed as the percent change found by the respective source.
[a] AP = ammonium phosphate; APAS = AP + ammonium sulfate; FCAP = flour chrome arsenate phenol; FRT = unidentified fire retardant; ZAB = zinc ammonium borate.
[b] SP = southern pine; DF = Douglas-fir.
[c] Treatment was by the pressure process; 1.0 lb/ft³ = 16.0 kg/m³.

scopic problems, but because they are proprietary, little is known of their nature.

Several studies have investigated the relationships between strength properties and fire-retardant salt treatments. The National Design Specifications (45) recommend a 10% reduction in all allowable design stresses for fire-retardant-treated lumber. Table VI shows losses in mechanical properties in wood treated with fire retardants. The effects on strength are greater in kiln-dried salt-treated wood than in air-dried salt-treated wood (46–49). Like the waterborne preservatives, this is due to hydrolytic degradation of the fiber, caused by the combination of high moisture content, high temperature, and acid salts or alkalies during the post-treatment redrying process.

Summary

Over the years the strength of wood has been, for the most part, studied by physical chemists and engineers and the chemistry of wood has been studied by organic chemists and biochemists. In a materials science approach to wood research, these disciplines must work together to relate the physical properties to the chemistry of the wood material.

In this chapter an explanation is presented of certain engineering aspects that are important in understanding the mechanical properties of wood. Individual factors such as growth, environment, chemicals, and use can greatly affect the physical and mechanical properties of the wood material. A theoretical model is presented to explain the relationship between physical properties and chemistry of wood at three distinct levels: macroscopic or cellular, microscopic or cell wall, and molecular or polymeric. These three levels and their implications on material properties must be understood to relate both wood chemistry and wood engineering from a materials science standpoint. When this is accomplished, the treatment and processing of wood and wood products can be controlled to yield more desirable and uniform properties.

The theories presented are unproven. They are offered as a starting point—a point to be verified, refined, and reconsidered, a point from which, through mutual cooperation between the fields of engineering and chemistry, the chemistry of wood strength may be truly explained.

Literature Cited

1. Larson, P. G.; Cox, W. J. "Mechanics of Material"; Wiley and Sons: New York, 1938; 408 p.
2. American Society for Testing and Materials; ASTM Stand. Desig. D 143–52 (reapproved 1978); Philadelphia, 1982.

3. American Society for Testing and Materials; ASTM Stand. Desig. D 198-76; Philadelphia, 1982.
4. American Society for Testing and Materials; ASTM Stand. Desig. D. 245-81; Philadelphia, 1982.
5. American Society for Testing and Materials; ASTM Stand. Desig. D 2555-81; Philadelphia, 1982.
6. American Society for Testing and Materials; ASTM Stand. Desig. D 2915-74 (reapproved 1980); Philadelphia, 1982.
7. U.S. Department of Agriculture, Forest Service. Wood handbook. USDA Agric. Handb. 72, rev.; Washington, D.C., 1974.
8. Wilson, T. R. C. USDA Tech. Bull. No. 282; Washington, D.C., 1932; 88 p.
9. U.S. Department of Agriculture, Forest Service. Wood handbook: Wood as an engineering material. USDA Agric. Handb. 72; Washington, D.C., 1955; 528 p.
10. Gerhards, C. C. *Wood Fiber* **1982**, *14*(1), 4-36.
11. Rowell, R. M. in "How the Environment Affects Lumbar Design;" Lyon, D. E. and Galligan, W. L.; Eds: *U.S., For. Serv., For. Prod. Lab. Rep.* **1980**.
12. Gerhards, C. C. *U.S., For. Serv. Res. Pap., FPL* 283 **1977**.
13. Wood, L. W. *U.S. For. Prod. Lab. Rep.* No. R-1916 **1951**.
14. Wangaard, F. F. "The Mechanical Properties of Wood"; John Wiley & Sons: New York, 1950; 377 p.
15. Freudenberg, K. *J. Chem. Ed.* **1932**, *9*(7), 1171-80.
16. Mark, R. E. "Cell Wall Mechanics of Tracheids"; Yale Press: New Haven, CT, 1967; 310 p.
17. Stamm, A. J. "Wood and Cellulose Science"; The Ronald Press Co.: New York, 1964.
18. Thomas, R. J. in "Wood: Its Structure and Properties"; Wangaard, F. F., Ed.; Penn State Univ. Press: University Park, PA, 1981; p. 101-46.
19. Lagergren, S.; Rydholm, S.; Stockman, L. *Sven. Papperstidn.* **1957**, *60*, 632-44.
20. Prentice, I. W. "Annual Report. East Malling Res. Stn." Kent, U. K., 1949; p. 122-25.
21. Ifju, G. *For. Prod. J.* **1964**, *14*(8), 366-72.
22. Kollmann, F.; Côté, W. A., Jr. "Principles of Wood Science and Technology. Vol. 1. Solid Wood"; Springer-Verlag: Berlin, 1968.
23. Alliott, E. A. *J. Soc. Chem. Ind.* **1926**, *45*, 463-66.
24. Koenigs, J. W. *Wood Fiber* **1974**, *6*(1), 66-80.
25. Erickson, H. D.; Rees, L. W. *J. Agric. Res.* **1940**, *60*, 593.
26. Siau, J. F. *Wood Sci.* **1969**, *1*(4), 250-53.
27. Kalnins, M. A. *U.S., For. Serv. Res. Pap., FPL* 57 **1966**.
28. Feist, W. C.; Hajny, G. J.; Springer, E. L. *Tappi* **1973**, *56*(8), 91-95.
29. MacLean, J. D. *U.S., For. Serv., FPL Mimeo* No. 1471 **1945**.
30. MacLean, J. D. *Proc.—Annu. Meet. Am. Wood-Preserv. Assoc.* **1953**, *49*, 88-112.
31. Millett, M. A.; Gerhards, C. C. *Wood Sci.* **1972**, *4*(4), 193-201.
32. Stamm, A. J.; Burr, A. K.; Kline, A. A. *Ind. Eng. Chem.* **1946**, *38*, 630-37.
33. Cowling, E. B. *U.S., For. Serv. Tech. Bull.* No. 1258 **1961**, p. 50.
34. Kennedy, R. W. *For. Prod. J.* **1958**, *8*, 308-14.
35. Betts, H. S.; Newlin, J. A. *USDA Bull.* No. 286. Washington, D.C., 1915; 15 p.
36. Hatt, W. K. *USDA For. Serv. Circ.* 39. Washington, D.C., 1906; 31 p.
37. Wood, M. W.; Kelso, W. L., Jr.; Barnes, H. M.; Parikh, S. *Proc.—Annu. Meet. Am. Wood-Preserv. Assoc.* **1980**, *76*, 22-37.
38. Dale, F. A. *For. Prod. Newsl.* **1966**, *328*, 1-4.

39. McCarthy, W. G.; Seaman, E. W.; DaCosta, B.; Bezemer, L. D. *J. Inst. Wood Sci.* **1972**, *6*(1), 24–31.
40. Koukal, M.; Bednarcik, V.; Medricka, S. *Drev. Vysk.* **1960**, *5*(1), 43–60.
41. U.S. Department of Defense, Department of the Navy, Civil Engineering Laboratory. CEL Tech. Data Sheet 79–07. USN, CEL: Port Hueneme, CA, 1979; 4 p.
42. Gillwald, W. *Holztechnologie* **1961**, *2*, 4–16.
43. MacLean, J. D. *USDA Agric. Handbook* No. 40. Washington, D.C., 1951; 160 p.
44. Holmes, C. A. in "Wood Technology: Chemical Aspects," Goldstein, I. S., Ed.; ACS SYMPOSIUM SERIES No. 43, ACS: Washington, D.C., 1977; p. 82–106.
45. National Forest Products Association. "National Design Specifications for Stress-Grade Lumber and Its Fastenings"; NFPA: Washington, D.C., 1982, 81 p.
46. Gerhards, C. C. *U.S., For. Serv. Res. Pap., FPL* **145, 1970**.
47. Brazier, J. D.; Laidlaw, R. A. *BRE Infor. Supp. IS 13/74* Princes Risborough Lab.: Aylesbury, Bucks., U.K., 1974; 3 p.
48. Adams, E. H.; Moore, G. L.; Brazier, J. D. *BRE Infor. Paper IP 24/79* Princes Risborough Lab.: Aylesbury, Bucks., U.K., 1979; 4 p.
49. Johnson, J. W. *Oregon State Univ. Rep.* T–23. Forest Res. Lab., School of Forestry, Oregon State Univ.: Corvallis, OR, 1967; 12 p.
50. Bendtsen, B. A.; Gjovik, L. R.; Verrill, S. P. *U.S., For. Serv. Res. Pap., FPL* **1983**.
51. Thompson, W. S. *For. Prod. J.* **1964**, *14*(3), 124–28.
52. Adams, R. D.; Mateer, S.; Krueger, G. P.; Lund, A. E.; Nicholas, D. D. *Res. Rep. EL-1745* Michigan Technological Univ.: Houghton, MI, 1981.
53. Terentjev, V., Jr. *Derevoobrab. Promst.* **1971**, *21*(8), 15–16.
54. Siemon, G. R. Res. Note No. 29. Dep. For., Univ. of Queensland: Brisbane, Queensland, Australia, 1979; 6 p.
55. Eggleston, R. C. *For. Prod. J.* **1952**, *2*(1), 3–24.
56. Burmester, A. *Holz Roh- Werkst.* **1970**, *28*(12), 478–85.
57. Pechman, H.; Aufsess, H. *Holz Roh- Werkst.* **1968**, *26*(12), 454–62.
58. Burmester, A.; Becker, G. *Holz Roh- Werkst.* **1963**, *21*(10), 393–409.
59. Hesp, T.; Watson, R. W. *Wood* **1964**, *29*(6), 50–53.
60. U.S. Department of Defense, Department of the Navy, Civil Engineering Laboratory; CEL Tech. Data Sheet 74–07. USN, CEL: Port Hueneme, CA, 1974; 2 p.
61. Winandy, J. E.; Bendtsen, B. A.; Boone, R. S. *For. Prod. J.* **1983**, *33*(6).
62. Gillwald, W. Translation by Pronin, D. *Holztechnologie* **1961**, *2*, 4–16.
63. Lutomski, K. *Zesz. Probl. Postepow Nauk Rolh.* **1976**, *178*, 143–52.
64. Zarudnaya, G. I.; Katayev, O. A.; Filipov, A. E. *Lesn. Zh. (Archangel, USSR)* **1980**, *1*, 53–55.
65. Jessome, A. P. Rep. No. 193. For. Prod. Res. Branch, Dep. Forestry: Ottawa, Ontario, Canada, 1962; 12 p.
66. Countryman, D. Douglas-Fir Plywood Assoc. Lab. Pap. No. 75. Douglas-Fir Plywood Assoc.: Tacoma, WA, 1957.
67. Brazier, J. D.; Laidlaw, R. A. *BRE–15 13/74* Princes Risborough Lab.: Aylesbury, Bucks., U.K., 1974; 3 p.
68. King, E. G.; Matteson, D. A. *Lab. Rep. No. 90* Douglas-Fir Plywood Assoc.: Tacoma, WA, 1961; 14 p.

RECEIVED for review May 17, 1983. ACCEPTED July 7, 1983.

6
Wood–Polymer Materials

JOHN A. MEYER

State University of New York, College of Environmental Science and Forestry, Syracuse, NY 13210

> *Treatment of solid wood over the years for increased utility included many chemical systems that affected the cell wall and filled the void spaces in the wood. Some of these treatments found commercial applications, while some remain laboratory curiosities. A brief description of the earlier treatments is given for heat-stabilized wood, phenol–formaldehyde-treated veneers, bulking of the cell wall with polyethylene glycol, ozone gas-phase treatment, ammonia liquid- and gas-phase treatment, and β- and γ-radiation. Many of these treatments led to commercial products, such as Staybwood, Staypak, Impreg, and Compreg. This chapter is concerned primarily with wood–polymer composites using vinyl monomers. Generally, wood–polymers imply bulk polymerization of a vinyl-type monomer in the void spaces of solid wood. This bulk polymerization takes place in the vessels, capillaries, ray cells, etc., but not in the cell wall or middle lamellas. The monomer is introduced into the solid wood by a vacuum process. The wood–monomer is then converted into the solid polymer by γ-radiation or a heat-sensitive catalyst dissolved in the monomer. The wood–polymer is fabricated into the final product.*

Background and History

Wood, as a renewable resource, has provided persons with tools, weapons, and shelter since the beginning of our coexistence on this planet. During the millennium of our development we learned how to make wood harder and stronger by drying and heat-tempering our wooden tools and weapons. As our knowledge of the world we lived in increased, we attempted other modifications to better fit our increased requirements. Over the years tars, pitches, creosote, resins, and salts have been used to coat wood or to fill its porous structure.

Treatment of wood to modify its properties has two objectives: dimensional stabilization due to moisture content and improvement in physical and mechanical characteristics. During the period from 1930 to 1960 research was carried out and many attempts were made to alter wood properties by the application of heat, pressure for surface densification, chemical addition to bulk the cell walls, impregnation with polymers, alteration of the chemical composition of wood components using liquid and gaseous reagents, irradiation of wood with γ- and β-radiation, as well as other techniques. Two excellent reviews of these processes have been published (1, 2). Each of the major systems is outlined in the following paragraphs. A similar outline has been published elsewhere (3, 4).

Heat-Stabilized Wood (Staybwood). In this process kiln-dried lumber is held at 150–300 °C for various lengths of time ranging from minutes to hours. This process gives up to 60% antishrink efficiency (ASE). This type of dimensional stabilization is accompanied by a serious loss in strength, toughness, and abrasion resistance (5, 6). Heating green wood in water also imparts some dimensional stabilization to wood (7). There is no commercial application for Staybwood.

Heat-Stabilized Compressed Wood (Staypak). Pressures of 400–4000 psi are applied to the wood after it has been heated. Both heat and pressure plasticize wood. At 160 °C and 12% moisture content, the maximum plastic yield per increment of pressure occurs at 1100 psi. Pressures of 1500–2500 psi are required to yield a specific gravity of 1.3. Highly densified wood must be cooled in the press. Some strength properties, such as impact strength and hardness, are increased in direct proportion to the density. Staypak finds limited application for silverware handles and desk legs (6).

Phenol–Formaldehyde Wood Composite (Impreg). In this process, kiln-dried veneer is impregnated with a water solution of phenol–formaldehyde prepolymer (trimers and tetramers) with a molecular size small enough that it can be carried into the cell wall by the water. This system has the advantage causing the cell wall to swell up to 25% beyond the swelling in water, so that after curing the composite has a final volume almost equal to that of water-swollen wood (6, 8). After impregnation using vacuum and pressure, the water is removed by drying and the prepolymer is polymerized and cross-linked to a high molecular weight by applying heat in a press or dry kiln. Green veneers can be treated directly. At a loading of 35% by weight of the dry wood there is no visible deposition of the polymer in the void spaces of the wood, and the composite has an ASE of 70–75%. The high ASE is an indication that most of the polymer is in the cell wall. Impreg is in commercial use, primarily for die molds and patterns in the automobile industry. The desired thickness is built from layers of impregnated veneer sheets; the assembly is then heat cured.

Phenol–Formaldehyde Compressed Wood Composite (Compreg). This composite is similar to Impreg in that the veneer, green or dried, is soaked in a water or alcohol solution of low molecular weight phenol–formaldehyde prepolymers (trimers and tetramers), and the prepolymer is carried into the cell wall by the water. The water is removed at a temperature low enough to prevent curing of the prepolymer. A stack of the treated veneer sheets is placed in a press with heated platens and, as the composite material is heated, pressure up to 1000 psi is applied to compress the wood and collapse the cell structure (6, 8, 9). The density of the final, cured (polymerized and cross-linked) composite approaches that of the cell wall (solid cell wall substance) with a specific gravity from 1.3 to 1.4. Incorporation of the polymer into the cell wall prevents springback in the presence of high relative humidities and imparts high dimensional stability. A retention in the wood of about 30% polymer gives optimum stability and a high ASE (90–95%). Compreg will absorb less than 1% moisture when immersed in water for 24 h. Strength (particularly in compression), hardness, and abrasion resistance are all increased, and the composite is quite resistant to decay and termites. Impact strength, however, is adversely affected by the process, and the composite is much more brittle.

Many specialty items, knife and cutlery handles in particular, are made from Compreg. This composite can be machined with great precision and the surface finish can be renewed by sanding and buffing because the polymer is throughout the wood. Compreg is also used for electrical insulators requiring high tensile strength.

Polyethylene Glycol Treatment. The polymers of dihydric alcohols are polyethers with an oxygen atom separating the hydrocarbon groups and reactive hydroxyl groups on the ends only; molecular weights up to 6000 are highly soluble in water (6, 10). Because of the low vapor pressure of polyethylene glycol (PEG) **1**, it remains in the cell walls when the wood is dried; this bulking action prevents the wood from shrinking. As water evaporates and increases the concentration of PEG in the solution, the rate of diffusion into the cell wall increases (6). This is evident by the swelling of the treated wood as it dries. Green cross-sectional disks of southern pine sapwood, 3.18 cm thick treated with PEG have bulked sufficiently to prevent checking during air-drying. Treatment consists of an overnight, or longer, soak in a 30% solution of PEG-1000 or two surface coats of molten PEG-1000 a day apart. For thicker disks, soak time should increase in proportion to the square of the thickness. Heartwood requires more soaking time or more coats than sapwood (11).

$$HO-CH_2CH_2-O-CH_2CH_2-(O-CH_2CH_2)_n-O-CH_2CH_2-OH$$
1

Green loblolly pine 12.7 cm long, 7.6 cm wide, and 0.95 cm thick can be dried in 10–40 min by immersion in molten PEG-1000 at 135

°C. (Drying time in molten PEG is shorter than in air.) Sufficient PEG diffuses into the samples during the immersion to give about a 35% ASE (*12*).

The PEG-bulked wood feels moist when relative humidity is above 70% because of its hygroscopicity, but certain polyurethane finishes tend to reduce this clammy feeling. The treated wood is highly stable toward changes in humidity, but in water, the PEG is leached out of the wood with time. Treatment causes a slight loss of abrasion resistance and bending strength. Toughness is generally unaffected at a 45% PEG loading with the ASE in the region of 80%. PEG is used where wood must have dimensional stability to prevent cracking and checking. Valuable art carvings have been preserved in this manner, and PEG treatment has permitted marine archeologists to preserve water-logged wooden ships brought up from lakes and oceans.

Ozone Gas-Phase Treatment. This process holds little promise for solid wood treatment because both the cellulose and lignin are degraded. The effect of ozone on small wood samples, ground wood, and chips is described elsewhere (*13, 14*).

Ammonium Liquid- and Vapor-Phase Treatment. Application of this ammonium treating process shows promise as an alternative to steam bending. Some of the disadvantages of steam bending include recovery of the original shape if exposed to high relative humidity, the necessity of holding the part in its final shape until the wood is set by reducing the moisture content to ambient conditions, and high breakage during the bending process. The use of anhydrous ammonia, in the liquid or vapor phase, permits the bending of many woods in much more complicated shapes than steam bending (*15–17*).

Complicated shapes can be formed without springback because the crystalline structure of the cellulose is changed and the hydrogen bonds reformed upon the evaporation of the ammonia. When ammonia-treated bent wood is exposed to liquid water it will change shape to some extent, but upon drying the treated wood will return to its new shape. X-Ray crystallography shows that the original cellulose I pattern changes to cellulose II upon treatment of cellulose with ammonia. The evacuated wood is exposed to ammonia vapor at a pressure of 150 psi at room temperature. The time of exposure is determined by the thickness of the wood and the moisture content; this time can range from minutes to hours. Bending of 180° has been accomplished with 3.8 cm thick wood.

Softwoods require a longer exposure to ammonia vapor than do hardwoods for the same degree of bending. After the ammonia-treated wood is bent into its final shape only mild restraint is required for a short period of time until the treated wood is set—that is, until some of the ammonia in the wood has evaporated.

In most cases the color of the ammonia-treated wood is darker and approaches a walnut color. Some color streaking is a drawback with certain wood species. In addition, the ammonia vapor reacts with some of the wood components to produce a liquid that drains from the wood upon release of the ammonia pressure. This process is not in commercial use at the present time.

Vinyl Composites

Radiation Catalysts. Because wood is essentially a mixture of high molecular weight polymers, high energy radiation exposure [in million electron volts (MeV)] will depolymerize the polymers by creating free radicals along the C–C backbone. If two free radicals are created on separate chains in close proximity, cross-linking will take place. Other types of reactions will take place when the free radical is created near a reactive or functional group. When the free radical is on a tertiary carbon, disproportionation will occur with chain scission. Surprisingly lignin is more resistant to γ-radiation than cellulose, but lignin is also more susceptible to natural UV rays when exposed to the atmosphere. (UV rays have about 20 eV per quanta compared to about 1.25 MeV for γ-ray quantas from the cobalt-60 source. Some slight increase in mechanical properties and a decrease in hygroscopicity were noted with radiation exposure up to 1,000,000 rd (rads) (*18*). Exposures above this level degraded the cellulose and impaired the mechanical properties; the wood was completely soluble above 3×10^8 rd (*20, 21*). The ultrastructure of γ-irradiated Douglas-fir and yellow poplar was studied by deLhoneux (*22*). The results were similar for both woods over the range of cobalt-60 γ-irradiation (1.25 MeV) of 32–1500 Mrd. The scanning electron microscope revealed that the middle lamella and the primary wall were the most resistant to the γ-radiation; the secondary wall depolymerized rapidly with low exposure. At 1500 Mrd the wood was completely soluble.

During the early 1960s a new class of chemicals containing one or more double bonds was used to treat wood—vinyl-type monomers that could be polymerized into the solid polymer by means of free radicals (*23*). This vinyl monomer polymerization was an improvement over the condensation polymerization reaction because the free radical catalyst was neither acidic nor basic, nor does the reaction leave behind a reaction product, such as water, that must be removed from the final composite. The acid and base catalysts used with the other treatments degrade the cellulose chain and cause a brittleness in the composite. Vinyl polymers have a large range of properties from soft rubber to hard brittle solids depending upon the groups attached to the C–C backbone. A few examples of vinyl monomers are shown here:

$$CH_2=CH_2 \quad \underset{\underset{CH_3}{|}}{CH_2=CH} \quad \underset{\underset{Cl}{|}}{CH_2=CH} \quad \underset{\underset{CN}{|}}{CH_2=CH} \quad \underset{\underset{O=C-CH_3}{|}}{CH_2=C-CH_3}$$

ethylene   propylene 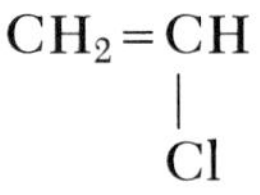   vinyl chloride 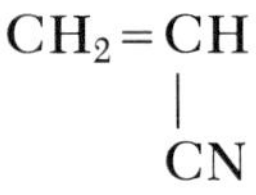   acrylonitrile 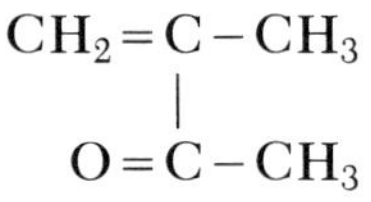   methyl methacrylate

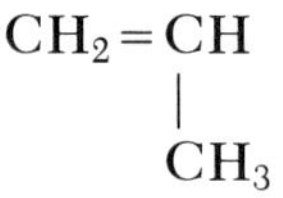

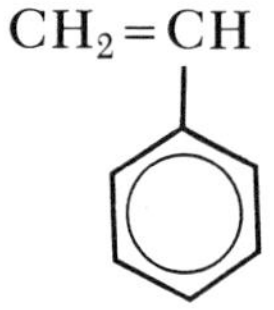

styrene

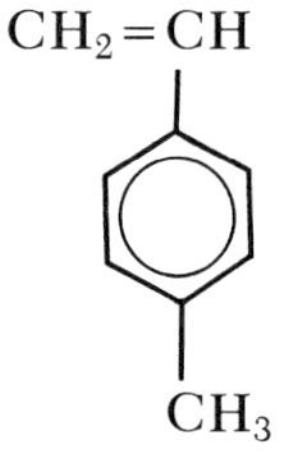

vinyltoluene

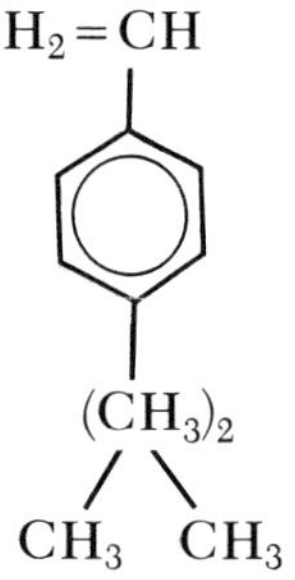

tert-butylstyrene

Most vinyl monomers are nonpolar; consequently, there is little if any interaction with the hydroxyl groups attached to the cellulose molecule. In general, vinyl polymers simply bulk the wood structure by filling the capillaries, vessels, and other voids in the wood structure.

The U.S. Atomic Energy Commission, in the early 1960s, sponsored research that used γ-radiation to develop wood–polymers. The γ-radiation generated free radicals in the wood–monomer, which in turn initiated the polymerization of the vinyl monomer (24). This support was expanded to other organizations, such as Lockheed–Georgia who supplied wood–polymer samples for industrial evaluation, North Carolina State University (25, 26) for the evaluation of the wood–polymer physical properties, and Arthur D. Little Company for economic evaluation of the wood–polymers. The first paper on the catalyst-heat process for making wood–polymers was presented at the 1965 annual meeting of the Forest Products Research Society in New York City (27). During the past 20 years the industrial development has been slow but steady for both γ-radiation and catalyst-heat processes.

In general, the free radicals used for the polymerization reaction come from two sources: temperature-sensitive catalysts and cobalt-

60 γ-radiation. In each process a free radical is generated from one of these sources. After free-radical generation the vinyl polymerization mechanism is the same. Each process for generating free radicals has its own peculiarities.

When γ-radiation is used as a source of free radicals many complications arise immediately, the least of which is the chemistry of the process. The use of radiation mandates compliance with government regulations and arouses the concern of environmentalists (28).

Safety requirements must be satisfied before a cobalt-60 source can be installed and licensed. Radiation-trained personnel must be on the staff before a license can be issued. At least 500,000–1,000,000 Ci (curies) of cobalt-60 are required in a production source for making wood–polymers, and at $1.00 or more per curie, a considerable investment must be made before production can begin. Besides cost considerations, the cobalt-60 radiation process does have some distinct advantages in making wood–polymer composites. Because the monomer is not catalyzed it can be stored at ambient conditions as long as the proper amount of inhibitor is maintained. The rate of free radical generation is constant for a given amount of cobalt-60 and does not increase with temperature.

When γ-radiation passes through a material such as wood or a vinyl monomer it leaves behind a series of ions and excited states as the energy of the γ-ray is absorbed through photoelectric, Compton, and pair production collisions (see Figure 1). (Cobalt-60 produces two γ-rays of 1.17 and 1.33 MeV. Approximately 30 eV is required to rupture a covalent bond and to cause ionization.) The ions and excited states generated in the absorbing material immediately rearrange to form free radicals, which in turn initiate the polymerization process.

$$(\text{Monomer})^* \rightarrow R^{\cdot} + H^{\cdot}$$
Excited State   Free Radicals

The free radicals usually consist of H· and the radical monomer. Once the free radical is generated, the polymerization reaction is the same as that of a normal free-radical-catalyzed, vinyl monomer bulk polymerization (27).

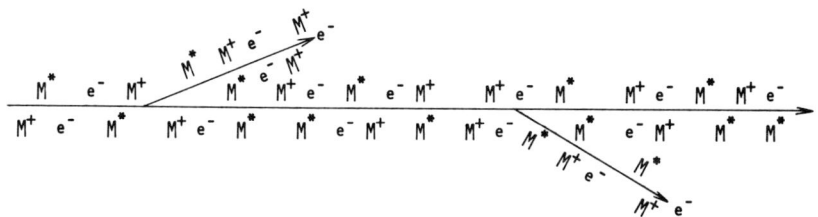

Figure 1. Ionized and excited molecules along the path of a γ-ray.

The initiation step can be represented in general by
$$R\cdot + M \text{ (monomer)} \to R-M\cdot$$
and the propagation step by
$$R-(M)_n-M\cdot + M \to R-(M)_{n+1}-M\cdot$$
The termination step recombines the growing radicals
$$R-(M)_n-M\cdot + R-(M)_n-M\cdot \to R-(M)_n-M-M-(M)_n-R$$
Because the wood's cell wall structure is not swollen by the vinyl monomer, there is little opportunity for the monomer to reach the free-radical sites, generated by the γ-radiation on the cellulose, to form a vinyl polymer branch. From this short discussion, it is reasonable to conjecture that there should be little if any difference in the physical properties of catalyst–heat initiated or γ-initiated in situ polymerization of vinyl monomers in wood.

Chemical Catalysts. Vazo or 2,2′-azobisisobutyronitrile catalyst (**2**) is preferred over peroxide catalysts because of its low decomposition temperature and its nonoxidizing nature. Vazo will not bleach dyes dissolved in the monomer during polymerization.

$$\begin{array}{c} CH_3 \quad\quad CH_3 \\ | \quad\quad\quad | \\ CH_3-C-N=N-C-CH_3 \\ | \quad\quad\quad | \\ CN \quad\quad CN \end{array} \to 2\; \begin{array}{c} CH_3 \\ | \\ CH_3-C\cdot \\ | \\ CN \end{array} + N_2$$

This first-order reaction is independent of the concentration of vazo and independent of the type of monomer (29).

The rapid decomposition of vazo catalyst with an increase in temperature (Table I) can be used to advantage in the bulk vinyl polymerizations in wood. A moderate temperature of 60 °C can be

Table I. Half-Life of Vazo Catalyst vs. Temperature

Temperature (°C)	$T_{1/2}$ (min)
0	4×10^7
7	1×10^7
18	1×10^6
30	1×10^5
46	1×10^4
70	270
100	5.5

(Reproduced from Ref. 30. Copyright 1977, American Chemical Society.)

used to initiate the reaction, and, because the half-life is more than 40,000,000 min or about 20 years at 0 °C, the catalyzed monomer can be stored safely for months. Catalyzed monomers have been stored for over a year at 5 °C. The nitrogen released during the vazo catalyst decomposition is normally soluble in the monomer–polymer solution. At high temperatures the nitrogen forms gas bubbles in the highly viscous monomer–polymer solution and the final wood–polymer will contain void spaces. In the wood–polymer this is of little consequence because the methyl methacrylate monomer (MMA) shrinks about 25% during the polymerization to create additional void spaces in the solid polymer. The cost of vazo catalyst is in the range of \$1–\$10 a pound depending upon the amount ordered. Theoretically, 1 g will produce 7.4×10^{21} free radicals and at \$10/lb; this is 3.3×10^{23} free radicals per dollar, or about \$0.02/g.

Impregnation Process. In both processes the first step in the impregnation of wood is carried out by evacuating the air from the wood vessels and cell lumens (27). Any type of mechanical vacuum pump is adequate if it can reduce the pressure in the apparatus to 750 mm of mercury or less. Some industrial producers only reduce the pressure to about 711 mm of mercury. Experience has shown that the air in the cellular structure of most woods is removed as fast as the pressure in the evacuation vessel is reduced. Pumping for 30 min at 1 mm pressure is sufficient to remove the air. The vacuum pump is isolated from the system at this point.

The catalyzed monomer containing cross-linkers, and on occasion dyes, is introduced into the evacuated chamber through a reservoir at atmospheric pressure. (*See* Figure 2, which illustrates the components necessary for making wood–polymers on both a laboratory and industrial scale.) The wood must be weighted so that it does not float in the monomer solution. In the radiation process, the catalyst is omitted from the monomer. A surge tank 10 times the volume of the treating vessel is included in the system to allow the air dissolved in the monomer to expand without greatly changing the pressure in the impregnation vessel. Alternatively, the system can be pumped as the monomer is admitted into the evacuated vessel. With this procedure much monomer is lost due to the high vapor pressure of MMA (40 mm at room temperature). After the wood is covered with monomer solution, atmospheric pressure is regained, or, in the case of radiation process, dry nitrogen is admitted into the evacuated vessel. Immediately the monomer solution begins to flow into the evacuated wood structure to fill the void spaces. Care must be taken to maintain enough monomer solution above the wood so that air is not readmitted to the cell structure.

The soaking period like the evacuation period depends upon the

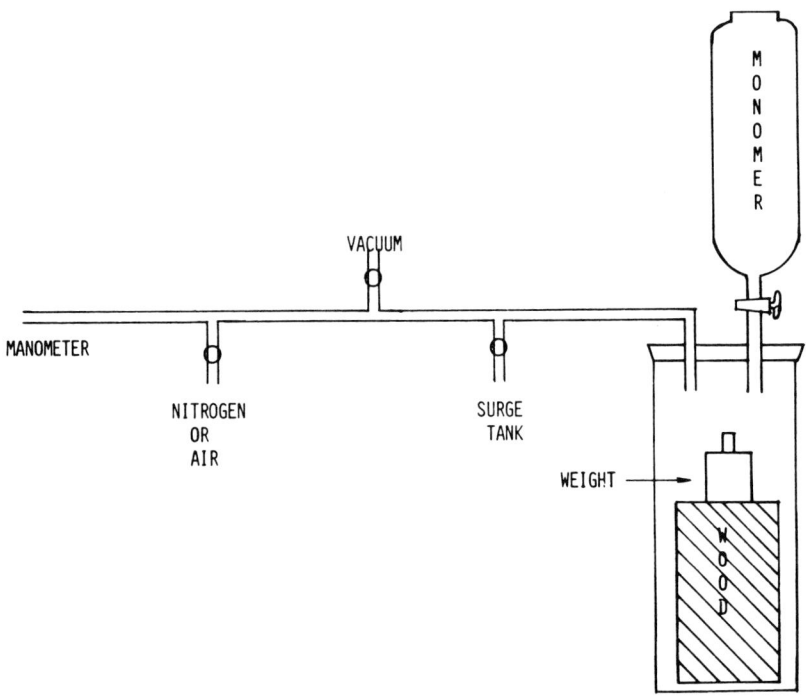

Figure 2. Apparatus used to impregnate wood.

structure of the wood: maple, birch, and other open-celled woods fill in about 30 min, and other woods require longer periods of time. A block of 7.6 × 7.6 × 17.8 cm maple absorbed 300 mL of monomer in less than 10 min.

After the monomer impregnation is complete the wood–monomer is removed and placed in an explosion-proof oven, or the cobalt-60 source for curing. On a laboratory scale or small production unit the wood–monomer is wrapped in aluminum foil before placing in the curing oven at 60 °C. In larger production operations the wood–monomer is placed directly into the curing oven, usually in the basket that held the wood during impregnation. In the radiation-cured procedure the thin metal can, in which the wood was impregnated, is flushed with nitrogen and is lowered into a water pool next to the cobalt-60 source. With high vapor pressure monomers, the wood surface is depleted to some extent by evaporation, but this depleted area is usually removed by machining to expose the wood–polymer surface. As mentioned previously, MMA has a vapor pressure of 40 mm at room temperature; *tert*-butylstyrene has a vapor pressure of only 0.8 mm at room temperature.

Monomers. Many different vinyl monomers (32) have been

used to make wood–polymers during the past 10 years, but MMA appears to be the preferred monomer for both the catalyst–heat and radiation processes. In fact, MMA is the only monomer that can be economically polymerized with γ-radiation. All types of liquid vinyl monomers can be polymerized with vazo or peroxide catalysts. In many countries styrene and styrene–MMA mixtures are used with vazo or peroxide catalysts.

All vinyl monomers contain inhibitors to prevent premature polymerization during transportation and storage. If these inhibitors, such as 1,4-benzenediol, monomethyl ether of 1,4-benzenediol, *tert*-butyl-1,2-benzenediol, and 2,4-dimethyl-6-*tert*-butylphenol, are not removed before use the catalyst or radiation must generate enough free radicals to use up the inhibitor before polymerization will begin. This induction period depends upon the amount and type of inhibitor present. In the case of radiation, the inhibitor must be kept to a minimum for efficient use of the γ-rays. The production of commercial polymethyl methacrylate rod or sheet stock, sold as Lucite or Plexiglas, requires 0.01% vazo catalyst with the inhibitor removed. With 50 ppm of 2,4-dimethyl-6-*tert*-butylphenol in MMA in wood, 0.25% vazo catalyst is required to obtain complete polymerization. Wood contains natural inhibitors, which is the reason for the high vazo content. Again, the amount of natural inhibition will depend upon the species of wood. Monomers extract the soluble fractions from the wood structure, and, with repeated use, the extractive content builds up in the monomer. Therefore, excessive foaming is produced under vacuum and the polymerization reaction is completely inhibited, and additional catalyst must be added.

The polymerization of vinyl monomers is an exothermic reaction in which a considerable amount of heat is released, about 75.3 kJ/mol. In both the catalyst–heat and γ-radiation processes the heat released during polymerization is the same for a given amount of monomer. The rate at which the heat is released is controlled by the rate at which the free-radical initiating species is supplied and at which the chains are growing. The vazo and peroxide catalysts are temperature dependent; consequently, the rate of decomposition, and thus the supply of free radicals, increases rapidly with an increase in temperature. Because wood is an insulator due to its cellular structure, heat flow into and out of the wood–monomer–polymer material is restricted. In the catalyst–heat process, heat must be introduced into the wood–monomer to start the polymerization, but once the exothermic reaction begins the heat flow is reversed. The temperature of the wood–monomer–polymer composite increases rapidly, because the heat flow out of the wood is much slower than the heat generation. Figure 3 illustrates the heat-transfer process (*31*).

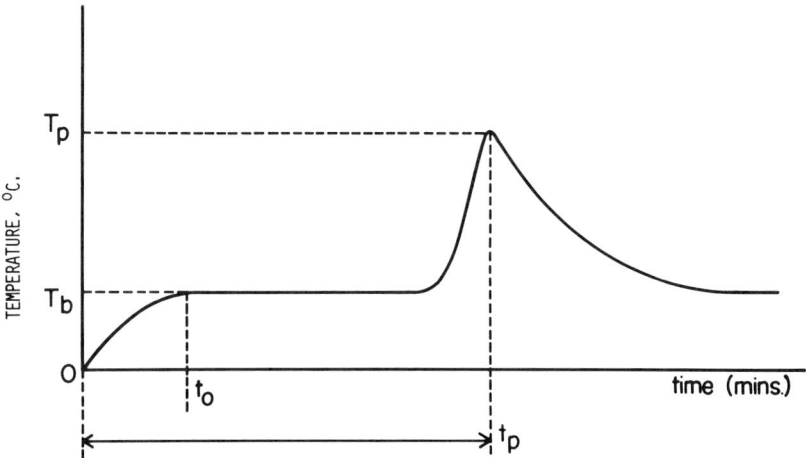

Figure 3. Idealized temperature–time exothermic curve.

The time to t_0 is the time for the wood–monomer mass to reach oven or curing temperature at T_b. During the period of constant temperature (the induction period), the inhibitor is being removed by the reaction with the free radicals. When the inhibitors are eliminated from the monomer and wood, the temperature rises to a maximum that corresponds to the peak of the exothermic polymerization reaction. Polymerization continues to completion although at a decreased rate, and the temperature returns to that of the curing chamber. The time to the peak temperature depends upon the amount of catalyst present, the type of monomer, the type of cross-linker, and the ratio of the mass of monomer to that of the wood. The wood mass acts as a heat sink.

The Gel Effect. The gel or Trommsdorff effect (33) is the striking autoacceleration of the vinyl polymerization reaction as the viscosity of the monomer–polymer solution increases. Chain termination involving the recombination of two free radicals becomes diffusion controlled; this results in a decrease in the rate of termination. The concentration of active free radicals therefore increases proportionally. To sum up the gel effect: the rate of vazo catalyst initiation increases with temperature, the rate of propagation or polymerization increases with the viscosity, and the rate of termination of the growing polymer chains decreases with the viscosity. The gel effect also results in an increase in the molecular weight of linear polymers, but this has no practical significance when cross-linking is part of the reaction.

As mentioned previously, a given γ-radiation-source geometry will supply free radicals at a constant rate for vinyl monomer polymerization. An increase in temperature would only affect the propagation and termination rates. The exothermic heat from the vinyl monomer polymerization is still released in the wood–monomer composite, but the temperature is much lower because of the slow rate of initiation. Complete radiation curing of wood–monomer composites usually requires 8 to 10 h depending upon the radiation source geometry; the vazo initiated reaction is over in 30 to 40 minutes. Thus, all the catalytic heat of polymerization of a given monomer mass is released in one-sixteenth of the time it is in the radiation process. Because the wood–monomer material in a thin metal can is immersed into a water pool for irradiation, the cooling by the water radiation shield also assists in lowering the temperature. Additional heat is added to the wood–monomer–polymer composite by the absorption of the γ-rays by the wood, although this heat is small compared to the exothermic heat from the polymerization.

When the heat of polymerization is released quickly in a wood–monomer composite the high temperature increases the vapor pressure of the moisture in the cell walls and distills the moisture out of the wood. The change in volume of the cell wall causes changes in dimensions which are manifested by shrinkage and distortion of the original wood shape. Wood–polymer composites cured by the catalyst–heat process must be machined to the final shape after treatment. Conversely, because the heat of polymerization by γ-rays is released over a longer period of time, the temperature of the wood–polymer remains low and not as much cell wall moisture is driven off. Therefore, the amount of distortion and dimensional change is somewhat less (30, 31).

Vazo Catalyst Effect on Polymerization. Figure 4 illustrates the temperature–time curves for the curing of basswood samples impregnated with MMA containing various concentrations of vazo catalyst. The data for each polymerization have been compiled, are presented in Table II, and are plotted in Figure 5 as the percent vazo catalyst per weight of monomer vs. the time to the exothermic peak (t_p). The data show that varying the amount of catalyst has a definite effect on the time to the exothermic peak, and on the exothermic peak temperature (T_p). There is also a decrease in the percent conversion at high catalyst concentrations (31, 34).

Increasing the concentration of catalyst brings about a reduction in the time to the exothermic peak. The exothermic peak temperature increases as the percentage of catalyst is increased. This increase in temperature is due to the autoacceleration effect that occurs when the viscosity of the monomer–polymer solution increases very rapidly with polymer formation. The percentage of conversion is approximately constant, except for a drop at the 1.2% and 1.5% vazo

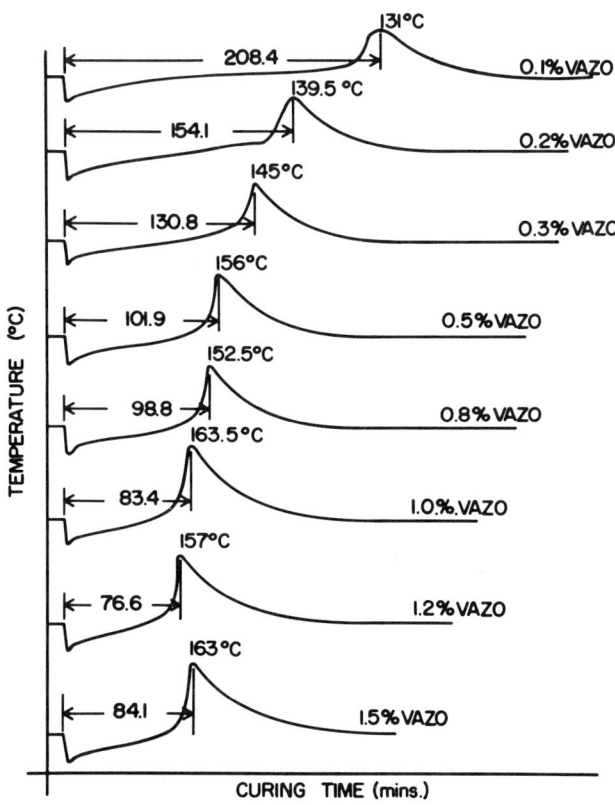

Figure 4. Temperature–time curves showing the effects of various concentrations of vazo catalyst on the polymerization exotherm of basswood–MMA composite. (Reproduced with permission from Ref. 31. Copyright 1972, Springer–Verlag.)

Table II. Effect of Vazo Catalyst on Polymerization of Basswood–MMA Composite

Basswood–MMA Sample	W_s (g)	W_m (g)	W_p (g)	M (%)	P (%)	C (%)	Vazo (%)	t_o (min)	t_p (min)	T_p (°C)
1	48.5	56.7	48.4	116.9	99.8	85.4	0.1	102.1	208.4	131.0
2	46.7	59.2	50.8	126.7	108.7	85.8	0.2	90.1	154.1	139.5
3	49.3	56.4	49.5	114.4	100.4	87.8	0.3	76.1	130.8	145.0
4	49.4	57.5	49.8	116.4	100.8	86.6	0.5	71.3	101.9	156.0
5	50.5	56.6	48.6	112.1	96.2	85.9	0.8	70.0	98.8	152.5
6	46.1	58.2	48.6	126.2	105.4	83.5	1.0	59.4	83.4	163.0
7	49.1	56.8	44.7	115.7	91.0	78.7	1.2	53.3	76.6	157.0
8	47.3	58.1	45.3	122.8	95.8	78.0	1.5	65.5	84.1	163.0

Key: W_s, weight of oven-dry wood; W_m, weight of monomer; W_p, weight of polymer; M, % of monomer; P, % of polymer; and C, % conversion of monomer to polymer.
(Reproduced with permission from Ref. 31. Copyright 1972, Springer–Verlag.)

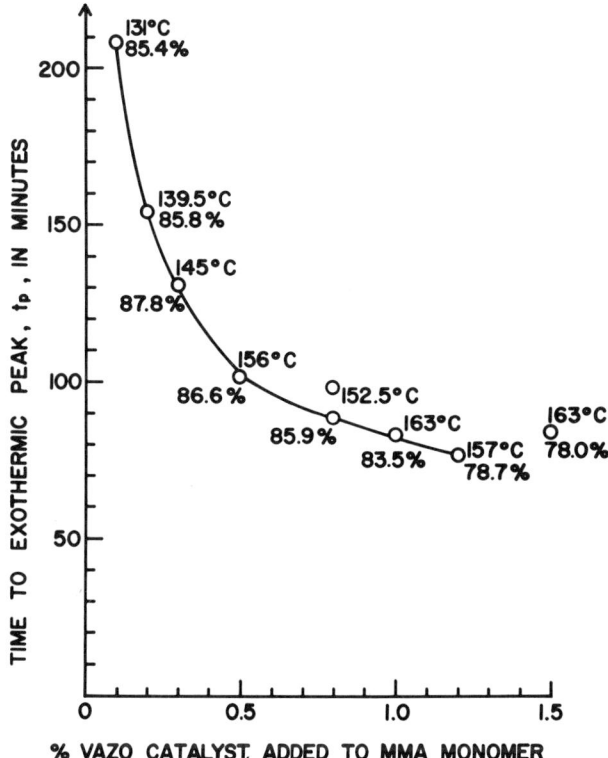

Figure 5. Time to exothermic peak (t_p) vs. percent vazo catalyst added to basswood–MMA composite. Exothermic peak temperature and percent monomer conversion are shown on the curve.(Reproduced with permission from Ref. 31. Copyright 1972, Springer–Verlag.)

concentrations. This can be explained because the monomer expands beyond the ends of the sample as the temperature rises and evaporates from the surface of the wood.

Individual monomers respond to catalytic polymerization in different ways. Figure 6 illustrates the time–temperature curves obtained when basswood was impregnated with several monomers containing 0.8% vazo catalyst. The time to the exothermic peak (t_p) was the shortest for MMA, followed by *tert*-butylstyrene, styrene, and vinyltoluene. Correspondingly, the exothermic peak temperature (T_p) for MMA was the highest at 152 °C, for *tert*-butylstyrene 145 °C, for styrene, 141 °C, and for vinyltoluene, 133 °C. A higher vazo content was used in order to assure polymerization of the four monomers. The effect of adding increasing amounts of vazo catalyst and crosslinkers to *tert*-butylstyrene, styrene, and vinyltoluene was similar to that for MMA (*see* Figures 4 and 7) (*35*).

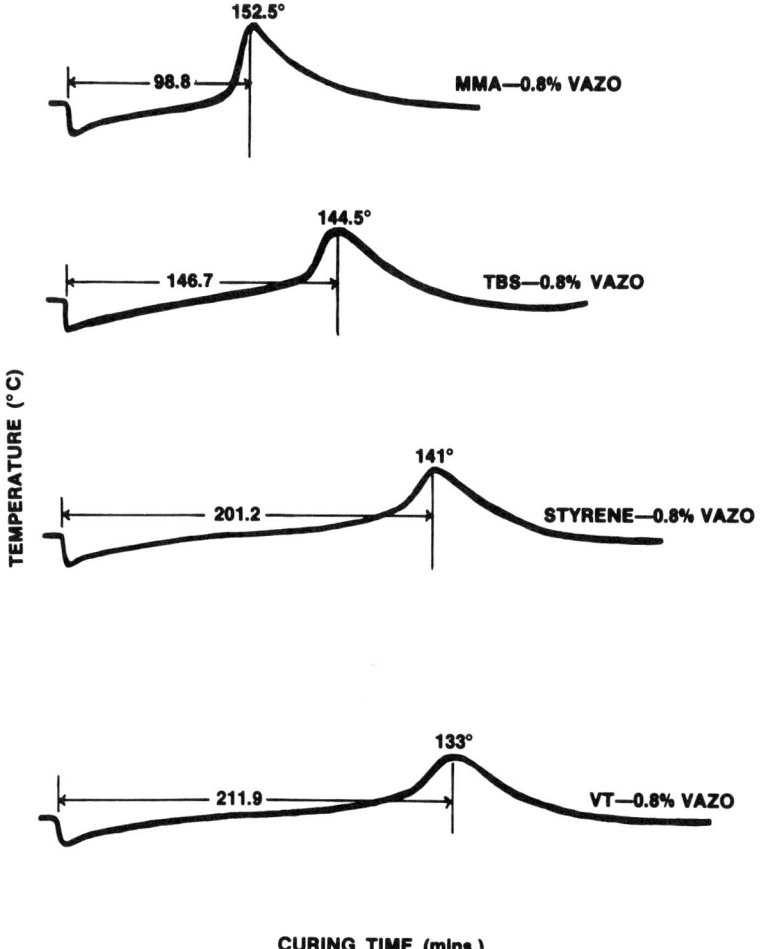

Figure 6. Temperature–time curves showing the effects of various monomers, catalyzed with the same vazo concentration, on the polymerization exotherms of basswood–monomer composites (35). Key: MMA, methyl methacrylate; TBS, tert-butylstyrene; and VT, vinyltoluene.

Cross-linking Effect on Polymerization. The general cross-linking reaction that occurs during polymerization and involves components with a functionality greater than one (two or more double bonds, divinyl monomer) has been studied extensively. The statistical analysis of molecular distributions in such reactions is due to Flory (36).

The cross-linking of unsaturated polymer chains by polymerization with divinyl monomers can be considered to occur in three stages. The first stage is equivalent to a linear copolymerization reaction with a low degree of conversion. The second stage is characterized by the formation of a three-dimensional gel structure and by an exponential increase in the rate of reaction resulting from diffusion control of the termination step (Trommsdorff–Norrish acceleration) (33, 37) as the viscosity increases. In the third stage, propagation, transfer, and even the initiation rate become diffusion controlled as the gel structure becomes increasingly dense.

Data are presented in Table III for the polymerization of basswood–MMA–trimethylol propane trimethacrylate (TMPTMA) impregnated samples containing 0.25% vazo catalyst. The melting point of the three-dimensional polymer is raised by adding TMPTMA to

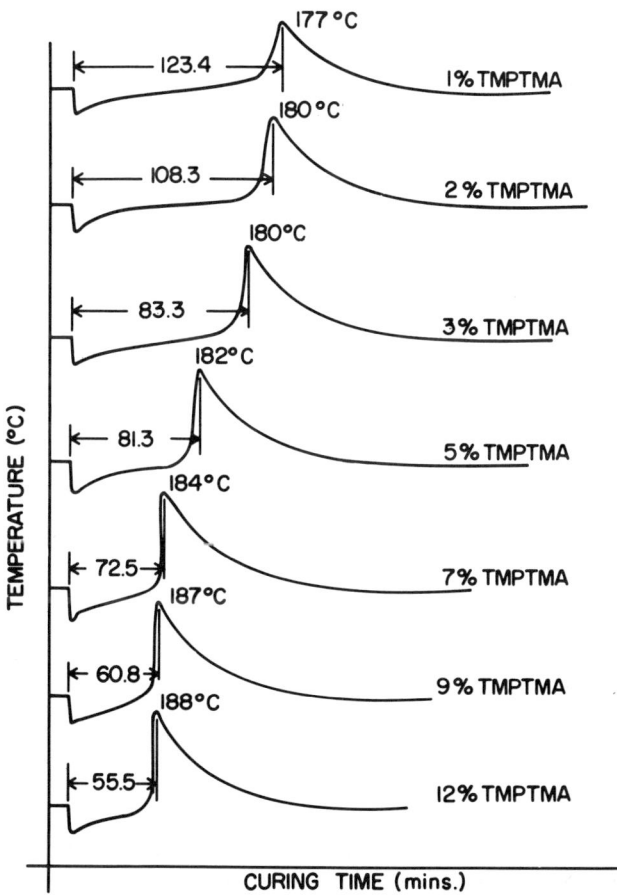

Figure 7. Temperature–time curves showing the effects of various concentrations of TMPTMA cross-linker on the polymerization exotherm of basswood–MMA composite (35).

Table III. Effect of TMPTMA Cross-linker on Polymerization of Basswood–MMA Composite

Basswood–MMA Sample	W_s (g)	W_m (g)	W_p (g)	M (%)	P (%)	C (%)	TMPTMA (%)	t_o (min)	t_p (min)	T_p (°C)
1	43.3	62.0	57.1	143.2	131.8	92.0	1.0	80.0	123.4	177
2	43.4	60.2	54.3	138.5	125.0	90.2	2.0	76.8	108.3	180
3	43.0	62.5	54.6	145.3	126.9	87.4	3.0	65.0	83.3	180
4	43.6	62.5	54.3	143.3	124.5	86.9	5.0	66.3	81.3	182
5	43.8	62.3	53.1	142.2	121.2	85.9	7.0	58.8	72.5	184
6	43.6	63.3	53.6	145.2	122.9	84.7	9.0	51.9	60.8	187
7	43.0	63.3	53.6	147.2	124.6	84.7	12.0	49.0	55.5	188
8	43.1	62.6	53.6	145.2	124.4	85.6	15.0	49.4	55.0	188
9	41.3	64.5	53.1	156.2	128.6	82.3	20.0	49.4	54.8	183

NOTE: The key is the same as in Table II.
(Reproduced from Ref. 35.)

MMA monomer, which causes cross-linking to the point where the polymer will decompose before it melts. On the average, the exothermic peak temperature remains about the same as the concentration of the TMPTMA is increased from 1.0 to 20% by weight of MMA.

The cross-linking agent affects the time to the exothermic peak dramatically as shown in Figure 7. Adding a cross-linking agent is comparable to increasing the amount of catalyst in the basswood–MMA system. One advantage of TMPTMA is that it forms a gel at the beginning of the polymerization, so that the monomer cannot expand out of the wood–monomer ends after the initial rise in temperature. Noncross-linked monomer will expand out of the ends of the wood–monomer composite where it will evaporate or polymerize into a foam. This decreases the polymer loading and wastes monomer. Cross-linking also raises the temperature high enough so that the polymer will not melt during sanding operations, and, thus, prevents loading of the sandpaper (38). The percent conversion of the monomer to polymer remains fairly constant and is comparatively higher than for basswood–MMA samples without cross-linker.

Figure 8 shows a plot of the time (t_p) to the exothermic peak as a function of the percentage of TMPTMA added to the basswood–MMA composite. At TMPTMA concentrations above 10%, the time to the exothermic peak remains essentially constant. The increase in the rate of reaction, i.e., the decrease in the time to the exothermic peak, primarily is due to the formation of the three-dimensional gel structure because the rate of reaction becomes diffusion controlled as the viscosity increases. Thus, the rate of termination decreases and the radical becomes effectively fixed in space and only adds more monomer molecules. The probability of encountering another free radical and terminating the chain growth is small.

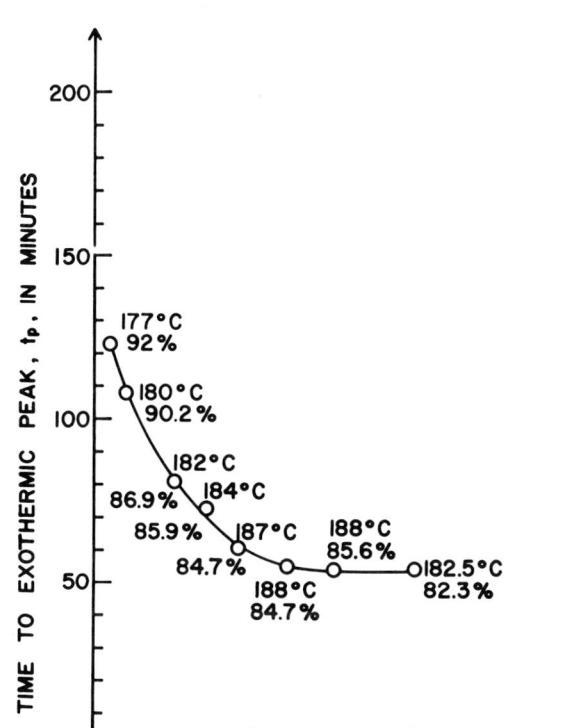

Figure 8. Time to exothermic peak (t_p) vs. percent TMPTMA cross-linking agent added to basswood–MMA composite. Exothermic peak temperature and percent monomer conversion are shown on the curve (35).

Wood Moisture Effect on the Polymerization Exotherm. The moisture content of wood used for wood–polymer composites is usually below 10%. On occasions liquid water is present after the polymerization of the wood–monomer is complete. This water is distilled out of the wood–monomer as the exotherm reaches temperatures above 100 °C, and it condenses on the cooler aluminum foil or oven walls. Water requires 2255 J/g for evaporation and could act as a heat sink in the polymerization reaction. Evaporation of the water in the wood could be accomplished by using the exothermic heat from the polymerization, and the maximum exothermic peak temperature could be depressed below 150 °C where wood cell wall materials begin to decompose. Some preliminary research has been carried out in this area (39) to show the effect of moisture content on the exothermic peak temperature. Table IV and Figure 9 illustrate this effect using methyl methacrylate, vinyltoluene, and styrene mono-

Table IV. Influence of Wood Moisture Content on the Polymerization Exotherm Temperature

Sample	Monomer[a]	Vazo Catalyst (%)	Moisture Content (%)	Polymer Loading (%)	t_p (min)	T_p (°C)
1	MMA	0.2	0.0	57.2	105	157
2	MMA	0.2	8.8	62.4	132	152
3	MMA	0.2	13.6	62.8	142	123
4	MMA	0.2	23.9	53.2	193	116
5	VT	1.0	0.0	56.5	175	128
6	VT	1.0	6.9	72.3	192	125
7	VT	1.0	11.1	58.6	206	97
8	VT	1.0	16.6	53.5	236	98
9	STY	1.0	0.0	68.0	165	138
10	STY	1.0	7.6	73.0	195	130
11	STY	1.0	15.8	69.0	228	127
12	STY	1.0	19.1	62.8	238	118

[a] Key: MMA, methyl methacrylate; VT, vinyltoluene; and STY, styrene. (Reproduced from Ref. 39.)

mers. The results show that the peak temperature T_p can be lowered by increasing the moisture content of the wood before impregnation.

Additives Effect on the Catalyzed Monomer Solution. Soluble dyes can be added to the catalyzed monomer solution to color the final wood–polymer composite. Any color of the visible spectrum can be added, browns to simulate black walnut, red and blues for national colors. The color emphasizes the grain structure of the particular species and combines with the polymer to add a three-dimensional depth not present in surface-finished wood. A dense black wood–polymer, so desirable for musical instruments, is difficult to obtain because of wood's light color and the tendency of the microstructure to chromatographically separate a dye of several components into its separate colors. Dyes have an inhibiting effect on the polymerization of wood–monomer composites, some more so than others. Additional catalyst can be added to overcome this inhibition, but in the radiation process of a given geometry additional time must be allowed for complete curing.

Fire retardants are added to wood–polymers used for flooring and other public installations. Polymethyl methacrylate is one of the few polymers that will depolymerize into the monomer when heated to the temperature of decomposition. The monomer in the gas phase burns with an intense blue flame similar to a natural gas flame.

Commercial fire retardants based on phosphorus and the halogens, (i.e., Phosgard C-22-R, SAF-707, Santicizer-140, triphenyl phosphate) are added directly to the monomer solution (40, 41). Many of the fire retardants are based on benzene ring structures such as triphenyl phosphate and Santicizer-140. Sooty smoke is a test for aromatic compounds in basic organic chemistry. The fire-tube tests and flame-spread tests produced huge amounts of smoke for wood–polymer composites containing aromatic fire retardants (40, 41). The presence of nonaromatic Phosgard decreased the smoke intensity. If chlorine is attached to the benzene ring of styrene to obtain monochlorostyrene, flame spread is reduced because of the fire-retardant

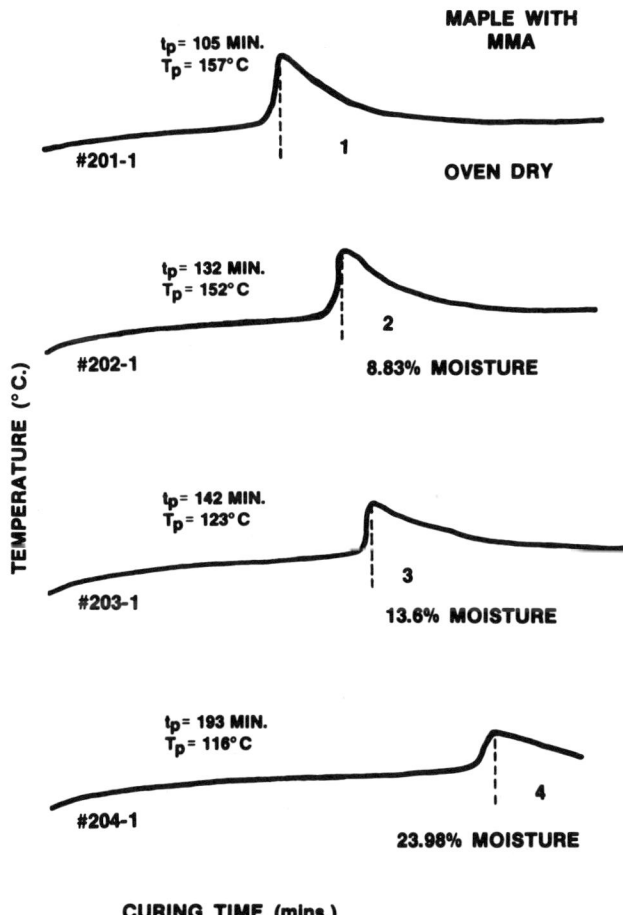

Figure 9. Temperature–time curves showing how the wood moisture content affects the polymerization exotherm of maple–MMA composite (39).

properties of the halogen. In this case the fire retardant is built into the vinyl monomer and is part of the polymer upon polymerization in the wood.

Some research has been done on the addition of polar solvents to the nonpolar monomer in an attempt to swell the cell wall structure and anchor it in a swollen state (32). This procedure can be done and the ASE does increase, but after the solvent evaporates the wood is only partially loaded, which decreases the physical properties. Wood–polymer composites normally have about a 10–15% ASE, which means that there is some penetration of the cell wall structure to reduce the swelling over that of untreated wood.

Water and Polymer Location. The mechanism of water absorption by dry wood proceeds in two steps. Water entering dry wood in vapor form is absorbed into the cell wall, and hydrogen bonds to the cellulose. As a result, the cell wall swells, and the overall dimensions of the wood increase. After 25–28% of the water is absorbed (based on the oven-dry weight of the wood) and the cell wall has swollen to its maximum, additional water will be condensed in the capillaries or other void spaces in the wood until it is filled. The fiber-saturation point is where the cell walls have absorbed the maximum amount of water and are swollen to the maximum extent, but no water has condensed in the capillaries. This point is surprisingly consistent, 26–30%, considering the large number of species of wood on this planet. As pointed out previously, normal wood–polymer material contains polymer only in the void spaces that are available, and little if any in the cell walls. This loading of the capillary vessels reduces the rate of water diffusing into the cell walls. But, given enough time (10–20-fold greater than in untreated wood) at high humidity, eventually water will reach the cell walls and cause the same volume swelling as untreated wood. Figures 10 and 11 show the differences in water absorption in basswood–polymer composites (32).

Water in a never-dried fresh-cut tree, when exchanged with a series of organic solvents, could be replaced with ^{3}H- (or tritium-) labeled MMA (42). After polymerization and the use of autoradiography the ^{3}H-labeled MMA was located in the cell wall and compound middle lamella. The polymer (polymethyl methacrylate) labeled with ^{3}H was removed from the capillaries and void spaces in the wood by solvent extraction, but the labeled polymethyl methacrylate in the wood structure remained. The same wood, after oven drying, was treated with the ^{3}H-labeled monomer by using the full cell method. Again, by autoradiography, the ^{3}H-labeled polymethyl methacrylate was located in the capillaries and void spaces. After solvent extraction of the polymethyl methacrylate no labeled polymer

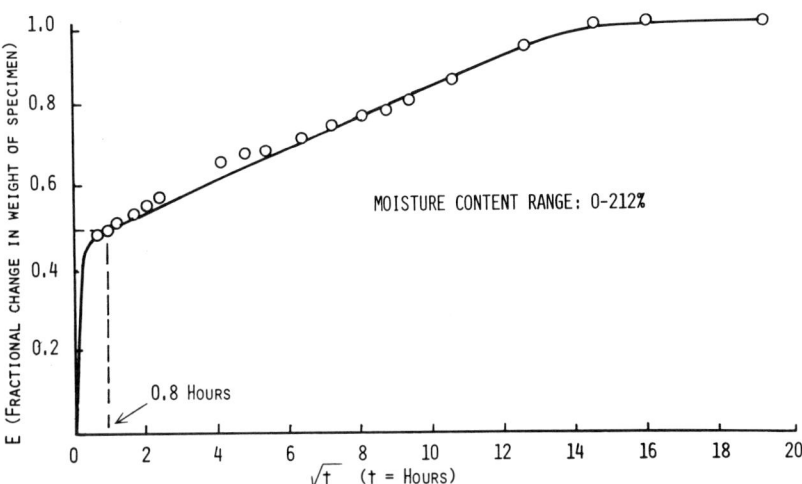

Figure 10. Fraction of total weight change vs. time for untreated wood. (Reproduced with permission from Ref. 32. Copyright 1969, Forest Products Research Society.)

was present in the cell walls. This points out how relatively impermeable the oven-dried cell wall is to vinyl monomers like MMA. Pentane solvent exchange dried cell walls contained 25% less polymer than the never dried wood. In all three cases the capillaries and void

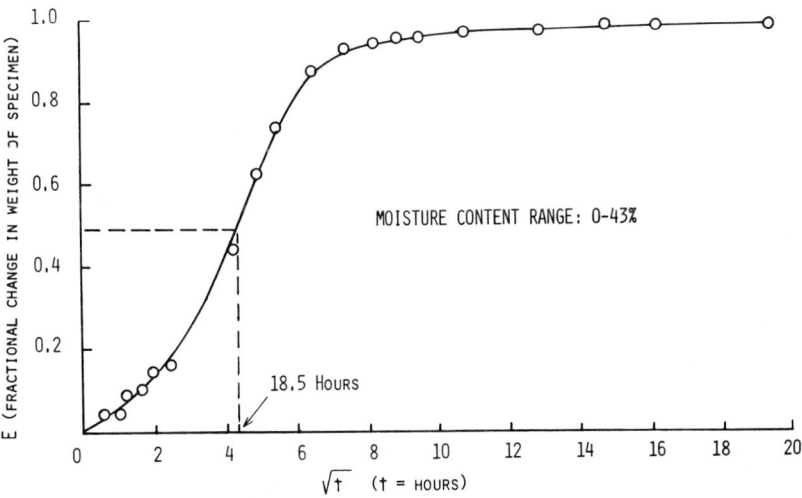

Figure 11. Fraction of total weight change vs. time for basswood treated with tert-butylstyrene. (Reproduced with permission from Ref. 32. Copyright 1969, Forest Products Research Society.)

spaces in the wood were filled with ^{3}H-labeled polymethyl methacrylate.

The flow into wood, especially of liquids, is along the grain. Siau permeability measurements were done on yellow birch and the permeabilities (cm^3cm/cm^2 s atm) observed were: longitudinal, 696; radial, 0.000177; and tangential, 0.000092 (43). These facts prevent the surface treatment with monomers from forming a wood–polymer surface shell around the wood. Ray cells in the radial direction are often connected to the longitudinal cell structure; this configuration makes shell loading improbable.

Physical Properties. Wood, when dry, has unique physical properties in that its tensile strength, bending strength, compression strength, impact resistance, and hardness per unit weight are the highest of all construction materials. The hydrogen bonding, the unique helical structure of the cell walls, the combination of the linear cellulose molecules impregnated with low molecular weight extractives, and all of the varying amounts of cross-linked lignin make wood an infinitely variable resource. All the unusual features of wood are the reason for the "art" of wood treatment.

The polymer loading of wood depends not only on the permeability of the wood species, but also on the particular piece of wood being treated (44). Because the void volume is approximately the same for the sapwood and heartwood of each species, it would be expected that the polymer would fill them to the same extent. Table V, however, shows that the sapwood is filled to a much greater extent than the heartwood for six of the eight species. This is contrary to what would be assumed from the measured void volume.

The sugar maple and the basswood are two exceptions; there is essentially the same retention of polymer in the sapwood and heartwood. The heartwood probably has less of the voids filled with polymer because of organic deposits and tyloses that block the penetration of the monomer into the capillaries. In the extreme case of red pine heartwood, visible amounts of resin exuded from the sample during drying. Table V also lists physical properties for a limited number of wood species of treated wood–polymer composites (30). Figures 12 and 13 illustrate typical test data for static bending and compression parallel to the grain for basswood–polymer composites (45). Table VI sums up some test results for static bending and compression parallel to the grain. The test data show that the variability amount in untreated test samples is high, but after polymer loading the coefficient of variability is reduced by 50% or more, thus producing much more uniform test data (45).

Langwig (32) used several different monomers to make wood–polymer composites. These same monomers were diluted with sol-

Table V. Physical Property Enhancements

Species	Void Volume (%)	Voids Filled (%)	Polymer in WPC (%)	Density Increase (%)	Compression Strength Increase (%)	Tangential Hardness Increase (%)	Permeability Ratio (Untreated: Treated)
Acer rubrum (Red Maple)							
Sapwood	64	65	46	82	171	280	270
Heartwood	63	56	71	67	82	209	40
Acer saccharum Marsh. (Sugar Maple)							
Sapwood	60	61	40	65	160	229	225
Heartwood	58	60	38	58	125	200	186
Prunus serotina Ehrh. (Black Cherry)							
Sapwood	64	63	45	78	202	289	717
Heartwood	63	46	37	56	86	124	428
Tilia americana L. (Basswood)							
Sapwood	80	61	63	168	435	626	107
Heartwood	77	66	62	160	288	505	104
Betula lutea Michx. F. (Yellow Birch)							
Sapwood	55	67	37	58	146	215	1047
Heartwood	52	60	31	43	56	120	1896
Liquidambar styraciflua (Red Gum)							
Sapwood	68	58	48	88	175	243	2884
Heartwood	65	35	33	46	33	95	110
Pinus resinosa Ait. (Red Pine)							
Sapwood	68	65	51	100	636	523	1395
Heartwood	68	6	8	7	1	1	14
Fagus grandifolia Ehrh. (Beech)							
Sapwood	59	53	36	53	201	261	213
Heartwood	55	34	24	30	30	112	19

(Reproduced with permission from Ref. 44. Copyright 1968, Forest Products Research Society.)

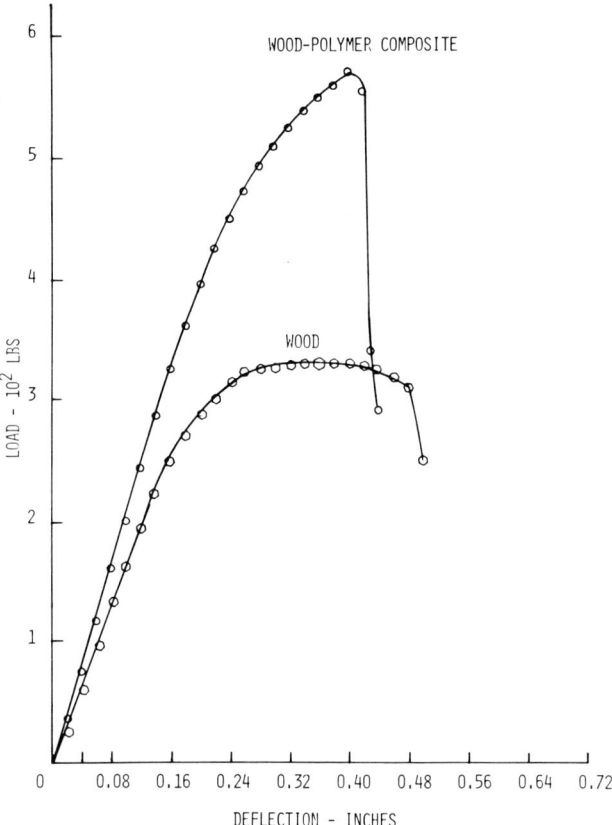

Figure 12. Example of typical bending test data. (Reproduced with permission from Ref. 45. Copyright 1968, Forest Products Research Society.)

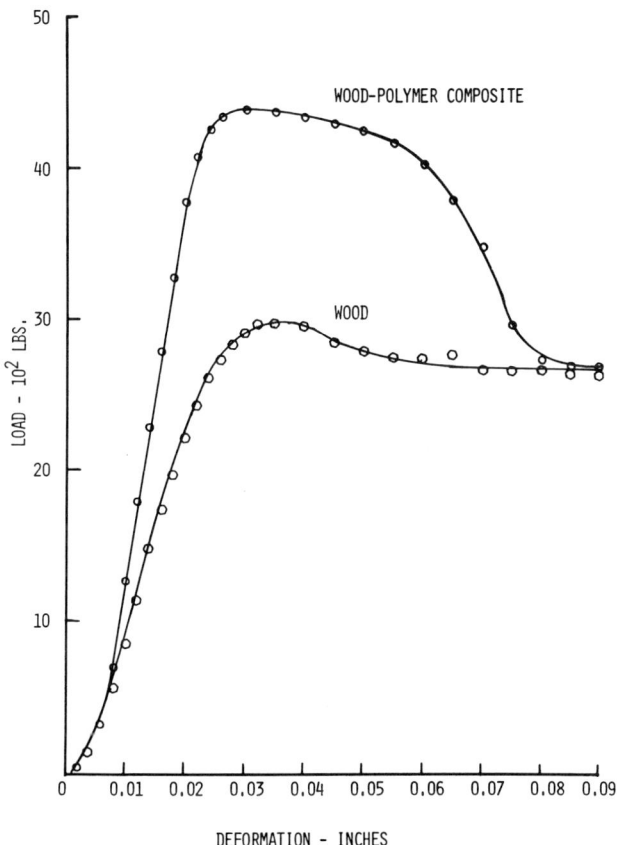

Figure 13. Example of typical compression test data. (Reproduced with permission from Ref. 45. Copyright 1968, Forest Products Research Society.)

Table VI. Test Results for Treated and Untreated Basswood

Strength Property	Units	Mean		% Change (based on un-treated wood)	Coefficient of Variation	
		Untreated (MC – 7.2%)	Treated		Untreated (%)	Treated (%)
Static Bending						
Modulus of elasticity	10^6 psi	1.356	1.691	25	13.3	10.4
Fiber stress at proportional limit	psi	6,387	11,587	81	17.8	6.3
Modulus of rupture	psi	10,649	18,944	78	15.7	5.4
Work to proportional limit	in.–lb/in.^{3}	1.66	4.22	154	25.0	11.6
Work to maximum load	in.–lb/in.^{3}	10.66	17.81	77	24.3	5.3
Compression Parallel to Grain						
Modulus of elasticity	10^6 psi	1.113	1.650	48	25.2	13.8
Fiber stress at proportional limit	psi	4,295	7,543	76	22.5	11.6
Maximum crushing strength	psi	6,505	9,864	51	18.9	14.1
Work to proportional limit	in.–lb/in.^{3}	11.28	21.41	90	25.1	17.5
Toughness	in.–lb/in.^{3}	41.8	62.6	50	17.2	9.5

(Reproduced with permission from Ref. 45. Copyright 1968, Forest Products Research Society.)

vents in order to obtain partial loading of the wood. In addition to MMA, *tert*-butylstyrene, and a Heilman–Gulf epoxy monomer were used. The monomers were diluted with 50% methanol and acetone. Dimethyl sulfoxide, 0.5%, was used as a transport medium to carry the monomer into the cell wall structure. Figure 14 shows the results of hardness tests on these wood–polymer composites of basswood and sugar maple. Table VII sums up some of the properties of the wood–polymer composites made with monomers other than MMA *(45)*. Figure 15 illustrates the load-deflection curves that were obtained from the experimental data. The Heilman–Gulf epoxy and the *tert*-butylstyrene monomers were an improvement over the MMA. When the *tert*-butylstyrene was diluted with 50% methanol the wood–polymer composite contained half of the maximum polymer loading, and the load-deflection data were the same as the untreated basswood. To obtain the maximum improvement in physical properties the wood must be fully loaded with polymer. In another study *(46)* hardness and hardness modulus tests were made on untreated and polymethyl methacrylate-treated red oak, aspen, and hard maple. Untreated hardness values for these woods were statistically significant when related to the untreated sample density. The hard-

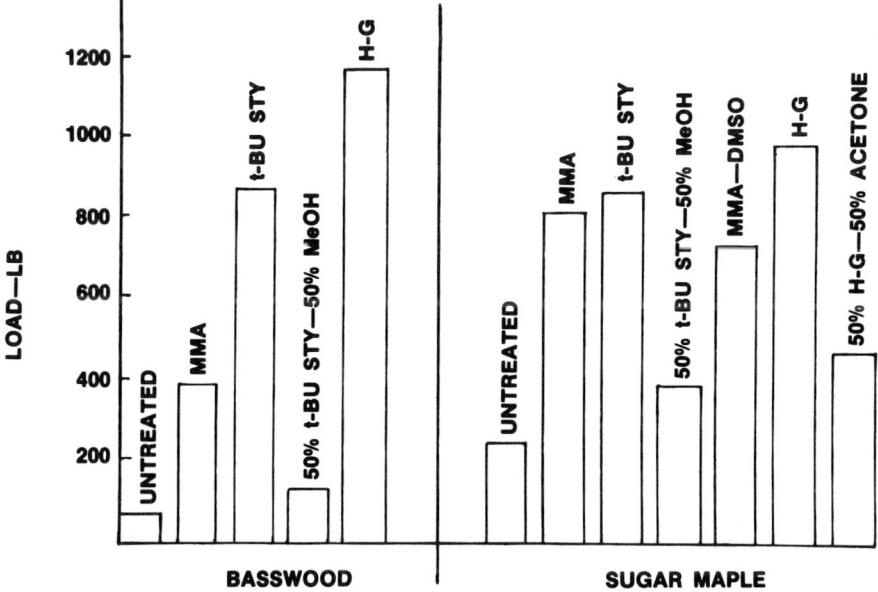

Figure 14. Hardness of various wood–polymer composites.(Reproduced with permission from Ref. 32. Copyright 1969, Forest Products Research Society.)

Table VII. Properties of Various Wood–Polymer Composites

Property	Methyl Methacrylate	tert-Butyl-styrene	tert-Butylstyrene, 50% Methanol	Methyl Methacrylate, 0.5% DMSO	Heilman–Gulf Epoxy Monomer	Heilman–Gulf Epoxy Monomer, 50% Acetone
Void spaces with polymer						
Basswood	65.3	87.7	35.3	—	76.5	—
Sugar maple	59.3	76.8	38.4	60.4	67.4	36.6
Volume change						
Basswood	2.2	−1.7	11.2	—	−5.8	—
Sugar maple	−1.0	−1.3	9.2	−0.8	−2.0	4.4
Dimensional stability (antishrink efficiency)						
Basswood	56.8	33.6	54.1	—	—	—
Sugar maple	4.8	0.0	39.8	12.7	4.2	21.1
Modulus of elasticity (increase)						
Basswood	0	11	0	—	31	—
Sugar maple	14	10	0	0	0	0
Fiber stress at proportional limit (increase)						
Basswood	17	38	2	—	66	—
Sugar maple	15	19	16	9	2	0
Modulus of rupture (increase)						
Basswood	54	87	7	—	107	—
Sugar maple	8	12	10	11	8	0

NOTE: All values in this table are expressed in percentages.
(Reproduced with permission from Ref. 45. Copyright 1968, Forest Products Research Society.)

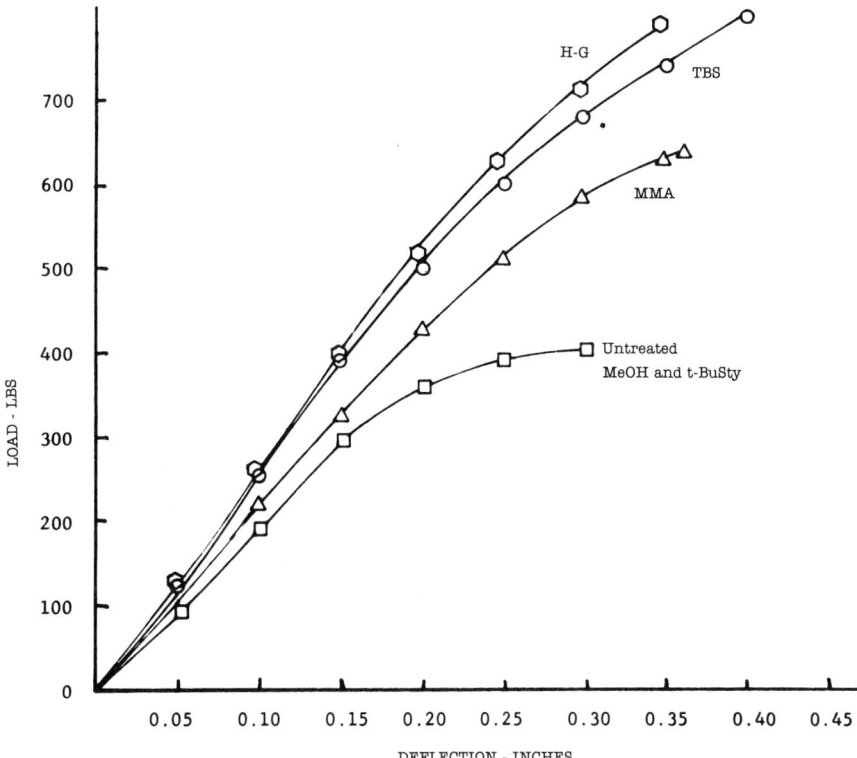

Figure 15. Load deflection curves for basswood–polymer composites. Key: H–G, Heilman–Gulf epoxy monomer; TBS, tert-butylstyrene; and MMA, methyl methacrylate. (Reproduced with permission from Ref. 32. Copyright 1969, Forest Products Research Society.)

ness values for the polymethyl methacrylate-treated woods had no statistical significance when related to the original untreated density. Highly significant relationships were found between treated hardness modulus and density and loading. Predicting equations were developed for hardness modulus based on species density and polymer loading.

Literature Cited

1. Seborg, R. M.; Tarkow, H.; Stamm, A. J. *U.S. For. Serv. Res. Rep.* FPL–2192 **1962**.
2. Tarkow, H. *Encycl. Polym. Sci. Technol.* **1966**, 5, 98–121.
3. Koch, P. *U.S. For. Serv. Agric. Handbook* No. 420 **1972**, 2, 1128–49.
4. Meyer, J. A.; Loos, W. E. *For. Prod. J.* **1969**, 19(12), 32–38.
5. Seborg, R.; Tarkow, H.; Stamm, A. J. *For. Prod. J.* **1953**, 3(3), 59–67.

6. Stamm, A. J. "Wood and Cellulose Science"; Ronald Press Co.: New York, 1964.
7. Yokota, T.; Tarkow, H. *For. Prod. J.* **1962**, *12*(1), 43–45.
8. Seborg, R. M.; Vallier, A. E. *For. Prod. J.* **1954**, *4*(5), 305–12.
9. Stamm, A. J.; Seborg, R. M. *U.S. For. Serv. Res. Rep.*, *FPL–1381* **1960**.
10. Stamm, A. J. *For. Prod. J.* **1956**, *6*(5), 201–4.
11. Stamm, A. J. *For. Prod. J.* **1959**, *9*(10), 375–81.
12. Stamm, A. J. *For. Prod. J.* **1967**, *17*(9), 91–96.
13. Lantican, S. M.; Cote, W. A. et al. *Ind. Eng. Chem.* **1965**, *4*(2), 66–70.
14. Schuerch, C. *J. Polym. Sci., Polym. Symp.* **1963**, *2*, 79–95.
15. Schuerch, C. *Ind. Eng. Chem.* **1963**, *55*(10), 39.
16. Pentoney, R. E. *Ind. Eng. Chem. Prod. Res. Dev.* **1966**, *5*, 105–10.
17. Davidson, R. W.; Baumgardt, W. G. *For. Prod. J.* **1970**, *20*(3), 19–25.
18. Kenega, D. L.; Cowling, E. B. *For. Prod. J.* **1959**, *9*(3), 112–16.
19. Schuerch, C., U.S. Patent No. 3 282 313, 1966.
20. Mater, J. *For. Prod. J.* **1957**, *7*(6), 208–9.
21. Siau, J. F.; Meyer, J. A.; Skaar, C. *For. Prod. J.* **1965**, *15*(4), 162–66.
22. deLhoneux, B. Masters thesis, "Ultrastructure Implications of Gamma Irradiated Wood," December 1981. SUNY College of Environmental Science and Forestry, Syracuse, NY, in press.
23. Siau, J. F.; Meyer, J. A.; Skaar, C. *For. Prod. J.* **1965**, *15*(4), 162–66.
24. International Atomic Energy Agency "Impregnated Fibrous Materials, Report of a Study Group, Bangkok, Thailand"; Vienna, 1968, STI/PUB/209.
25. Elwood, L.; Gilmore, R.; Merrill, J. A.; Poole, W. K. U.S. Atomic Energy Commission Contract AT(40–1)–25–13, Task 20; Report No. ORO–638 (RTI–2513–T13) Division of Isotope Development, 1971.
26. Elwood, L.; Gilmore, R. C.; Stamm, A. J. *Wood Sci.* **1972**, *4*(3), 137–41.
27. Meyer, J. A. *For. Prod. J.* **1965**, *15*(9), 362–64.
28. Riddle, E. H. "Monomeric Acrylic Esters"; Reinhold Publishing Co.: New York, 1954; p. 29.
29. DuPont Product Bulletin "DuPont Vazo"; Industrial and Biochemicals Dept., Wilmington, DE.
30. Meyer, J. A. in "Wood Technology: Chemical Aspects," Goldstein, Irving S., Ed.; ACS SYMPOSIUM SERIES No. 43, ACS: Washington, D.C., 1977; pp. 301–25.
31. Duran, J. A.; Meyer, J. A. *Wood Sci. Technol.* **1972**, *6*, 59–66.
32. Langwig, J. E.; Meyer, J. A.; Davidson, R. W. *For. Prod. J.* **1969**, *19*(11), 57–61.
33. Trommsdorff, E.; Kohle, H.; Lagally, P. *Makromol. Chem.* **1948**, *1*, 169–98.
34. Beall, F. C.; Witt, A. E. *Wood Fiber J.* **1972**, *4*(3), 179–84.
35. Duran, J. A.; Meyer, J. A., unpublished senior research problem; "Exothermic Heat Released During Catalytic–Heat Polymerization of Basswood and Various Vinyl Monomers"; SUNY Coll. Environ. Sci. and For., Chem. and Wood Prod. Eng. Depts., Syracuse, NY, 1971.
36. Flory, P. J. *J. Am. Chem. Soc.* **1941**, *63*, 3083–100.
37. Norrish, W. G. W.; Smith, R. R. *Nature (London)* **1942**, *150*, 336–37.
38. Meyer, J. A. *For. Prod. J.* **1968**, *18*(5), 89.
39. Campos, G.; Meyer, J. A., unpublished senior research problem; "The Influence of Wood Moisture Content on Wood–Plastic Exotherms"; SUNY Coll. Environ. Sci. and For., Chem. Dept., Syracuse, NY, 1973.
40. Siau, J. F.; Meyer, J. A.; Kulik, R. S. *For. Prod. J.* **1972**, *22*(7), 31–36.
41. Siau, J. F.; Campos, G. S.; Meyer, J. A. *Wood Sci.* **1975**, *8*(1), 375–83.
42. Timmons, T. K.; Meyer, J. A.; Cote, W. A., Jr. *Wood Sci.* **1971**, *4*(1), 13–24.
43. Siau, J. F.; Meyer, J. A. *For. Prod. J.* **1966**, *16*(8), 47–56.
44. Young, R. A.; Meyer, J. A. *For. Prod. J.* **1968**, *18*(4), 66–68.

45. Langwig, J. E.; Meyer, J. A.; Davidson, R. W. *For. Prod. J.* **1968,** *18*(7), 33–36.
46. Beall, F. C.; Witt, A. E.; Bosco, L. R. *For. Prod. J.* **1973,** 23(1), 56–61.
47. Meyer, J. A. *Wood Sci.* **1981,** *14*(2), 49–54.
48. Meyer, J. A. *For. Prod. J.* **1982,** 32(1), 24–29.

RECEIVED for review May 17, 1983. ACCEPTED August 16, 1983.

7
Bioactive Wood–Polymer Composites

R. V. SUBRAMANIAN
Department of Materials Science and Engineering, Polymeric Materials Section, Washington State University, Pullman, WA 99164

The preparation and properties of bioactive wood–polymer composites are discussed. The basic, effective approach to bring about simultaneous improvements in decay resistance, dimensional stability, and mechanical behavior of wood is in situ polymerization and copolymerization of organotin monomers carrying the bioactive tributyltin group. Tri-n-butyltin methacrylate–maleic anhydride and tri-n-butyltin methacrylate–glycidyl methacrylate are examples of suitable monomer combinations for in situ copolymerization. Comonomers that carry anhydride or epoxy functional groups graft to wood through esterification or etherification of wood hydroxyls. Electron microprobe analysis for tin atoms shows that a detectable portion of tin copolymer is located in cell walls. The treated wood is effective in providing resistance against white rot and brown rot fungi, and against marine organisms as determined by laboratory and ocean tests. Notable improvements in flexural and impact strengths, and significant reduction in moisture absorption are also observed.

THE POLYMERIZATION OF VINYL MONOMERS in the void spaces of bulk wood results in wood–polymer composites of increased strength properties and dimensional stability (see Chapter 6). Because the different environmental conditions expose in-service timber to attack by numerous wood-deteriorating microorganisms, it is desirable to enhance the biodegradation resistance of wood, with simultaneous improvements in mechanical behavior. This chapter summarizes the formation of bioactive wood–polymer composites (1–4). The basic approach is still in situ polymerization of vinyl monomers in wood, with the appropriate choice of a bioactive, toxic, functional group incorporated in the monomer, and with other modifications based on wood–polymer reactions.

Biological Activity

The biodegradation of wood, whether it is above ground, by soil contact, or in marine applications, is brought about by fungi, bacteria, insects, and marine borers (5). Both toxic and nontoxic preservative treatments have been adopted in protecting wood. The chemical modification of wood has been investigated as a nontoxic preservative treatment (6, 7). The basis for such nontoxic treatments is the assumption that the enzymes exuded by the microorganisms must come directly in contact with the wood and that the substrate must have a specific configuration in order for the highly selective enzyme-initiated wood-degrading reaction to take place. Therefore, if the woody substrate is chemically modified, even by the use of nontoxic chemicals, these reactions cannot take place, and chemically modified wood should become unrecognizable as a food source to support microbial growth.

Toxic preservatives function by disrupting the cellular organization of microorganisms so that the organism dies. Thus, numerous organic salts of copper, zinc, arsenic, and boron have been used as wood preservatives, generally with added chromium compounds, to reduce rapid leaching of the water-soluble compounds. Among organic compounds, coal-tar creosote and pentachlorophenol are important toxic preservatives in wide commercial use. The impregnation of wood with toxic compounds that are leached out during actual use of wood represents the most widely used, and frequently, the only practical method of preserving wood. Details of the various aspects of wood and biodegradation protection are discussed in Chapters 8 and 12.

Biotoxicity of Organotin Compounds and Polymers

The trialkyltin group was chosen for incorporation in monomers used for in situ polymerization in wood because trialkyltin compounds have emerged as broad spectrum toxicants having high toxicity toward marine fouling organisms (8), as well as wood-destroying organisms (9–11). In addition, they possess a tolerable toxicity toward mammals. They are also approximately 10 times more toxic against wood-destroying fungi than pentachlorophenol (9). Trialkyltin compounds eventually degrade to harmless inorganic oxides of tin by the action of UV light, microbes, etc., and thus present a minimal environmental hazard (12).

A further advantage to using vinyl monomers with trialkyltin functional groups for in situ polymerization is the possibility of a controlled release of the toxic trialkyltin group from the treated wood. In the conventional technique of wood treatment with organic pesticides, the preservative is dispersed in wood, and the loading level of the biocide is maximized by various timber preconditioning techniques. This preconditioning at maximum loading levels results in initial overkill through the high leaching rate of the dispersed toxicant and consequent reduction in the duration of protection afforded. On the contrary the chemical anchoring of the toxic trialkyltin group to the polymer impregnated in wood requires the dissociation of the chemical bond linking the toxicant to the polymer, and the subsequent diffusion of the toxic group through the polymer matrix. Hence, a method is available for increasing the duration of protection significantly through uniform and optimum release of toxicant throughout the lifetime of the preserved wood. Further, the environmental hazards associated with rapid release of toxicants to the environment is also minimized.

The biocidal effect of the trialkyltin groups depends on the nature of the alkyl group, and the lower homologues (methyl, propyl, etc.) possessing the highest activity (8). However, for a safe balance between tolerable toxicity to mammals and high activity against microorganisms, $(n\text{-}C_4H_9)_3Sn-$, the tri-n-butyltin functional group (TBT), is the preferred choice. The synthesis and properties of polymers with TBT functional groups have been investigated extensively and their biocidal effect established and related to their chemical structure in both laboratory and field tests (13–16). The polymers are formed by polymerization of organotin monomers such as tri-n-butyltin methacrylate (TBTMA) or by esterification of polymers carrying carboxylate functional groups, such as styrene–maleic anhydride copolymers with tri-n-butyltin oxide (TBTO).

The release of the TBT toxicant from these polymers is controlled by the polymer matrix properties. For example, copolymers of TBTMA and glycidyl methacrylate (GMA) are effective against *Pseudomonas nigrifaciens* (marine bacterium), *Sarcina lutea* (soil fungus); and *Giomeralla cingulata* (soil fungus); the leaching rate determined by inhibition zones against these organisms is modified by the nature and extent of cross-linking of the polymers (17, 18). Thus it is possible to vary the biotoxicity of these organotin polymers by modifications of chemical structure.

Organotin Polymers in Wood

The impregnation of wood by polymers with TBT functional groups offers a viable route to the preparation of bioactive wood–polymer composites with a number of advantages. Polymer impregnation may be expected to improve the strength properties and dimensional stability of wood in water. The controlled release of the toxic TBT moiety, chemically linked to the polymer located within the wood, can increase the service life of wood while ensuring minimal adverse impact on the surrounding environment. By decreasing the amount of water absorbed by the wood, polymer impregnation minimizes the swelling and shrinking of wood in cyclic moisture exposure, and thereby slows the loss of toxicant from wood. Thus, improvement in biodegradation resistance and mechanical behavior may be achieved.

Although it is possible to incorporate organotin polymers in wood by vacuum or pressure impregnation with solutions of preformed polymer, it would be preferable to utilize monomer impregnation followed by in situ polymerization because the smaller molecular size, as well as the low viscosity of monomers, is conducive to efficient penetration of wood. Therefore, vinyl monomers in which the TBT group is chemically bonded, such as TBTMA, are used for in situ polymerization in wood (2).

In the selection of monomers for in situ polymerization, the possibility of chemical reaction with hydroxyl groups in wood must be considered. This reaction possibility is achieved by selecting comonomers such as maleic anhydride (MAnh) or GMA for copolymerization with the monomer containing TBT. In reaction with TBTMA MAnh can react with wood hydroxyls by esterification as shown in Reaction 1, as well as copolymerizing through the double bonds. Similarly, GMA can react by etherification of hydroxyls, through its epoxide group. As a result of these reactions, the copolymer does not just fill in the voids of the wood matrix, but it is chemically bonded, i.e., *grafted* to the wood structure. Therefore, in situ copolymerization in these systems involves copolymerization through the

Wood–OH + ~~–CH–CH–CH$_2$–C(CH$_3$)~~ →
(TBTMA–MAnh Copolymer, with anhydride ring and C=O / O–Sn(Bu)$_3$ groups)

TBTMA–MAnh Copolymer

Wood–O–C(=O)–CH–CH–CH$_2$–C(CH$_3$)~~
with O=C–OH and C=O / O–Sn(Bu)$_3$ groups

TBTMA–MAnh Copolymer grafted to wood

Reaction 1

double bonds of the comonomers, and grafting through the additional functional groups present (the anhydride or epoxide). Homopolymerization of epoxide groups may also occur. Thus, the monomer pairs TBTMA-MAnh or TBTMA-GMA are interesting systems to study (1, 2). Conversion of wood hydroxyls by acetylation or etherification can improve dimensional stability and biodegradation resistance in wood (7, 9, 20).

Copolymerization of TBTMA–MAnh and TBTMA–GMA was conducted in situ by free radical initiators benzoyl peroxide or azobisisobutyronitrile after vacuum impregnation of grand fir (*Abies grandis*) wood specimens (0.64 × 2.54 × 11.4 cm) containing wood fibers either parallel to length or parallel to width. Working with 10–40% (by weight) of monomers in solution in benzene or acetone, it was found that 15–60% (by weight) loading of copolymer in wood can be obtained. The amount of polymer incorporated into treated wood was much higher when acetone, a moderate wood swelling solvent, was used than from benzene solutions. (Benzene is a nonswelling solvent for wood.) The viscosity of benzene solutions of the monomers is significantly higher than the viscosity of acetone solutions.

The results of solvent extraction of polymer-impregnated wood samples point to a high degree of grafting of the polymer to wood (Table I) as revealed by the unextractable fraction of the impregnated polymer. Comparison of these data with those obtained when only the individual monomers were homopolymerized showed that MAnh, or GMA, still gives a high degree of grafting but TBTMA did not. Therefore, under the polymerization conditions used, the acylation and etherification of wood hydroxyls by MAnh or GMA, respectively, readily can occur in addition to homopolymerization reactions.

Polymer Distribution. The macrodistribution of polymer in treated wood specimens can be examined by scanning electron microscopy (SEM) of transverse and longitudinal sections of the speci-

Table I. Fractions of Various Polymers and Copolymers Grafted to Wood

Polymer	Polymer in Wood (wt %)	Grafting (%)
TBTMA–MAnh	16.2	87.3
	37.9	93.5
	57.6	91.5
TBTMA–GMA	17.6	86.0
	32.5	88.3
	51.5	83.4

NOTE: The solvent used for extraction was benzene.

mens (2). Small sections from edge and end surfaces could thus be compared with sections from the body center of the specimens. In the case of GMA polymerized from acetone solutions, wood cells fill to the same extent in all parts of the specimen; therefore, the polymer is distributed uniformly throughout the treated specimen. However, when TBTMA–GMA or TBTMA–MAnh are copolymerized from benzene solutions, polymer distribution in treated wood was not uni-

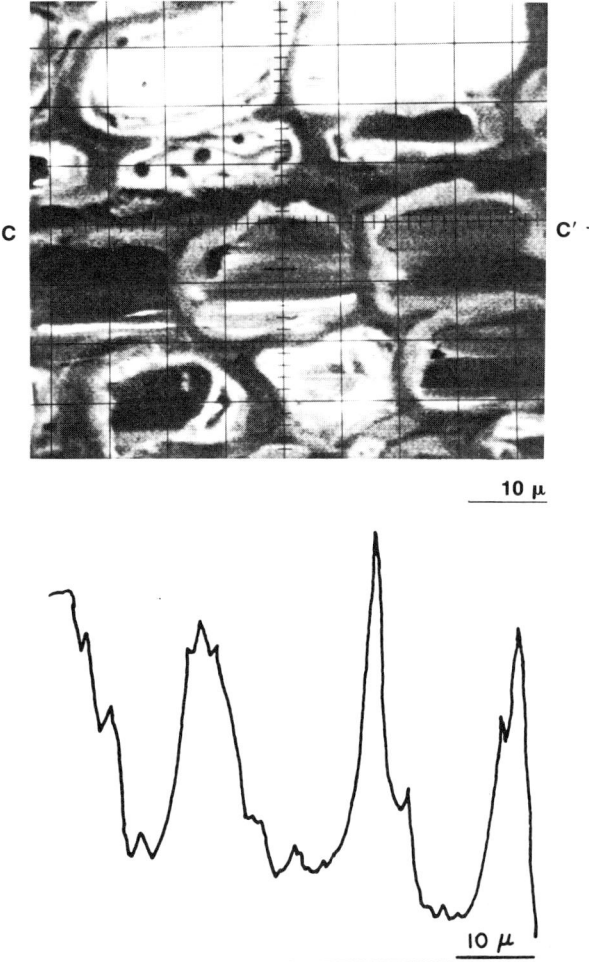

Figure 1. Concentration of tin atoms in the cell walls of TBTMA–GMA-treated wood. Key: top, SEM using secondary emission electrons of treated wood; and bottom, concentration of tin determined by electron microprobe moving along line cc′ (top).

(Reproduced with permission from Ref. 2. Copyright 1981, Springer–Verlag.)

form. Samples taken from the body center of the treated wood showed the highest fraction of empty wood cells, and those taken from the short edge of treated wood showed the highest fraction of filled cells. Thus, the polymer distribution is uniform at the surface but declines toward the center of the sample. Both the nonswelling nature of benzene and the higher viscosity of monomer solutions in benzene contribute to the observed nonuniformity in polymer distribution.

Polymer migration into the cell walls of treated wood can be expected to have important consequences on improvement in mechanical properties, dimensional stability, and resistance to wood decay. Because the copolymer in the TBTMA systems contains tin, the microdistribution of the copolymer in treated wood can be examined by determining the location of tin atoms with the aid of electron microprobe analysis. The results for TBTMA–GMA copolymer are shown in Figure 1. A substantial amount of tin atoms and, presumably, the TBTMA–GMA copolymer are distributed within the cell walls of treated wood. Although benzene is a nonswelling solvent, the copolymer is formed in wood cell walls in detectable amounts.

Organotin polymers are incorporated uniformly into wood by in situ polymerization of organotin monomers. The use of comonomers containing functional groups capable of reacting with wood hydroxyls leads to the grafting of the copolymer to wood. The effects of organotin polymer impregnation on the properties of wood may be examined now.

Dimensional Stability. As expected, the antishrink efficiency (ASE) is increased, and the amount of water absorbed is decreased because of polymer impregnation. Table II summarizes the results of

Table II. Antishrink Efficiency of Wood Treated with Various Polymers

Soaking Time (h)	Antishrink Efficiency[a]					
	GMA (wt %)		TBTMA–MAnh (wt %)		TBTMA–GMA (wt %)	
	25.4	59.6	16.2	57.6	17.6	51.5
1.5	67.0	73.0	68.0	82.0	66.0	77.5
6	62.0	66.0	60.0	76.0	60.5	74.0
18	61.5	62.5	57.5	71.5	56.0	72.5
48	61.0	62.0	52.0	67.0	52.0	71.0

NOTE: Samples had wood fibers parallel to sample length.
[a] Antishrink efficiency = (1 − % swelling of treated wood/% swelling of control) × 100. (Reproduced, with permission from Ref. 3. Copyright 1981, Springer–Verlag.)

Table III. Weight Percent of Water Absorbed by Untreated Wood and Various Treated Woods

Soaking Time (d)	Untreated	GMA (wt %)		TBTMA–MAnh (wt %)		TBTMA–GMA (wt %)	
		25.4	59.6	16.2	57.6	17.6	51.5
0.50	51	25	13	12	4	14	10
1.50	60	39	24	20	13	26	21
4.50	76	58	35	34	23	37	32
8.50	88	64	38	45	28	42	38

(Reproduced with permission from Ref. 3. Copyright 1981, Springer–Verlag.)

ASE determined by thickness (tangential) swelling of treated and untreated grand fir wood specimens immersed in distilled water. The effectiveness of the copolymers TBTMA–MAnh and TBTMA–GMA is slightly better than that of the GMA homopolymer in improving ASE.

Similarly, water absorption determined by weight gain following immersion in water is decreased considerably by polymer incorporation (Table III). For all three polymer types, water absorption decreases with increasing polymer content initially, but levels off to a constant value at higher polymer contents. Thus, all accessible sites for water absorption are not completely sealed off even at the highest levels of polymer loading. Furthermore, the polymers themselves will have some levels of water absorption. The effectiveness in decreasing water absorption decreases in the order TBTMA–MAnh > TBTMA–GMA > GMA. This is the order of decreasing hydrophobicity of the three polymers. The degree of reaction between epoxy groups of GMA and wood hydroxyls is enhanced by TBT groups in the TBTMA comonomer compared to reaction of GMA alone in wood (3). With larger numbers of wood hydroxyls left unreacted, the GMA–wood system shows more water absorption.

Mechanical Properties. Significant improvements in strength properties are observed to varying extents in grand fir specimens treated with the different polymer systems. In specimens with fibers parallel to the length, moderate increases in flexural strength, between 50 and 65%, are observed. The flexural strength is increased much more drastically in the transverse direction, by 317%, 61%, and 243% for GMA-, TBTMA–MAnh-, and TBTMA–GMA-treated wood specimens, respectively (see Figure 2). Such significant improvement in the weak transverse direction is highly desirable.

Similarly, the improvements in flexural modulus of elasticity

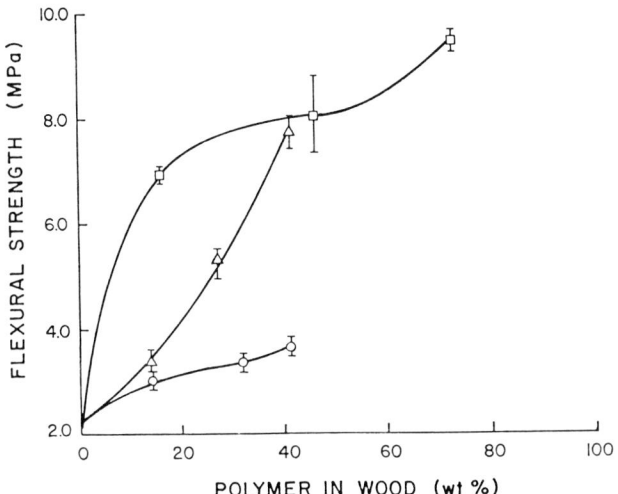

Figure 2. Dependence of flexural strength on polymer content of wood treated with GMA (□), TBTMA–MAnh (○), and TBTMA–GMA (△). The specimens had fibers parallel to their width.
(Reproduced with permission from Ref. 3. Copyright 1981, Springer–Verlag.)

were also moderate in the longitudinal direction. For example, at 40% (by weight) polymer content, the flexural moduli were: 1460 MPa for TBTMA–MAnh, compared to 1265 MPa for TBTMA–GMA, 1110 MPa for GMA, and 1000 MPa for untreated wood. As seen in Figure 3, GMA, TBTMA–MAnh, and TBTMA–GMA bring about improvements to the extent of 535%, 80%, and 456%, respectively, in the transverse flexural modulus. The TBTMA–MAnh copolymer is hard and brittle compared to the other two copolymers. Considering that the TBTMA–MAnh copolymer produces the highest increase in modulus in the longitudinal direction, the observed low efficiency for this polymer in the transverse direction could be due to the formation of longitudinal microcracks in wood tracheids during the treatment process. This hypothesis is consistent with the observation that the tensile strength of specimens with fibers parallel to the width actually decreases with polymer content for TBTMA–MAnh, although increasing measurably for the other two systems (Figure 4). The impact strengths of specimens with fibers parallel to the length and treated with TBTMA–MAnh increase uniformly with polymer content, whereas well-defined maxima are observed for the other two polymers (Figure 5). The complex microstructure and submicroscopic ultrastructure of wood combine very effectively in contributing to the fracture toughness of wood (21).

Currently, only some general features of mechanical property

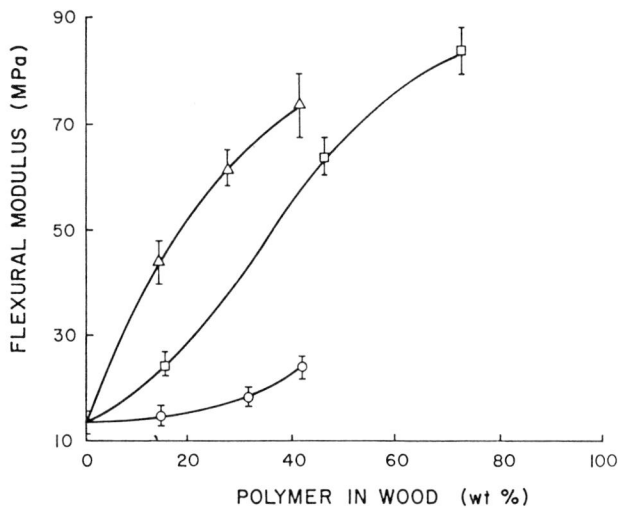

Figure 3. Dependence of flexural modulus on polymer content of wood treated with GMA (□), TBTMA–MAnh (○), and TBTMA–GMA (△). The specimens had fibers parallel to their width.

(Reproduced with permission from Ref. 3. Copyright 1981, Springer–Verlag.)

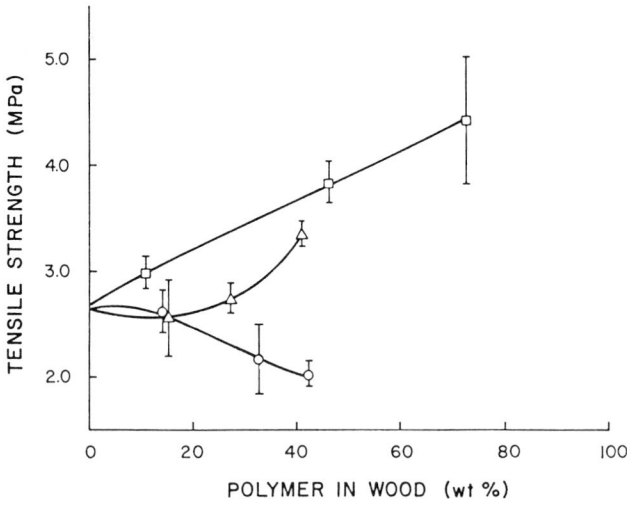

Figure 4. Dependence of tensile strength on polymer content of wood treated with GMA (□), TBTMA–MAnh (○), and TBTMA–GMA (△). The specimens had fibers parallel to their width.

(Reproduced with permission from Ref. 3. Copyright 1981, Springer–Verlag.)

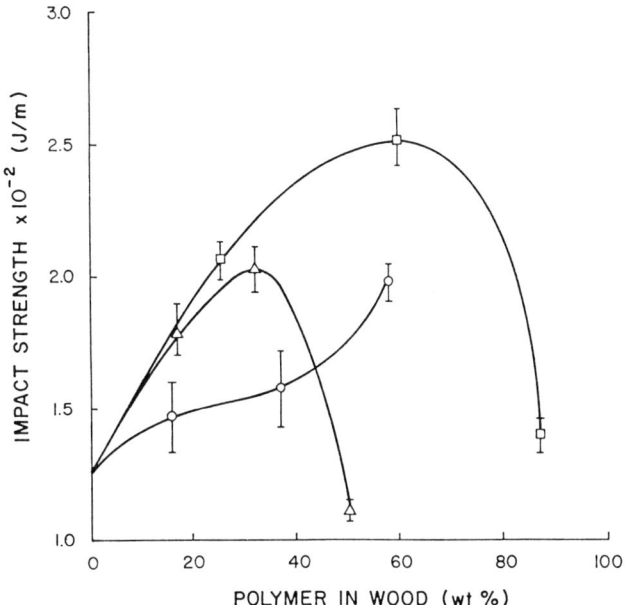

Figure 5. Dependence of impact strength on polymer content of wood treated with GMA (□), TBTMA–MAnh (○), and TBTMA–GMA (△). The specimens had fibers parallel to their length.

(Reproduced with permission from Ref. 3. Copyright 1981, Springer–Verlag.)

improvement by polymer impregnation can be inferred. In addition to filling up voids and microcracks, and thus reducing defects in the structure, the impregnated polymer, because of its high cohesive strength, acts as a strong binder matrix in the wood structure. Better stress transfer within the structural elements of wood may also be facilitated by polymer impregnation. Although it is not easy to interpret the details of the observed effects, it should be rewarding to study fracture and the details of crack initiation and propagation in these systems. These studies should be coupled with an examination of failure modes, as revealed in instrumented impact tests and scanning electron micrographs of fracture surfaces, in order to lead to a better understanding of the changes caused by polymer impregnation of wood.

Biodegradation Resistance. Organotin polymers increase the decay resistance of grand fir wood to attack of several microorganisms example, *Coniophora puteana*, a brown-rot fungus, and *Polyporus versicolor*, a white-rot fungus, completely covered untreated and *versicolor*, a white rot fungus, completely covered untreated and GMA-treated wood blocks in only 4 weeks, while the organotin polymer-treated specimens were, to visual examination, completely free of fungal attack even after 8 weeks of exposure. The untreated control samples exposed to *Coniophora puteana* for 8 weeks suffered a weight loss of 5.2%; the GMA-treated samples showed a 1% weight loss. But the TBTMA–MAnh- and TBTMA–GMA-treated samples showed no weight loss at all. Thus, based on visual appearance and weight loss, the TBTMA–MAnh and TBTMA–GMA treatments provide superior protection against decay as compared to either untreated or GMA-treated wood. This confirms the expectations based on the presence of the bioactive tri-n-butyltin group in TBTMA–MAnh and TBTMA–GMA copolymers and the slow leaching of the toxicant from the polymer-treated wood.

The decay resistance provided by organotin polymers has been confirmed by field tests conducted in the ocean (4, 22). The TBTMA–GMA polymer treatment and other similar polymers carrying TBT functional groups were applied to southern yellow pine (*Pinus* sp) specimens designed for exposure at Key West Naval Station, Key West, Florida. At each semiannual inspection, in June and again in December, the racks of specimens were pulled up, the surface fouling on the coupons was scraped off, and the exposed surface was inspected carefully and rated by degree of attack on a scale from 0 to 10. A rating of 10 signifies no attack; 9, trace attack; 7, moderate attack; 4, heavy attack; and 0, no structural integrity left. Heaviest attack takes place between June and December when the water temperature is higher. This test method corresponds to ASTM D2381 ("Accelerated Evaluation of Wood Preservatives for Marine Services by Means of Small-Size Specimens").

The field test data are given in Table IV, and confirm the evidence obtained in laboratory tests of wood decay with grand fir specimens discussed earlier. After the first 2 years of exposure, none of the specimens with organotin treatments showed any attack. The total failure of untreated samples in this period demonstrates the presence of sample inoculum for attack. After 24 months, the marine creosote-treated controls had an average rating of 8.3 and the special marine creosote treatment resulted in a rating of 9.7. In 36 months, however, they have been attacked and damaged severely, whereas the organotin polymer-impregnated systems do not suffer any deterioration.

The long-term effectiveness of TBTMA as a biocide has been confirmed (23). The deterioration of southern pine wood specimens impregnated with TBTMA has been compared with specimens that were treated with pentachlorophenol methacrylate (PCPMA) or pentabromophenol methacrylate (PCBMA). The three esters were incorporated as the copolymer by in situ copolymerization with methyl methacrylate (MMA) at different molar ratios of the comonomers.

Table IV. Field Test Data for Marine Decay of Wood

Sample Number	Treatment	12 Months	18 Months	24 Months	30 Months	36 Months
A1	P-13 Marine Grade		10	10		10
A2	Creosote (368 kg/m^3)		10	4		—
A3			10	10		4
A4			10	9		—
A5			9	7		—
A6			10	10		4
A13	TBTMA–MMA Copolymer		10	10		10
A14	(223 kg/m^3)		10	10		10
A15			10	10		10
A16			10	10		10
A17			10	10		10
A18			10	10		10
A31	Tributyl Tin Ester of		10	10		10
A32	Methyl Vinyl Ether/Maleic		10	10		10
A33	Anhydride (52.5 kg/m^3)		10	10		10
A34			10	10		10
A35			10	10		10
A36			10	10		10
A37	P-13 Marine Grade		9	7		—
A38	Creosote (254 kg/m^3)		10	10		—
A39			10	10		10
A40			10	10		10
A41			10	10		—
A42			10	10		0
A43	Special Marine Creosote,		10	9		7
A44	40% Naphthalene		10	9		4
A45	(435 kg/m^3)		9	10		4
A46			10	10		10
A47			10	10		10
A48			10	10		9
B44	TBTMA–MMA, in situ	10	10		10	
B45	Polymerization	10	10		10	
B46	(365 kg/m^3)	10	10		10	
B77		10	10		10	
B78		10	10		10	
B79		10	10		10	
B57	TBTMA–GMA in situ	10	10		10	
B58	Polymerization, 10%	10	10		10	
B59	Polymer in Wood	10	10		10	
B60		10	10		10	

(Reproduced from Ref. 4. Copyright 1982, American Chemical Society.)

Both soil block samples and marine exposure samples of southern pine were impregnated thus by MMA, or copolymers of PCPMA, PCBMA, and TBTMA with MMA.

Soil blocks were exposed to a brown rot fungus *Gloeophyllum trabeum* for 12 weeks to determine weight loss by biodecay. The untreated control showed 42% weight loss, and the MMA-treated specimens suffered 17% weight loss. The weight loss of blocks treated with PCPMA–MMA and PCBMA–MMA was close to that observed with MMA treatment alone. The inference was that the reduction in weight loss in these cases was due to the effect of the impregnated polymer as a moisture barrier, and not due to release of the incorporated toxic chemicals, PCPMA or PCBMA. In marked contrast to these results, the weight loss in the presence of TBTMA–MMA was negligible, and shows that TBTMA was indeed effective in preventing attack by the brown rot fungus.

The failure of esters, PCPMA and PCBMA, in these tests to afford any level of protection is probably due to the stability of these esters, which precludes release of the toxic pentachlorophenol and pentabromophenol (23). These observations serve to emphasize the importance of achieving controlled release of toxicants for protection against biological degradation.

In marine tests conducted with specimens treated as described previously, TBTMA–MMA was found effective in preventing attack by marine organisms (23) for over 12 months. These results confirm observations reported earlier (3, 4). The untreated control specimens were largely destroyed by limnoria and teredine borers, while specimens treated with PCPMA–MMA and PCBMA–MMA already showed signs of incipient attack. (These tests are being continued.)

Conclusion

The feasibility of simultaneously improving decay resistance, dimensional stability, strength, and toughness seems well established by the results of in situ polymerization and copolymerization of organotin polymers in wood. Proven techniques of wood impregnation by monomers can be adopted in these experiments to prepare wood–polymer composites stronger than untreated wood. Tributyltin functional groups, incorporated through chemical attachment to the polymerizing monomers, have proved effective in rendering the wood–polymer composites bioactive, with decay resistance superior to that obtained by conventional treatments of wood by creosote. Scientific and technological advances in this field can be expected to proceed rapidly by exploring and exploiting the concepts underlying these early bioactive wood–polymer composites.

Literature Cited
1. Mendoza, J. A. "Wood Preservation by In Situ Polymerization of Organotin Monomers"; Masters thesis, Materials Science and Engineering, Washington State University; Pullman, WA, 1977.
2. Subramanian, R. V.; Mendoza, J. A.; Garg, B. K. *Holzforschung* **1981**, 35, 253.

3. Subramanian, R. V.; Mendoza, J. A.; Garg, B. K. *Holzforschung* **1981**, *35*, 263.
4. Andersen, D. M.; Mendoza, J. A.; Garg, B. K.; Subramanian, R. V. in "Biological Activities of Polymers"; Carraher, C. E., Jr.; Gebelein, C. G., Eds.; ACS SYMPOSIUM SERIES No. 186, ACS: Washington, D.C., 1982; p. 27.
5. Becker, G. *Wood Sci. Technol.* **1974**, *8* (3), 163.
6. Rowell, R. M.; Gutzner, D. I. *Wood Sci.* **1975**, *7* (3), 240.
7. Rowell, R. M.; *Proc.—Annu. Meet. Am. Wood-Preserv. Assoc.* **1975**, *1*, 1.
8. Phillip, A. T. *Prog. Org. Coat.* **1973/1974**, *2*, 159.
9. Hof, T. *J. Inst. Wood Sci.* **1969**, *4* (5), 19.
10. Levi, M. P. *J. Inst. Wood Sci.* **1969**, *4* (19), 45.
11. Cockroft, R. *J. Inst. Wood Sci.* **1974**, *6* (6), 2.
12. Sheldon, A. W. *J. Paint Technol.* **1975**, *47* (600), 54.
13. Dyckman, E. J.; Montemarano, J. A. "Antifouling Organometallic Polymers: Environmentally Compatible Materials"; NSRDC Report 4136, David W. Taylor Naval Ship R&D Center: Annapolis, MD, February 1974.
14. Montermoso, J. C.; Andrews, J. M.; Marinelli, L. P. *J. Polym. Sci.* **1958**, *32*, 523.
15. Subramanian, R. V.; Somasekharan, K. N. *J. Macromol. Sci., Chem.* **1981**, *A16*, 73.
16. Subramanian, R. V.; Garg, B. K. *Polym.–Plast. Technol. Eng.* **1978**, *11*, 81.
17. Subramanian, R. V.; Garg, B. K.; Corredor, J. "Organometallic Polymers"; Carraher, C. E., Jr.; Sheats, J. E.; Pittman, C. U., Eds.; Academic Press: New York, 1978; p. 181.
18. Somasekharan, K. N.; Subramanian, R. V. in "Controlled Release of Bioactive Materials"; Baker, R., Ed.; Academic Press: New York, 1980; p. 415.
19. Stamm, A. J.; Tarkow, H. *J. Phys. Colloid Chem.* **1947**, *51*, 493.
20. Goldstein, I. S.; Dreher, W. A.; Jeroski, E. B.; Nielson, J. F.; Oberley, W. J.; Weaver, J. W. *Ind. Eng. Chem.* **1959**, *51*, 1315.
21. Borgin, K. *New Sci.* **1974**, November 21, 556.
22. Anderson, D. M. Report #DTNSRDC/SME–78/41, David W. Taylor Naval Ship R&D Center, Annapolis, MD, 1979.
23. Rowell, R. M. "Controlled Release Delivery Systems"; Roseman, T. J.; Mansdorf, S. Z., Eds.; Marcel Dekker, Inc.: New York, 1983; Chap. 23.

RECEIVED for review May 17, 1983. ACCEPTED July 15, 1983.

8
Interaction of Preservatives with Wood

DARREL D. NICHOLAS
Mississippi State University, Forest Products Utilization Laboratory, Mississippi State, MS 39762

ALAN F. PRESTON
Michigan Technological University, Institute of Wood Research, Houghton, MI 49931

> *Many of the chemicals used in wood-preservative formulations interact both physically and chemically with the wood substrate. The inorganic salts—principally copper and chromium—are the most chemically reactive components. These compounds undergo complex reactions to form organometallic and inorganic complexes that render them nonleachable. Organic compounds undergo a simple physical interaction with wood, which can be either confined to the gross structure or can involve microdistribution within the cell wall. The type of distribution can significantly affect the performance of preservatives.*

W OOD PRESERVATION IS TOTALLY DEPENDENT on the treatment of wood with toxic chemicals. Because wood contains chemically active functional groups, reactions between the wood components and chemical preservative systems are possible in some cases. In this chapter these interactions will be explored and their significance in the wood-preserving process and resulting treated products will be discussed.

Major-Use Wood Preservatives

Before proceeding with potential interactions between preservative chemicals and the wood substrate, it will be necessary to describe briefly the pertinent chemical and physical properties of the commercial wood preservatives as well as some new biocides that may have future potential in wood preservation.

Pentachlorophenol. In the pure state pentachlorophenol (penta) **1** is a white, needlelike crystalline solid. It is only slightly soluble in water, 14 ppm at 20 °C, but is readily soluble in a number of organic solvents (*1*). Under alkaline conditions, penta can be converted to alkali metal salts that are readily soluble in water. Because of its chlorinated ring structure, pure penta is considered to be inert

$$\underset{1}{\text{C}_6\text{Cl}_5\text{OH}}$$

chemically and does not decompose when heated at temperatures up to its boiling point (1). Unlike most phenols, penta is not subject to easy oxidative coupling or electrophilic substitution reactions. Commercially available penta is not pure and usually contains about 83–84% penta, 6% of the three isomeric tetrachlorophenols, 6% other chlorinated phenols, and the remainder consists of other chlorine compounds and inert materials.

Creosote. The major portion of creosote is derived from coal and is a complex mixture of aromatic hydrocarbons with condensed ring systems. The remaining components are *tar acids*, which are phenolic derivatives of these compounds, and *tar bases*, which are heterocyclic compounds containing nitrogen plus some neutral oxygenated compounds. At least 200 chemical compounds have been identified in coal-tar creosote, but many of these are present in small amounts. The chemical composition is variable, but some idea of a typical creosote is given in Table I (2).

Inorganic Salts. A number of the metal salts have fungicidal activity and are used to formulate commercial wood preservatives. The principal metal salts used are compounds of arsenic, chromium, copper, and zinc. In order to provide the desired fungicidal activity, leach resistance, and low corrosivity, combinations of these compounds are used. All of the formulations discussed later are waterborne solutions. Only a brief description of these preservative systems will be presented here; more detailed presentations can be found elsewhere (3).

AMMONIACAL COPPER ARSENATE (ACA). This preservative is a combination of copper carbonate, arsenic trioxide, and ammonia. The ammonia serves as both a solubilizing and anticorrosion agent.

CHROMATED COPPER ARSENATE (CCA). This preservative is formulated either as a *salt-type* or *salt-free* system, with the latter being

Table I. Chemical Composition of a Typical U.S. Creosote

Component	%
Naphthalene	3.0
Methylnaphthalene	2.1
Acenaphthene	9.0
Dimethylnaphthalene	2.0
Dibenzofuran	5.0
Fluorene-related compounds	10.0
Methylfluorenes	3.0
Phenanthrene	21.0
Anthracene	2.0
Carbazole	2.0
Methylphenanthrene	3.0
Methylanthracenes	4.0
Fluoranthene	10.0
Pyrene	8.5
Benzofluorene	2.0
Chrysene	3.0

the most frequently used. The salt-type is formulated from potassium dichromate, copper sulfate, and arsenic pentoxide, whereas the salt-free system is formulated from chromic acid, copper oxide, and arsenic acid.

ACID COPPER CHROMATE (ACC). This preservative is formulated by mixing copper sulfate, sodium dichromate, and chromic acid.

CHROMATED ZINC CHLORIDE. This preservative is formulated by mixing sodium dichromate and zinc chloride.

Minor-Use Wood Preservatives

There are several organometallic compounds used commercially on a limited basis for wood preservation. The chemical and physical properties of these compounds are discussed here.

Copper Naphthenate. When naphthenic acids are neutralized with alkali, they react with a number of metal salts to form naphthenates. Because of its fungicidal properties, copper is normally the metal used. The naphthenic acids 2 are structurally variable (3). The

$$\text{cyclopentane}-(CH_2)_3-COOH$$
$$\phantom{\text{cyclopentane}}-(CH_3)_2$$

2

copper salts are viscous, waxy compounds and generally are diluted with petroleum solutions.

Copper 8-Quinolinolate (Copper-8). Copper 8-quinolinolate **3** is derived by reacting copper with quinolinol. This compound is

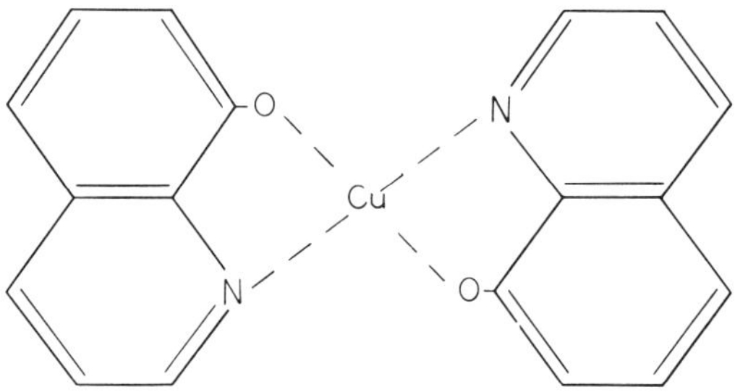

3

formulated in both water and hydrocarbon solvents. Because of its poor solubility, cosolvents such as nickel 2-ethylhexanoate and dodecylbenzenesulfonic acid are used in commercial formulations.

Tributyltin Oxide. Under ambient conditions, tributyltin oxide (TBTO) **4** is a liquid and is soluble in many organic solvents. Solutions

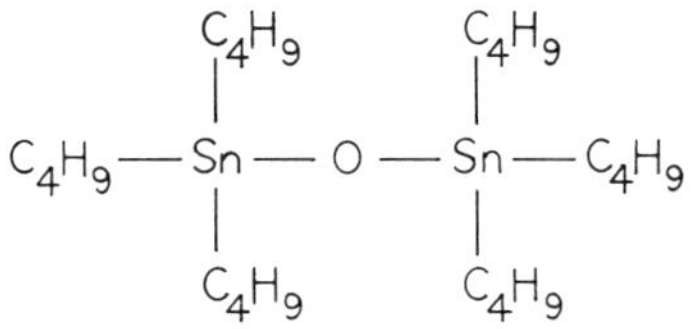

4

of mineral spirits are used for formulating treating solutions, but waterborne systems can be obtained by solubilizing TBTO with alkylammonium compounds (4).

New Wood Preservatives

Currently, the most promising biocides for potential new wood preservatives are the alkylammonium compounds (AAC). These compounds have been subjected to extensive studies (5, 6) and the dialkyldimethylammonium compounds are suggested to be the best (7). Didecyldimethylammonium chloride **5** appears to be particularly effective.

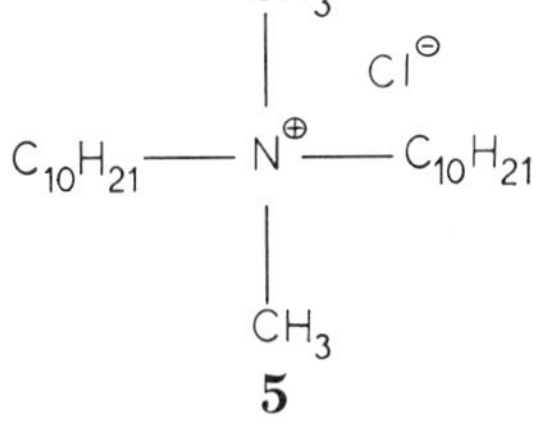

5

Other compounds under consideration include ammoniacal copper-fatty acids, isothiazolinones, benzothiazoles, sulfonamides, tetrachloroisophthalonitrile, salicylanilide derivatives, and 3-iodo-2-propynyl butylcarbamate (8–13).

Preservative Distribution in Wood

When wood is impregnated with wood-preservative solutions, the initial flow is throughout the macrostructure—mainly the cell lumens—and this is termed macrodistribution. Following this, some solutions are capable of undergoing further movement into the cell wall structure, which results in microdistribution of the preservatives. The effect of these two types of distribution on treated wood will be discussed in this section.

Macrodistribution. The ability of any wood preservative to control biodegradation is affected by the macrodistribution of the chemical within the wood product being protected. The macrodistribution of a preservative is influenced by three basic factors: wood characteristics, treating process, and characteristics of the treating solution. Consideration of the principles of flow in wood and of the factors that influence the treatment of wood are covered in Chapters 3 and 4 (14, 15). Suffice it to say that when the preservative has been distributed through the wood, fixation will occur either through chemical interaction between the preservative and the wood structure, between the preservative components themselves, or by physical deposition as a result of solvent loss. These fixation mechanisms are covered in the section on microdistribution.

The macrodistribution of a preservative has an obvious effect on the ability of the treatment to protect the wood product. The longevity of the protection is also influenced by changes in the macrodistribution with time and exposure to weathering. With multicomponent preservatives such as CCA, movement of just one of the components, usually copper, can have a marked effect on performance. Depending on the severity of exposure, redistribution effects can have either a positive or a negative effect on field performance. For instance, the movement of the copper component in CCA-treated pine stakes upon leaching to give a higher loading in the outer zones will give an apparent enhanced performance over that predicted in laboratory tests (16). However, this same type of phenomenon, which also occurs in marine exposures of CCA-treated pine, may eventually cause major problems through sudden failure of CCA-treated marine piling (17).

One problem that concerns macrodistribution occurs with the use of Douglas-fir in situations where treatment is necessary. The heartwood of this species is nondurable and very difficult to treat.

The abundance of the resource and the highly desirable strength properties have led to widespread use of treated Douglas-fir with subsequent biodegradation problems occurring in poorly treated products (18). Extensive research has been carried out to develop fumigant treatments for internal decay in poles (19). In Hawaii the problems of termite attack are being considered now (20, 21).

Microdistribution. Although studies on the macrodistribution of preservatives have diminished in number over the last decade, research on microdistribution effects has increased considerably. Two major reasons for this increase are the availability of scanning electron microscopy–energy dispersion X-ray analysis (SEM–EDXA) techniques for monitoring the position of preservatives in the wood structure, and the attention being given to the poor performance of in-ground hardwoods, particularly when treated with CCA preservatives. It is routine practice to analyze the treatment properties of any new candidate preservative by the SEM–EDXA technique. A notable example in the United States is in the development of waterborne pentachlorophenol preservatives, both solutions and emulsions. Analysis by SEM–EDXA shows that these formulations penetrate into and through the cell wall (22). Apparently the aqueous emulsions only break on deposition of the pentachlorophenol as the water evaporates.

The greatest interest in the effects of distribution on preservative efficacy has centered recently on the variable performance characteristics of treated hardwoods in ground contact. Attention has centered around the waterborne preservatives of the CCA type. Many theories have been promulgated to explain why wood treated seemingly to levels considered sufficient for long-term protection can suffer severe soft-rot degradation and rapid failure in-service. One explanation of the phenomenon is known as the microdistribution theory (23–25). This theory suggests that the preservative is distributed poorly in the wood, both at the macroscopic level and microscopically within cell walls; that is, the fungitoxic elements of the preservative do not penetrate into the bulk of the fiber walls during treatment. Although the overall preservative loading is adequate for protection of the wood, most of the preservative remains close to the original penetration pathways in the vessels and rays. Soft-rot microfungi are thus able to grow in the relatively unprotected wood cells remote from the main concentrations of preservatives.

Undoubtedly, distribution patterns do account for soft-rot attack in a number of cases but they cannot be the sole explanation. Butcher (26) has shown that there is a correlation between the natural susceptibility of timber to soft-rot and the level of copper required to prevent attack. This concept is based on the idea that soft-rot fungi are able to tolerate higher levels of toxicants if the substrate is comparatively rich in nutrients. Support for this idea comes from studies that showed that an increase in carbohydrates and nitrogen levels in the substrate increases the tolerance of the fungi to toxicants (27, 28).

The highly susceptible hardwoods have a higher than average content of cellulose and hemicellulose that can be utilized for energy required for processes such as colonization, microbial transport of nitrogen into wood, and enzyme production. However, along with a high carbohydrate content, the highly susceptible hardwoods are low in lignin. This last factor is believed to play a major role in the soft-rot susceptibility of treated hardwoods and it also explains and is compatible with the microdistribution theory (29).

In high lignin content wood species the lignin impeded enzymatic degradation of the microfibrils through a barrier effect (30). Both the level and type of lignin are important as the guaiacyl lignin predominant in softwoods appears to present a greater barrier to decay than syringyl–guaiacyl lignin found in susceptible hardwoods. This hypothesis also assumes that lignin is the major fixation site for copper—which appears to be the case—and that copper can be fixed to retention levels in the S_2 layers of the cell walls of high-lignin wood species which are in excess of toxic thresholds. Conversely, in low-lignin species, the level of copper fixed will be below the toxic threshold. Generally, this observation is compatible with the earlier microdistribution theory because microdistribution will be influenced by the position and levels of the lignin within the wood cell wall.

The lignin hypothesis implies that the active copper fungitoxicant is fixed by ion exchange to the lignin. The importance of the fixation mechanisms has not been accounted for, but soft-rot attack in hardwood treated with CCA was accelerated when the wood was subjected to rapid fixation of the preservative at elevated temperatures after treatment (31). Undoubtedly, the change in fixation mechanisms caused by the heat treatment has a marked effect on the control of soft-rot decay. This observation is also compatible with the fixation and lignin hypothesis theories.

The understanding of decay and protection processes in studies on the soft-rot problem of CCA-treated hardwoods will be of considerable value in the development of the next generation of wood preservatives.

Chemical Reactions of Preservatives with Wood

Because wood contains several different types of functional groups, chemical reactions with wood-preservative chemicals are a possibility. Depending on the circumstances, such reactions can be either advantageous or detrimental. For example, in some cases these reactions can result in the fixation of water-soluble chemicals and prevent their depletion from the wood in-service. Conversely, some chemical reactions can inactivate preservatives and, therefore, are undesirable.

Inorganic Salt Preservatives. Inorganic compounds used to formulate wood preservatives are normally water soluble. As a result, the salts deposited in the wood are susceptible to leaching unless they are transformed to insoluble compounds or are chemically fixed to the wood substrate. Indeed, such conversions do occur as a result of interactions between some of the salts and the wood substrate. Both single element and multicomponent reactions are involved in the fixation mechanisms; copper and chromium are the most reactive of the possible components.

Probably the most significant effect of the interaction of inorganic salts with wood is the ultimate fixation or insolubilization of these compounds which makes them resistant to leaching. The overall scheme of these complex reactions was initially elucidated by Dahlgren (32–36) and Dahlgren and Hartford (37–39). Subsequently this scheme was modified by Pizzi (40); Figure 1 is a schematic diagram based on his work.

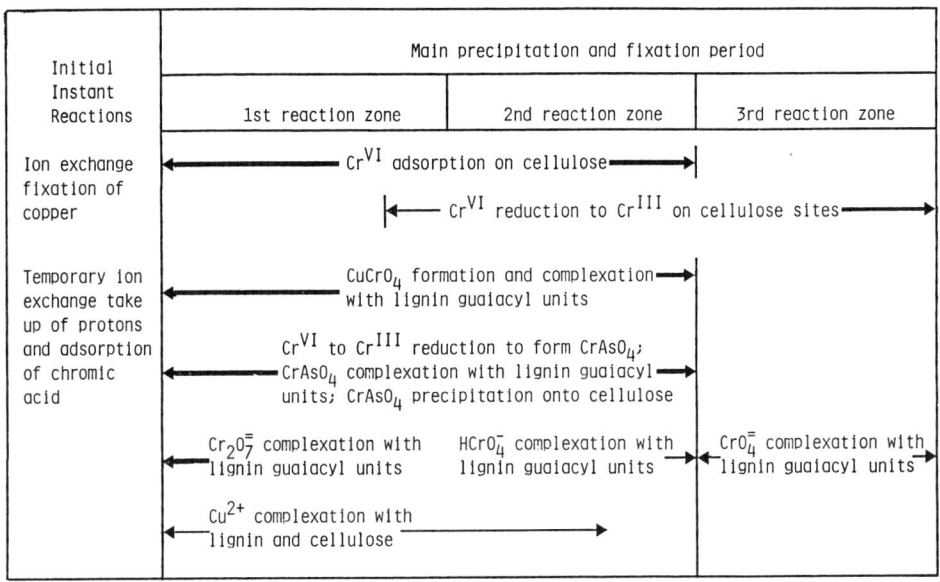

Bold line arrows indicate predominant reactions in reaction zone.

Figure 1. Schematic diagram of fixation of CCA wood preservatives. (Reproduced with permission from Ref. 40. Copyright 1982, John Wiley & Sons, Inc.)

The initial instant reactions of CCA with wood have been shown to result in a rapid decrease in the pH (41). The magnitude of this pH drop is dependent on the strength of the treating solution, where a higher CCA concentration produces a lower pH value. The reason for this drop is unknown but may be attributable to the ion-exchange fixation of copper which releases protons. However, other reactions may also be involved. Following this initial decrease, the pH gradually rises as the fixation reactions continue. This increase is attributable to the formation of chromic acid–lignin and copper chromate ($CuCrO_4$)–lignin complexes, and to the reduction of chromium.

A summary of the products formed when CCA reacts with wood is as follows (40):

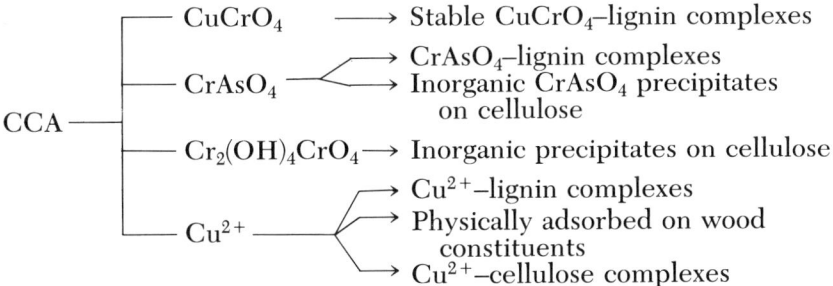

The products formed when CCA is reacted with wood are composed mainly of the $CuCrO_4$–lignin complexes, $CrAsO_4$–lignin complexes, and $CrAsO_4$ precipitates on cellulose. However, the actual proportions of various chemical species are dependent on the relative proportions of Cu, Cr, and As present in the formulation (40).

Copper reacts with both the cellulose and lignin components of wood (29, 30, 40, 42–46). In one study (45), the majority of the copper was located in the primary wall and S_1 layer. This result suggests that copper is associated with lignin because the bulk of this component is located in these layers (47). However, a large part of the copper in wood treated with a CCA solution has been demonstrated to be closely associated with the cellulose, and probably bound to the hydroxyl groups (44, 48). The exact nature of the copper–wood reaction is not known, but a drop in pH was observed (43) during the adsorption of metals on cellulose. The pH drop suggests removal of protons to form complexes.

Only about 10% of the copper in the system is bound to the chromium. The rest of the copper is bound, independent of other metals, to the cellulose and lignin. Copper is present in the treated wood in four different forms (40): (1) $CuCrO_4$ bound to guaiacyl units of lignin by Cr^{6+} (10–15%); (2) Cu^{2+} directly bound to carbohydrates and lignin guaiacyl units (10–22%); (3) Cu^{2+} directly bound to lignin functional groups other than guaiacyl units (40–70%); and (4) $CuSO_4$ physically adsorbed by the various wood constituents (5–20%).

The majority of the copper (83–90%) is associated with the lignin in wood (40).

Chromium also reacts with wood to form complexes. The exact nature of these complexes is unknown but probably occurs in both the hexavalent and trivalent states. In the hexavalent state, the reaction is probably with the lignin component where it apparently forms a complex with the guaiacyl units (49). This reaction is probably similar to that of chromium with an o-hydroxyphenol such as 1,2-benzenediol.

With chromium trioxide, the following polymeric complex is postulated (49):

$$\left[\left[\underset{2}{\text{guaiacyl-OH, OCH}_3} \right] \cdot \left[CrO_3 \right] \right]_n \quad \text{WITH } n \geq 2$$

6

The formation of this complex provides a significant amount of water repellency, which undoubtedly contributes to the efficacy of CCA-treated wood against degradative elements.

Distribution studies of chromium within the cell wall show high concentrations in the primary wall and S_1 layer (areas of high lignin content), which again suggests preferential association with lignin (50).

When chromium is reduced from the hexavalent to the trivalent state it reacts readily with arsenic to form $CrAsO_4$, which, in turn, has the ability to complex with the lignin and cellulose. In treated wood, approximately 85% of the arsenic reacts with chromium; the remaining arsenic forms fairly soluble complexes with lignin and cellulose (51).

Although the interaction of CCA with wood results in the desirable conversion of the chemicals from a water-soluble mixture to a highly water-insoluble complex, it also has an undesirable effect, namely, a decrease in strength properties (see Chapter 5). For example, the toughness of wood has been shown to be reduced significantly after treatment and redrying (41). Interestingly, the degree of strength loss is dependent on the CCA retention, with a higher loss being incurred in material treated to higher retentions. The exact reason for this strength loss is unknown, but may be associated with the decrease in pH associated with the first stage of fixation. This contention is supported by the fact that a lower pH and higher strength loss was obtained at higher CCA retentions (41).

The chemical reactions and mechanism of fixation of the ammoniacal preservatives such as ACA have not been studied extensively. The main mechanism of fixation is believed to be the formation of insoluble copper arsenate upon evaporation of the ammonia. However, the overall mechanism is undoubtedly more complex because cuprammonium ions react by ion exchange with functional groups, such as carboxyl, in wood (52). In addition, copper complexes can be formed with cellulose (52).

Needless to say, the fixation of inorganic chemicals in wood by interaction with the wood substrate and extractives is beneficial and greatly improves the durability of these preservatives. Contrarily, other interactions provide less desirable reactions. For example, under certain circumstances copper and zinc can become so tightly bound to the wood that their efficacy as wood preservatives is reduced. This result occurs when copper acetate and zinc acetate are used to treat wood (53). In this form, these elements are salts of acetic acid and they form ion-exchange bonds with the wood components that are stable in the weak acid environment and cannot be ionized readily by water. Conversely, this reduction in efficacy does not occur when copper sulfate and zinc chloride are used because they are salts of strong acids and the pH of the environment prevents insolubilization of these elements by the wood.

Another detrimental effect of preservative–wood interaction is the formation and buildup of colloidal material in the treating solutions as it is recycled. This particulate matter arises from an interaction between the preservative chemicals and wood components and often results in a significant decrease in the penetrability of the treating solution (54).

Organic Preservatives. In contrast to inorganic salts, the organic preservatives are much less reactive with wood. This is particularly true for the standard preservative—creosote and pentachlorophenol (penta). Nevertheless, some of the other minor-use or new preservatives are more reactive and undergo various reactions with the wood substrate. This reactivity has a significant influence on their performance.

TBTO undergoes a condensation reaction with cellulose to form a tributyltin alkoxide (57). This reaction with cellulose could prevent metabolism by inhibiting the extracellular enzymes of the wood-destroying fungi. However, such a reaction is unlikely to be initiated in situ because the alkoxide is highly susceptible to hydrolysis (58).

If a reaction between cellulose and TBTO does occur, it is far more likely that it would take place via electron donation from, or hydrogen bonding with, the oxygen atom of the compound (58). However, such interactions are still speculative and must be verified by experimentation.

Over a period of time, the TBTO in treated wood is degraded (59). This degradation involves the carbon–tin bonds and results in a progressive formation of dibutyl, monobutyl, and, in some cases, inorganic tin derivatives, all of which are far less toxic to wood-decay fungi. The exact mechanism for this degradation is not known but appears to involve molecular oxygen because organic coatings of various types retard the reaction (59).

Because of these detrimental reactions, the future of TBTO as a commercial wood preservative is in question.

When alkylammonium compounds are in the salt form, such as alkyldimethylammonium chloride, they are readily soluble in water. However, after they are impregnated into wood, they become fixed and very little can be leached out with water. It has been hypothesized that the fixation mechanism is an ion-exchange reaction with the carbonyl groups in wood. If this hypothesis is correct, the main reaction sites would be in the hemicellulose and lignin components. However, in recent unpublished studies at the Institute of Wood Research on the adsorption of alkylammonium compounds and copper salt formulations by wood, it has been found that alkylammonium compounds and copper from a mixed solution compete for the same ion-exchange sites, with the alkylammonium compound being preferentially adsorbed. As is the case with the fixation of CCA, the site of ion exchange is presumably within the lignin structure (29, 30). In addition to ion exchange, the alkylammonium compounds also will be fixed within the wood by ion-pair sorption (60) as is the case with their sorption onto cotton (61).

Literature Cited

1. Benvenue, A.; Beckman, H. *Residue Rev.* **1967**, *19*, 83–134.
2. Lorenz, L. F.; Gjovik, L. R. *Proc.—Annu. Meet. Am. Wood-Preserv. Assoc.* **1972**, *68*, 32–41.
3. Hartford, W. H. "Wood Deterioration and Its Prevention by Preservative

Treatments"; Nicholas, D. D., Ed.; Vol. II; Syracuse Univ. Press: Syracuse, NY, 1973; pp. 1–20.
4. Richardson, B. A.; Cox, T. R. G. *Tin Its Uses* **1974**, *102*, 6–10.
5. Butcher, J. A.; Preston, A. F.; Hedley, M. E.; Cross, D. J. *Proc. N. Z. Wood Preserv. Assoc.* **1978**, *17*, 36–45.
6. Nicholas, D. D.; Preston, A. F. *Proc.—Annu. Meet. Am. Wood-Preserv. Assoc.* **1980**, *76*, 13–21.
7. Preston, A. F.; Nicholas, D. D. *Wood Fiber* **1981**, *14*(1), 37–42.
8. Butcher, J. A.; Hedley, M. E.; Preston, A. F.; Cross, D. J. "Alternative Chemicals for Protection of Wood: A New Zealand Perspective." Internat. Union of For. Res. Organiz. World Congress, Div., Proc., 1981, 311–13.
9. Miller, G. A.; Lovegrove, T. *J. Coat. Technol.* **1980**, *52*(661), 69–72.
10. Hedley, M. E.; Preston, A. F.; Cross, D. J.; Butcher, J. A. *Wood-Preserv. Int. Biodeterior. Bull.* **1979**, *15*(1), 9–18.
11. Greaves, H.; Adams, N.; McCarthy, D. F. *Holzforschung* **1982**, *36*, 225–31.
12. Preston, A. F.; McKaig, P. A.; Walcheski, P. J. *Proc.—Annu. Meet. Am. Wood-Preserv. Assoc.* **1983**, in press.
13. Winebrenner, L. I. *Plant Eng.* **1982**, *6*, 58–59.
14. Nicholas, D. D.; Siau, J. F. "Wood Deterioration and Its Prevention by Preservative Treatments"; Nicholas, D. D., Ed.; Vol. II; Syracuse Univ. Press: Syracuse, NY, 1973; pp. 299–344.
15. Siau, J. F. "Flow in Wood"; Syracuse Univ. Press: Syracuse, NY, 1972.
16. Butcher, J. A.; Greaves, H. *Int. Res. Group Wood Preserv.* **1982**, Document No. IRG/WP/3188.
17. McQuire, A. J. *N. Z. For. Serv.* **1976**, Reprint No. 1083.
18. Graham, R. D.; Corden, M. E. *Spec. Rep.—Electr. Power Res. Inst. (Palo Alto, Calif.)* **1977**, Report No. EL–366.
19. Graham, R. D. *For. Prod. J.* **1979**, *29*(9), 21–30.
20. Botsa, E. E. *Build. Ind. Dig. Hawaii* **February 1982**, A–4.
21. Black, J. *Build. Ind. Dig. Hawaii* **March 1982**, A–7.
22. Hatcher, D. B. *Proc.—Annu. Meet. Am. Wood-Preserv. Assoc.* **1980**, *76*, 308–320.
23. Dickenson, D. J.; Sorkoh, N. A. A. H.; Levy, J. F. *Rec. Ann. Conv. Br. Wood Preserv. Assoc.* **1976**, 25–40.
24. Greaves, H. *Holzforschung* **1974**, *28*, 193–200.
25. Greaves, H.; Nilsson, T. *Holzforschung* **1982**, *36*, 207–13.
26. Butcher, J. A. *Mater. Org.* **1979**, *14*(3), 215–34.
27. Henningson, B. *Mater. Org.* **1975**, *3*, 175–85.
28. Hulme, M. A.; Butcher, J. A. *Mater. Org.* **1977**, *12*, 223–34.
29. Nilsson, T. *Int. Res. Group Wood Preserv.* **1982**, Document No. IRG/WP/1167.
30. Butcher, J. A.; Nilsson, T. *Int. Res. Group Wood Preserv.* **1982**, Document No. IRG/WP/1151.
31. Preston, A. F.; McKaig, P. A. *For. Prod. J.* **1983**, *33*, in press.
32. Dahlgren, S. E. *Rec. Ann. Conv. Br. Wood Preserv. Assoc.* **1972**, 109–28.
33. Dahlgren, S. E. *J. Inst. Wood Sci.* **1973**, *6*(4), 28–30.
34. Dahlgren, S. E. *Holzforschung* **1974**, *28*(2), 58–61.
35. Dahlgren, S. E. *Holzforschung* **1975**, *29*(3), 84–95.
36. Dahlgren, S. E. *Int. Res. Group Wood Preserv.* **1976**, Document No. IRG/WP/358.
37. Dahlgren, S. E.; Hartford, W. H. *Holzforschung* **1972**, *26*(2), 62–69.
38. Dahlgren, S. E.; Hartford, W. H. *Holzforschung* **1972**, *26*(3), 105–13.
39. Dahlgren, S. E.; Hartford, W. H. *Holzforschung* **1972**, *26*(4), 142–49.
40. Pizzi, A. *J. Polym. Sci., Polym. Chem. Ed.* **1982**, *20*, 739–64.
41. Winandy, J. E.; Bendtsen, B. A.; Boone, R. S. *For. Prod. J.* **1983**, *33*(6), 53–58.
42. Bayley, C. H.; Rose, G. R. F. *Nature (London)* **1960**, *185*(4709), 313–14.
43. Belford, D. S.; Meyers, A.; Preston, R. D. *Biochim. Biophys. Acta* **1959**, *34*, 47–58.

44. Belford, D. S.; Cook, C. D. *Wood* **1960**, *25*(8), 330–32.
45. Bland, D. E. *Nature (London)* **1963**, *200*(4903), 267.
46. Pizzi, A. *J. Polym. Sci., Polym. Chem. Ed.* **1982**, *20*, 707–24.
47. Panshin, A. J.; deZeeuw, C. "The Textbook of Wood Technology," 4th ed.; McGraw-Hill: New York, 1980; 705 pp.
48. Belford, D. S.; Preston, R. D.; Cook, C. D.; Nevard, E. H. *Nature (London)* **1957**, *180*(4595), 1081–83.
49. Pizzi, A. *Holzforschung und Holzrerwertung* **1979**, *31*(6), 128–30.
50. Yata, S.; Mukudai, J.; Kajita, H. *Mokuzai Gakkaishi* **1982**, *28*(1), 10–16.
51. Pizzi, A. *J. Polym. Sci., Polym. Chem. Ed.* **1982**, *20*, 725–38.
52. Hulme, M. A. *Rec. Ann. Conv. Br. Wood Preserv. Assoc.* **1979**, 38–50.
53. Goldstein, I. S.; Dreher, W. A.; Jeroski, E. B. *For. Prod. J.* **1961**, *11*(3), 128–30.
54. Nicholas, D. D. *For. Prod. J.* **1972**, *22*(5), 31–36.
55. Arsenault, R. D. "Wood Deterioration and Its Prevention by Preservative Treatments"; Nicholas, D. D., Ed.; Vol. II.; Syracuse Univ. Press: New York, 1973; pp. 121–278.
56. Resch, H.; Arganbright, D. G. *For. Prod. J.* **1971** *21*(1), 38–43.
57. Richardson, B. A. "Action Mechanism of Some Organometallic Preservatives. In Biodeterioration of Materials, Microbiological and Allied Aspects"; Walters, A. H.; Elphick, J. J., Eds.; Elsevier: New York, 1971; pp. 498–505.
58. Beaumont, H. G.; MacKay, C. A. *Int. Pest Control* **1974**, *16*(3), 8–11.
59. Henshaw, B. G.; Laidlaw, R. A.; Orsler, R. J.; Carey, J. K.; Savory, J. G. *Rec. Ann. Conv. Br. Wood Preserv. Assoc.* **1978**, 19–29.
60. Rosen, M. J. *J. Am. Oil Chem. Soc.* **1975**, *52*, 431–35.
61. Bender, M.; Carmello, L. *J. Colloid Interface Sci.* **1982**, *86*(1), 266–73.

RECEIVED for review May 19, 1983. ACCEPTED July 7, 1983.

SURFACE CHEMISTRY

9
Chemistry of Adhesion

R. V. SUBRAMANIAN
Department of Materials Science and Engineering, Polymeric Materials Section, Washington State University, Pullman, WA 99164

The basic features of the structure, properties, and modification of polymeric adhesives are discussed as they relate to adhesive behavior with special reference to wood substrates. Adhesive properties and bond performance, including durability, are derived from chemical composition (functional groups), molecular organization (branching, molecular weight distribution, crosslinking) and physical state (elastomer, thermoplastic, thermoset, crystalline). The various types of adhesives include those derived from naturally occurring polymers—carbohydrates, proteins, and natural rubber— and those derived from the multitude of synthetic polymers—phenolics, epoxies, acrylics, elastomers, urethanes, etc. The transformation of liquid adhesive to a solid bond calls for curing reactions of reactive components or film formation from emulsion, solution, or melt. The properties of adherend surfaces play a critical role in the wetting and bonding of substrates as do modifications of interfacial interactions by chemical surface treatments and coupling agents. Resistance to environmental effects of humidity, temperature, microbial attack, etc. determines the durability of bonds.

ADHESIVE BONDING PLAYS A VITAL ROLE in materials engineering—the design, fabrication, and application of new and improved products from a variety of materials. The science of adhesion has developed into a multidisciplinary study involving the chemistry and physics of adherend surfaces and adhesives and the fracture mechanics of adhesive joints. Because the adhesives in use are virtually all polymers, an understanding of polymer science is required, including the chemistry of polymerization reactions, and the rheology, deformation, and fracture behavior of polymers (1–3). This chapter discusses the chemistry of adhesion as it relates to the many aspects of adhesive bond formation and performance.

0065–2393/84/0207–0323/$07.50/0
© 1984 American Chemical Society

In focusing attention on wood bonding, we are dealing with nature's own unique material whose sophisticated structure and complexities are at the same time baffling and challenging. Thus, there are several complicating factors in the study of wood as an adherend: the species; heartwood; sapwood; earlywood; latewood; surface planes in radial, tangential, longitudinal, or intermediate directions; pH; porosity; moisture content; and extractives are all capable of modifying the bonding properties of wood (4, 5). Of these, some, like pH, moisture content, or extractives, have a more direct influence on adhesive chemistry and need to be considered carefully.

Adhesive Bond Formation

Adhesives wet, flow, and set to a solid during bond formation. The transformation from liquid adhesive to solid bond can be achieved in a number of ways. Where the adhesive is a polymer, the initial starting material is a liquid monomer or prepolymer that, under the conditions of bonding with heat, pressure, and/or catalyst, polymerizes to the solid polymer in the glue line. It is also usual to apply solutions of preformed polymers in suitable solvents to the faces of adherends, and allow bond formation to take place with evaporation of solvent. Alternatively, polymers that can be melted or softened to flow at elevated temperatures can be applied as *hot-melt* adhesives to form the bond on cooling. With porous adherends like wood, penetration of the pores by liquid or molten adhesives is an important factor in bond formation.

The transformation of liquid monomers to solid adhesive involves an increase in molecular size and molecular weight through polymerization. The increase in molecular weight of the polymeric adhesive is responsible for the attainment of adequate mechanical properties, cohesive strength, impact strength, etc., and should, therefore, be allowed to proceed to the required levels during bond formation.

In polymer adhesives synthesized separately for adhesive application from solution or as hot melt, the required solubility, softening temperature, viscosity of melts, and flow properties are dependent upon the molecular weight of the polymer and further, on the molecular weight distribution. The presence of low molecular weight polymer among the polymer molecules of higher molecular weight facilitates easier melting and flowing of the adhesive. Furthermore, because wood is a porous adherend, low molecular weight polymers achieve better penetration of the pores to enhance the resulting adhesion.

Thermoplastic and Thermosetting Polymers. Polymer adhesives that dissolve in solvents, or that soften and flow on heating and solidify on cooling, are able to do so because the long-chain polymer molecules are essentially linear, perhaps with occasional branches, but are not chemically bonded to each other. Because of their long chains and high molecular weights, the polymer molecules have physical entanglements with each other. As the temperature of the polymer is raised, various segments of the long polymer molecule acquire mobility at a temperature called the glass transition temperature, T_g. Below this temperature, large-scale segmental mobility is frozen, and the amorphous polymer is in the glassy state; above T_g, the onset of segmental mobility enables the polymer to respond quickly to any applied stress, move into the rubbery state, and exhibit the typical elasticity of elastomers. Polymer flow at still higher temperatures occurs as a consequence of the vigorous movements of various segments that shift the center of gravity of the polymer molecule progressively. Such a polymer is thermoplastic, and forms the basic constituent of soluble or hot-melt adhesives.

A polymer is in the glassy state at room temperature if the T_g is above room temperature, and it behaves as an elastomer if the T_g is considerably below room temperature. The T_g of a polymer depends on the freedom of rotation available about bonds in the polymer chain. Flexible polymers are made by the introduction of groups about which rotation is facile, like $-O-$, $-Si-O-Si-$, CH_2-S-, $=CH-CH_2-$. Discussions of different adhesive systems provide a number of examples of such groups being introduced in otherwise intractable polymers to make them flexible and tough. *Plasticizers* are also employed to make rigid polymers pliable. Plasticizer molecules, by their interaction with polymer chains, reduce interchain interactions and consequently, their rigidity.

In contrast to the linear thermoplastic polymers, which are soluble and fusible, the cross-linked network polymers are insoluble and infusible. They are formed from polymerizing systems containing monomers or prepolymers with a functionality of three or more. A good example is the phenol–formaldehyde resin systems. The cross-linking reaction takes place in the bond under applied pressure and heat, and the whole adhesive bond might consist of only one super giant molecule. Such resins are, therefore, called thermosetting resins.

The cross-link density and the spacing of cross-links along the polymer chain determine the relative flexibility of the rigid network structures, which are naturally more resistant to heat, solvents, and chemical attack than uncross-linked polymers. Flexibilizing polymers or fillers are also added to thermosetting resins to improve toughness and reduce brittleness of the adhesive.

An important aspect of forming a cross-linked structure is the extent of reaction of the available functional groups. As the polymer molecule grows in size its T_g may rise above the temperature of cure. Segmental mobility is therefore lost, and there remain a few functional groups in the polymer chain immobilized and incapable of reaction at that temperature. However, on raising the temperature again, molecular mobility is restored and more functional groups can react to form more cross-links and a stronger network structure. The increase in bond strength observed on post-curing is attributed to fuller utilization of the reactive groups present.

Molecular Forces Between Adherend and Adhesive. The various theories of adhesion invoke the occurrence and interplay of physical and chemical interactions across the adherend–adhesive interface, as well as the deformation behavior of the adhesive (6, 7). Therefore, bond formation depends upon the development of intermolecular attraction, both within the bulk of the polymer and between adhesive and adherend.

Of the different types of forces responsible for intermolecular attraction, the foremost are the London or dispersion forces that act between all atoms and account for virtually all of the molecular attraction or cohesion in all molecules except the very polar molecules (described later). Dispersion forces are short-range interactions, effective at about 4 Å, and rapidly decrease with the sixth power of the distance between molecules. Therefore, the adhesive polymer molecule must be flexible enough to come within this range of interaction with the rigid adherent surface under the conditions of bond formation.

Another interaction occurs between dipoles in molecules. Dipoles arise when the electrons of a chemical bond between atoms are not shared equally, thus creating positive and negative charge centers in the molecule. The interaction forces between permanent dipoles of polar molecules depend on the strength of the two dipoles, and decrease with the sixth power of the distance between their centers. Clearly, the dipolar interaction of polymeric adhesives will be strong when they carry polar chemical groups.

A particularly strong type of dipolar attraction results when the positive center is associated with the hydrogen atom attached to an electronegative atom, most commonly nitrogen or oxygen, as in the following examples:

$$\overset{\delta+~\delta-}{-C=O}\ldots\ldots\overset{\delta+~\delta-}{H-N-} \qquad \overset{\delta-~\delta+}{-O-H}\ldots\ldots\overset{\delta-~\delta+}{O-H}$$

This proton sharing between electronegative atoms is called the hydrogen bond. It occurs in polymers carrying amide (–CONH–), carboxyl (–COOH), or hydroxyl (–OH) groups, and contributes significantly to the adhesion to polar substrates of adhesives such as proteins, starch, polyvinyl alcohol, epoxy resins, phenolics (8), and especially to wood which has an abundance of hydroxyl groups.

The molecular forces are secondary or van der Waals forces. It is also conceivable that primary valence forces form chemical bonds, either covalent or ionic, between adhesive and adherend. The contribution of covalent bonds to bond strength is a subject of great, if sometimes controversial, interest (6).

Adhesives for Wood

The variety of wood products has increased enormously over the past decades as more adhesives have been found for bonding. The most important products in terms of volume are plywood, particle board, and fiber board. But adhesively bonded products range from tiny articles of jewelry to giant laminated timbers spanning hundreds of feet (5). The modification of adhesive properties to suit the different application requirements requires a sound understanding of the basic chemistry of adhesives.

In the following discussion, only the most widely used adhesive types are described. These are the urea–formaldehyde (UF) resins, melamine–formaldehyde (MF) resins, phenol–formaldehyde (PF) resins, diisocyanates, polyisocyanates, polymers and copolymers of vinyl acetate, and polyamides. These are all predominantly thermosetting resin systems.

Some general observations can be made regarding the above thermosetting resin systems before discussing their chemistry (9). The cohesive strength of these thermosetting resins exceeds the tensile strength of wood. The adhesive prepolymer resins have high enough polarity and low enough viscosity to penetrate into the micropores of wood and bring about mechanical anchorage of the adhesive at the outset of the bonding process. The polar groups are capable of forming strong hydrogen bonds with the hydroxyl groups in wood. Thus strong dipole interactions are formed in addition to secondary van der Waals forces. Primary chemical bonds can be formed by chemical reactions between functional groups in wood and those in PF, MF, or UF resins or diisocyanates. Resin types vary markedly in their water resistance which could affect their application in exterior or interior environments. The PF resins are the most durable, and the UF resins, the least. The correlation of these properties with their chemical structure will be made clear in the study of their chemistry.

Apart from the synthetic thermosetting polymer adhesives, other polymers such as polyvinyl acetate and polyamides are used in much smaller quantities. Adhesives of natural origin, such as animal, casein, soybean, starch, and blood glues, are now largely supplanted by the more versatile synthetic polymers. However, casein is used for structural laminating (5). Phenolic compounds derived from trees, such as tannins, are increasingly used for wood bonding, mostly in combination with synthetic adhesive components (10, 11).

Phenolic Resin Adhesives. PF resin adhesives are low molecular weight prepolymers formed from phenol and formaldehyde, and are of the thermosetting type. The polymerization is controlled by the acid or alkaline conditions, i.e., pH, at which it is conducted. Another important reaction variable is the molar ratio of phenol to formaldehyde. By varying the time and temperature of reaction, and by choosing catalysts and phenols of varying reactivity, a wide variety of adhesives can be prepared for many applications and conditions.

RESOLES. Resoles are phenolic resins produced under alkaline conditions with a molar excess of formaldehyde, HCHO, over phenol in the reaction mixture. The initial reaction is the substitution of phenol with methylol ($-CH_2OH$) groups, as shown in Figure 1, both at the *ortho* (**I**) and *para* (**II**) positions. Furthermore, because more than 1 mol of formaldehyde is used for each mole of phenol, products carrying two or three methylol groups (**III, IV**) are also formed. The *ortho:para* substitution ratio depends on the type of catalyst and pH, and decreases from 1.1 at a pH of 8.7 to 0.38 at a

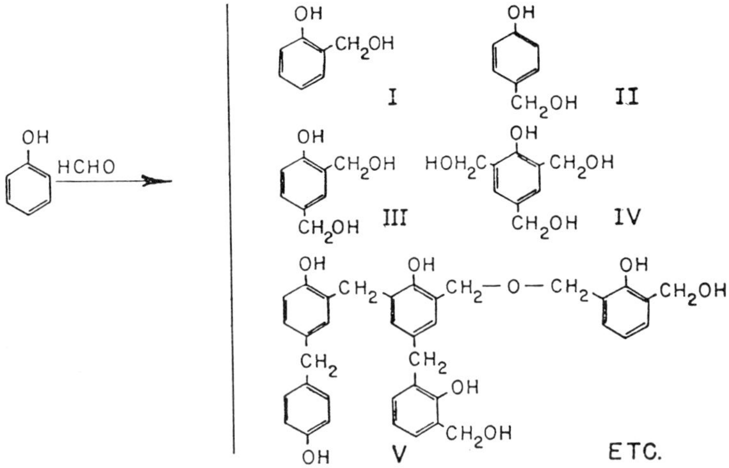

Figure 1. Mixture of reaction products in resole formed from phenol and formaldehyde.

pH of 13.0 (9). Alkali and alkaline-earth metal hydroxides considerably enhance *ortho* substitution in the order Mg > Ca > Sr > Ba > Li > Na > K (9). The chelating strength of the metal cation has a pronounced catalytic effect on directing *ortho* substitution, and cations of transition metals Fe, Co, Mn, etc. are even more effective.

The initial formation of these methylol-substituted phenols is followed by their reaction with each other and with unreacted phenol to give a complex mixture of molecules of different sizes and degrees of branching in which the phenolic nuclei are connected to each other through methylene links, $-CH_2-$, or methylene ether bridges, $-CH_2-O-CH_2-$, as exemplified in Figure 1 (V).

The extent to which the methylol phenols link and grow depends upon the conditions of reaction, pH, time, and temperature. In addition, the mole ratio of formaldehyde to phenol in the reaction mixture, is an important controlling factor. The properties of resole adhesive resins can be varied or reproduced only by careful control of these experimental conditions. Methylol formation is accelerated by increasing the pH. Methylol groups, which subsequently form the methylene links leading to the growth of the macromolecule, increase with increasing molar ratio of formaldehyde to phenol. Under any one set of conditions, continued heating or heating at higher temperature results in products of higher viscosity and lower water solubility. Resoles can thus be formed as water-soluble liquid resins of low molecular weight, about 150, or as grindable solids of molecular weight around 1000.

An alkaline catalyst like NaOH is used, with formaldehyde to phenol ratio varied from 1:1 to 3:1 (usually 1.8–2.4:1) for particle boards. Apart from catalyzing the hydroxymethylation of phenol, NaOH also serves to provide the required water solubility of the resin, even at a high molecular weight, through the formation of sodium phenoxides. Furthermore, it accelerates the curing of the resin.

Resins are prepared commercially that have extraordinarily low monomer content (less than 0.1% of either phenol or formaldehyde) (9). The molecular weight distribution of the typical resin shows very small amounts of bis(hydroxymethyl) and tris(hydroxymethyl) phenols, with the bulk of the product being larger condensation products (9).

Resole represents an intermediate stage in the progress of the reaction of phenol and formaldehyde. The final product, idealized in the cross-linked network shown in Figure 2, is formed by heating the resole to bring about cross-linking through the methylol groups already present in the resin. Further addition of reactants to resole is therefore unnecessary. The curing reaction can also be brought

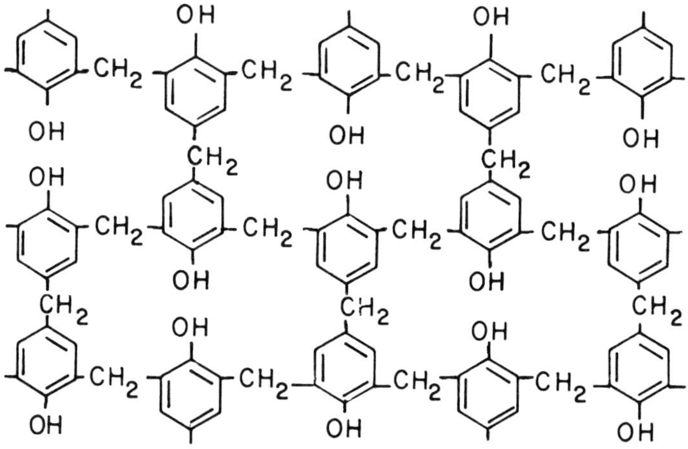

Figure 2. Illustration of the cross-linked structure of phenol–formaldehyde resin.

about by the addition of strong acids to the resole instead of by heating it. Mostly hydrochloric, phosphoric, p-toluenesulfonic, or phenolsulfonic acids are used. Corrosion of metal substrates and long-term attack on wood adherends are problems associated with the use of acid hardeners for phenolic adhesives. Although the acid-curing of resole adhesives is technically not important, the reaction of phenol and formaldehyde, under initially acidic conditions, is used to produce another class of phenolic resin intermediates, novolaks.

NOVOLAK. In the acidic pH range, the reaction of phenol and formaldehyde occurs as an electrophilic substitution reaction. Initially the reaction produces methylolphenol (**I–IV,** Figure 1) as under alkaline conditions. However, the methylol group is unstable under acid conditions and quickly forms methylene bridges by further reaction with phenol. By using phenol in excess, i.e., at a formaldehyde:phenol ratio less than one, linear molecules with terminal phenol groups are formed (**VI**). The molecular weight of these polymeric chains is usually less than 2000. Because they are uncross-linked, novolaks are soluble and thermoplastic. Special catalysts, such

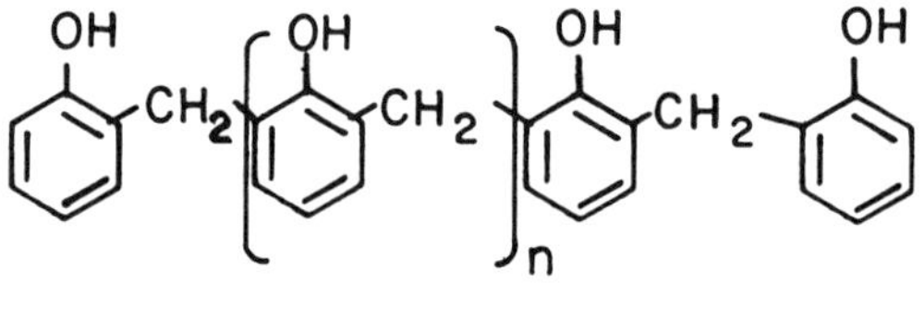

VI

as bivalent metal acetates, need to be used to obtain high-*ortho*-substituted novolaks, in which the phenolic nuclei are linked predominantly through the *ortho*-position (*see* **VI**) (*12*). In the absence of these special catalysts, *para* links are also formed. The mechanism of the selective *ortho*-hydroxymethylation is formation of metal ion chelates.

The novolak resin, unlike resoles, does not contain any free methylol groups. Therefore, it is incapable of further reaction without the addition of more formaldehyde. Consequently, hardening of novolaks to the infusible, cross-linked product shown in Figure 2 can be achieved only by further addition of formaldehyde, or formaldehyde donors. The usual formaldehyde donors are paraformaldehyde, or almost invariably, hexamethylenetetramine (methenamine) **VII**, both of which decompose to formaldehyde under the reaction conditions. Either of these can cross-link the linear novolaks to form methylene bridges that produce the network structure shown in Figure 2. The high-*ortho*-substituted novolaks are more reactive in this cross-linking step because the relatively more reactive *para* position has been left free during their preparation, and is now available for reaction with formaldehyde or hexamethylenetetramine.

RESORCINOL RESINS. The reactivity of phenol with formaldehyde is greatly increased with two hydroxyl groups on its nucleus (resorcinol **VIII**). Room temperature polymerization is observed without the need for any catalyst. The rate of reaction goes through a minimum at a pH of 3.5 and increases at lower or higher pH values. To make a useful adhesive, prepolymers, similar to novolaks, are pre-

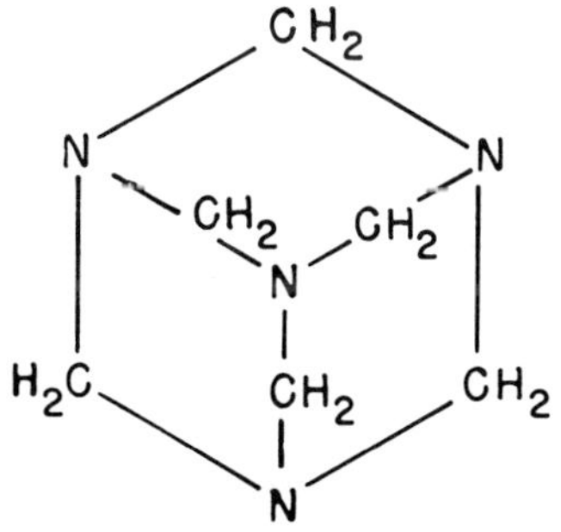

VII

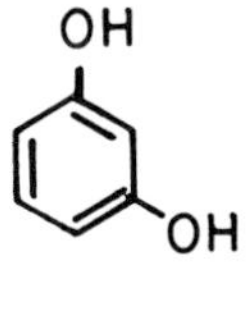

VIII

pared from resorcinol and formaldehyde at a low mole ratio, ~0.5–0.7, of the aldehyde. The prepolymer with the reactive resorcinol nuclei can then be cured, even at ambient conditions, with further addition of formaldehyde, usually as paraformaldehyde, to form the adhesive bond.

The sensitivity of resorcinol resin curing to changes in pH has interesting consequences in the variability of bond strength with species of wood bonded. Specimens of English oak, for example, which gave low strength when bonded by resorcinolic resins, were found to give an aqueous extract of pH 3.7, close to the point where the reactivity of the resin is a minimum (3). This retardation of curing by the acidity of oak could be overcome by washing the specimen with sodium acetate after which the strength produced by resorcinolic resin adhesive was excellent. The effect of extractives on gelation of curing resins will be discussed again later.

Modifications can be made in phenolic resins by using alkyl substituted phenols to produce less reactive resins, to improve compatibility with rubber, and to improve moisture resistance, electrical properties, etc.

APPLICATION OF PHENOL–FORMALDEHYDE RESIN ADHESIVES. The wood industry is a major outlet for phenol–formaldehyde (PF) adhesives and uses about 25% of the PF resin produced in making particle boards and plywoods, and in structural bonding. In these applications, resoles are used almost exclusively, as dry films, liquid resins, or powdered resins. Wood contains a variety of reactive functional groups, in addition to hydroxyls, capable of both hydrogen bonding and chemical reaction with PF resins. For example, lignin, which constitutes about 20–25% of wood, is phenolic in structure, and can coreact with PF adhesives (13). Furthermore, resin viscosity is low enough to enable penetration of micropores of wood, and cause mechanical anchorage of the adhesive. Finally, the cohesive strength of the resin surpasses that of wood. All of these factors contribute to the strength of the adhesively bonded wood products.

In preparing dry films, special paper is impregnated with a resole solution and dried. The dry film form has long storage life, up to 12 months and is particularly well adapted to gluing thin and crotch veneers because problems of controlling spread are avoided. The moisture needed to enable the resole to become a liquid under applied pressure has to come from the wood; therefore, the moisture content should be controlled carefully and maintained above 6%, preferably in the range of 8–12% depending on the species glued and the veneer thickness. Liquid resoles are prepared as aqueous or alcoholic solutions of the PF resin. Resin powders, preferred for long storage life, can be dissolved at the time of application to yield liquid adhesives. The hot-setting phenolic adhesives are used mostly for bonding exterior softwood plywood and require pressing at 130–150 °C for a few minutes. Usually a filler such as coconut or walnut shell flour is used to adjust wetting, avoid excessive penetration, and to obtain uniform joint thickness. The fillers reduce cost and, when used in reasonable proportions, the brittleness of the adhesive joint.

As indicated previously, a significant feature of these resins is the very low content (<0.1%) of the monomers, phenol and formaldehyde. The molecular weights of resins used in plywood manufacture are higher, and the viscosity is in the range of 700 (wet process) or 450 mPa s (dry process) (9). In the manufacture of particle boards, which consist of ~95% by weight of wood chips bonded together by the adhesive, the viscosity of the resin used is much lower, about 130 mPa s (9). In both cases, the resin has a dry solids content of 40–50%.

The cold-setting resole adhesives require an acid hardener, such as p-toluenesulfonic acid which is added at the time of application, for curing. These adhesives are particularly useful in assembly gluing of wood, as in building construction for bonding sandwich panels, prefabricated house panels, truck panels, etc. In such applications, the resin content of the adhesive is also higher than in hot-press adhesives, about 65–75%. Acid-catalyzed adhesives offer some disadvantages. Wood fibers are attacked by the acid in the long run, and the joints formed under acid conditions are not as durable as joints formed from resorcinol–phenol glues. Therefore, with the development of reactive resorcinol-based adhesives that may be cured at low temperatures, the use of acid-catalyzed phenolic adhesives has declined.

DURABILITY AND FRACTURE TOUGHNESS. The durability of joints prepared from resorcinols as well as the hot-setting phenolic adhesives is excellent. The joints are as durable as wood, and properly made joints can withstand, without significant delamination or loss in strength, prolonged exposures to cold and hot water, alternate soaking and drying, temperatures up to those that seriously damage wood, and outdoor weathering. They are also resistant to solvent, wood preservatives, fire-retardant chemicals, and acid or alkali. Thus well-made joints bonded by phenol, phenol–resorcinol, and resorcinol–resin glues are difficult to destroy without destroying the wood itself. The properties are derived from the highly cross-linked structure of aromatic nuclei (Figure 2) that confers thermal stability, solvent resistance, and moisture resistance to the cured adhesive.

The use of accelerated aging tests to predict the durability of phenolic adhesive joints has been critically examined to establish the dependence of fracture toughness of the joints on cure chemistry

(*14*). In this study, the composition and reactivity of the phenol–resorcinol resin, time and temperature of cure, and filler types were varied while measuring or following reaction exotherms, soluble resin fractions, the formation and disappearance of reactive intermediates and, finally, the fracture toughness of the joints. The fracture energy measurement was based on earlier studies (*15, 16*) of the effects of wood anisotropy (grain angle) and wood processing (surface roughness) on fracture toughness of adhesive joints. These studies used double cantilever beam cleavage specimens. By establishing that fracture toughness measurements can be used to follow progress of cure of the resin, some important conclusions have been reached. The observed maximum in the variation of fracture energy with curing time was shown to correspond to the initial formation and decomposition of the dimethylene ether linkages, $-CH_2-O-CH_2-$, (Figure 1, **V**). As more of these flexible, ether linkages are formed, the fracture energy increases; and with the decomposition of these links to the more rigid methylene bridges, $-CH_2-$, (Figure 1, **V**) a drop in fracture energy is observed.

The dependence of fracture toughness on cure temperature showed that the maximum fracture toughness was obtained at a cure temperature that coincided with the temperature of peak exotherm of the resin–formaldehyde reaction. Thus, the highest fracture energy is observed on curing at the temperature of maximum thermal response of the adhesive resin. As was mentioned earlier, the mobility of polymer chains decreases as curing proceeds with more and more of the available functional groups reacting, until curing ceases at the T_g of the cross-linked network. Therefore, at temperatures above and below the optimum cure temperature, polymer mobility, which controls cross-linking, is substantially increased or inhibited. Optimum cross-linking is observed at the optimum cure temperature, and provides the optimum combination of network strength and flexibility leading to the observed maximum in fracture toughness.

Because the fracture toughness depends both on cure time and temperature, the arbitrary selection of time and temperature for accelerated tests may not be appropriate for reliable prediction of long-term service life of joints (*17*). In order to reduce test variability and improve the durability prediction of adhesive joints, it would be necessary first to control the cure temperature and time required to produce a level of fracture toughness that does not change further (*14*). The study is thus an excellent example of a multidisciplinary approach combining chemistry, fracture mechanics, and wood science in the investigation of the adhesive bonding of wood.

Urea–Formaldehyde and Melamine–Formaldehyde Resins (*2*,

18). Urea (NH_2CONH_2) reacts with formaldehyde similarly to phenol to produce methylol derivatives that then condense further to yield a cross-linked network (Scheme 1). Actually, at a mole ratio of 1.5–2 mol of formaldehyde to urea and a pH of 7.5, a mixture of the monomethylol, dimethylol, trimethylol, and tetramethylol ureas are formed. For further extensive condensation to take place, the pH of the system must be made acidic. Thus, it is possible to concentrate the initial resin solution or spray-dry it to a soluble powder that can be dissolved and mixed with an acid catalyst at the time of application to induce the curing reaction. The ratio of formaldehyde to urea used in commercial resins varies with the manufacturer, but is always less than 2:1.

For efficient curing of urea–formaldehyde (UF) resins at low temperatures, strong acids are needed as *cold hardeners*. A drawback of such use of strong acids is that the acid attacks wood. Water-soluble organic acids like citric or tartaric acids reduce the danger of attack on wood. *Hot hardeners* are ammonium salts of strong acids like ammonium chloride or ammonium sulfate that liberate the acid at high temperatures.

About 85% of UF resin production is used in bonding wood materials. The lower cost of the resin, faster curing compared to PF resins at the same temperature, and the formation of a colorless glue line have favored its adoption in plywood and particle board manufacture. The disadvantage is that the UF resin bonds are not resistant to weathering. The lack of durability is attributed to the presence of the hydrolyzable $-C(O)-N-$ groups. Therefore, the quick-setting UF resin adhesives are preferred for gluing furniture and architectural plywood for interior use where the greater durability of PF resins is not required. Recently, increasing concern has been expressed over the release of unreacted, as well as hydrolytically formed UF resins (*19–21*).

Fortification or upgrading of UF to improve its weathering resistance is practiced by adding melamine **IX**, which provides amino groups for reaction without introducing easily hydrolyzable groups in the network.

IX

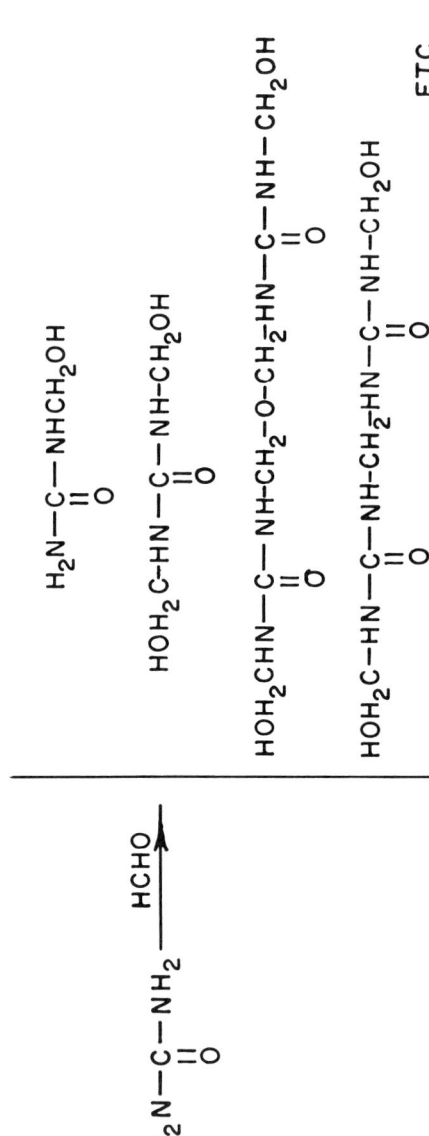

Scheme 1. Low molecular weight methylol derivatives formed from urea and formaldehyde.

Melamine, when used by itself, also forms adhesive resins with formaldehyde. However, melamine–formaldehyde (MF) resins are more expensive than UF and PF resins and have not gained much importance as adhesives. Current commercial applications of MF adhesives in structural wood laminating include bonding interior or finger joints and laminated decking with 60:40 melamine–urea combination. They have also been used successfully in high-frequency edge gluing where a durable, colorless glueline is desired.

Isocyanate-Based Adhesives. Generally diisocyanates or polyisocyanates are used in the preparation of polyurethane adhesives. The urethane link (–O–C(O)–NH–) is formed by the reaction of an alcohol with an isocyanate (–N=C=O) group:

$$R'-OH + O=C=N-R \rightarrow R'-O-CO-NH-R$$

Thus, when diisocyanates are reacted with dihydroxy compounds (diols) a polyurethane is formed, which is composed of repeating urethane links:

$$HO-R-OH + O=C-N=R'-N=C=O \rightarrow \sim\sim -O-R-O-C(O)-NH-R'-NH-C(O)\sim\sim \quad (1)$$

The structural variations possible in R and R' make it possible to vary the toughness and elasticity of the polyurethane adhesive. Polyesters or polyethers are prepared with terminal hydroxyl groups that can then be reacted with difunctional or polyfunctional isocyanates. Polyurethane prepolymers can be formed in Reaction 1, with the desired terminal group produced by using one or other of the diol and diisocyanate reactants in excess.

The two-part adhesive systems make use of the reactive terminal groups of the prepolymers to bring about chain extension and cross-linking with suitable curing agents. Whereas diisocyanates or polyisocyanates are needed to cure hydroxyl-terminated polymers, the terminal isocyanate groups can cross-link by reacting with themselves in the presence of absorbed moisture. In the latter case a one-part adhesive results. Other one-part polyurethane adhesives comprise a high molecular weight polyurethane dissolved in a suitable solvent. The polarity and hydrogen bonding capability of the urethane group give the polyurethanes enhanced adhesion to a variety of surfaces.

In bonding wood, the reaction of isocyanate groups with the numerous hydroxyl groups that are present in the various components of wood—cellulose, hemicellulose, and lignin—is possible. The product of this chemical reaction with wood is the urethane bond as shown in Reaction 1. Methyl, ethyl, propyl, and butyl isocyanates react quickly with dry wood to form urethane links with cell wall component hydroxyls (22, 23).

In a study concerned with the decay resistance provided by isocyanate bonding to wood, the distribution of the methyl isocyanate reaction in southern pine showed that 60% of the lignin hydroxyls and 12% of the holocellulose hydroxyls are substituted at the point where resistance to biological attack occurs. Therefore, it can be surmised that the chemical bonding of wood by isocyanates through the urethane link can contribute significantly to the excellent performance of diphenylmethane–diisocyanate **X** or polymeric isocyanates **XI** as adhesive binder in particle boards (24, 25).

There are a few other chemical reactions on the wood surface that could make important contributions. One is that of moisture on the surface of wood to form an unstable carbamic acid group that quickly decomposes to form a primary amine with evolution of carbon dioxide. The primary amine formed has active hydrogens reactive to isocyanate. Other successive reactions ensue leading first to disubstituted ureas and then to biurets. Furthermore, isocyanate reaction with urethane to form allophanates, and trimerization of isocyanates to form isocyanurate are also possible to variable extents, under the conditions of bonding. The different reactions are summarized in Scheme 2.

Although all the reactions discussed in this section can contribute to the cross-linking of the adhesive and the chemical bonding of the wood particles, the initial reaction of isocyanate with adsorbed water molecules on the wood surface results in immediate wetting of wood with the formation of polyurea. Even on impervious surfaces like glass or metal, the reaction of isocyanate with adsorbed water results in intimate contact after chemical reaction, including poly-

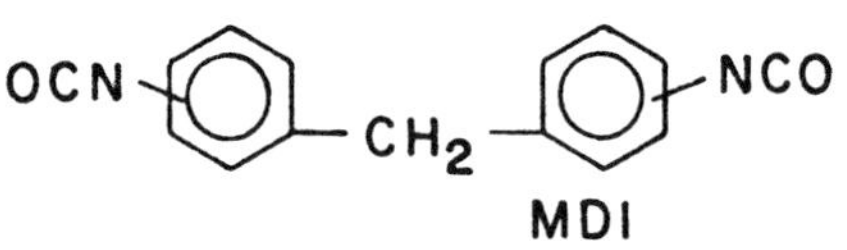

MDI

X

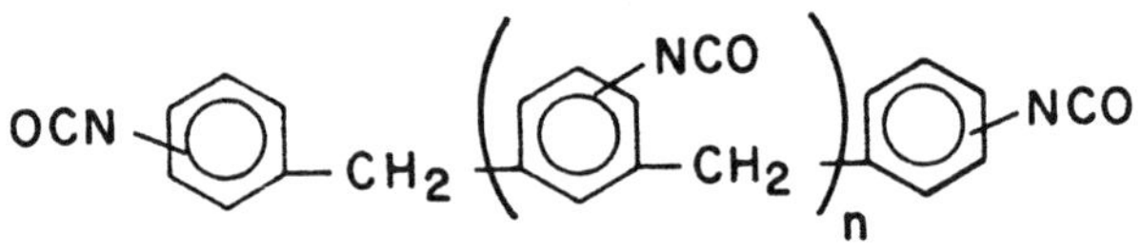

PMDI

XI

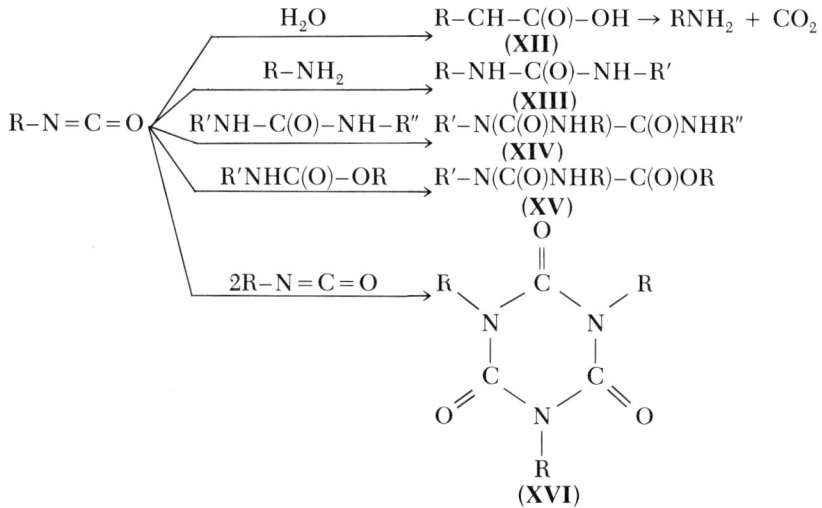

Scheme 2. *Typical reactions of isocyanates leading to the formation of substituted carbamic acid* (**XII**), *substituted urea* (**XIII**), *biuret* (**XIV**), *allophanate* (**XV**), *and isocyanurate* (**XVI**).

merization, and produces excellent adhesion (26). In the case of wood surface, the excellent wetting achieved by isocyanates leads to immediate penetration of exposed cells by the isocyanate adhesive.

The wetting and permeation of polymeric isocyanates mixed with a small amount of dye were observed in optical and scanning electron micrographs (27). The ready formation of an interphase by the permeation of polymeric isocyanates has important consequences for mechanical anchorage of adhesive as well as for the fracture behavior of the bonded wood composite.

Tannin-Based Adhesives. Condensed tannins—obtained as extracts of the barks of trees such as wattle, hemlock, or pine—consist of flavonoid units that have undergone varying degrees of condensation. The phenolic rings of tannins are important to adhesive chemistry because they have reactive sites available for condensation with formaldehyde. The main polyphenolic pattern is represented by flavonoid analogs based on resorcinol–pyrogallol, resorcinol–catechol, phloroglucinol–pyrogallol, and phloroglucinol–catechol rings shown in Figure 3. Wattle extracts, rich in the resorcinolic tannins, have been exploited more commercially than tannins of the phloroglucinolic type obtained from pine trees.

The major effort in understanding the chemistry of these complex derivatives in attempts to use them most effectively in adhesives is exemplified by the work of Pizzi (*10, 11, 28*). A detailed study of the composition of tannin extracts and their reaction characteristics

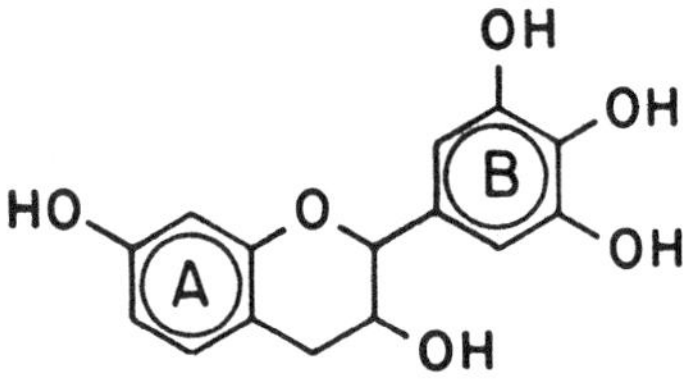

Figure 3. Polyphenolic patterns occurring in tannins. Key: **XVII**, *resorcinol A–pyrogallol B;* **XVIII**, *resorcinol A–catechol B;* **XIX**, *phloroglucinol A–pyrogallol B; and* **XX**, *phloroglucinol A–catechol B ring systems.*

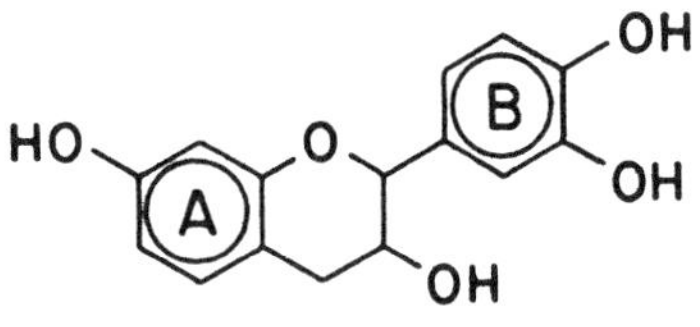

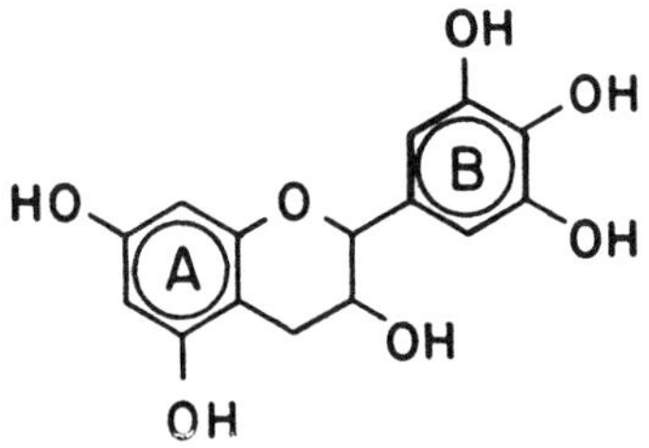

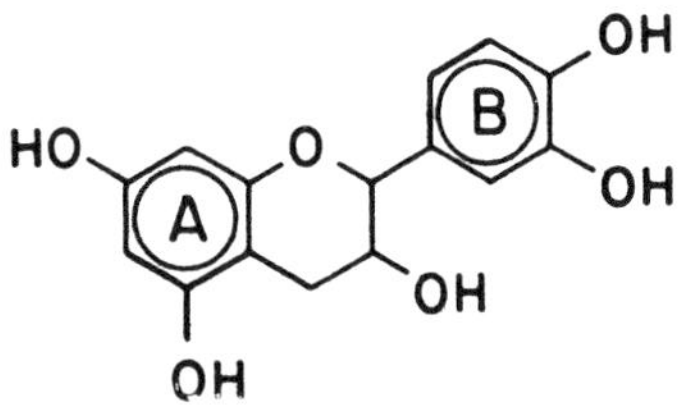

and macromolecular structure (*10, 11, 28*) has led to the formulation of processes for adhesives from tannin extracts that may be commercially useful. The basic reactions were discussed in the section on PF resins; experimental details may be found in References 10, 11, and 28.

Tannins are most useful when applied in combination with synthetic resins. In addition to the cross-linking reaction utilizing the phenolic rings, the reaction of phenolic hydroxyls with isocyanate groups can bring about additional cross-linking and improve the results considerably. Exterior-grade particle board adhesives have been made by combinations of tannins with commercial diphenylmethane–diisocyanate and formaldehyde (*10, 28, 29*). The isocyanate group is deactivated by water less readily in the presence of a phenol; therefore, the reaction of isocyanate with a water solution of tannin extract becomes quite practicable (*10*). Continued research on using tannins for adhesives can further the goal of substituting plant chemicals for expensive petrochemical intermediates.

Thermoplastic Adhesives. The polymer adhesives described in the following sections are not used as extensively with wood as the thermosetting adhesives discussed earlier. However, they do illustrate many interesting principles of the chemistry of adhesion through organic polymers.

HOT-MELT ADHESIVES. Thermoplastic polymers, it may be recalled, are essentially linear polymers that soften and flow when heated and solidify on cooling. Therefore, many of them are suitable for use as hot-melt adhesives. Because the cooling occurs quickly and there is no curing reaction involved, fast assembly on a mass scale, as in paperback book binding, is possible. Some problems that follow solvent use are avoided because no solvent is required. One such problem is solvent pollution. Another is swelling by the solvent, like swelling of wood by water-based systems. However, because hot-melt adhesives are based on polymers that soften on heating, the joints made with hot-melt adhesives are susceptible to softening whenever they are exposed to sufficiently high temperature.

Low molecular weight polyethylene, $(CH_2\text{-}CH)_n$, is used widely as a hot-melt adhesive. To improve the adhesive properties of this hydrocarbon polymer, especially to polar substrates, ethylene is copolymerized with vinyl acetate (VAc), which introduces the polar acetate groups into the polymer, to make ethylene–vinyl acetate copolymers (EVAc), $\{CH_2\text{-}CH_2)_x\{CH_2\text{-}C(OCOCH_3)H\}_y$. These copolymers are a versatile class of hot-melt adhesives whose properties can be modified over a wide range by varying the proportion of VAc comonomer from 18 to 40% (8). They are random copolymers, whose molecular weight can be varied to alter flow properties. The ease of flow, an important factor in wetting, is measured by the melt index, which is the weight of polymer flowing through a given orifice in 10 min at a fixed temperature and pressure prescribed by ASTM standards (30). The higher the molecular weight of the polymer, the lower the melt index. Copolymers with melt indices varying from 2 to 200, such as EVAc copolymers, and ethylene–acrylic acid, are employed in various applications.

More widely used in wood bonding are hot-melt adhesives based on polyamides. A polyamide is formed when a monomer carrying two amine groups ($-NH_2$) is reacted with another carrying two carboxylic acid groups ($-COOH$). Each amine group, at either end of the molecule, can condense with a carboxylic acid group of the other monomer to form an amide link ($-CONH-$) with the elimination of a water molecule:

$$-NH_2 + HOOC- \xrightarrow{-H_2O} -NHCO-$$

By such step wise condensation of the amine and carboxylic groups of many monomer molecules, a long chain polyamide, known as nylon **XXI** is formed.

$$H_2N(CH_2)_6NH_2 + HOOC(CH_2)_4COOH \xrightarrow{-H_2O} (NH-(CH_2)_6-NHCO(CH_2)_4-CO-)_n$$

(XXI)

The polarity of polyamides that are used as thermoplastic, hot-melt adhesives can be varied by altering the proportion of hydrocarbon ($-CH_2-$) and amide ($-CONH-$) groups in $-CO(CH_2)_x(CONH-(CH_2)_y-NH-$. They are more expensive than EVAc copolymers, but set more rapidly, and are preferred as edge veneer adhesives in the furniture industry.

ADHESIVE EMULSIONS. Thermoplastic, synthetic polymers can be prepared as emulsions for use as adhesives. For example, while EVAc hot-melt adhesives described in the previous section contain less than 40% VAc, when the content of VAc in the copolymer is increased to 60%, and the copolymer is prepared in the form of aqueous emulsions, a very useful and versatile adhesive polymer is obtained. Although the VAc homopolymer, poly(vinyl acetate), is a brittle solid, with a T_g = 28 °C, the ethylene units present in the EVAc copolymer act as an *internal plasticizer*, and lower the T_g to below room temperature. The plasticization results from the reduction of interchain interaction of the VAc polymer chains by the ethylene units interspersed among the strongly interacting VAc units. This reduction of the T_g has important consequences because the formation of a flexible adhesive film from the emulsion depends upon the T_g of the polymer.

This emulsion is not a liquid–liquid system, but is an aqueous dispersion of solid polymer particles. Therefore, if the T_g were above room temperature (at which the emulsion is applied), the polymer segments, lacking segmental mobility, would not diffuse readily from one particle of the emulsion into another after evaporation of the water medium in which the copolymer emulsion is prepared. The result would be a powdery film. Conversely, when the T_g is reduced below room temperature, segmental mobility in the copolymer leads to diffusion and formation of a flexible, strong adhesive film from the latex by coalescence of the emulsified particles during drying at room temperature.

Poly(vinyl acetate) (PVAc), $\{CH_2-CH(OCOCH_3)\}_n$, is also prepared as an emulsion for adhesive applications, and is familiar to users as *white glue*. As mentioned already, the T_g of PVAc is above room temperature, which makes the polymer rigid and brittle at room temperature. For adhesive application, therefore, an *external plasticizer*, such as dibutyl phthalate, is added to lower the T_g below room temperature and to facilitate film formation from emulsions.

The function of added plasticizer molecules is to reduce the interchain interaction of the VAc polymer chains and facilitate the mobility of the polymer segments.

PVAc emulsions have excellent adhesion to cellulosic materials and find extensive use in bonding paper and in wood assembly in a wide variety of products—paper bags, milk cartons, envelopes, books, pencils, etc. High bond strength, fast set, and colorless glue lines combined with ease of application are advantages of PVAc emulsions in wood bonding. As mentioned earlier, the polymer emulsion is an aqueous dispersion of swollen solid particles. One obvious advantage of an emulsion is the use of water as the medium from which the polymer is applied. The other advantage is that the emulsion enables the solids content of the polymer adhesive to be higher than in solution. Because the viscosity of a polymer solution rises rapidly with the molecular weight and concentration of polymer, the required concentration may reduce fluidity and decrease the ease of application of the adhesive. However, the high molecular weight of the polymer has little effect on the viscosity of the emulsion. Therefore an aqueous dispersion with a high ratio of high molecular weight solids to viscosity can be prepared and applied to produce adhesive films of desirable strength and thickness.

The lowering of the T_g of PVAc does result in some creep of the adhesive bond. In order to prevent sliding of macromolecules, which produces creep, VAc is copolymerized with small amounts of comonomers that can be cross-linked during curing of the adhesive. Thus, a significant reduction in creep is achieved by cross-linking the polymer chains and reducing their susceptibility to displacement relative to each other. Such curable PVAc emulsions are used in the bonding of nonwoven fabrics, i.e., to impart wash and dry-clean resistance. Copolymerization of VAc with a flexible comonomer results in tacky, soft, low-T_g copolymer emulsions suited for pressure sensitive adhesives.

PVAc is also the precursor to a few other types of adhesives. For example, poly(vinyl alcohol) (PVA), $\{CH_2-CH(OH)\}_n$, is formed by the hydrolysis of PVAc. Low molecular weight grades of PVA are used in PVAc emulsions to improve colloidal stability of the emulsions, i.e., as a protective colloid. Higher molecular weight grades, only partially hydrolyzed, modify other properties such as adhesion, viscosity, and film formation.

With its high hydroxyl content and excellent binding capacity, PVA is used widely as a water-soluble adhesive with excellent adhesion to paper, and natural and synthetic fibers. When the $-OH$ groups in partially hydrolyzed PVAc are reacted with aldehydes, acetal units are formed so that the polymer now contains acetal groups in addition to the alcohol and acetate units already present (Reaction 2).

$$
\begin{array}{c}
---CH_2-CH-CH_2-CH-CH_2-CH-CH_2-CH--- \\
O\!H H\!O OH O \\
\diagdown\!O\!\diagup C=O \\
+ \|| \\
C-H CH_3 \\
| \\
R
\end{array} \xrightarrow{-H_2O}
$$

$$
\left[\begin{array}{c}CH_2-CH-CH_2-CH \\ \diagdown\diagup \\ OO \\ \diagdown\diagup \\ C \\ \diagup\diagdown \\ RH\end{array}\right]\left[\begin{array}{c}CH_2-CH \\ | \\ OH\end{array}\right]\left[\begin{array}{c}CH_2-CH \\ | \\ O \\ | \\ CO \\ | \\ CH_3\end{array}\right] \quad (2)
$$

ACETAL ALCOHOL ACETATE

Poly(vinyl acetals) can be applied from solution in organic solvents or as hot-melt adhesives. The properties of these polymers depend on their molecular weights, on the degree of hydrolysis of the acetate, and on the type and proportion of acetal units. Poly(vinyl butyral) and poly(vinyl formal) formed by reaction with butyraldehyde and formaldehyde, respectively, are examples of such acetal resins. Poly(vinyl acetals) are capable of cross-linking by heat or mineral acids. In addition, the residual hydroxyl groups can condense with methylol derivatives in phenol–formaldehyde and melamine–formaldehyde resin systems, with isocyanates, and with epoxy resins. Therefore, it is no mere coincidence that poly(vinyl formal), cured with phenolic resins, was the first material to find major structural adhesive use with metals. Likewise, many commercial adhesives based on poly(vinyl acetals) involve curing with thermosetting resins, mainly, phenolics; the types of acetal, and the ratios of the two components in the adhesive provide a range of performance. The poly(vinyl acetal)–PF resin adhesive systems are excellent for wood bonding (3).

Poly(vinyl acetals) are also compounded with appropriate plasticizer and other additives for use as hot-melt adhesives. As hot-melt adhesives, they provide moderate strength as structural adhesives for bonding wood to metal.

Poly(vinyl butyral) is well known in its application as the interlayer in laminated automotive safety glass. Laminated architectural glass is made similarly with poly(vinyl butyral), with controlled transmission of light and heat, to provide aesthetic appeal as well as reduce glare, heat loss, and UV-light transmission. Multiple laminates are useful in making transparent bulletproof shields. As a component of washcoats and sealers, poly(vinyl butyral) finds use in wood finishing.

Acidity of Wood

The marked acidity of some species of wood such as oak is well known. Before completing the discussion of the chemistry of adhesion to wood, it is necessary to consider the interaction between wood acidity and adhesive chemistry. It was pointed out earlier that the bond strength of resorcinolic adhesives to oak surfaces was significantly reduced by the acidity of the latter. That this is a general phenomenon is apparent from several studies of the chemical reactivity of phenolic and urea resins in the presence of several species of wood (*31*, *32*). The curing reactions of the resins could be retarded by some woods and others had hardly any effect. It has been suggested that some low molecular weight substances migrate from the wood into the adhesive phase to retard the curing reaction in some cases.

There have been many attempts to investigate the effect of extractives on cure chemistry and bonding to wood (*33*–*35*). For example, the effect of extractives from pressure-refined hardwood fiber on urea-formaldehyde resin was studied (*34*, *35*) and it was found that the ethanol-soluble extractives decreased the gel time as much as 41%, and the sequentially extracted water-soluble extractives increased the gel time in excess of 65%. There was little correlation between the extractive content and gel time; however, an empirical relation between the pH of the extractives and the gel time was observed (*35*). The effect of several species of wood on the gel time of urea–formaldehyde resin have also been studied (*36*). In these studies the gel time was correlated with the pH and acid buffering capacity of the extract.

However, it appears that better correlation with gel times is obtained when the amounts of extractable and unextractable wood acids are considered separately (*37*). Although several procedures do exist in the literature for the determination of acid content in wood, many of these involve extraction of wood with various solvents, and they estimate the acid content in the extractives only. In contrast, we developed a procedure that acknowledges the presence of soluble as well as insoluble acids in wood. The method is based on the reaction of wood acids with aqueous sodium acetate to liberate an equivalent amount of acetic acid. Subsequent pH titration gives the total acid content. The soluble acids are determined by water extraction and titration. Subtraction of soluble acids from total acid content gives the insoluble or bound acid in wood. The reaction of the bound, i.e., unextractable, wood–carboxylic acid can be written in a simplified manner as follows:

$$\text{Wood}\sim\text{COOH} + \text{CH}_3\text{-COONa} \rightarrow \text{Wood}\sim\text{COONa} + \text{CH}_3\text{-COOH}$$

Our estimates of the acid contents of heartwoods and sapwoods of red oak, hickory, southern pine, white fir, and Douglas-fir are given in Table I. The acid contents were correlated with the gel times of urea–formaldehyde resins (36) in contact with the same sample lots of the different species of wood. The correlations are presented in Table II. It is seen from Table II that the best correlation of gel times is obtained with insoluble acid content for each of the empirical fits that were tried. Similarly, the pH of the sodium acetate extracts showed better correlations with the gel times than the pH of the water extracts (37).

The unextractable acid in wood plays a major role in the catalysis of the urea–formaldehyde polycondensation reaction. The significance of this indication must be viewed in contrast to previous investigations which have attempted to correlate properties of wood with the properties or amounts of extractives. It would not be prudent to generalize regarding the effect of unextracted acids because only seven species were studied. However, in future studies these observations may be found to be generally true for most, if not all, species.

The marked influence of bound acids in wood on the curing

Table I. Acid Content in Woods

Species of Wood	Acid Content (meq/100 g wood)			Gel Time[c] (min)
	Water-Soluble	Total[a]	Water-Insoluble[b]	
Red oak				
Heartwood	0.118	2.577	2.459	19.00
Sapwood	1.845	4.183	2.338	19.18
Hickory				
Heartwood	1.651	2.869	1.218	25.00
Sapwood	1.982	3.937	1.955	15.00
Southern pine	1.390	4.824	3.434	12.08
White fir				
Sapwood	0.588	2.071	1.483	16.38
Douglas-fir	1.930	9.130	7.200	7.75

[a] By sodium acetate extraction
[b] By difference
[c] Values reported in Ref. 36 for urea–formaldehyde reaction.
(Reproduced with permission from Ref. 37. Copyright 1983, Springer–Verlag.)

Table II. Empirical Correlations Between Acid Content and Gel Time of Urea–Formaldehyde (36)

Empirical Correlation[a]	r	a	b
Linear: $y = a + b \cdot x$			
Soluble	−0.219	18.608	−1.670
Total	−0.770	23.948	−1.799
Insoluble	−0.816	22.691	−2.213
Exponential: $y = a \cdot e^{bx}$			
Soluble	−0.288	18.936	−0.150
Total	−0.863	27.681	−0.138
Insoluble	−0.899	24.936	−0.167
Logarithmic: $y = a + b \cdot \ln x$			
Soluble	−0.220	16.361	−1.171
Total	−0.727	27.304	−8.238
Insoluble	−0.851	23.464	−8.018
Power: $y = a \cdot x^b$			
Soluble	−0.277	15.470	−0.101
Total	−0.805	35.434	−0.624
Insoluble	−0.895	25.799	−0.578

[a] The acid content (x) is in milliequivalents per 100 g wood, and gel time (y) is in minutes.
(Reproduced with permission from Ref. 37. Copyright 1983, Springer-Verlag.)

reaction of urea–formaldehyde resins has strong implications for the adhesive bonding of wood. It must not be construed, however, that the pH value and buffering capacity of wood are the only factors that affect its bonding.

Further experiments have shown the applicability of this analytical technique to the determination of bound acid generated by HNO_3 treatment of wood (38). Thus the kinetics of the reaction can be followed. The type of acid groups generated are carboxylic acids. Their ability to initiate the in situ polymerization of furfuryl alcohol in wood has been observed. Detailed studies of the reactions of wood acid groups with adhesive systems and their implication for adhesive bonding of wood are subjects deserving of continued investigation.

Literature Cited

1. Kaelble, D. H. "Physical Chemistry of Adhesion"; Wiley-Interscience: New York, 1971.
2. Wake, W. C. "Adhesion and the Formulation of Adhesives"; Applied Science Publishers Ltd.: London, 1976; 2nd ed., 1982.
3. Houwink, R.; Salomon, G. "Adhesion and Adhesives," Vol. 1; "Adhesives," 2nd ed.; Elsevier Publishing Company: New York, 1965.
4. Blomquist, R. F., Ed. "Adhesive Bonding of Wood and Other Structural Materials"; Educational Module for Materials Science and Engineering (EMMSE), The Pennsylvania State University, University Park, PA, 1983.

5. "Adhesive Bonding of Wood," Technical Bull. No. 1512, U.S. Department of Agriculture, Forest Service, 1975.
6. Kinloch, A. J. *J. Mater. Sci.* **1980**, *15*, 2141.
7. Kinloch, A. J. *J. Mater. Sci.* **1982**, *17*, 617.
8. Skeist, Irving "Handbook of Adhesives," 2nd ed.; Van Nostrand Reinhold: New York, 1977.
9. Knop, A.; Scheib, W. "Chemistry and Application of Phenolic Resins"; Springer-Verlag: New York, 1979.
10. Pizzi, A. *Ind. Eng. Chem. Prod. Res. Dev.* **1982**, *21*, 359.
11. Pizzi, A. *J. Macromol. Sci., Rev. Macromol. Chem.* **1980**, *C18* (2), 247.
12. Peer, H. G. *Rec. Trav. Chim.* **1960**, *79*, 825.
13. Marton, J.; Marton, T.; Falkehag, S. I.; Adler, E. in "Lignin Structure and Reactions," Marton, J., Ed.; ACS ADVANCES IN CHEMISTRY SERIES No. 59, ACS: Washington, D.C., 1966; 125.
14. Ebewele, R. O.; River, B. H.; Koutsky, J. A. *J. Adhes.* **1982**, *14*, 189.
15. Ebewele, R. O.; River, B. H.; Koutsky, J. A. *Wood Fiber* **1979**, *11*, 197.
16. Ebewele, R. O.; River, B. H.; Koutsky, J. A. *Wood Fiber* **1980**, *12*, 40.
17. Ebewele, R. O. "The Fracture Mechanics Approach to the Assessment of Adhesive Joint Performance in Bonded Wood Products," Ph.D. thesis, Univ. of Wisconsin: Madison, WI, 1980.
18. Meyer, Beat "Urea Formaldehyde Resins"; Addison-Wesley Publishing Company: New York, 1979.
19. Myers, G. *For. Prod. J.* **1983**, *33* (5), 27.
20. Myers, G. *For. Prod. J.* in press.
21. Nestler, F. H. Max "The Formaldehyde Problem in Wood-Based Products—An Annotated Bibliography"; *USDA Forest Service General Technical Report* 1977, FPL–8.
22. Rowell, R. M.; Ellis, W. D. in "Urethane Chemistry and Applications"; Edwards, K. N., Ed.; ACS SYMPOSIUM SERIES No. 172; ACS: Washington, D.C., 1981, 263.
23. Rowell, R. M.; Ellis, W. D. *Wood Sci.* **1979**, *12*, 52.
24. Frink, J. W.; Sachs, H. I. in "Urethane Chemistry and Applications"; Edwards, K. N., Ed., ACS SYMPOSIUM SERIES No. 172; ACS: Washington, D.C., 1981, 285.
25. McLaughlin, A.; Alberino, L. M.; Farrissey, W. J.; Waszeciak, D. P. in "Proc. Symposium on Wood Adhesives—Research, Applications and Needs," Madison, WI, Sept. 23–25, 1980, p. 112.
26. Schallenberger, C. S. in "Handbook of Adhesives"; Skeist, I., Ed.; 2nd ed.; Van Nostrand Reinhold: New York, 1977; Chap. 7.
27. Johns, W. E.; Plagemann, W., paper presented at the Annual Adhesion Society Symposium, Savannah, GA, Feb. 20–23, 1983.
28. Pizzi, A. *Holz Roh- Werkst.* **1982**, *40*, 293.
29. Pizzi, A. *J. Macromol. Sci., Chem.* **1981**, *A16*, 1243.
30. ASTM, Standard D1238–70, "Measuring Flow Rates of Thermoplastics by Extrusion Plastometer," 1970.
31. Mizumachi, H. *Wood Sci.* **1973**, *6* (1), 14.
32. Mizumachi, H.; Morita, H. *Wood Sci.* **1975**, *7* (3), 256.
33. Jain, N. C.; Gupta, R. C.; Chauhan, B. R. S. *Holzforsch. Holzverwert.* **1974**, *26* (6), 129.
34. Albritton, R. O.; Short, P. H. *For. Prod. J.* **1979**, *29* (2), 40.
35. Slay, J. R.; Short, P. H.; Wright, D. C. *For. Prod. J.* **1980**, *30* (3), 22.
36. Johns, W. E.; Niazi, K. E. *Wood Fiber* **1981**, *12* (4), 255.
37. Subramanian, R. V.; Somasekharan, K. N.; Johns, W. E. *Holzforschung* **1983**, *37*, 117.
38. Subramanian, R. V.; Balaba, W. M.; Somasekharan, K. N. *J. Adhes.* **1982**, *14*, 295.

RECEIVED for review May 19, 1983. ACCEPTED August 22, 1983.

Activation of Wood Surface and Nonconventional Bonding

EUGENE ZAVARIN

Forest Products Laboratory, University of California, Berkeley, CA 94804

> *Nonconventional bonding includes many different methods of bonding wood, all of them radically different from the conventional phenol–formaldehyde and urea–formaldehyde and related methods. In many cases the methods rely, at least in principle, on formation of covalent bonds to wood surfaces. Some of the systems involve direct covalent bonds between the wood surfaces, some employ bifunctional monomers for joining the surfaces, and others covalently bridge the surfaces by polymeric chains. The last methods appear to bridge the gaps between wood surfaces with the least difficulty. The methods include gluing by spent sulfite liquor at low pH; gluing by a mixture of spent sulfite liquor, furfuryl alcohol, and maleic anhydride with oxidative surface activation; gluing by water-soluble carbohydrates with a catalyst; and gluing by isocyanates. Some methods are at the pilot plant stage, some are at the laboratory stage, while gluing by isocyanates has been in industrial use for some time. The products often exhibit improved dimensional stability and water resistance, but tend to suffer from abnormally high variability in the mechanical properties. Progress is handicapped by insufficient knowledge of the chemical composition of wood surfaces as well as of the chemical processes involved in bonding. The reacting wood surfaces are commonly richer in lignin than the bulk of the wood and are covered with a layer of polar and nonpolar extractives. This coverage as well as chemical transformations during surface preparation and history can influence the formation of covalent bonds to wood. Acid or oxidant activators can be required for bond formation. Such activators could promote cross-linking of the introduced polymers without formation of covalent bonds to wood surface, could*

change the polymer and enable it to form covalent bonds with the wood surface, or could change the wood surface and enable it to form covalent bonds with the polymer. Some systems involve the lignin portion of wood for formation of covalent bonds (lignophilic systems), others preferentially form covalent bonds with cellulose or hemicelluloses (cellophilic systems).

THE TERM NONCONVENTIONAL BONDING has been in use for some time, but its choice can be hardly termed as fortunate. In the first place it is negative, i.e., it is based on concepts outside of the scope of definition. Secondly, it lacks time stability; what is nonconventional today might be conventional tomorrow. Thirdly, it is too broad, as it can relate to nonconventional glues, nonconventional practices, or even to nonconventional equipment. Probably the only reason for using this term is the lack of appropriate alternatives.

In common usage the term nonconventional bonding of wood has been applied somewhat imprecisely to a group of bonding procedures involving a wide variety of chemical monomeric or polymeric reagents. These reagents are different from the conventionally used adhesives, such as phenol–formaldehyde and urea–formaldehyde. The word "different" is rather ambiguous and allows for an appreciable gray area. In this chapter only those bonding agents that involve completely new ways of bonding and cross-linking are included, and the agents that bear appreciable similarity to phenol–formaldehyde and urea–formaldehyde resins (e.g., phenol–formaldehyde resins including tannin as a partial phenol substitute) are excluded.

Most nonconventional bonding systems share the idea of covalently bonded wood surfaces. In conventional bonding the wood surface represents, or is thought to represent, a secondary reaction partner only, with covalent bonding restricted mainly to cross-linking reactions of the bonding agents (1–3). Some of the methods involving covalently bonded surfaces require activation of the wood surfaces as a necessary prerequisite for successful bonding. Activation of wood's external surfaces[1] causes a change in the chemical behavior of the wood and enables the components of wood either to undergo new

---

[1] External wood surfaces, occasionally simply designated as wood surfaces, are commonly artificially created and comprise the interfaces between wood and the external world. Internal wood surfaces comprise the interfaces between cell walls and cell lumena. The depth of the surface layer is not restricted to monomolecular thickness, but is regarded as the depth necessary to produce a certain surface effect. The definition of depth is thus relative and depends on the type of interaction. Consequently, analytical results arrived at by different methods of surface analysis are not strictly comparable.

reactions or to react at an increased rate. Such increased reactivity results in certain external effects (e.g., improved adhesion). Because of the variety of these effects, activation is a relative concept and depends upon the nature of the effect and cannot be discussed per se. Thus, in addition to nonconventional bonding, activation of lignocellulosic materials forms the basis of grafting of organic monomers to lignocellulosic surfaces (4) and is responsible for certain improvements in the performance of the surface coatings of wood (5–8).

Advantages of nonconventional bonding are associated with covalently bonded wood surfaces (external and to some extent internal) and include dimensional stability of the products. Occasionally increased brittleness and a loss in mechanical properties due to acidic degradation of carbohydrates are observed.

Some nonconventional bonding methods are based on the use of agricultural by-products, i.e., on nonpetroleum-based materials; this use constitutes another advantage. Some nonconventionally bonded materials produce reduced amounts of toxic gaseous materials, such as formaldehyde, that make them preferable to phenol–formaldehyde products and urea–formaldehyde resins. Economically, the nonconventional methods do not offer any particular advantages, although they appear to be competitive with the conventional methods.

This chapter includes discussions on the chemical composition of the wood surface prior to interactions with bonding agents—a topic often neglected in discussions of the surface reactions of wood; nonconventional bonding methods based on direct, covalent, or wood-to-wood bonding; bonding through intermediacy of bivalent molecules; bonding through intermediacy of a cross-linked polymer, commonly covalently attached to wood surfaces; and fundamental research in these areas.

Wood Surface Composition Prior to Activation or Bonding

The chemical composition of a wood surface does not necessarily correspond to the chemical composition of the bulk of the wood and is a function of the conditions and methods of surface formation; the redistribution of extractives following or during the surface formation; the incorporation of foreign materials during surface formation and thereafter; and the chemical changes in time due to interactions with air–oxygen, light, and other chemical and physical reagents. Although the knowledge of the chemical composition of the surface is of great importance in understanding the surface performance in nonconventional bonding, the amount of information currently available on the above areas is scarce.

Conditions and Methods of Wood Surface Formation. The

conditions and methods of wood surface formation can strongly influence the percentages of lignin, hemicelluloses, and cellulose in the surface layer of wood. With transverse surfaces (cuts made perpendicular to wood trunk, i.e., perpendicular to the length of the tracheids), the percentages of above wood constituents should not deviate much from those of total wood. Such surfaces are however less practically important than other surfaces. However, the situation is different with radial and tangential surfaces. Morphological evidence from microscopic studies [including scanning electron microscopy (SEM)] of wood surfaces and wood fibers has been provided for spruce and birch wood (9–15) and on black spruce.

In the work on black spruce [*Picea mariana* (Mill.) BSP] (14, 15) the wood surfaces were produced at temperatures ranging between −190 and 250 °C by tangential and radial tensile failures. The low softening point of hemicelluloses (50–60 °C) and of lignin (90–100 °C), the two materials that bind the microfibrils of the cell wall, strongly influenced the morphology of the surfaces. Thus with both radial and tangential failures, the percent of tracheids broken by transwall failure decreased between 0 and 200 °C from 40–50% down to ∼0%. Furthermore, with tangential surfaces the fiber faces produced at or below 100 °C revealed mainly the S_1 surface structure; above 150 °C a predominantly primary wall structure, heavily embedded in, or covered by an amorphous matrix of lignin and hemicelluloses was produced. The results suggested that with an increase in temperature the wood fibers are more easily separated, then broken, from parent wood (Figure 1). Because the ratio of hemicelluloses to cellulose percentages and the lignin percentage increases from the secondary wall to the middle lamella (16–18) the results also suggest that wood surfaces of different chemical composition are produced under different conditions. This is important in respect to the reactions connected with the activation of wood surfaces; for example, oxidizing, activating agents such as hydrogen peroxide (H_2O_2) and certain nitrates react preferentially with lignin (19, 20).

Methodologies allowing direct assessment of the chemical composition of the wood surface are those based on Auger spectroscopy and particularly on electron spectroscopy for chemical analysis (ESCA). Except for one marginally successful attempt to determine the sulfur distribution on the surface of a sulfonated cross section of a black spruce wood sample (21), Auger spectroscopy has not been used on wood.

Appreciably more work has been done with ESCA, however. From the results of ESCA a ratio of oxygen to carbon atoms on the surface (N_O/N_C) can be calculated. Theoretically for pure cellulose this ratio is 0.83, for milled wood lignin from conifers it is around

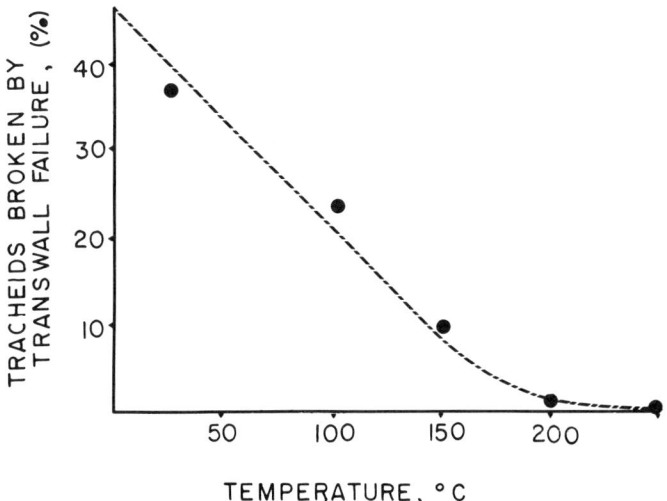

Figure 1. Percentage of tracheids broken by transwall failure on the tangential fracture surface as a function of temperature (15).

0.37, and from hardwoods around 0.43. For nonpolar extractives (monoterpenoids, resin acids, fatty acids) this ratio is around 0.10 or lower. The ESCA analyses of lignocellulosic materials available include those on lignin and lignocellulosic fibers (22–24) on lignocellulosic fibers and chips of pine wood (25), and on maple wood (26).

The N_O/N_C values for lignin and pure cellulosic fibers (cotton, filter paper) agreed reasonably well with the calculated values, particularly if the fibers were extracted with EtOH, acetone, or similar solvents. Assuming that the surface of lignocellulosic materials was composed after extraction of only cellulosics and lignin, Dorris and Gray (22, 23) estimated 30–45% of lignin on the surface of ground wood pulp fibers. This is appreciably higher than the lignin content of wood. In a related work 7% lignin in the bulk and 17% on the surface of extracted sulfite pulps were reported (27, 28). The amount of lignin on the surface increased with temperature of defibration (25) which agrees with the work of Koran (15, 16). The extracted pine chips had an N_O/N_C ratio similar to pure lignin. These results suggest that lignin was probably the main chemical component of the wood surface (Table I).

Without extraction with polar solvents the N_O/N_C ratios were regularly much lower. These ratios were well below those of cellulose and lignin for maple wood and pine wood, amounting to 0.15 and 0.26, respectively. The extraction effect was due to nonpolar wood extractives; no effort has been made to identify these extractives. Although the likelihood of extractives covering the surface cannot be

Table I. ESCA N_O/N_C Ratios for Various Lignocellulosic Materials

Lignocellulosic Material	N_O/N_C Ratio
Calculated	
Cellulose	0.83
Milled wood lignin	
Softwood	0.37
Hardwood	0.43
Nonpolar extractives	0.10
Found	
Filter paper	0.79–0.83
Milled wood lignin (spruce)	0.39
Pine chips	0.26
Acetone extracted	0.42
Maple wood	0.15
HNO_3 treated	0.43

dismissed, it is also possible that part of the observed effect relates to chemical changes during surface preparation. Such changes could involve dehydration, could lower the N_O/N_C ratio, and could render some of the products of decomposition more soluble in polar solvents.

Although electronic absorption spectroscopy in its reflectance variation has been used to assess the distribution of lignin across the cell walls of tracheids (16, 17), no method based on electronic or IR spectroscopy has been developed to enable quantitative determination of the lignin content of wood surfaces.

Redistribution of Extractives. Redistribution of the extractives during or after surface preparation could result in their deposition on the surface in larger amounts. If a surface was created prior to the removal of moisture from wood, or if the wood subsequently was wetted (e.g., aqueous solution of a reagent used in wood activation), the water-soluble, polar extractives are likely to migrate and become deposited on the wood surface during the process of drying. A number of studies discussed discolorations during kiln or air drying of wood (29–32). Some nonpolar, water-insoluble extractives, such as fatty and resin acids, also migrate during drying and become distributed on the wood surface. Several studies on adhesion inactivation of wood surfaces are available. The vapor-phase translocation of stearic acid as a model compound was studied (34) at temperatures between 25 and 105 °C by using wood pulp handsheets. Although the rate of translocation is strongly dependent upon temperature, it can take place even at the ambient temperature. The redistribution of extractives during wood drying is strongly influenced by the drying method (*34*). The nonpolar toluene-soluble extractives of *Pinus taeda* L. (resin acids, fatty materials, turpentine components, steroids, and other unsaponifiables) were found to become strongly enriched in the outer layer of wood during kiln drying, although practically no enrichment took place in air-drying of wood. However, on the basis of wood anatomical considerations (*Betula alleghaniensis* Britton— yellow birch) it was concluded (*35*) that vapor transport of fatty acids from inside of wood to the surface was concluded to be highly unlikely, except for the region close to the surface. In Hemingway's opinion (*35*), the amount of free saturated fatty acids (i.e., not occurring as glycerides) was too small in this region to interfere with gluing through surface deposition. He favored the deposition of the products of air-oxidation of linoleic acid (free and bound as glyceride) on the surface.

Additional work dealing with surface deposition of nonpolar extractives is that of Suchland and Stevens (*36*), Hancock (*37*), Chen (*38*), Chow (*39*), and Troughton and Chow (*40*). In the last work, the surface of *Picea glauca* (Moench) Voss veneer was covered with a layer of silicic acid powder, the veneer was heated at 150 °C for various amounts of time, and the migrant acetone extractives that became adsorbed to silicic acid were isolated. The time of heating (20–80 min) did not influence the percent of total acetone solubles that migrated, which amounted to an average of 0.11% for heartwood and 0.18% for sapwood (dry wood percent basis). Conversely, the amount of free fatty acids on the surface increased from 65 ppm for 20 min to 74 ppm for 80 min of drying time.

This supports the previously mentioned work based on ESCA that nonpolar extractives can be deposited on wood surface during its preparation and subsequent history.

Deposition of Foreign Materials. Foreign materials are deposited on the surface of wood during and after surface formation. Some deposits, such as dust and water of condensation, are related to wood storage, and their amount can be controlled in principle. The others are, however, connected with the methods of surface formation. Thus in various machining operations small amounts of metal, primarily iron, from cutting parts are likely to be transferred to wood surfaces. Although, due to their minute amounts, such materials are not generally likely to influence the chemical behavior of the wood surfaces greatly, in some instances their presence can be felt. Thus, in interactions of lignocellulosic materials with H_2O_2, traces of iron can exert an appreciable catalytic effect on the rate of H_2O_2 decomposition. In laboratory experiments, traces of iron can be removed easily by treatment with chelating agents, such as sodium salts of ethylenediaminetetraacetic acid (EDTA) and iron reintroduced in controlled quantities as desired. In industrial practice, however, the quantitative control of the traces of iron or other catalytically acting materials would represent a more difficult problem.

Chemical Changes. The chemistry of the wood surface can be changed during and after preparation of the surface due to interaction with various physical and chemical natural reagents. Some sawing conditions, commonly with excessive saw vibrations, cause the wood surface to burn. Although not much is known about the temperatures of the wood surface during sawing, the temperatures of plane circular saws were found to be about 40–60 °C, occasionally 100 °C, and even 160 °C, above the ambient temperature, depending upon the distance from the teeth. The temperatures of the saw teeth are generally appreciably higher, reaching as high as 774 °C (41, 42). Thus, the possibilities are given for pyrolytic and oxidative changes on wood surface, although the times of exposure are very short and the effects correspondingly less. Additional cases of solid-wood exposure to elevated temperatures are met in drying wood particles and veneer.

The degradation of wood at moderately elevated temperatures or at short exposures to higher temperatures in the presence of air is composed of pyrolytic and oxidative changes. At longer exposures to higher temperatures the combustion process sets in—autocatalytic pyrolytic decomposition coupled with oxidation of the produced volatiles and char. Pyrolytic and oxidative changes of cellulose, hemicelluloses, and lignin at moderate temperatures proceed independently of each other, i.e., wood behaves like a mixture of these materials (43). Pyrolysis of cellulose was a subject of numerous investigations and has been reviewed several times (44–48). The process begins with depolymerization of the polysaccharides by transglycosylation to yield glucosan and other monosaccharide and oligosaccharide derivatives. Concurrently, dehydration to unsaturated compounds takes place (46). With lignin the low temperature decomposition is dominated by condensations and formation of ether linkages between the n-propyl side chains, and by generation of alkyl–aryl bonds, which is paralleled by dehydration reactions that form double bonds in the side chains (49, 50). Oxidation of cellulose apparently takes place at or above 140 °C and is accompanied by depolymerization and formation of carbonyl and carboxyl groups, followed by some decarboxylation. Moisture strongly catalyzes the process (51, 52, 53).

The information on the pyrolytic and oxidative changes that occur on wood surfaces resulting from the history of their formation is unsatisfactory. To a large extent such information is connected with investigations of surface inactivation toward conventional gluing (54), or with the dimensional stabilization of wood by exposure to moderately elevated temperatures.

A loss of hygroscopicity by prolonged heating of solid cellulosic materials to 100 °C or higher was explained by gradual loss of hydroxyls (55). A quantitative correlation was obtained between the loss of hygroscopicity and loss of weight by using yellow poplar and loblolly pine wood samples heated to 200 °C for 5 min. This correlation was explained by formation of intramolecular epoxy groups between hydroxyls 2 and 3 of anhydro units of cellulose (56); intermolecular ether linkages apparently do not form (57). The increased dimensional stabilization of wood after heating it to 300 °C was explained by the decomposition of hygroscopic hemicelluloses and other carbohydrates, followed by condensation or polymerization of the resulting furan-type compounds (58). Changes in chemistry of microsections of wood used as models for surface layers of wood of *Picea glauca* (Moench) Voss were studied by Chow and Mukai (39, 59) between 100 and 240 °C in air and in nitrogen. Below 180 °C the changes were associated mainly with oxidation, and above 180 °C they were of mixed pyrolytic and oxidative nature. The hydroxyl absorption in IR spectra decreased with time at 180 °C, the color of wood darkened, and both crystallinity and degree of polymerization (DP) of cellulose decreased. The IR C=O absorption of ester and carboxyl groups first decreased and then increased with temperature. Extractives were found to catalyze the rate of oxidation. This catalysis is probably why refined fibers give better medium-density fiberboard through an increase in oxidation moieties at the surface.

Increased temperatures are also likely to lead to changes in extractive makeup on the wood surface. Polymerization of tannins and monomeric phenolic materials to synthetic phlobaphenes and similar materials is likely to take place. Unsaturated fatty acids such as linoleic acid apparently can cleave oxidatively, and the resulting smaller molecular weight fragments attach themselves to the surface of wood (35).

Aside from special circumstances the changes in the chemistry of the wood surface due to exposure to light during surface formation and thereafter (drying, storage) are of little importance to nonconventional bonding (excepting light as a potential activator). The reactions are rather complicated and depend upon the wavelength and intensity of light, temperature, time of exposure, moisture content of wood, atmospheric composition, and presence of light-absorbing substances (activators) (51, 60–64). The surface changes include formation of free radicals, chain scission, dehydrogenation and dehydroxymethylation of cellulose and splitting of double bonds, formation of phenoxy radicals and quinone structures, and polymerization of lignin. In the presence of oxygen and water, H_2O_2 and peroxy groups also form. With solid wood (45–50 °C, 50% relative humidity, 75 d of exposure, and Xe arc as light source) there is a loss of lignin and hemicelluloses from the surface (61); water strongly increases the rates of loss. The color of wood surface can lighten or darken depending upon the wavelength of light to which the surface was exposed to (65). The changes in color depend strongly upon the amount and kind of extractives present. A photooxidative reaction operates to transform flavonoids taxifolin (I) and aromadendrin (III) into quercetin (II) and kaempferol (IV), respectively, and results in a general decrease in flavonoids and increase in vanillin-related compounds (65).

Direct Covalent Wood-to-Wood Bonding

Attempts to induce bonding between lignocellulosic surfaces by formation of direct surface-to-surface covalent bonds were made as early as 1945 when Linzell patented a process for making fiber products by compressing and heating a mixture of lignocellulosic fibers and a ferric compound as oxidant (66). Schur and Levy noted an improvement in the wet strength of paper upon oxidation of the pulp with sodium periodate or sodium hypochlorite (67). Additional patents involved interaction of lignocellulosic materials with acids in the absence of oxidants (68–72). More recently the area has drawn attention following the experiments of Stofko (73). His initial work was connected with the manufacture of particle board and plywood from white fir [*Abies concolor* (Gord. and Glend.) Lindl.] and incensecedar [*Calocedrus decurrens* (Torr.) Florin] wood in combination with various oxidants including aqueous H_2O_2/ferrous sulfate, aqueous HNO_3, and ethanolic ferric chloride.

It was originally hoped that free radicals formed by oxidative activation of the wood surface would join via oxidative coupling to form covalent bonds (Figure 2). The particle board was prepared by

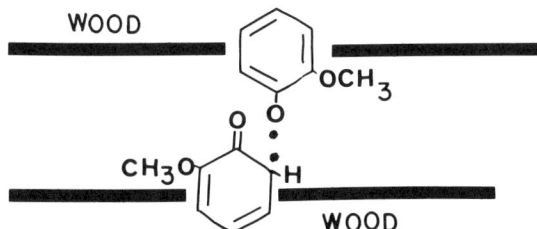

Figure 2. Hypothetical mechanism for the direct wood-to-wood bonding through oxidative phenolic coupling.

spraying the particles with activators, followed by pressing at 130 °C for 2 min. The resulting particle board had a specific gravity of 0.70–0.72 g/cm^3, internal bond (IB) of 63.5–74.0 psi, 24-h water absorption of 44.0–55.5%, and 24-h thickness swelling of 10.4–26%. Addition of hammer-milled bark exerted a beneficial effect (wood-to-bark ratio, 50:50; specific gravity, 0.725 g/cm^3; IB, 81.9 psi; 24-h water absorption, 59.9%; 24-h thickness swelling, 5.6%). Incense-cedar plywood had a shear strength of 405 or 385 psi (measured on a Globe plywood tester) (73).

Later experiments concentrated on plywood. Douglas-fir veneer [*Pseudotsuga menziessi* (Mirb.) Franco] was used under acidic or alkaline conditions, in combination with an expanded list of activators including H_2O_2 with ferric chloride or zirconium tetrachloride as catalysts; sodium chlorate; sodium hypochlorite; potassium persulfate; ammonium nitrate; potassium permanganate; and potassium ferricyanide; or combinations of the above; and occasionally in the presence of cobaltous chloride and sulfate, manganese dioxide, and cupric nitrate. Sodium chlorate under alkaline conditions produced plywood with dry shear strength of 246 psi, but the bonds were not water-resistant. Acceptable water resistance was attained, however, under acidic conditions. Dry shear was acceptable although an increase in acidity beyond a certain point tended to decrease dry shear, probably by surface hydrolysis. Percent of wood failure reached values between 90 and 100% but not consistently; it tended to be higher with more acidic mixtures and in some cases was inversely proportional to shear strength. Addition of wheat flour or of ammonium lignosulfonate to reduce penetration of the bonding reagents into the interior of wood resulted in only small improvements and small reduced variability. The main problems were connected with reproducibility and variability of the plywood characteristics as well as with the anticipated time effect of acid on the strength of the product (74, 75). The process was finally patented (76) and included examples of making plywood and particle board by using incense-cedar and Douglas-fir veneer and white fir shavings. Activators used for plywood were ferric chloride in EtOH, H_2O_2 with ferric chloride or with zirconium tetrachloride as catalysts in water, and sodium chlorate in water. In some cases H_2SO_4 or HCl was added. The mean plywood dry shear strength for the examples cited ranged from 210 to 385 psi. The bond was resistant to 4 h of boiling. Particle board was made with H_2O_2/ferrous sulfate in the presence of HCl and gave an IB of 65 psi at a density of 0.70 g/cm^3.

The work of Stofko et al. was continued by Pohlman et al. (77), mainly in the area of particle board and using primarily H_2O_2/ferrous sulfate as oxidant. It was thought that improvement in mechanical properties and decreased variability of particle board properties could be achieved by stabilizing the H_2O_2 reagent. Use of phosphoric acid or of pyrophosphoric acid as stabilizers or preoxidation of wood with sodium hypochlorite failed to produce any positive results, however. As expected, the stability of H_2O_2 increased as the amount of added ferrous sulfate decreased. Despite all these approaches the most important parameter determining the IB was still the density; thus, at densities of 0.88 g/cm^3, IB values above 70 psi were reached, although at 0.81 g/cm^3, IB values dropped to 38 psi (150 °C, 5 min presstime, 0.2% FeSO$_4$, 4.0% H_2O_2, 0.5% H_3PO_4, 0.5% HCl). This is exemplified in Figure 3. Addition of bark exerted a favorable effect on IB, although at least 50% bark had to be added to obtain an IB of about 75 psi at 0.75 g/cm^3 density. Improvement in IB and water resistance were directly related to the amount of H_2O_2 reagent used.

A remarkable improvement in particle board properties was achieved using branchwood. Thus at densities of 0.78–0.80 g/cm^3 the IB of ponderosa pine Bauer-refined material (*Pinus ponderosa* Laws.) was 76 psi for normal heartwood, 47 psi for normal sapwood, 163 psi for nondebarked branches, and 228 psi for debarked branches with satisfactory water resistance (0.01% FeSO$_4$, 6.0% H_2O_2, 0.5% HCl). Unfortunately, not enough attention has been given to this result. The heat necessary for bonding is provided by the exothermic reactions taking place with the heat from the press platens used only for initiation of these reactions; in conventional bonding heating is provided mainly by the platens (77).

Attempts to nonconventionally bond veneer of white fir, sugar pine (*Pinus lambertiana* Dougl.) and aspen (*Populus tremuloides* Michx.) were made using peroxyacetic acid under acidic conditions (78). The mean dry shear values were determined according to ASTM D 905 on a Baldwin Universal testing machine and ranged from 107 to 972, from 204 to 916, and from 363 to 918 psi, with phenol–formaldehyde boards giving dry shear values of 1367–1638 psi. Increase in peroxyacetic acid concentration or in acid strength had a positive effect on shear strength. Shear strength varied within a wide

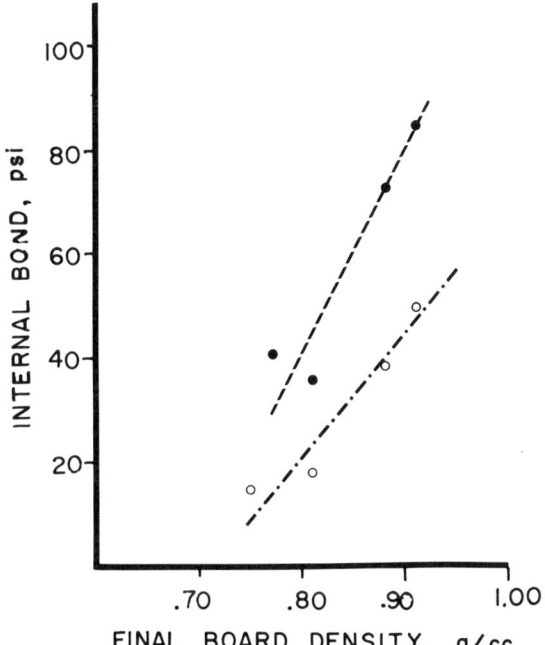

Figure 3. Relationship between board density and internal bond strength in particle board made by direct oxidative wood-to-wood bonding. Key: ●, 0.2% $FeSO_4$–4.0% H_2O_2–0.5% H_3PO_4–0.5% HCl; and ○, 0.2% $FeSO_4$–4.0% H_2O_2–1.0% H_3PO_4 (77).

range within any of the boards made. The general conclusions drawn were that variability in the bond quality of the particle board and plywood, and the excessive dependence on density of IB in the case of particle board (satisfactory products could be obtained at industrially impractical high densities only) were the main obstacles to the acceptance of the process; both were related to the lack of gap-filling properties of the reagents.

Investigations in the area of liner boards indicated that treatment of ground wood, of Kraft fiber, or of their mixtures with H_2O_2 or sodium chlorate in the presence of acids improved wet strength of the resulting liner boards. This was related to the formation of interfiber bonds (79). The process was patented 4 years later under inclusion of a fiber confrication step (80).

A process involving preoxidation of lignocellulosic materials in particle form with HNO_3, oxygen, or nitrous gases, and molding them into various shapes has been patented (81). The covalent bonding arises by formation of ester links between carboxyls and hydroxyls on the surface.

Bonding Through Intermediacy of Bifunctional Molecules

Covalent bonding of wood by means of bifunctional molecules appears to offer additional possibilities through more efficient bridging of the gaps between the wood surfaces, i.e., the wood surfaces do not need to be as near as about one bond length as in the case of direct bonding, but could be separated by gaps of several bond lengths (Figure 4).

Schorning et al. (82) attempted to make particle boards by using ethylenediamine and 1,6-hexanediamine as bonding agents. These amines are known to interact with wood surface by condensation with lignin. Addition of 15% of ethylenediamine imparted noticeable strength to particle board, which was still insufficient for commercial considerations. 1,6-Hexanediamine was more efficient with the particle board having a bending strength of 2176 psi and an IB of 48.0 psi at 7% addition (density, 0.85 g/cm^3, pressing at 140 °C for 12 min); however, the water resistance was low. The better results obtained with 1,6-hexanediamine are explained by its higher reactivity, although the more efficient gap bridging ability of the amine might be a more realistic explanation in view of its longer bridging chains.

Collett et al. (83, 84) attempted to improve the method of Schorning et al. by preoxidizing wood particles either with HNO_3 in the presence of oxygen, or with nitrogen oxides in the presence of oxygen at controlled time and temperature conditions. The bifunctional agents 1,6-hexanediamine, ethylenediamine, phenylenediamine, ethylene glycol, and 1,6-hexanediol as well as the monofunctional ammonia were used. Overall, diamines gave the best IB values, followed by ammonia, and glycols performed poorly. As with

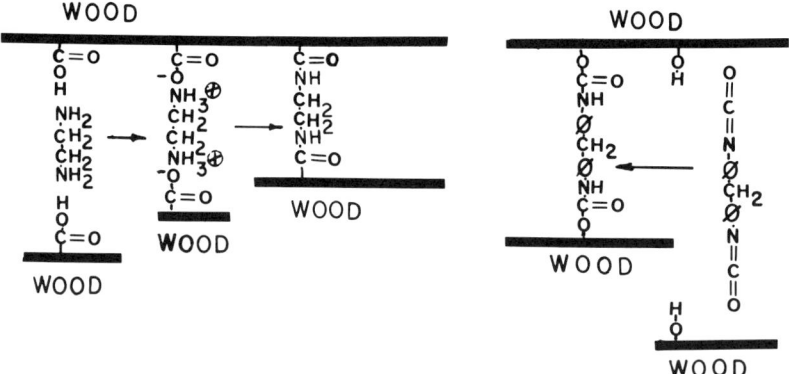

Figure 4. Hypothetic mechanism for bonding of preoxidized wood with ethylenediamine (left), and bonding of wood with a bifunctional isocyanate (right).

Schorning et al., 1,6-hexanediamine proved to be better than ethylenediamine. At densities of 0.81–0.88 g/cm^3, the 10% dry wood basis 1,6-hexanediamine board gave IB values between 101 and 142 psi, i.e., appreciably above the values reached by Schorning et al., which demonstrated the value of preoxidation (Table II). This is, however, still well below IB values of about 175 psi obtained with 6% phenol–formaldehyde board at 0.72 g/cm^3 density, although for some water resistance properties the 1,6-hexanediamine board proved to be superior to phenol–formaldehyde board. Increased preoxidation with nitrous gases or higher amine levels resulted in less swelling and an increase in IB. The results suggest the formation of water-resistant covalent bonds. Formation of amide and ester linkages was used to explain the bond formation. The results of this research were patented (81).

Bifunctional molecules were studied (84, 85), including maleic anhydride, maleic acid, succinic anhydride, azelaic acid, and saccharinic acid as cross-linking agents, in combination with surface activators including HCl, hydrobromic acid, perchloric acid, H_2SO_4, ferric chloride, zinc chloride, ferric nitrate, oxalic acid, and formic acid. The press temperatures used were 150–165 °C and the pressing time was 15 min. The first three cross-linking reagents produced board with the best properties. With 6% maleic acid, IB values of 131–144 psi at densities of 0.92–0.94 g/cm^3 were reached as compared with phenol–formaldehyde board with an IB of 229 psi at 0.95 g/cm^3 or of 165 psi at 0.70 g/cm^3. Although superior in water resistance, overall the board was appreciably inferior to phenol–formaldehyde board. Extraction experiments indicated that between 97 and 99% of monomers interacted with surface. Ester linking as well as

Table II. Some Properties of 1,6-Hexanediamine (HMDA) and Phenol–Formaldehyde-Bonded Particle Board

Binder	Amount (% dry wood)	Density (g/cm^3)	IB (psi)	Thickness Swelling (%)
Phenol–formaldehyde	6	0.65	110	36.8
	6	0.72	175	26.0
	6	0.79	179	38.2
HMDA (wood NO/O$_2$ preoxidized)	10	0.81	104	18.6
	10	0.88	142	19.0
HMDA (wood not preoxidized)	7	0.85	43	80.7
	10	0.85	48	127.8

NOTE: Data for HMDA with nonpreoxidized wood are from Ref. 82; all other data are from Refs. 83 and 84. Thickness swelling refers to 1-h boil test, except for the last two numbers which refer to 2-h immersion (temperature not given).

partial depolymerization of carbohydrates under the acid conditions used, and condensation and polymerization of the resulting sugar monomers explain the cross-linking.

Overall it appears that cross-linking of wood using bifunctional monomers and involving either preoxidation or acidic surface transformations can lead to boards of acceptable water resistance. The respective IB depends strongly on the amount of cross-linking agents and particularly on the density of the boards with the acceptable IB values reached again only at impractically high density values. The results suggest that use of monomeric bifunctional amines or carboxylic acids as cross-linking agents is insufficient for solving the problems connected with bridging of the surface-to-surface gaps in wood.

Conifer wood particles were bonded, preoxidized with nitrogen oxide–oxygen mixture, and cross-linked with furfuryl alcohol in the presence of small amounts of HCl. Acceptable particle board was obtained at relatively high densities (density, 0.74–0.78 g/cm^3; IB, 102–154 psi; swelling in 2-h boil test, 26–42%). Good results were also obtained by using furfuryl alcohol/formaldehyde cross-linking mixtures with maleic acid instead of HCl (*84, 86*). Both of these procedures involve acid polymerization of furfuryl alcohol and are discussed in the next section. The nature of covalent bonding to the wood surface is somewhat obscure and could involve in part ester linkages.

Bridging wood surfaces with bifunctional isocyanate monomers or resins under formation of substituted urethane linkages has been known for some time (*87*) and is only marginally nonconventional. Particle board with an IB of 105 psi at 0.65 g/cm^3 density and 24-h thickness swelling of 14% (5% resin content) has been reported (*88, 89*), and IB values of 59–72 psi at 0.65 g/cm^3 density and with 24-h thickness swelling of 12.9–20.0% and water absorption of 32.6–45.1% (4% resin content) have also been reported (*90*). Similar results were reported by Loew and Sachs (*91*). In addition, polyurethane foam has been used for these studies (*92*). Experiments with isocyanate resin bonded particle board of IB of 68–96 psi and with average linear expansion of 0.1% at relative humidities of 50% and 90% have been conducted. Water was detrimental to strength properties at the mat moisture content of 18% (*93–95*). The advantage of isocyanate methodology resides in good strength properties at relatively low densities. Unfortunately water resistance appears to be a matter of some concern, at least at the economically viable levels of isocyanates.

Bonding by Intermediacy of a Covalently Attached Polymer

Use of Oxidants. The logical follow-up of the efforts to overcome gap-bridging problems is the use of polymers covalently attached to the surface of wood (86). The longer chains of polymeric macromolecules are better suited for bridging larger distances than direct bonds or bridges of monomeric length (Figure 5). One process was conceived by Philippou and subjected to extensive experimentation. It consisted in application of an oxidant to the wood surface, followed by a mixture including in most cases various percentages of furfuryl alcohol, lignosulfonates, and maleic acid, and pressing the furnish at elevated temperatures (86, 96–99). The original idea resided in oxidative activation of the wood surfaces during pressing under formation of active free radical sites to initiate the chain polymerization of furfuryl alcohol under ultimate bridging of the wood surfaces by poly(furfuryl alcohol) chains. The roles of the other ingredients of the mixture were more nebulous.

Of the oxidants used by Philippou (H_2O_2, peroxyacetic acid, HNO_3, potassium ferricyanide, and sodium dichromate) the peroxy compounds acted best at densities of 0.70–0.75 g/cm^3 so that most experimentation was conducted using H_2O_2. Generally, ammonium lignosulfonate was used as lignosulfonate. White fir, Douglas-fir, sugar pine (*Pinus lambertiana*, Dougl.) and Bishop pine (*Pinus muricata* D. Don) were used as furnish. Increases in the amount of H_2O_2 up to 4.0% at 0.71–0.76 g/cm^3 density increased the IB and modulus of elasticity (MOE), had an indefinite effect on modulus of rupture (MOR), and increased water resistance. IB values of 130 psi for white fir, of 146 psi for Douglas-fir, and of 211 psi for Bishop pine were reported with 24-h thickness swelling of 4.6%, 5.4%, and 3.7%, respectively for 4% H_2O_2 (Table III). The difference between the performance of various species was explained in part by extractives and in part by differences in wood density; less dense woods allow for more compression and contact to reach the particle board density given above. Increase in particle board density from 0.58 to 0.80

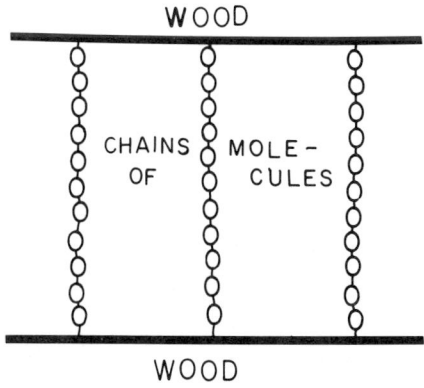

Figure 5. Bonding of wood through polymeric chains.

Table III. Effect of Species and Hydrogen Peroxide Concentration on the Properties of Particle Board

Species	Hydrogen Peroxide concentration (% in oven-dried wood)	Density (g/cm³)	IB (psi)	2-h Boil Test Thickness Swelling (%)	2-h Boil Test Water Absorption (%)
Coastal Douglas-fir *Pseudotsuga menziesii* var. *menziesii* (Mirb.) Franco	1.0	0.72	100	48.0	105.0
	2.0	0.72	122	28.0	72.4
	4.0	0.74	146	7.7	35.9
Interior Douglas-fir *Pseudotsuga menziesii* var. *glauca* (Beissn) Franco	2.0	0.73	168	13.4	46.8
Bishop pine (*Pinus muricata* D. Don)	1.0	0.72	136	17.2	48.0
	2.0	0.74	183	9.9	30.7
	4.0	0.75	211	5.5	14.5
Sugar pine (*Pinus lambertiana* Dougl.)	1.0	0.74	148	10.5	29.7

Note: Particle board was made with H_2O_2/ammonium lignosulfonate–furfuryl alcohol–maleic anhydride. Conditions: 7% total polymerizing solids (oven-dried basis); ammonium lignosulfonate solids to furfuryl alcohol, 7:3 by weight; and 2% of maleic anhydride (polymerizing solids basis).

Selected values were taken from Ref. 97.

g/cm³ improved the IB, MOE, MOR, and water resistance. Increasing the amount of polymerizing chemicals from 3 to 11% improved all parameters. Addition of wax decreased IB and MOR but increased resistance to water. Although furfuryl alcohol used alone performed well, addition of ammonium lignosulfonate improved practically all properties of the particle board with a 6:4 ratio of lignosulfonate to furfuryl alcohol representing the best mixture. An increase in the amount of catalytically active maleic acid (0–10% crosslinking chemicals basis) increased water resistance, and had only a small negative effect on MOR. Increase in pressing time and temperature (4–10 min and 120–205 °C) had a positive effect on water resistance. Substitution of furfuryl alcohol by formaldehyde produced acceptable board; substitution by acid prepolymerized furfuryl alcohol gave appreciably higher IB values with lower water resistance. The extent of experimentation was not sufficient, however, for forming final conclusions.

Philippou also experimented with wood particles preoxidized with nitrogen oxide–oxygen by using furfuryl alcohol acid polymer as cross-linking agent and maleic acid as catalyst. He obtained particle board with IB values of 43–159 psi and 9–20% swelling in a 2-h boil test at 0.70–0.77 g/cm³ densities and 7.5% resin usage (84, 86). The procedure does not differ much in principle from the one where furfuryl alcohol monomer was used.

A limited amount of experimentation was made with plywood and laminated panels. These experiments used the same ammonium lignosulfonate/poly(furfuryl alcohol) resin or monomer systems, with H_2O_2 activation, maleic acid as catalyst, and Douglas-fir veneer. A dry wood failure of 100% was obtained with plywood (83% and 95% wet, after vacuum-pressure soak), with dry shear strength of 1135 psi for parallel laminated panel made with ammonium lignosulfonate–furfuryl alcohol (525 psi wet), and of 1640 psi for the one made using ammonium lignosulfonate/furfuryl alcohol resin (650 psi wet).

The work of Philippou et al. was continued by others in the direction of developing acceptable particle board with 0.65-g/cm³ density, maximizing economics, extension to large-scale production, and exploration of various ancillary, industrially important factors (78, 100–102). Comparison of HNO_3 with H_2O_2 as oxidative activators indicated that H_2O_2 gives products with mechanical properties equal to or slightly superior to those of HNO_3 (e.g., IB of 77.3 psi vs. 62.9 psi at 0.65 g/cm³ density and 1.5% of oxidant). However, HNO_3 produced board with appreciably better water resistance (e.g., water absorption after 2 h of boiling: HNO_3, 90.3%; H_2O_2, fail; density, 0.65 g/cm³; oxidant, 1.5%). At a density of 0.75 g/cm³ H_2O_2 boards were water resistant, with a water absorption of 118.7%. Because the pH of the polymerizing medium strongly influences the mechanical and water resistance parameters of the particle board produced by the Philippou process (pH should be below 2.3 for good wet strength), the difference between H_2O_2 and HNO_3 oxidants was likely to reside in the difference in acidity of the two systems. However, because economic aspects favored the use of HNO_3, practically all experimentation was based on the use of this oxidant.

On the basis of extensive experiments it was concluded that the optimum composition of the cross-linking agent including ammonium lignosulfonate, furfuryl alcohol, and maleic acid was in the proportion of 4.2:1.8:1.0. The cross-linking agent was used in an amount of 7–9% in combination with 1.5% HNO_3. The furnish was pressed at 177 °C and 445 psi for 7 min. In a pilot plant run the resulting white fir flakeboard (7% solids, 0.65-g/cm^3 density) gave an IB of 68.0 psi and swelling after 2 h boiling of 14.0% vs. an IB of 65.0 psi and swelling of 39.4%, respectively for 5.0% solids phenol–formaldehyde board (Table IV). Attempts to use different lignosulfonates (sodium or calcium), to replace maleic acid with other acids, or to replace furfuryl alcohol met with no success or with indefinite results. Use of hardwood or of bagasse resulted in particle board of lesser quality. Influence of the moisture content of the furnish was studied, and it was concluded that 9.5% moisture represents the best compromise.

A thorough investigation of the organic emissions during pro-

Table IV. Properties of Nonconventionally Bonded Full-Size Panels Compared to Laboratory-Made Boards and Phenol–Formaldehyde (PF) Boards

Board Type	IB (psi)	MOR (psi)	MOR Retention (%)	MOE × 1000 (psi)	2-h Boil Swelling (%)	24-h Soak Swelling (%)
Fir, flake						
Full size	68	2572	36	631	22.5	14.4
Lab size	68–72	2600–3100	38–46	600–700	22.0	12.0–14.0
5% PF[a]	65	4054	58	493	39.4	27.0
Pine, flake						
Full size	69	2510	30	625	19.5	13.5
Lab size	100	2446	36	527	20.5	12.8
5% PF[a]	128	3690	48	405	32.6	24.8
Mill residue						
Full size	74	860	29	227	13.2	8.5
Lab size	82	852	23	283	15.0	11.3
5% PF[a]	105	1329	44	24.4	20.2	14.2

Conditions: Activator, 1.5% HNO_3 (72%) (oven-dry basis); cross-linking agent, 7% (oven-dry basis), composed of 60% ammonium lignosulfonate, 25% furfuryl alcohol, and 15% maleic anhydride; target density, 0.65 g/cm^3; bonding, 7 min at 177 °C with 9–11% moisture content of the material.

[a] On the basis of oven-dried wood

duction of particle board indicated that treatment with HNO_3 results in the production of nitric oxide, methyl nitrite, and methyl formate. Addition of polymerizing mixture and pressing results in emission of furfural, difurylmethane, difurfuryl ether, and furfuryl alcohol. Furfural was the only abundant compound found in emissions from newly pressed boards. The toxicity hazard from these materials was judged to be moderate to high if they are present in significant concentrations (101).

In terms of the performance of coatings, the developed system proved to be comparable to phenol–formaldehyde. The board appeared to be less susceptible to fungal decay in preliminary studies, as compared to phenol–formaldehyde board, but had high mold susceptibility. According to ASTM 631 72 corrosion test, the board was initially highly corrosive to metals but after 16 d the corrosiveness dropped and was only slightly higher than that of untreated wood flakes (102).

Economically the Philippou process appears to be somewhat more expensive than phenol–formaldehyde. In December 1981, the cost of binder raw material for 1000 ft^2 of HNO_3 board was estimated to run \$22.60, vs. \$28.30 for H_2O_2 board, and \$21.00 for phenol–formaldehyde board. Further work, particularly on alternative ingredients, can change the situation (103).

The other cross-linking mixtures studied and involving oxidative activation included ammonium and magnesium lignosulfonates in combination with H_2O_2/HCl. The boards produced were, however, inferior to the above (77).

Procedures bearing some similarity to the Philippou process were patented by Emerson in 1953 and 1963 (104, 105). The first patent consisted of pretreatment of lignocellulosic particulate material with urea (0.7–11%), followed by addition of furfural (0.8–16%) and an acid or acid salt catalyst (1.0–16%), and pressing at elevated temperatures. The second involved a mixture of urea, furfural, lignin, petroleum resin, drying oil, wax, and an oxidation catalyst as a cross-linking mixture.

Systems conceptually involving only oxidative cross-linking reactions of lignosulfonates were studied (see Scheme 1). The cross-linking material included spent sulfite liquor, potassium ferricyanide, and H_2O_2 in percentages of 25%, 0.5–1.0% and 3.0%, respectively (calculated as water-free reagents, dry wood percent basis), and gave particle board of IB of 102–122 psi at a density of 0.73–0.78 g/cm^3 (106), or an IB of 115 psi at a density of 0.56 g/cm^3 (107). The water resistance was generally not great unless paraffin emulsion was added. The advantages of the process seem to be in the relatively high pH (such as 4.2). Besides potassium ferricyanide, potassium

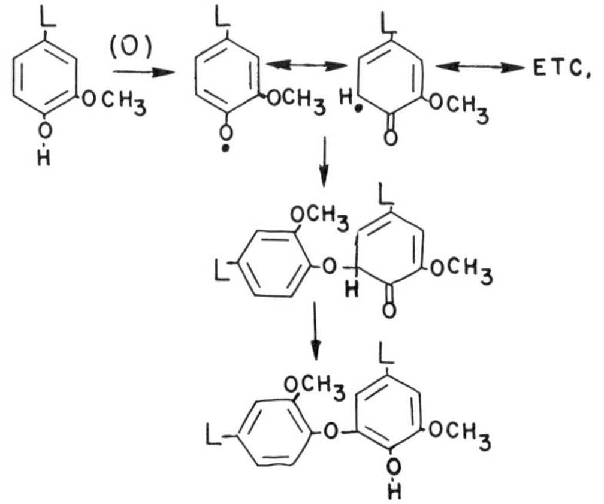

Scheme 1. Partial mechanism of oxidative cross-linking of lignosulfonates.

ferrocyanide could be used as catalyst. Other catalysts such as ferrous sulfate, mercurous oxide, silver oxide, peroxidase, and potassium persulfate were inactive as they led to no gelation of the spent sulfite liquor.

Substitution of potassium ferricyanide for 1% of sulfur dioxide dissolved in calcium-base spent sulfite liquor was possible at a pH of about 2.0. An increase in pH stabilized the cross-linking mixture unless 4% of ammonium chloride (dry wood basis) was added. In this case particle board could be produced at a pH of 4.5, with properties equivalent to those produced at a pH of 2.0, i.e., with IB values of up to 82.0 psi and acceptable water resistance (*108, 109*).

Use of Acids. Several gluing procedures employ acid-catalyzed cross-linking agents without oxidants. One procedure (*110*) involves the use of a mixture of a carbohydrate such as sucrose, glucose, or starch, and a catalyst—a weak acid in most cases (e.g., ammonium salt of an inorganic acid)—in an amount of 2–32 g/1000 cm^2 and pressing the particle board at a temperature of 140–250 °C and a pressure of 70–360 psi for 5 to 10 min. The bonding mechanism was explained by transformation of carbohydrates into furan-type compounds, polymerization, and condensation with lignin of wood. The results included an IB of 180.6 psi at a density of 0.80 g/cm^3 and an IB of 76.8 psi at a density of 0.56 g/cm^3, with very good water resistance and dimensional stability. A particular advantage of the system resides in the high pH during chemical transformations, which lies between 3.5 and 5.5.

A method using H_2SO_4 in combination with spent sulfite liquor as a cross-linking agent was developed (*111–14*). Initially dilute H_2SO_4 was sprayed on wood wafers, followed by addition of calcium lignosulfonate powder; later H_2SO_4 was mixed with calcium, ammonium, sodium, or magnesium lignosulfonate binder to an extent of 10–12% (*112*). The material was pressed at 166 °C at 400–500 psi and gave IB values of up to 95 psi. The strength of the board tended to deteriorate rapidly after heating in humid air. Although the pH of the board was reported to be as high as 3.55–3.87, it is likely that this deterioration was connected with a low pH localized around the glue lines.

Fractionated ammonium lignosulfonate of pH 1.5–4.3 was used without H_2SO_4 for making waferboard (*115–17*). The board was pressed at a temperature of 210 and 220 °C and a pressure of 500 psi for 6 to 12 min (*115*). The amount of binder was 6%. The IB values reported were up to 80 psi with water absorption of 123–157% and thickness swelling of 27–48% (2-h boil) at a density of 0.64–0.67 g/cm^3. Interestingly, the strength properties were improving with the carbohydrate content of the binder tested dry and reached a maximum at around 60% carbohydrates when tested wet. This observation suggests that the water-soluble carbohydrates of the spent sulfite liquor are more important binding ingredients than lignin (*110*).

Other Methods. Difunctional amines were used in combination with poly(vinyl chloride) (PVC) aqueous dispersion and occasionally with epichlorhydrin as cross-linking binders (*82*). Cross-linking was assumed to take place through reaction of the amine moiety with PVC, although some amine–wood surface interaction probably also took place in view of what was mentioned before. The material was pressed at 140 °C to give particle board with IB values as high as 153.6 psi at the density of 0.76 g/cm^3. Addition of hydrophobic materials improved water resistance. A combination of maleic acid and poly(vinyl alcohol) (PVA) was used as a cross-linking binder for particle board and plywood (*118*). After pressing at 140 °C the particle board exhibited IB values as high as 142.2 psi at densities of 0.74–0.78 g/cm^3; the resistance to water was poor, however.

A mixture of maleic anhydride, styrene, and diethylene glycol diacrylate was used, partially prepolymerized, and reacted with untreated or with H_2O_2/ferrous sulfate-pretreated wood. The IR spectra indicated the probable formation of covalent bonds between the introduced mixture and the hydroxyls of wood. Details of the procedure were, however, unavailable (*119*).

Several reports on the activation of lignocellulosic surface by corona discharge, microwave plasma, and ozone treatments have been published (*120–22*). Strong bonds were produced between thermoplastics and wood by corona treatment of wood surfaces and particularly by treatment of polyethylene and polystyrene surfaces. With 5-min treatment wood–polymer–wood bonds of over 569 psi were obtained using low-density polyethylene sheets at a pressing temperature of 115 °C. The results suggested good moisture resistance (120). Marked improvement in bonds between cellulose strips was reported after microwave discharge or ozone treatment (121, 122). Treatment with corona discharge improved bonding between fibers and polyethylene powder in hardboard composites and decreased the water uptake by a factor of 2–4. The effect was connected partially to oxidation of surface and partially to formation of long-lived charges (electrets) on the surface of materials.

Fundamental Studies

The material discussed in the last sections indicates that wood surface activation involves the interaction of lignocellulosic surfaces with relatively simple reagents, generally acids or oxidants. Basic reagents and reducing materials were rarely used. The introduced organic polymers represented also rather common materials and the general nature of the reactions in question was also known. For these reasons there was no lack of hypotheses proposed to explain the chemical mechanisms of activation and bonding. This is contrasted, however, with the small number of fundamental studies aimed at substantiating or at refuting the hypotheses advanced, i.e., dealing with what actually takes place during activation or bonding. In view of the voluminous nature of the material and the available reviews, this section is not going to cover in depth the entire field of the reactions coming potentially in question and involving the activation reagents and the cross-linking substances used; instead only more fundamental work specifically aimed to substantiate the hypotheses presented or to uncover new interactions connected with wood surface activation or nonconventional bonding will be presented.

Surface Activation. ACID ACTIVATION. Acid treatment of cellulose and hemicelluloses generally leads to hydrolysis to monosaccharides, which can subsequently dehydrate and condense to form furan-type compounds such as furfural and its 5-hydroxymethyl adduct. Further reactions lead to polymeric materials of dark color as well as to monomers such as levulinic acid, formic acid, and angelica lactones. Various condensation and solvolysis reactions also accompany the acid treatment of lignin (123). The hydrolysis, dehydration, and condensation reactions have been used to explain formation of covalent bonds between surfaces (85), increase in water resistance (85, 124), and weakening of wood (75) in nonconventional plywood or particle board production. However, very little factual information is available on how far, in terms of the consecutive reactions mentioned, and in what direction, in terms of the parallel reactions mentioned, does the surface of lignocellulosic materials actually change in acidic surface reactions connected with nonconventional wood bonding, i.e., at low moisture conditions and at temperatures between 100 and 200 °C.

Submitting cellulose powder–H_2SO_4 mixture to 150 °C at 1200 psi for 10 min resulted in the appearance of bands at 1720 (carbonyl) and 1620 cm^{-1} (aromatic) in the IR spectrum (124). This observation was interpreted as an indication of the formation of furfural-type compounds. Cellulose treated with acidic salts such as aluminum chloride, zinc chloride, and sodium bisulfate was reported to show an endothermic nadir at 180–250 °C in differential thermal analysis (DTA) (125). The thermogravimetric (TG) experiments indicated, however, a markedly lower temperature at which weight loss (presumably dehydration) begins, with acceleration taking place between 100 and 200 °C. With DTA and TG, H_2SO_4 destabilized cellulose with an endotherm occurring between 180 and 260 °C, depending upon the amount of acid added (126). Thermal analysis of lignocellulosic materials has been recently reviewed (44, 127).

HYDROGEN PEROXIDE ACTIVATION. More work has been done on the oxidative activation of wood. The reactions of H_2O_2 and organic peroxides in acid media with various organic materials, including lignocellulosics, have been investigated and reviewed many times (128–37). In order to obtain information on the hydrogen peroxide activation of wood surface, Jenkin (138) studied the reaction of cellulose, xylan, lignin, and wood with H_2O_2 at 100 °C in the presence and absence of Fe^{2+} and Cu^{2+} catalysts. The reaction was followed by attenuated total reflectance (ATR) (Figure 6) and by transmission IR spectroscopy.

In the case of cellulose the main reaction products were carboxylic groups and a smaller number of keto, aldehyde, or ester groups. The reaction between noncatalyzed H_2O_2 and cellulose was slow at the beginning, but its rate increased with time. In the case of xylan and lignin the reaction rate decreased with time. The aberrant behavior of cellulose was related to the crystallinity of cellulose, and the time increase in accessibility connected with gradual swelling. Oxidation proceeded much faster with catalyzed H_2O_2. Carboxylic absorption reached a maximum in 5–10 min but decreased afterward. This decrease was explained by a Ruff-type decarboxylation.

Thermal analysis provides information on changes in physical properties of a substance as a function of increasing temperature. These methods are fruitful for investigations of wood activation and transformations leading to bonding of wood as they allow for the study of changes in the properties of solids, liquids, or solid–liquid mixtures under temperature and pressure conditions approximating those in the press. The physical properties predominantly studied

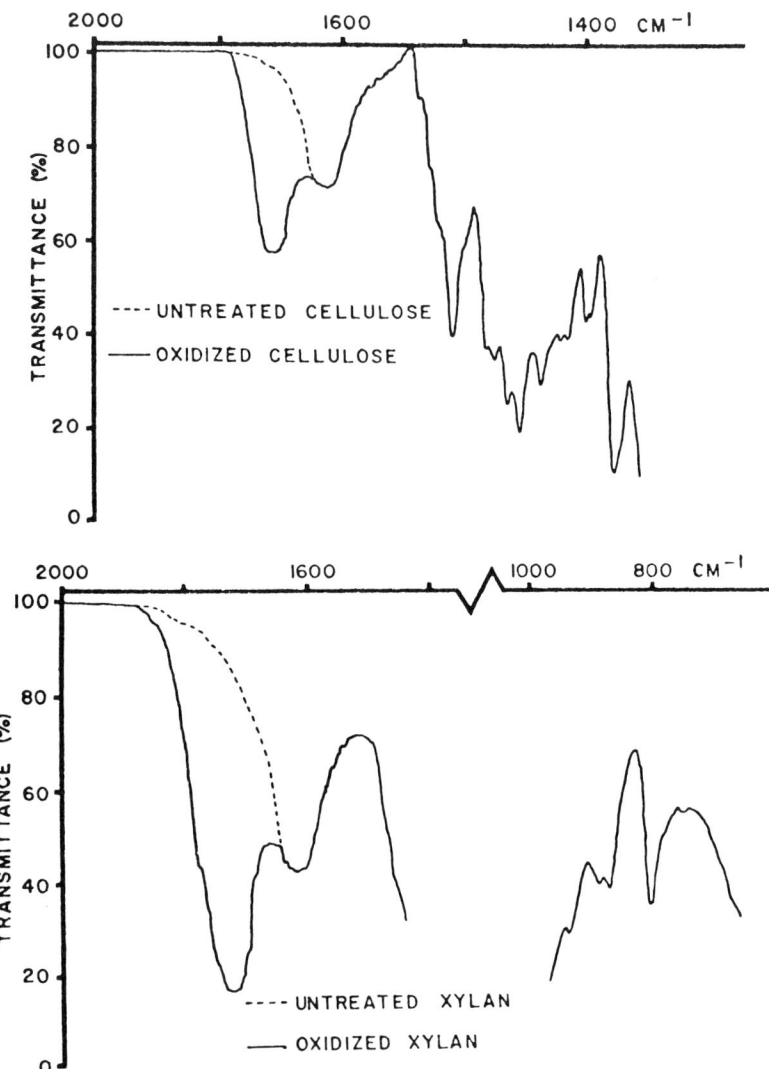

Figure 6. IR reflectance (ATR) spectrum of cellulose before and after oxidation with H_2O_2 (top) and IR transmission spectrum of xylan before and after oxidation with H_2O_2 (bottom). (Reproduced with permission from Ref. 138. Copyright 1976, John Wiley & Sons, Inc.)

include heat absorption or release [differential scanning calorimetry (DSC)], temperature of the sample [differential thermal analysis (DTA)], weight [thermogravimetry (TG) or differential thermogravimetry (DTG)], and certain mechanical properties [thermomechanical analysis (TMA)]. The methods allow determination of many parameters describing the transformations taking place. These include melting points, glass transitions, reaction temperatures, enthalpies of reactions, and kinetic parameters and are particularly efficient if coupled with other analytical approaches, such as gas chromatography (GC) of the gases formed and measurement of spectra and pH.

The reaction between H_2O_2 and wood powder, cellulose, and lignin, was studied by DSC using alodined aluminum pans without removing catalytically acting metal ions from lignocellulosic materials. The alodining process consists of the treatment of aluminum surface with pyrophosphate–fluoride and makes the surface catalyze the decomposition of H_2O_2. In all cases two exotherms were observed—one between 75 and 90 °C and the other between 115 and 130 °C (shoulder at 95 °C with lignin). The first exotherm was thought to stem from H_2O_2 decomposition and the other from decomposition of the organic peroxides formed. The IR spectra of the reacted materials showed carbonyl bands between 1700 and 1850 cm^{-1} in the case of wood and lignin with the intensity of the bands and complexity of the spectra generally increasing with reaction temperature. Only a slight increase in carbonyl absorption was noted with cellulose (86). The results suggested preferential lignin oxidation in wood.

Continuing the above work the reaction of H_2O_2 with lignocellulosic materials in powder form was studied under varying pressure, heating rate, particle size, H_2O_2 concentration, and atmospheric composition, by using much less catalytically active pans made of pure, untreated aluminum, and lignocellulosic substrates free of heavy metal ions (139–144). Two exotherms, at 130–190 °C, and at 200–250 °C were observed. The first exotherm was assigned to decomposition of H_2O_2 and/or its reaction with lignocellulosic substrates; the other exotherm was assigned to the reaction of oxygen with lignocellulosic hydroxyls. Thermodynamic and kinetic constants were determined for these reactions. Arabinoxylan and lignin were found to be more reactive than wood or cellulose. Most substrates reacted with H_2O_2 to form oxygen in various amounts. By using model compounds it was found that oxygen was formed under catalytical influence of aldehydic or keto groups in the lignocellulosic materials. Compounds yielding such groups by acid hydrolysis, e.g., monosaccharides, disaccharides, or polysaccharides, were also active. Esters or carboxyls were inactive. A mechanism that involved intermediacy of hydroperoxides was proposed for the formation of oxygen.

Experimenting with cellulose–lignin mixtures showed that the position of both exotherms shifted to higher temperatures with an increase in cellulose percentage (Figure 7); the heat of the reaction increased with an increase in the lignin percentage. There is a pos-

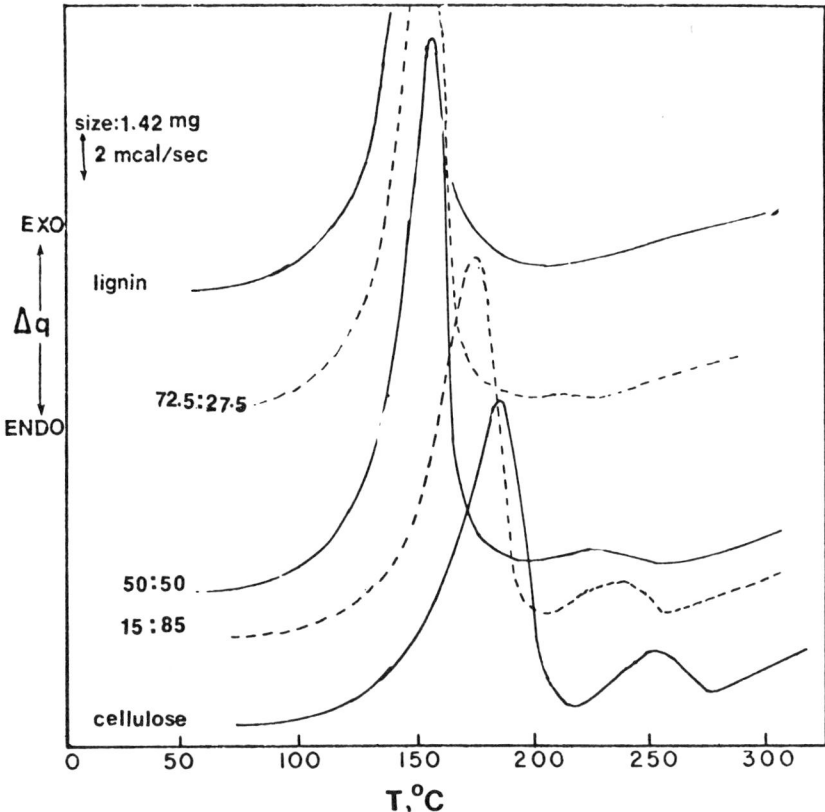

Figure 7. DSC curves of H_2O_2-treated lignin–cellulose mixtures. Conditions: closed aluminum pan with a pinhole; 975 psi N_2 pressure; 20 °C min^{-1} heating rate. (Reproduced with permission from Ref. 142. Copyright 1982, Gordon and Breach, Science Publishers.)

itive correlation between the tensile strength of the cellulose–lignin mixtures (125) and the heat of reaction. All of the above suggested that H_2O_2 activation of wood surfaces is connected primarily with the lignin moiety.

NITRIC ACID AND NITRATE ACTIVATION. The reactions of HNO_3, nitrates, and nitrogen oxides with lignocellulosic materials have been a subject of numerous publications. The specific transformations can be subdivided into acid reactions, oxidation reactions, and nitration–esterification reactions. With HNO_3 and more acidic salts the acid reactions predominate at higher dilutions, and parallel those of aqueous solutions of H_2SO_4 and other similar acids. Oxidation of cellulose results primarily in oxidation of primary hydroxyls to carboxyls, with the secondary hydroxyls oxidized less. Nitration leads to the formation of nitric acid esters. Lignin is oxidized to quinone structures, and later, under ring opening, to carboxylic acids. Nitration introduces nitro groups at the C-6 position of the aromatic ring; position C-4 is nitrated at higher acid concentrations only by displacement of the side chain. For details the reader is referred to the original literature and reviews (145–55).

The reactions between about 20 nitrates including nitric acid and various lignocellulosic materials were investigated using DSC, TG, and DTG as well as IR spectroscopy (145, 156). The lignocellulosic materials included lignins from a softwood and a hardwood, microcrystalline cellulose, precipitated cellulose, xylan, and wood powder. Reaction with HNO_3 produced an exotherm whose position depended strongly upon the moisture content of the sample; below 10% moisture content the position of the exotherm was relatively stable, however. This result agrees with the well-known increase in reactivity of HNO_3 with concentration. The exotherm peaked between 78 °C (wood flour) and 172 °C (microcrystalline cellulose) with lignin samples reacting below 100 °C and carbohydrate samples reacting well above 100 °C. The pH of lignocellulosic materials increased from 0.9–1.4 to 1.8–2.8 following the appearance of the exotherm. Thermodynamic and kinetic reaction parameters were determined for all lignocellulosic materials mentioned.

Oxidation of microcrystalline and precipitated celluloses resulted in the appearance of 1765- and 1740-cm^{-1} bands in the IR spectrum. These bands were explained by formation of five-membered lactones (possibly between C-6 carboxyls and C-3 hydroxyls) and of six-membered lactones or esters, respectively. The formation of ester linkages could be responsible for the observed interfiber or intersurface bonding mentioned previously. Both of the above bands were exhibited also by oxidized xylan. Oxidation of lignin resulted in a band at 1725 cm^{-1} (ester) and at 1625 cm^{-1} (double bonds or nitrate ester) as well as in weak bands at 1520 and 1350 cm^{-1}, which suggested the formation of nitrate groups. No bands at 1765 and 1740 cm^{-1} were produced in the case of wood, suggesting the preferential oxidation of lignin.

Thermal decomposition of nitrate salts goes through formation of NO_2, O_2, and of metal oxides in most cases; ammonium nitrate produces chiefly N_2O. In the presence of lignocellulosic materials the reactions might take another course, however. The reaction between nitrates and various lignocellulosics produced an exotherm in DSC peaking within a wide temperature range, depending upon the cation and the nature of the substrate. A positive linear correlation was found to exist between the pH of the aqueous solutions of the nitrate salts and the peak temperature (Figure 8). The metal salts of

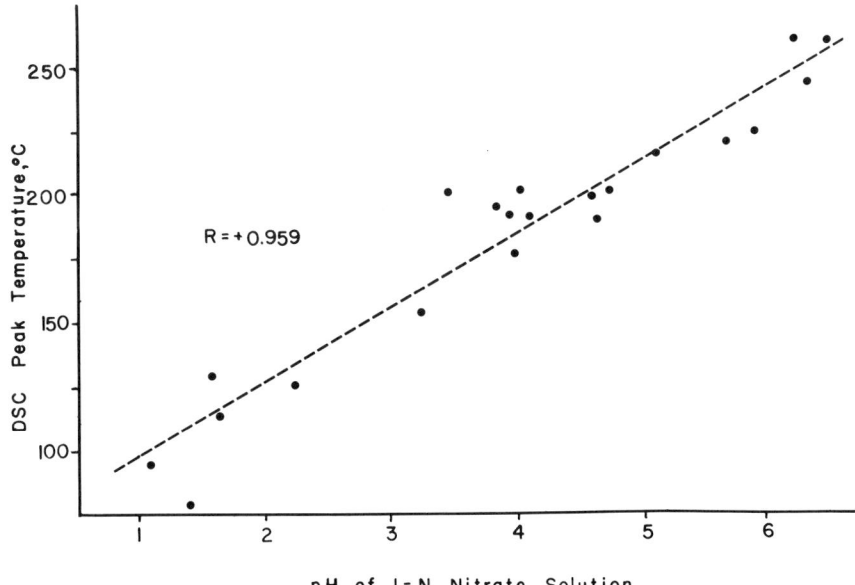

Figure 8. Relationship between DSC exotherm peak of wood flour oxidation by various nitrates and pH of the nitrates (145).

the weaker organic acids formed in the course of the oxidation coupled with disappearance of the nitrates resulted in a marked increase in pH of the reaction mixtures after the reaction. Ammonium nitrate represented an exception, however, due to the oxidation of the ammonium cation. In spite of the pH increase the acidity of the reaction mixtures involving aluminum, ferric, chromic, beryllium, and ammonium nitrates was judged to be still too high (pH < 3.5) to eliminate the danger of acid hydrolysis in particle board. Precipitated cellulose reacted at appreciably lower temperatures than microcrystalline cellulose, indicating the influence of accessibility. Between pH values of 1.0 and 5.5 the heat of the reaction was independent of pH; above pH 5.5 the heat of reaction became lower, most likely due to incomplete oxidation, because pH increases and the reaction slows down as it progresses. The ester bands were absent in IR when nitrate salts were used as oxidants probably because of the formation of the metal salts of carboxylic acid groups.

Electron spectroscopy for chemical analysis (ESCA) is a valuable method for studying the oxidation state of the wood surface. The carbon (1s) peak of the ESCA output consists of several overlapping components. Mathematical computer methods can resolve this peak into four component peaks corresponding to carbons with no oxygen substituents (C_1), to carbons carrying one single carbon-to-oxygen bond [alcohols, ethers (C_2)], to carbons carrying one double or two single carbon-to-oxygen bonds [aldehydes, ketones, acetals (C_3)], and to carbons carrying one double and one single carbon-to-oxygen bond [esters, carboxyls (C_4)] (22, 26).

Oxidative chemical changes on the surface of maple wood (*Acer saccharum* Marsh.) were studied by using ESCA (26). Treatment with HNO_3 at ambient temperature for 24 h resulted in a strong increase in carboxylic groups (C_4 carbons) and some increase in keto, aldehyde, ketal or acetal groups (C_3 carbons), while the number of carbons attached to a single noncarbonyl oxygen [alcohols, ethers (C_2 carbons)] and of carbons carrying no oxygen (C_1 carbons) decreased (Figure 9). These observations substantiated the results obtained by Wu (145). Treatment of wood with sodium periodate resulted in a disappearance of C_1 carbons, an increase in C_2 and C_3 carbons and a slight increase in C_4 carbons. High percentages (50%) of carbons carrying no oxygen in nontreated wood (C_1 carbons) and disappearance of these upon periodate treatment, as well as a strong increase in C_2 carbons, are difficult to interpret. It is possible, in view of the tendency of extractives to accumulate at the surface, that these changes do not involve skeletal lignocellulosic materials, but rather the deposited extractives or environmental contaminants. A strong increase in C_3 carbons is in keeping with the known periodate oxidation of glycol units to aldehyde groups.

Oxidation of wood by HNO_3 at ambient temperature that results in the formation of carboxylic acid groups on the surface has been substantiated (157). These carboxylic groups are sufficiently acidic to catalyze polymerization of furfuryl alcohol.

PLASMA ACTIVATION. Treatment by microwave plasma has been used for surface activation of cellulose. The involvement of free radicals in the activation is unlikely; surface oxidation and degradation under formation of a gluelike surface layer is favored as the source of cellulose-to-cellulose adhesion (122). Electron spin resonance (ESR) and free radical titration have shown that action of radio-frequency argon plasma on rayon results in the formation of free radicals (158). The number of paramagnetic centers increases with the power applied. Concurrently hydroxyl (OH) groups decrease, and degree of polymerization (DP), water solubility, and the number of carbonyl groups increase. Nitrogen and air radio-frequency plasma can modify OH groups as well as break the chains of cotton cellulose (159). The concentration of paramagnetic centers was greatly influenced by the pressure in the reactor.

Treatment of corrugating medium with corona discharge and analysis of the surface by ESCA indicated an increase in oxygen, with N_O/N_C ratio of 0.57 vs. 0.42 and 0.41 for the untreated surface, not

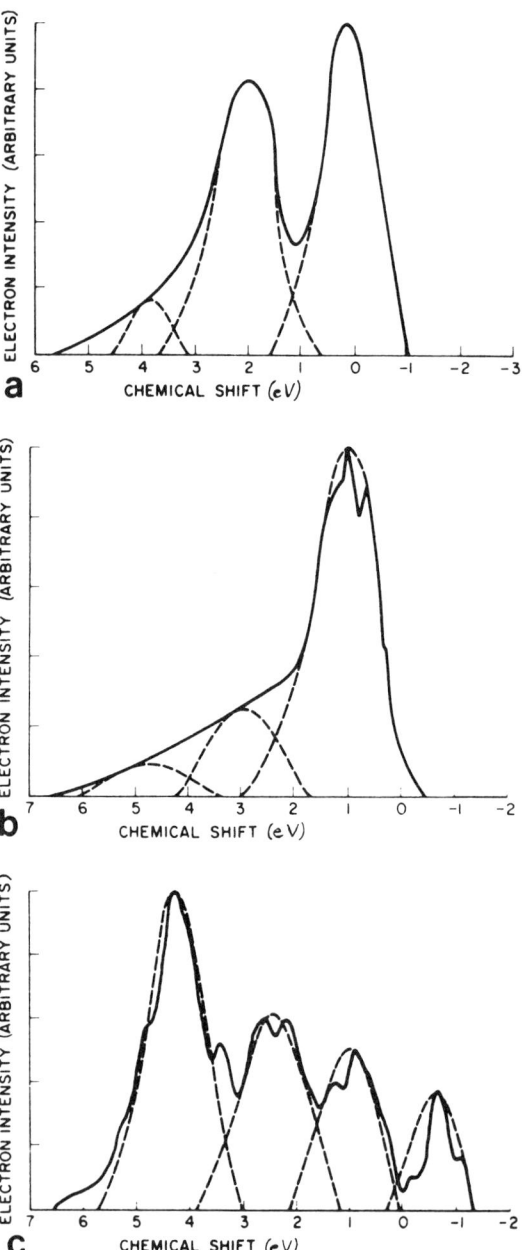

Figure 9. Depiction of separation of smoothed, expanded, deconvoluted, and Gaussian curve-fitted ESCA C(1) peak of untreated maple wood (a), periodate-treated maple wood (b), and HNO_3-treated maple wood (c). (Reproduced with permission from Ref. 26. Copyright 1982, Forest Products Research Society.)

extracted and extracted with acetone, respectively. Concurrently corona discharge decreased the number of C_1 carbons and increased the number of carboxylic C_4 carbons, which also increased the water absorbency of the surface (160).

OTHER OXIDIZING ACTIVATORS. Chlorates have been used at low pH as surface activators (75). The oxidation mechanism most likely involves decomposition to chlorine or chlorine dioxide which are likely to react preferentially with lignin. Other compounds used include chromates, nitrates, organic peroxides and peroxide salts, hypochlorites, chlorites, perchlorates, halogens, ozone, permanganates, and transition metal salts, but little or no fundamental work related to nonconventional bonding has been done with them (76). Ceric salts, Mn^{3+}, ozone, γ-radiation, and UV radiation often have been used as activators in graft polymerization of lignocellulosic fibers. The pertinent literature has been reviewed (4, 161–63).

Nonconventional Bonding. Studies of the reactions of chemical agents with lignocellulosic surfaces, resulting or thought to result in an ability of such surfaces to adhere to each other under formation of covalent bonds, do not generally give an answer to the question whether such bonds do actually form or how they form. The results of such experiments tell us only what might happen and what is not likely to happen. To gain a better understanding of the bonding process requires an insight into the chemical processes actually taking place during bonding, i.e., into the interactions between lignocellulosic surfaces and the introduced materials (if present). The direct experimentation in this area is commonly difficult, and indirect evidence involving studies of interactions of the materials in question must be obtained under conditions approaching those of the practice.

DIRECT BONDING. The formation of covalent bonds between wood surfaces in the direct wood-to-wood bonding method (124) is easier to accept for the lack of alternative explanations, although the nature of such bonds is still obscure (see Figure 2). In an attempt to inquire deeper into the bonding processes Stofko prepared small composite boards using his process, temperatures between 70 and 170 °C, pressures of 1200 psi, and pressing times of 4–20 min. Lignocellulosic substrates such as wood powder, microcrystalline cellulose, refined brown ("pecky") rot resulting from the attack of *Polyporus amarus* Hedgcock on incense-cedar wood (81% Klason lignin) (164), and mixtures thereof were used. The activating agents included $H_2O_2/ZrCl_4$, H_2O_2/Fe^{2+}, $NaClO_3/H_2SO_4$, $NaClO_3/NaOH$, H_2SO_4, and NaOH. Bonding was evaluated by tensile and Izod-impact tests, water absorption, thickness swelling, boiling tests, and scanning electron microscopy (SEM). Lignin and wood bond easily with the products resistant to boiling in water. Cellulose bonded under nonoxidizing, acidic conditions, apparently by hydrolytic degradation, dehydration to furan compounds, and subsequent acidic repolymerization. Permanent bonding between cellulose and lignin apparently results from both oxidative and hydrolytic reactions. The bonding strength achieved with wood and lignin was about that of the tensile strength of Douglas-fir wood perpendicular to the grain. Overall the level of bonding in cellulose was significantly lower than in lignin or in wood, and the oxidative bonding produced materials with significantly higher bonding strength than acidic bonding. Oxidative bonding resides primarily in lignin-to-lignin and possibly in lignin-to-cellulose bonds. These results were supported by the SEM results that demonstrated that in bonding using H_2O_2/Fe^{2+} the cellulose particles occasionally split lengthwise in preference to separation of the bonded lignin–cellulose particles (Figure 10).

BIFUNCTIONAL AMINES. The obvious nature of the assumed reactions makes the general explanations of the processes involved in bonding by bifunctional monomers rather attractive, too. The bonding of wood by ethylenediamine and 1,6-hexamethylenediamine can be explained by the known condensation reactions between amines and lignin (166). In experiments of diamine bonding of wood and fiber preoxidized with the HNO_3 or nitric oxides, the disruption of the cellular components at the wood surface was assumed (83), followed by consolidation and condensation with diamines in the press under formation of ester and amide linkages. This mechanism was substantiated by SEM photographs of the oxidized wood surface as well as by quantitatively determining the retention of the diamines by the carboxylic groups on the surface (see Figure 4).

BIFUNCTIONAL ACIDS. Bonding of wood by bifunctional acids and anhydrides has been explained by ester formation during pressing and heating (85). Extraction of the boards with acetone indicated that between 97.0 and 99.2% of maleic anhydride participated in bond formation. Removal of extractives improved internal bond. Cellulose (bleached pulp) gave drastically lower IB values, particularly at higher maleic anhydride percentages. These results could indicate the preferential participation of lignin in bond formation.

BIFUNCTIONAL ISOCYANATES. The bridging reaction between isocyanates and wood involves the formation of urethanes with hydroxyls of wood. Presence of water exerts a negative influence by formation of polyureas, carbon dioxide, and amines. Although some controversy still exists (87), the indirect evidence for covalent bond formation is strong, as cellulose reacts readily with isocyanates (162–64); however, because only a few reactive groups are available, lignin reacts poorly (165). In isocyanate bonding the reaction involves chiefly the cellulosic part of wood (see Figure 4).

POLYMERS. The general involvement of surface, activated or

Figure 10. Scanning electron micrograph of a composite of cellulose powder–lignin powder mixture. The big, fibrous cellulose particle (right) appears to be bonded to the big amorphous lignin particle (left). A split in the cellulose particle suggests that bonding between lignin and cellulose particles was stronger than the tensile strength of cellulose perpendicular to the fiber axis (124).

nonactivated, in direct bonding or in bonding involving intermediacy of bifunctional monomers is often relatively easy to understand from the chemical point of view, but the situation is more complicated in case of bonding by polymers allegedly forming covalent bonds with the surface. If the activator is introduced with the polymer (mixed or separately applied to the surface) several alternatives present themselves.

(1) Role of activator is restricted mainly or only to polymer cross-linking. In this case the action of the adhesive becomes equivalent to that of phenol–formaldehyde or urea–formaldehyde.

(2) Activator is changing the polymer, enabling it to react with wood surface. In this case the surface participates in covalent bonding, but it is not activated.

(3) Activator is reacting with wood surface, enabling it to form covalent bonds with the polymer.

(4) Combination of all three or of any two of the above.

The situation is simpler with wood preactivated in a separate step (83, 85), although in this case a question might be raised whether changes on the wood surface (e.g., formation of carboxyls) merely catalyze cross-linking of the polymer or whether covalent bonds form involving surface. Not much experimentation has been done so far aimed at elucidation of the mechanisms of nonconventional bonding methods involving intermediacy of a polymer. Most of the available evidence is restricted to the Philippou process.

In the Philippou process, bonding is achieved by pressing wood containing an oxidant (e.g., H_2O_2) on the surface and treated with a mixture of furfuryl alcohol, maleic acid, and ammonium lignosulfonate (or closely related replacements). Acid polymerization of furfuryl alcohol has been known for some time and is known to involve intermolecular dehydration of alcoholic groups under formation of a polymer composed of furan nuclei linked by methylene bridges. Some of the furan nuclei open up during the reaction to form 1,4-diketo units (Scheme 2). In addition formation of furfuryl–furfuryl ether groups, furfuryl–furyl ether groups, splitting of methylol groups as formaldehyde as well as some less well understood reactions leading to cross-linking and darkening of the polymeric product take place. Although oxidation leads to a variety of products, free radical chain polymerization apparently does not occur (170–76).

In addition to the obvious ester formation with hydroxyl groups, maleic anhydride is known to react with furan compounds via Diels–Alder reaction. The double bond of maleic acid is susceptible to free radical attack and is oxidizable, with oxidation by H_2O_2 leading to tartaric acid (129). The chemistry of maleic anhydride has been reviewed thoroughly (183). The reactions of ammonium lignosulfonate with acid or oxidizing agents are complicated and have been mentioned previously (122, 177). Because of these alternatives the transformations forming the chemical basis of the Philippou process are difficult to negotiate.

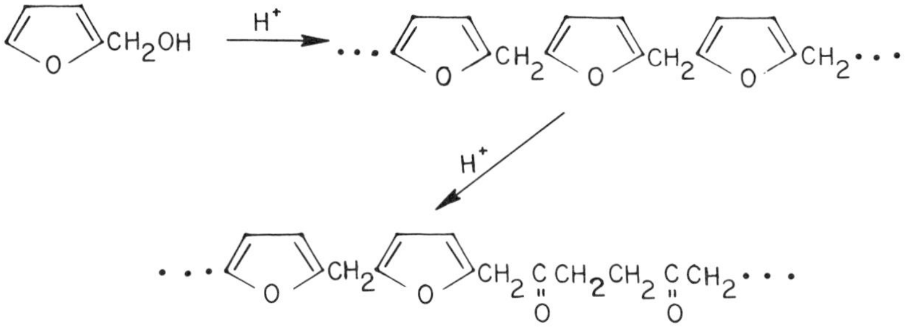

Scheme 2.

Philippou reported (86, 96) that water absorption and thickness swelling of his particle boards strongly decreased with increased density of the product in contrast to the conventionally bonded products (Table V). As is known from polymer science, the swelling of a three-dimensional polymer in solvents is reciprocally related to the cross-linking density of the polymeric network. The behavior of the nonconventionally bonded particle boards mentioned before, thus, suggested that wood was bonded by a continuous network of covalent bonds extending to and including the surface. To demonstrate the covalent bonding between wood and the polymers additionally, Philippou investigated the grafting of his polymeric mixture to 60–80 mesh wood sawdust. Although some grafting took place in the absence of H_2O_2, addition of this chemical strongly increased the percent of graft. Strongest grafting was exhibited by ammonium lignosulfonate–furfuryl alcohol mixtures, followed by furfuryl alcohol alone, while ammonium lignosulfonate alone grafted poorly. The results were in agreement with his particle board experiments, in which mixtures of furfuryl alcohol with ammonium lignosulfonate performed best.

Additional experiments based on DSC, IR, and UV spectroscopy were made. Furfuryl alcohol by itself or in the presence of milled wood lignin or ammonium lignosulfonate and with maleic acid as catalyst produced an exotherm between 135 and 155 °C in DSC. In the presence of H_2O_2, or in the presence of wood or lignin prereacted with H_2O_2, the exotherm shifted to lower temperatures. Prereaction of cellulose had no major effect on the position of the exotherm peaks, however. The IR and UV spectra of the cross-linking polymers prepared in the presence of H_2O_2 differed from those prepared only by acid polymerization. The results were interpreted as indicating that grafting of the polymeric mixture to lignocellulosic surfaces takes place during particle board pressing with the lignin portion of wood playing the dominant role in formation of covalent bonds to the wood surface (86).

Grafting experiments give considerable insight into the nature of covalent bonding of nonconventional adhesives to the lignocellulosic surfaces by allowing for the assessment of whether and to what extent such bonds do form. The experiments can be run in a solvent medium (e.g., water) or in the condensed phase; the latter corresponds better to what takes place in the heated press. Reliable distinction between grafted polymer and homopolymer (generally performed by dissolving the homopolymer in an appropriate solvent) is very important and often subject to considerable difficulties because of insolubilization of the homopolymer by cross-linking and other reactions.

Table V. Relationship Between Board Density, Thickness Swelling, and Water Absorption for the Particle Board Produced by the Philippou Process

Density (g/cm³)	IB (psi)	2-h Boil Test		24-h Water Soak Test	
		Thickness Swelling (%)	Water Absorption (%)	Thickness Swelling (%)	Water Absorption (%)
0.58	73	84.3	116.1	33.1	92.1
0.64	102	46.7	86.6	17.8	59.4
0.68	121	30.2	69.2	11.5	45.7
0.74	130	14.6	44.7	5.8	28.2
0.80	134	5.8	14.6	3.4	18.4
R^a	0.936	−0.959	−0.999	−0.944	−0.974

Conditions: Activator; 3% H_2O_2 (dry wood basis); adhesive, 7% (dry wood basis) with ammonium lignosulfonate–furfuryl alcohol in 7:3 ratio; catalyst, 2% (adhesive basis) maleic acid.
NOTE: For phenol–formaldehyde board see Ref. 184.
[a] Correlation coefficient with density

The results of the grafting experiments are commonly reported as degree of grafting (%) or percent grafting = $(A - B)100/B$, and as grafting efficiency, (%) = $(A - B)100/(A - B) + C$, where A is the weight of the grafted sample after extraction, B is the weight of the original, nongrafted sample, and C is the weight of the homopolymer (178).

The grafting experiments of Philippou were continued by C. Nguyen and Zavarin (178). In the presence of H_2O_2/Fe^{2+} furfuryl alcohol grafted to cellulose at a pH of 2.0 and 90 °C in aqueous medium to 14%; acid prepolymerized furfuryl alcohol grafted to 91% under the same conditions. No grafting was observed in the absence of H_2O_2 in spite of extensive polymerization of 2-furfuryl alcohol. Use of monomers other than 2-furfuryl alcohol indicated that monomers able to acid polymerize, such as 3-furfuryl alcohol and furfural, grafted in aqueous medium to cellulose in presence of H_2O_2 and Fe^{2+}. Monomers unable to acid polymerize, such as 5-methyl-2-furfuryl alcohol, acetonylacetone, acetylacetone, and tetrahydrofurfuryl alcohol, did not graft. The results suggested that acid polymerization of furfuryl alcohol is an indispensable step in oxidative grafting to cellulose (Figure 11). Oxidation of cellulose with H_2O_2/Fe^{2+} until disappearance of the oxidant ($TiCl_4$ test), followed by addition of furfuryl alcohol acid polymer resulted in no grafting; similar preoxidation of the acid polymer followed by addition of cellulose resulted in 72% graft. An oxidative change in furfuryl alcohol acid polymer apparently is necessary for formation of covalent bonds to cellulose; activation of cellulosic surface probably plays a negligible role.

The chemistry of nonconventional bonding involving a cross-linking mixture of tannin, furfuryl alcohol, and maleic acid in conjunction with surface activation by HNO_3 were studied by DSC, and IR, UV, and NMR spectroscopies (179). The UV and NMR experiments gave results of limited importance. In DSC furfuryl alcohol gave an exotherm between 50 and 100 °C in the presence of maleic acid and HNO_3, which tended to shift to lower temperatures with an increase in the acidity of the catalysts. A sharp exotherm was exhibited at about 180 °C with mixtures containing HNO_3. On the basis of DSC results and IR spectra it was concluded that HNO_3 promoted condensation reactions of the cross-linking mixture and that furfuryl alcohol became nitrated with the nitro groups degrading to give the 180 °C exotherm.

Participation of the wood surface has not been stressed much in nonconventional bonding based on cross-linking of carbohydrates in the presence of weak acids (110), cross-linking of lignosulfonates in the presence of strong acids (111–17), and based on cross-linking of lignosulfonates in the presence of oxidants (106–9). At the same time

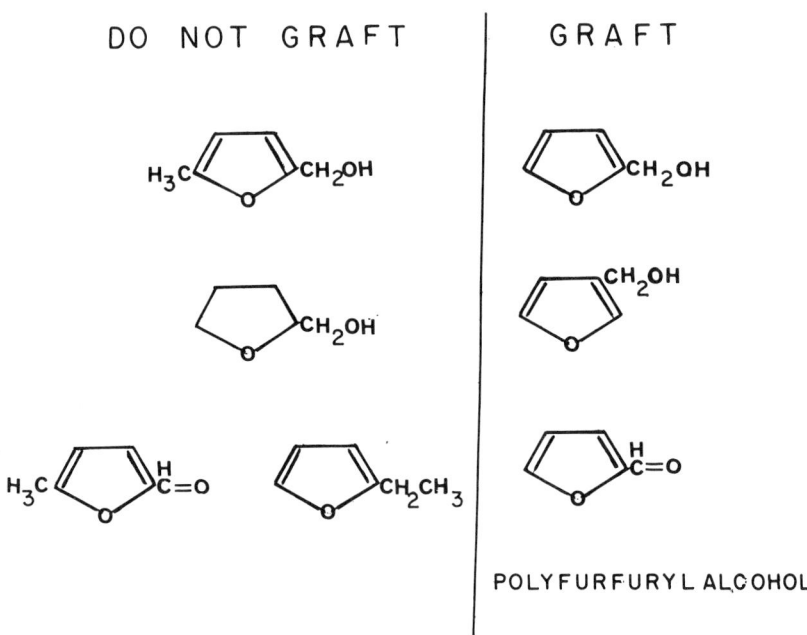

Figure 11. Cellulose graftable and cellulose nongraftable derivatives of furan CH_2O_2-ferrous ion as activator (178).

it would be surprising if various acid dehydration–condensation cross-linking reactions taking place with carbohydrates and lignosulfonates or by phenolic oxidative coupling of lignosulfonates would not extend to the related constituents—cellulose, hemicelluloses, and lignin—on the wood surface. The question of the surface participation in these processes awaits additional experimentation.

The mechanism of nonconventional bonding based on mixtures of diamines with PVC was explained (82) on the basis of the reported ability of the two materials to partially cross-link (180) (Scheme 3). Epichlorohydrin, used occasionally as an additive, polymerizes with polyfunctional amines (181) (Scheme 4) and reacts with lignin in the presence of amines or carboxylic acids (182). Even simpler to visualize is the cross-linking of the mixture of maleic anhydride and PVA by formation of ester linkages (118) that attach to the hydroxyls of the wood surface.

Conclusions

The research in the area of nonconventional wood bonding has not reached its zenith as yet and much applied and fundamental work remains to be done. So far, in spite of an appreciable number of nonconventional bonding systems in existence, only a few are com-

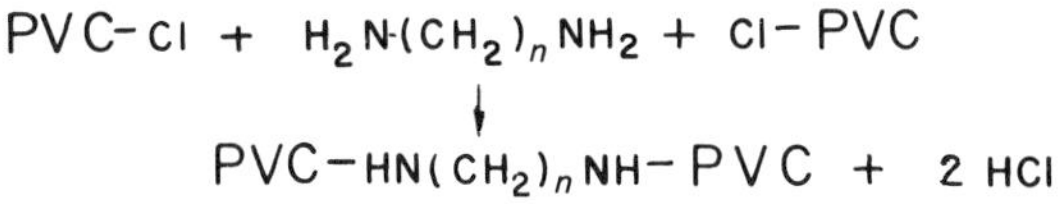

$$PVC-Cl + H_2N(CH_2)_nNH_2 + Cl-PVC$$
$$\downarrow$$
$$PVC-HN(CH_2)_nNH-PVC + 2 HCl$$

Scheme 3.

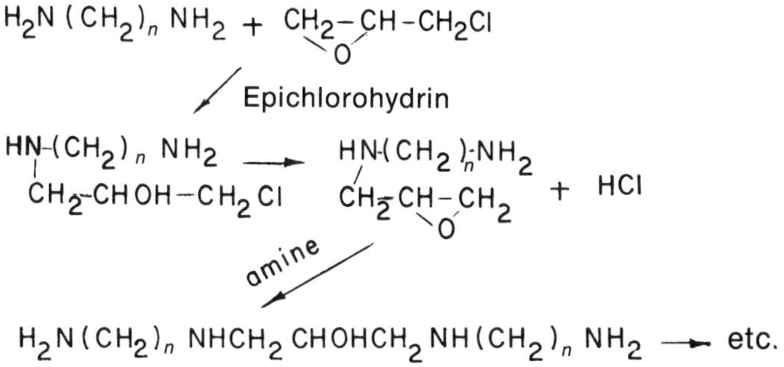

$$H_2N(CH_2)_nNH_2 + CH_2-CH-CH_2Cl \atop \diagdown O \diagup$$

Epichlorohydrin

$$\underset{CH_2-CHOH-CH_2Cl}{HN-(CH_2)_n NH_2} \longrightarrow \underset{CH_2-CH-CH_2 \atop \diagdown O \diagup}{HN-(CH_2)_n NH_2} + HCl$$

amine

$$H_2N(CH_2)_n NHCH_2 CHOHCH_2 NH(CH_2)_n NH_2 \longrightarrow \text{etc.}$$

Scheme 4.

petitive with the conventional ones in terms of economics and properties of the products, e.g., bonding by spent sulfite liquor at low pH (Shen); bonding by a mixture of spent sulfite liquor, furfuryl alcohol, and maleic anhydride with an oxidizing activator (Philippou); and bonding by water-soluble carbohydrates with a catalyst (Stofko). Apparently, only one (isocyanates) is in industrial use. The methods of application of activators, bifunctional monomers, or polymers, and of pressing the treated particles at elevated temperatures correspond to those used with the conventional systems and generally do not present difficulties. Exceptions are the methods requiring pretreatment of the particles with gases such as nitrogen oxides as halogens that require special treatment chambers. If strong acids or oxidants are employed, corrosion of the equipment must be considered and corrosion-resistant materials used. Conversely, use of iron equipment (press platens, spray bottles) could lead to catalytical decomposition of certain chemicals, such as H_2O_2, if these are used as activators. Storage of some strong oxidants such as chlorates could involve safety hazards. The methods of testing the produced particle board correspond to those conventionally used.

The nonconventionally produced particle board can offer several advantages. Some systems offer better water absorption and thickness swelling characteristics and/or low formaldehyde emissions. Reliance on agricultural by-products and independence from the international oil market represent additional attractions. The problems yet to be solved are mainly connected with abnormally high variability in the mechanical properties and generally inferior performance in the low density area.

The chemistry of the bonding processes must be understood better for further progress to be made. Although there is no lack of hypotheses explaining the chemical transformations taking place, only a limited amount of work has been done to allow an insight into what actually happens. In many cases even such items as the chemical nature of surface activation and existence of covalent bonds to wood surface are not understood.

Addenda

Wood Surface Studies. Attenuated total reflectance (ATR) was used to study the wood surface by IR spectroscopy. Aromatic lignin bands were less intensive in ATR spectra than in transmittance spectra. This was explained by the preferential exposure of the S_2 layer during cutting (185).

It was proposed on the basis of the results of ESR and iodometric peroxide determinations that photodegradative modification of the wood surface starts with formation of free radicals that interact with oxygen to form hydroperoxide groups. The latter decompose rapidly to form carbonyl, carboxyl, and similar groups (186).

Nonconventional Bonding with Acid Activation. Concentrated or spray-dried, spent sulfite liquor, with or without previous ultra-filtration, was used as a binder for waferboards. The press temperatures used were between 210 and 230 °C and the press times between 5 and 10 min, with the amount of resin between 4 and 5%. The boards produced were better or comparable to the boards made using phenol–formaldehyde resin according to IB, MOE, MOR, hardness, and linear expansion tests; the cost of the binder was twofold or threefold less (187).

The torsion shear test was used to evaluate the role of carbohydrates in the polymerization of spent sulfite liquor. It was demonstrated that the lower molecular weight fraction of spent sulfite liquor, which contained large amounts of reducing sugars, was reactive and had good binding properties. The higher molecular weight fraction, containing no carbohydrates, would not thermoset. Addition of glucose to the latter fraction resulted in good bonding properties, however. At a temperature of 210 °C best results were obtained with a mixture of crude spent sulfite liquor containing 40–45% glucose. At 230 °C the activity of spent sulfite liquor increased and even pure glucose, with ammonium sulfate as the acidic component, gave excellent results. Other sugars behaved similarly to glucose. Furaldehyde derivatives were far less reactive than glucose; glucitol and gluconic acid gave no bonding. Positive results were obtained with unsaturated aldehydes such as acrolein and cinnamaldehyde. All these results suggested that formation of an unsaturated aldehyde function plays an important role in the polymerization of sugars. Study of the various polymerization catalysts indicated a clear correlation between acid strength and catalytic activity (188).

Nonconventional Bonding with Oxidant Activation. Pretreatment of wood flakes with HNO_3, followed by application of a mixture of furfuryl alcohol, ammonium lignosulfonate, and maleic anhydride (weight ratio 4.2:1.8:1.0, respectively, 7% total of oven-dried wood), gave boards with a thickness swelling of 14–16% over a range of 0.45–0.75 g/cm^3 densities. With H_2O_2 instead of HNO_3 the boards failed in the 2-h boiling test at 0.65 g/cm^3 density and suffered severe thickness swelling at 0.75 g/cm^3 density. The optimum board-making parameters were established as 180 °C, 7-min press time, and 9.5% mat moisture content (189).

Improvement in the particle board properties made by the process based on HNO_3 surface activation and ammonium lignosulfonate–furfuryl alcohol–maleic anhydride binder was reported to result from drying of the HNO_3 treated chips (190).

Direct wood-to-wood bonding was investigated using wood panels of *Acer saccharum* Marsh., *Betula alleghaniensis* Britton, *Quercus rubra* L., *Pseudotsuga menziesii* (Mirb.) Franco, and *Pinus palustris* Mill in combination with activating agents including HNO_3 (40%), H_2SO_4 (30%), H_2O_2 (30%), potassium persulfate (0.3 M), and potassium periodate (0.3 M). In terms of shear strength of the resulting HNO_3-bonded products, *Acer saccharum* performed best and *Pinus palustris* performed worst (8 g of reagents per 12.7 × 17.8-cm panels, 100 °C, 290 psi, 1 h). With *Acer saccharum*, HNO_3 activation performed best (shear of above 2031 psi). None of the products were water-resistant, however.

Urea, hexamethylenediamine, maleic and phthalic anhydrides, vanillin, benzoic acid, phenol, resorcinol, and tannin (from *Acacia mearnsii* De Wild) were used as bridging agents in combination with activation by HNO_3 (170 °C, 290 psi, 30 min, and 6 g of bridging material per panel). In nearly all cases the shear values improved upon acid treatment and were highest with maleic anhydride, benzoic acid, resorcinol, and tannin (2705, 2379, 2583, and 2768 psi, respectively); the same bridging agents had wet shears of 708, 743, 726, and 1334 psi, respectively, and the other agents in most cases did not give any water resistance to wood panels. The combination of tannin with HNO_3 pretreatment produced panels comparable to those bonded with phenolic adhesives (191).

Extractive-free wood flour of *Acer saccharum* Marsh. (sugar maple), α-cellulose, birch acetyl-4-*O*-methylglucuronoxylan, and spruce-milled wood lignin were treated with 40% HNO_3 and the resulting products were investigated by IR and UV spectroscopy in combination with wet chemical analytical methods. Cellulose was oxidized at 100 °C to form carboxyls; evidence of cellulose nitration and of the presence of HNO_3 absorbed by cellulose at ambient temperature was obtained. Xylan was strongly changed by hydrolysis of acetyl groups at the ambient temperature; at 100 °C hydrolysis of the chains, recondensation reactions, and formation of uronic and aldonic acids, followed by their lactonization, took place. The lignin portion of wood was extensively modified at the ambient temperature with the reactions including oxidation, nitration, degradation, and condensation. The major degradation product isolated from HNO_3 treatment of wood was 2,4-dinitroguaiacol (*192*).

Isocyanates. The influence of various manufacturing parameters on the properties (IB, MOE, MOR, boiled MOR) of waferboard bonded by polymeric isocyanates was examined. The amount of adhesive used was 1.5–2.25%, the press time was 1–3 min, the temperature was 177–204 °C, and the density of the boards was 0.679 g/cm^3. The experiments indicated that under the conditions used the National Bureau of Standards (NBS) 2-B-2 standards can be achieved (*193*).

IR spectroscopic evidence for covalent urethane bond formation in the reaction between isocyanates and wood has been obtained. Isolation of holocellulose by the sodium chlorite method, isolation of lignin by the H_2SO_4 procedure, and subjecting both to IR spectroscopy indicated that isocyanates reacted with both cellulose and lignin (*194*).

In three-layer flakeboards based on five softwood and hardwood species of 4, 10, and 18% moisture content, with press temperature of 177 °C, 6-min press time, and 3% isocyanate binder, the results showed that moisture content of wood was the most important variable; at 18% moisture level, IB and bonding properties were lowest. Species of wood influenced strongly the bonding efficiency. In almost all cases the bending properties were the key characteristic of the panel performance. Southern pine produced the boards with lowest IB (81–116 psi), and red oak gave the highest IB values (98–213 psi) (*195*).

The use of mixtures of diphenylmethane diisocyanate and various polyols as particle board binders was recommended, and IB values up to 157 psi at 1.5% binder content (0.80–0.86-g/cm^3 density) were reported (*196*).

Several reviews and discussions of bonding by isocyanates are available (*197–201*).

Nonpolar Nonconventional Binders. Wood cross sections and veneer specimens of *Betula maximowicziana* Regel were methacrylated, acrylated, propionated, and isobutyrated and glued with styrene monomer−benzoyl peroxide. The specimens that were esterified with acids containing an active double bond bonded properly, and all specimens failed in which one or both partners were esterified with saturated acids. Upon alkaline saponification of the bonded wood the copolymers of styrene with methacrylic acid were isolated. Bonding was explained by bridging through graft copolymerization. Shear strength of the three-ply plywood prepared by the above methods was small, however, and the values strongly scattered; this observation was explained by the roughness of the wood surface and by incomplete copolymerization, inhibited by lignin and residual extractives (202−5).

Experimentation with polypropylene and modified polypropylene as hot-melt adhesives gave plywood samples that satisfied specifications of the Japan Agricultural Standards. According to SEM data, molten polypropylene made good contact with the veneer surface and penetrated into the lumena of the cells and other spaces. The glue-joint strengths of acetylated and silylated wood glued with polypropylene were nearly independent from the degree of acetylation or silylation. With modified polypropylene the strength decreased slightly with increased acetylation. Deuterium exchange indicated that accessibility of hydroxyls was nearly the same before and after gluing in the case of polypropylene (53% vs. 51%, respectively). In modified polypropylene the accessibility decreased to 45% after gluing. It was concluded that mechanical adhesion dominated with nonpolar polypropylene, and with modified polypropylene some adhesion due to primary and secondary valence forces was contributing to the bonding (206, 207).

Dry rubber or latex was used as a binder for particle board. Particle board met specifications and was superior to the boards prepared using thermosetting binders in the steam and hot water resistance as well as in price (208).

Other Methods. *Pentacme contorta* Merr. and Rolfe (white lauan) and *Swietenia macrophylla* King (large-leaf mahogany) wood formed end-grain joints when pressed at elevated temperatures. Increase in temperature from 100 to 156 °C increased bond strength. Pressure had some effect. Bond strength obtained amounted to about 52% of that obtained with urea−formaldehyde adhesives. Bonding was explained by softening of lignin at the temperatures used (209).

Aqueous hexamethylenetetramine solution adjusted to pH 2.0 by addition of H_2SO_4 was used as a binder for particle board. The amount of hexamethylenetetramine solids was 10% of wood weight. Heating was by press platens and by high frequency; temperatures ranged from 180 to 220 °C with 5 min of press time. The boards pressed at 220 °C had a density of 0.75 g/cm^3 and exhibited an IB of

45 psi, thickness swelling of 5.7% (2 h), and water absorption of 15.9%. The amount of formaldehyde given off was higher than with the phenol–formaldehyde boards, but lower than with urea–formaldehyde boards (210).

Patents. A number of patents on nonconventional binders appeared, including aqueous dispersion of epoxy resin (211); a mixture of diisocyanates or polyisocyanates, alkylene oxides or halogenated alkylene oxides, and lignin (212); alkali-treated chlorolignin, optionally including phenolic or amino resins (213); mixtures of polyisocyanates with lignin (lignosulfonates or Kraft lignin) (214); and ammonium lignosulfonate (215).

Acknowledgments

We are indebted to D. G. Arganbright and G. Grozdits for reviewing the manuscript. Most of the work done at the University of California Forest Products Laboratory was supported by grants-in-aid from Weyerhaeuser Company, Boise Cascade Corporation, Potlatch Corporation, Tablopan de Venezuela, Crown Zellerbach Corporation, U.S. Plywood Corporation, Masonite Corporation, Quaker Oats Company, Simpson Timber Company, and Blandin Wood Products, for which appreciation is extended.

Literature Cited

1. Schmidt, A. X.; Marlies, C. A. "Principles of High Polymer Theory and Practice"; McGraw-Hill Book Co.: New York, 1948; p. 655.
2. Allan, G. G.; Neogi, A. N. *J. Adhes.* **1971,** 3, 13.
3. Ramiah, M. V.; Troughton, G. E. *Wood Sci.* **1970,** 3, 120.
4. "Graft Copolymerization of Lignocellulosic Fibers," Hon, D. N.-S., Ed.; ACS SYMPOSIUM SERIES No. 187, ACS: Washington, D.C., 1982.
5. Black, J. M.; Mraz, E. A. *U.S., For. Serv., Res. Pap. FPL* 232, 1974.
6. Black, J. M. *U.S., For. Serv., Res. Note FPL* 134, 1969.
7. Black, J. M.; Mraz, E. A. *U.S., For. Serv., Res. Pap. FPL* 271, 1976.
8. Feist, W. C.; Mraz, E. A.; Black, J. M. *For. Prod. J.* **1977,** 27(1) 13.
9. Lagergren, S.; Rydholm, S.; Stockman, L. *Sven. Papperstidn.* **1957,** 60, 632.
10. Garland, H. *Ann. Mo. Bot. Gard.* **1939,** 26, 1.
11. Wardrop, A. B. *Aust. J. Sci.* **1951,** B-4, 4, 391.
12. Wardrop, A. B.; Addo-Ashong, F. W. in Proc. Melbourne Univ. Engin. Dept.; Symp. on Fracture, 1963.
13. Atack, D.; May, W. D.; Morris, E. L.; Sproule, R. N. *Tappi* **1961,** 44, 555.
14. Koran, Z. *Tappi* **1967,** 50, 61.
15. Koran, Z. *Sven. Papperstidn.* **1968,** 71, 567.
16. Fergus, B. J.; Goring, D. A. I. *Holzforschung* **1970,** 24, 118.
17. Fergus, B. J.; Procter, A. R.; Scott, J. A. N.; Goring, D. A. I. *Wood Sci. Technol.* **1969,** 3, 117.
18. Janes, R. L. in "Pulp and Paper Manufacture. The Pulping of Wood"; MacDonald, R. G.; Franklin, J. N., Eds.; McGraw-Hill: New York, 1969; Vol. I, Chap. 2, pp. 33.
19. Nguyen, T. *J. Adhes.* **1982,** 14, 283.
20. Wu, K.-T. "Investigation of the Reactions between Lignocellulosic Mate-

rials and Nitrates by Differential Scanning Calorimetry," Ph.D. thesis, Berkeley, CA, 1981.
21. Gancet, C.; Heitner, C.; Beatson, R. P.; Gray, D. G. *Tappi* **1980**, *63*, 139.
22. Dorris, G. M.; Gray, D. G. *Cellul. Chem. Technol.* **1978**, *12*, 9.
23. Dorris, G. M.; Gray, D. G. *Cellul. Chem. Technol.* **1978**, *12*, 721.
24. Gray, D. G. *Cellul. Chem. Technol.* **1978**, *12*, 735.
25. Mjoeberg, P. J. *Cellul. Chem. Technol.* **1981**, *15*, 481.
26. Young, R. A.; Rammon, R. M.; Kelley, S. S.; Gillespie, R. H. *Wood Sci.* **1982**, *14*, 111.
27. Takeyama, S.; Gray, D. G. Int. Pap. Phys. Conf., Harrison Hot Springs, B. C., Canada, Preprint Book (CPPA Montreal), 1979, 179.
28. Takeyama, S.; Gray, D. G. *Cellul. Chem. Technol.* **1982**, *16*, 133.
29. Anderson, A. B.; Ellwood, E. L.; Zavarin, E.; Erickson, R. W. *For. Prod. J.* **1960**, *10*, 212.
30. Ellwood, E. L.; Anderson, A. B.; Zavarin, E.; Erickson, R. W. *For. Sci.* **1960**, *6*, 315.
31. Evans, R. S.; Halvorson, H. N. *For. Prod. J.* **1962**, *12*, 367.
32. Zavarin, E.; Anderson, A. B.; Berolzheimer, C. P. *For. Prod. J.* **1965**, *15*, 73.
33. Swanson, J. W.; Cordingly, Sh. *Tappi* **1959**, *42*, 812.
34. Huffman, J. B. *For. Prod. J.* **1955**, *5*, 135.
35. Hemingway, R. W. *Tappi* **1969**, *52*, 2149.
36. Suchsland, O.; Stevens, R. R. *For. Prod. J.* **1968**, *18*, 38.
37. Hancock, W. V. *For. Prod. J.* **1963**, *13*, 81.
38. Chen, C. M. *For. Prod. J.* **1970**, *20*, 36.
39. Chow, S.-Z. *Wood Sci. Technol.* **1971**, *5*, 27.
40. Troughton, G. E.; Chow, S.-Z. *Wood Sci.* **1971**, *3*, 129.
41. Mote, C. D.; Szymani, R. *Holz Roh-Werkst.* **1977**, *35*, 189.
42. Zaitsev, N. A. *Derevoobrab. Promst.* **1968**, *17*(4), 15.
43. Allan, G. G.; Mattila, T. in "Lignins"; Sarkanen, K. V.; Ludwig, C. H., Eds.; Wiley-Interscience: New York, 1971; p. 575.
44. Nguyen, T.; Zavarin, E.; Barrall, E. M., II *J. Macromol. Sci., Rev. Macromol. Chem.* **1981**, *C20* (1), 1.
45. Tillman, D. A.; Rossi, A. J.; Kitti, W. D. "Wood Combustion"; Academic: New York, 1981, p. 74.
46. Shafizadeh, F.; DeGroot, W. F. in "Thermal Uses and Properties of Carbohydrates and Lignins"; Shafizadeh, F.; Sarkanen, K. V.; Tillman, D. A., Eds.; Academic: New York, 1976; p. 1.
47. Kilzer, F. J. in "Cellulose and Cellulose Derivatives"; Bikales, N. M.; Segal, L., Eds.; Wiley-Interscience: New York, 1971; p. 1015, Pt. V.
48. Shafizadeh, F.; Chin, P. P. S. in "Wood Technology: Chemical Aspects," Goldstein, I. S., Ed.; ACS SYMPOSIUM SERIES No. 43; ACS: Washington, D.C., 1977; p. 57.
49. Domburg, G. E.; Sharapova, T. E. *Khim. Drev.* **1978**, (3) *39*, 31, 46.
50. Domburg, G. E.; Skripchenko, T. N. *Khim. Drev.* **1982**, (5), 81.
51. McBurney, L. F. in "Cellulose and Cellulose Derivatives"; Ott, E.; Spurlin, H. M.; Grafflin, M. W., Eds.; Interscience: New York, 1954; p. 168, Pt. I.
52. Tryon, M.; Wall, L. A. in "Autoxidation and Autoxidants"; Lundberg, W. O., Ed.; Interscience: New York, 1966; p. 963, Vol. II.
53. El-Rafie, M. H.; Khalil, E. M.; Abdel-Hafiz, S. A.; Hebeish, A. *J. Appl. Polym. Sci.* **1983**, *28*, 211.
54. Nguyen, T. Report, Univ. Calif. For. Prod. Lab., 1978.
55. Stamm, A. J.; Hansen, L. A. *Ind. Eng. Chem.* **1937**, *29*, 831.
56. Salehuddin, A. B. M. Ph.D. thesis, North Carolina State Univ., 1970.
57. Seborg, A. M.; Tarkow, H.; Stamm, A. J. *For. Prod. Res. Soc. J.* **1953**, *3*(3), 59.
58. Mitchell, R. L.; Seborg, R. M.; Millett, M. A. *For. Prod. Res. Soc. J.* **1953**, *3*(4), 38.

59. Chow, S.-Z.; Mukai, H. N. *Wood Sci.* **1972**, *4*, 202.
60. Kringstad, K. P. *Papier (Darmstadt)* **1973**, *27*, 462.
61. Feist, W. C.; Rowell, R. M. in "Graft Copolymerization of Lignocellulosic Fibers," Hon, D. N.-S., Ed.; ACS SYMPOSIUM SERIES No. 187; ACS: Washington, D.C., 1982; p. 349.
62. Hon, D. N.-S. in "Development in Polymer Degradation: 3"; Grassie, N., Ed.; Appl. Sci. Publ.: Essex, England, 1981; p. 229.
63. Baugh, P. J.; Phillips, G. O. in "Cellulose and Cellulose Derivatives"; Bikales, N. M.; Segal, L., Eds.; Wiley-Interscience: New York, 1971; p. 1047, Pt. V.
64. Hon, D. N.-S.; Chan, H.-Ch. in "Graft Copolymerization of Lignocellulosic Fibers," Hon, D. N.-S., Ed.; ACS SYMPOSIUM SERIES No. 187; ACS: Washington, D.C., 1982; p. 101.
65. Minemure, N.; Umehara, K. *Rep. Hokkaido For. Prod. Res. Inst.* **1979**, *68*, 92.
66. Linzell, H. K. U.S. Patent 2 388 487, 1945.
67. Schur, M. O.; Levy, R. M. *Pap. Trade J. (Tappi Sect.)* **1947**, *124* (20) 221 (43).
68. Glab, W. T. U.S. Patent 3 033 695.
69. Olson, E. T.; Plow, R. H. U.S. Patent 2 156 160.
70. Philips, M.; Goss, M. J. U.S. Patent 2 190 909.
71. Ehrlich, J. U.S. Patent 2 430 922.
72. Wilson, W. E. U.S. Patent 2 639 994.
73. Stofko, J.; Zavarin, E. Patent Disclosures: "A new bonding system for particleboard," 1970; and "New bonding system of lignocellulosic materials," 1971.
74. Stofko, J; Zavarin, E.; Schniewind, A. Abstract, Amer. Chem. Soc., Div. of Cellulose, Wood, and Fiber Chemistry, 167th Nat. Meeting, Los Angeles, CA, 1974, No. 14.
75. Stofko, J.; Zavarin, E.; Schniewind, A.; Dickinson, F. E. Univ. of Calif., For. Prod. Lab., Techn. Rep. 35.01.98, 1972–73.
76. Stofko, J.; Zavarin, E. U.S. Patent 4 007 312, 1977.
77. Pohlman, A.; Philippou, J.; Wu, J.; Nguyen, T.; You, Y.-S.; Johns, W. Univ. of Calif., For. Prod. Lab., Techn. Rep. 35.01.134, 1974–75.
78. Johns, W. E.; Nguyen, T. *For. Prod. J.* **1977**, *27*, 17.
79. Jenkin, D.; Zavarin, E. Univ. of Calif., For. Prod. Lab. Techn. Rep. 35.01.113, 1973.
80. Goheen, D. W.; Barton, J. S. U.S. Patent 4 022 965, 1977.
81. Brink, D. L. U.S. Patent 3 900 334, 1975.
82. Schorning, P.; Roffael, E.; Stegmann, G. *Holz Roh-Werkst.* **1972**, *30*, 253.
83. Collett, B. M. Ph.D. thesis, Univ. of Calif., Berkeley, 1973.
84. Brink, D. L.; Collett, B. M.; Pohlman, A. A.; Wong, A. F.; Philippou, J. in "Wood Technology: Chemical Aspects", Goldstein, I. S., Ed.; ACS SYMPOSIUM SERIES No. 43; ACS: Washington, D.C., 1977; p. 169.
85. Pohlman, A. A. M. S. dissertation, Berkeley, Calif., 1974 (*see also* Ref. 81).
86. Philippou, J. L. Ph.D. dissertation, Univ. of Calif., Berkeley, 1977.
87. Zicherman, J. B. *For. Prod. J.* **1975**, *25* (6), 21.
88. Deppe, H. J. "Proc. 11th Wash. State Univ. Symp. on Particleboard"; Maloney, Th.M., Ed.; Wash. State Univ.: Pullman, WA, 1977; p. 13.
89. Deppe, H. J.; Ernst, K. *Holz Roh- Werkst.* **1971**, *29*, 45.
90. Roffael, E.; Rauch, W. *Holz Roh- Werkst.* **1973**, *31*, 402.
91. Loew, G.; Sachs, H. I. "Proc. 11th Wash. State Univ. Symp. on Particleboard"; Maloney, Th. M., Ed.; Wash. State Univ.: Pullman, WA, 1977; p. 473.
92. Deppe, H. J. *For. Prod. J.* **1969**, *19* (7), 27.
93. Johns, W. E. "Proc. 14th Wash. State Univ. Symp. on Particleboard"; Maloney, Th. M., Ed.; Wash. State Univ.: Pullman, WA, 1980; p. 177.

94. Johns, W. E.; Maloney, Th. M.; Huffaker, E. M.; Saunders, J. B.; Lentz, M. T. "Proc. 15th Wash. State Univ. Symp. on Particleboard"; Maloney, Th. M., Ed.; Wash. State Univ.: Pullman, WA, 1981; p. 213.
95. Johns, W. E. *For. Prod. J.* **1982**, *32*(11), 47.
96. Philippou, J. L. *J. Wood Chem. Technol.* **1981**, *1*, 199.
97. Philippou, J. L.; Johns, W. E.; Nguyen, T. *Holzforschung* **1982**, *36*, 37.
98. Philippou, J. L.; Johns, W. E.; Zavarin, E.; Nguyen, T. *For. Prod. J.* **1982**, *32*(3), 27.
99. Philippou, J. L.; Zavarin, E.; Johns, W. E.; Nguyen, T. *For. Prod. J.* **1982**, *32* (5), 55.
100. Johns, W. E.; Layton, H. D.; Nguyen, T.; Woo, J. K. *Holzforschung* **1978**, *32*, 162.
101. Brink, D. L.; Johns, W. E.; Zavarin, E.; Kuo, M. L.; Nguyen, T.; Layton, D.; Wong, A.; Birnbach, M.; Merriman, M. M.; Breiner, T.; Grozdits, G.; Wu, K. T. Univ. of Calif., For. Prod. Lab. Techn. Rep. 35.01.193, 1977–80.
102. Bibal, J. N.; Grozdits, G. A.; Dao, L. T.; Tee, L. B.; Nguyen, Ch.; Chen, H.; Zavarin, E.; Wu, K. T.; Saunders, R. S. Univ. of Calif., For. Prod. Lab. Techn. Rep. 35.01.272, 1981–82.
103. Grozdits, G. private communication.
104. Emerson, R. W. U.S. Patent 2 764 569, 1953.
105. Emerson, R. W. U.S. Patent 3 097 177, 1963.
106. Nimz, H. H.; Mogharab, I.; Gurang, I. *Appl. Polym. Symp.* **1976**, *28*, 1225.
107. Nimz, H.; Razvi, A.; Mogharab, I.; Clad, W. *Offenlegungsschrift (Fed. Repub. Ger.)* **1974**, *2*, 211, 353.
108. Nimz, H. H.; Hitze, G. *Cellul. Chem. Technol.* **1980**, *14*, 371.
109. Nimz, H.; Gurang, I.; Mogharab, I. *Liebigs Ann. Chem.* **1976**, 1421.
110. Stofko, J. U.S. Patent 4 183 977, 1980.
111. Shen, K. C. *For. Prod. J.* **1974**, *24*(2), 38.
112. Shen, K. C. "Proc. of 8th Wash. State Univ. Symp. on Particleboard"; Maloney, Th. M., Ed.; Wash. State Univ.: Pullman, WA, 1974; p. 231.
113. Shen, K. C.; Fung, D. P. C. *For. Prod. J.* **1979**, *29* (3), 34.
114. Shen, K. C. *Adhes. Age* **1978**, *21* (3), 31.
115. Shen, K. C.; Calvé, L. *Adhes. Age* **1980**, *23*(8), 25.
116. Shen, K. C. U.S. Patent 4 193 814, 1981.
117. Shen, K. C.; Calvé, L.; Lau, P. "Proc. 13th Wash. State Univ. Symp. on Particleboard"; Maloney, Th. M., Ed.; Wash. State Univ.: Pullman, WA, 1979; p. 369.
118. Schorning, P.; Stegmann, G. *Holz Roh- Werkst.* **1972**, *30*, 329.
119. Nicholas, D.D.; Dziubak, M. Annu. Rep. Res. Div., Mich. Tech. Univ., Houghton, Mich., 1978, p. 151.
120. Kim, C. Y.; Goring, D. A. I. *Pulp Pap. Mag. Can.* **1976**, *72* (11), T363.
121. Wertheimer, M. R.; Suranyi, G.; Goring, D. A. I. *Tappi* **1972**, *55*, 1707.
122. Goring, D. A. I. in "The Fundamental Properties of Paper Related to Its Uses"; Bolam, F., Ed.; Techn.Div., The British Paper and Board Industry Federation: London, 1976; Vol. I, p. 172.
123. Wallis, A. F. A. in "Lignins"; Sarkanen, K. V.; Ludwig, C. H., Eds.; Wiley-Interscience: New York, 1971; p. 345.
124. Stofko J. Ph.D. dissertation, Univ. of Calif., Berkeley, 1974.
125. Košik, M.; Lužakova, V.; Reiser, V. *Cellul. Chem. Technol.* **1972**, *6*, 589.
126. Parks, E. J. *Tappi* **1971**, *54*, 537.
127. Nguyen, T.; Zavarin, E.; Barrall, E. M., II *J. Macromol. Sci., Rev. Macromol. Chem.* **1981**, *C21*(1), 1.
128. Schumb, W. C.; Satterfield, Ch. N.; Wentworth, R. L. "Hydrogen Peroxide"; ACS MONOGRAPH SERIES, Reinhold Publ. Corp.: New York, 1955.
129. Wallace, J. G. "Hydrogen Peroxide in Organic Chemistry"; DuPont, 1962.

130. "Organic Peroxides"; Swern, D., Ed.; Wiley-Interscience: New York, 1971; Vol. II.
131. Moody, G. J. *Adv. Carbohydr. Chem.* **1964**, *19*, 149.
132. Ivanov, V. I.; Kaverzneva, E. D.; Kuznecova, Z. I. *Dokl. Akad. Nauk. SSSR* **1952**, *86*, 301.
133. Latosh, M. V.; Alekseev, A. D.; Reznikov, V. M. *Khim. Drev.* **1980** (2), 43.
134. Walling, Ch.; Goosen, A. *J. Am. Chem. Soc.* **1973**, *95*, 2987.
135. Ingles, D. L. *Aust. J. Chem.* **1972**, *25*, 87.
136. Ingles, D. L. *Aust. J. Chem.* **1972**, *25*, 97.
137. Ingles, D. L. *Aust. J. Chem.* **1972**, *25*, 105.
138. Jenkin, D. J. *Appl. Polym. Symp.* **1976**, *28*, 1309.
139. Nguyen, T.; Zavarin, E.; Barrall, E. M., II *Thermochim. Acta* **1980**, *41*, 107.
140. Nguyen, T.; Zavarin, E.; Barrall, E. M., II *Thermochim. Acta* **1980**, *41*, 269.
141. Nguyen, T.; Zavarin, E.; Barrall, E. M., II *J. Appl. Polym. Sci.* **1982**, *27*, 1019.
142. Nguyen, T. *J. Adhes.* **1982**, *14*, 283.
143. Nguyen, T.; Zavarin, E.; Barrall, E. M., II *J. Appl. Polym. Sci.* **1983**, *28*, 647.
144. Nguyen, T. Ph.D. dissertation, Univ. of California, Berkeley, 1979.
145. Wu, K.-T. Ph.D. dissertation, Univ. of Calif., Berkeley, 1981.
146. Dence, C. W. in "Lignins"; Sarkanen, K. V.; Ludwig, C. H., Eds.; Wiley-Interscience: New York, 1971; p. 373.
147. Barsha, J. in "Cellulose and Cellulose Derivatives"; Ott, E.; Spurlin, H. M.; Grafflin, M. W., Eds.; Interscience: New York, 1954; p. 713, Pt. II.
148. Sergeeva, L. L.; Shorygina, N. N. *Izv. Akad. Nauk SSSR, Ser. Khim.* **1972**, (4) 924, ABIPC *44*, 8220, 1974.
149. Grinshpan, O. D.; Kaputskii, F. N.; Ermolenko, I. N. *Dokl. Akad. Nauk BSSR* **1973**, *17* (12), 117; ABIPC **1975**, *45*, 7842.
150. Laisha, G. M.; Sharkov, V. I. *Vysokomol. Soedin.* **1974**, *16A* (8), 1703; ABIPC **1975**, *45*, 11464.
151. Poller, S.; Patscheke, G. *Holztechnologie* **1979**, *20* (4), 216.
152. Rashin, M. N.; Sokolov, O. M. *Khim. Drev.* **1979**, (4), 50.
153. Kaputskii, F. N.; Bobrovskii, A. P.; Gert, E. V.; Bashmakov, I. A. *Zh. Prikl. Khim. (Leningrad)* **1979**, *52* (4), 900.
154. Lužakova, V.; Macincinova, T.; Blažej, A. *Vysk. Pr. Odboru Pap. Celul. 25/Pap. Celul.* 35 (11) V73, V79, 1980. ABIPC 52, 1196, 1981.
155. Lužakova, V.; Macincinova, T.; Blažej, A. *Vysk. Pr. Odboru Pap. Celul. 25/Pap. Celul.* 35 (11) V86, V95, 1980. ABIPC 52 1197, 1981.
156. Wu, K.-T.; Zavarin, E. *Proc. Natl. Sci. Counc., Repub. China Part A: Applied Sciences* **1982**, *6* (2), 102.
157. Subramanian, R. V.; Balaba, W. M.; Somesekharan, K. N. *J. Adhes.* **1982**, *14*, 295.
158. Simionescu, Cr. I.; Macoveanu, M. M.; Olaru, N. *Cellul. Chem. Technol.* **1976**, *10*, 197.
159. Simionescu, Cr. I.; Macoveanu, M. M. *Cellul. Chem. Technol.* **1977**, *11*, 87.
160. Suranyi, G.; Gray, D. G.; Goring, D. A. I. *Tappi* **1980**, *63*(4), 153.
161. "Modified Cellulosics"; Rowell, R. M.; Young, R. A., Eds.; Academic; New York, 1978; p. 171.
162. Hebeish, A.; Guthrie, J. T. "The Chemistry and Technology of Cellulosic Copolymers"; Springer: New York, 1981.
163. Arthur, J. C., Jr. "Proceedings of the Symposium on Graft Polymerization onto Cellulose"; Interscience: New York, 1972.
164. Zavarin, E.; Nguyen, C.; Worster, J. R.; Romero, E. *J. Wood Chem. Technol.* **1982**, *2*, 343.

165. Pichler, R. M.; Hruschka, A.; Prey, V. *Holzforschung* **1969**, *23*, 192.
166. Simionescu, Cr.; Butnaru, R.; Rozmarin, Gh. *Cellul. Chem. Technol.* **1973**, *7*, 153.
167. Mack, C. H.; Hobart, S. R. *J. Appl. Polym. Sci.* **1966**, *2*, 133.
168. Ellzey, S. E., Jr.; Mack, C. H. *Text. Res. J.* **1962**, *32*, 1023.
169. Kratzl, K.; Buchtela, K.; Gratzl, J.; Zauner, J.; Ettingshausen, O. *Tappi* **1962**, *45*, 113.
170. Dunlop, A. P.; Peters, F. N. "The Furans"; American Chemical Society, Reinhold: New York, 1953; p. 791.
171. Barr, J. B.; Wallon, S. B. *J. Appl. Polym. Sci.* **1971**, *15*, 1079.
172. Wewerka, E. M.; Loughran, E. D.; Walters, K. L. *J. Appl. Polym. Sci.* **1971**, *15*, 1137.
173. Milković, J.; Myers, G. E.; Young, R. A. *Cellul. Chem. Technol.* **1979**, *13*, 651.
174. Schmitt, C. R. *Polym.–Plast. Technol. Eng.* **1974**, *3*(2), 121.
175. Gandini, A. *Adv. Polym. Sci.* **1977**, *25*, 47.
176. Gandini, A.; Rieumont, J. *Tetrahedron Lett.* **1976**, *25*, 2101.
177. Chang, H.-M.; Allan, G. G. in "Lignins"; Sarkanen, K. V.; Ludwig, C. H., Eds.; Wiley-Interscience: New York, 1971; p. 471.
178. Nguyen, Ch.; Zavarin, E. to be published.
179. Kelley, S. S.; Young, R. A.; Rammon, R. M.; Gillespie, R. H. *J. Wood Chem. Technol.* **1982**, *2*, 317.
180. Kainer, H. "Polyvinylchlorid und Polyvinylchlorid-Mischpolymerisate"; Springer: Berlin, 1965; pp. 121, 202.
181. Lee, H.; Neville, K. in "Encyclopedia of Polymer Science and Technology"; Mark, H. E.; Gaylord, N. G.; Bikales, N. M., Eds.; Interscience: New York, 1967; p. 209.
182. Ball, F. J.; Dougherty, W. K.; Moorer, H. H. U.S. Patent 3 149 085, 1964.
183. Trivedi, B. C.; Culbertson, D. M. "Maleic Anhydride"; Plenum Press: New York, 1982.
184. Moslemi, A. A. "Particleboard"; So. Illinois University Press: London, 1974; p. 141.
185. Hse, Ch.-Y.; Bryant, B. S. *Mokuzai Gakkaishi* **1966**, *12*, 187.
186. Hon, D. N.-S., Chang, Sh.-T.; Feist, W. C. *Wood Sci. Technol.* **1982**, *16*, 193.
187. Go, T. A. in "Proc. Canadian Waferboard Symposium"; Szabo, T.; Gribble, H. W., Eds.; Forintek Canada Corp., 1980, Spec. Publ. SP 505E, p. 259.
188. Calvé, L.; Fréchet, J. M. J. *J. Appl. Polym. Sci.* **1983**, *28*, 1969.
189. Brink, D. L.; Kuo, M. L.; Johns, W. E.; Birnbach, M. J.; Layton, H. D.; Nguyen, T.; Breiner, T. *Holzforschung* **1983**, *37*, 69.
190. Johns, W. E.; Jahan-Latibari, A. *J. Adhes.* **1983**, *15*, 105.
191. Kelley, S. S.; Young, R. A.; Rammon, R. M.; Gillespie, R. H. *For. Prod. J.* **1983**, *33* (2), 21.
192. Rammon, R. M.; Kelley, S. S.; Young, R. A.; Gillespie, R. H. *J. Adhes.* **1982**, *14*, 257.
193. McLaughlin, A.; Farrissey, W. J.; Albertino, L. M.; Waszeciak, D. P. in "Proc. Canadian Waferboard Symposium"; Szabo, T.; Gribble, H. W., Eds.; Forintek Canada Corp., 1980, Spec. Publ. SP 505E, p. 271.
194. Rowell, R. M.; Ellis, W. D. in "Urethane Chemistry and Applications"; Edwards, K. N., Ed.; ACS SYMPOSIUM SERIES No. 172; ACS: Washington, D.C., 1981; p. 263.
195. Johns, W. E.; Maloney, Th. M.; Saunders, J. B.; Huffaker, E. M.; Lentz, M. T. "Proc. 16th Wash. State Univ. Intern. Symp. on Particleboard"; Maloney, Th. M., Ed.; Wash. State Univ.: Pullman, WA, 1982; p. 71.
196. Gallagher, J. A. *For. Prod. J.* **1982**, *32* (4), 26.
197. Ball, G. W.; Redman, R. P. *FESYP Internat. Particleboard Symp. (Hamburg)* **1979**, 121.
198. Frink, J. W.; Sachs, H. I. in "Urethane Chemistry and Applications"; Ed-

wards, K. N., Ed.; ACS SYMPOSIUM SERIES No. 172; ACS: Washington, D.C., 1981; p. 285.
199. Adams, A. D. "Proc. 14th Wash. State Univ. Symp. on Particleboard"; Maloney, Th. M., Ed.; Wash. State Univ.: Pullman, WA, 1980; p. 195.
200. Johns, W. E. "Proc. 14th Wash. State Univ. Symp. on Particleboard"; Maloney, Th. M., Ed.; Wash. State Univ.: Pullman, WA, 1980; p. 177.
201. Johns, W. E. *J. Adhes.* **1983**, *15*, 59.
202. Nakagami, T.; Yokota, T. *Bull. Kyoto Univ. Forests* **1979**, *51*, 274.
203. Nakagami, T.; Yokota, T. *Mokuzai Gakkaishi* **1981**, *27*, 32.
204. Nakagami, T.; Yokota, T. *Mokuzai Gakkaishi* **1981**, *27*, 87.
205. Nakagami, T.; Oishi, T.; Yokota, T. *Mokuzai Gakkaishi* **1983**, *29*, 248.
206. Goto, T.; Saiki, H.; Onishi, H. *Wood Sci. Technol.* **1982**, *16*, 293.
207. Onishi, H.; Goto, T.; Saiki, H. *Holzforschung* **1983**, *37*, 29.
208. Zamani Bin Abdul Wahid, A. Proc. Natural Rubber Technol. Sem. (Kuala Lumpur), 1979, p. 289. Rubber Res. Inst. of Malaysia. ABIPC 52(11) (1982) p. 1288, No. 12213.
209. Cassilla, R. C. *Philippine Lumberman* **1977**, *23* (11), 16. *For Prod. Abstr.* 3 (11) (1980) No. 3256.
210. Roffael, E.; Parameswaran, N. *Adhaesion* **1981**, *25*, 286.
211. Wistuba, E.; Wittmann, O. German Patent 3 025 522, 1982.
212. Gaul, J. M.; Nguyen, T. U.S. Patent 4 361 662, 1981.
213. Eka, A. B. Jap. Patent Kokai 76 056 82, 1982.
214. Lambuth, A. L. U.S. Patent 4 279 788, 1981.
215. Shen, K. C.; Fung, D. P. C.; Calvé, L. Can. Patent 1 101 625, 1981.

RECEIVED for review May 16, 1983. ACCEPTED August 2, 1983.

11
Chemistry of Weathering and Protection

WILLIAM C. FEIST
U.S. Department of Agriculture, Forest Service, Forest Products Laboratory, Madison, WI 53705

DAVID N.-S. HON
Department of Forest Products, Virginia Polytechnic Institute and State University, Blacksburg, VA 24061

> *Wood exposed to the outdoors undergoes photodegradation and photooxidative degradation in the natural weathering process. UV light interacts with lignin to initiate discoloration and deterioration. Deterioration of wood in the natural weathering process involves a very complex, free radical reaction sequence. Light does not penetrate wood past 200 µm; therefore, degradation reactions are a surface phenomenon. The free radicals generated in wood by light rapidly interact with oxygen to produce hydroperoxides which in turn are easily decomposed to produce chromophoric groups. In this chapter the influence of outdoor weathering on the performance of wood and wood-based materials is discussed in detail. Macroscopic, microscopic, chemical, and physical changes are described. The mechanisms of weathering and methods of protection of exposed wood surfaces are summarized.*

WOOD IS A NATURALLY DURABLE MATERIAL that has been recognized for centuries throughout the world for its versatile and attractive engineering and structural properties. However, like other biological materials, wood is susceptible to environmental degradation. When wood is exposed to the outdoors above ground, a complex combination of chemical, mechanical, and light energy factors contribute to what is described as *weathering* (1). Weathering is not to be confused with decay, which results from decay organisms (fungi) acting in the presence of excess moisture and air for an extended period of time (2). Under conditions suitable for the development of decay,

This chapter not subject to U.S. copyright.
Published 1984, American Chemical Society

wood can deteriorate rapidly and the result is far different than that observed for natural outdoor weathering.

The degradation of wood by any biological or physical agent modifies some of its organic components. The organic components in wood are primarily polysaccharides and polyphenolics: cellulose, hemicelluloses, and lignin. Extractives are also present in relatively small quantities and their concentration determines color, odor, and other nonmechanical properties of a wood species. A change in these components may be caused by an enzyme, a chemical, or electromagnetic radiation, but invariably, the net result is a change in molecular structure through some chemical reaction. Stalker (3) conveniently divided the environmental agencies that bring about wood degradation into categories. *Physical* forms of energy were used to describe all factors other than fungi, insects, or animals. The importance of the various physical destructive agents on wood can be considered by comparing two situations, inside and outside the wood structures (Table I). The most serious risk to wood indoors comes from fire. Outdoors, the most important factor is weathering.

This chapter updates and consolidates past literature on the weathering and protection of wood, and emphasizes recent and new research in this area.

Table I. Relative Effect of Various Energy Forms on Wood

Energy Form	Indoor		Outdoor	
	Result	Degree of Effect	Result	Degree of Effect
Thermal				
Intense	fire	severe	fire	severe
Slight	darkening of color	slight	darkening of color	slight
Light				
Visible and UV	color change	slight	large color changes	severe
			chemical degradation (especially lignin)	severe
Mechanical	wear and tear	slight	wear and tear	slight
			wind erosion	slight
			surface roughening	severe
			defiberization	severe
Chemical	staining	slight	surface roughening	severe
	discoloration	slight	defiberization	severe
	color changes	slight	selective leaching	severe
			color changes	severe
			strength loss	severe

Background

Perhaps the earliest record of the sun's effect on wood materials can be found in Exodus 15:23 when Moses led the Israelites into the wilderness of Shur:

> And when they came to Marah, they could not drink the waters of Marah, for they were bitter; therefore it was named Marah. So the people grumbled at Moses, saying "What shall we drink?" Then he cried out to the Lord, and the Lord showed him a *tree;* and he threw *it* into the waters, and the waters became sweet.

Marah is an area of desert located near the Red Sea on the Sinai Peninsula. The desert water is bitter due to high alkalinity. Weathering of wood on the desert by the sun causes the alcohol groups of cellulose and hemicellulose to be oxidized to carboxyl groups. By throwing a piece of weathered wood into the alkaline water, an acid–base reaction takes place in which the alkalinity of the water is reduced. Thereby, the water becomes *sweet*.

In addition to this incident, man no doubt was aware of the environment's degradative effect on wood since he first began using such materials. However, it was not until 1827 that the chemical phenomenon of wood weathering was reported by Berzelius (4), followed by Wiesner (5) in 1846, and Schramm (6) in 1906. However, systematic studies on weathering reactions in wood did not begin until the 1950s (1).

General Aspects of Wood Weathering

In outdoor weathering of smooth wood, original surfaces become rough as the grain raises, the wood checks, and the checks grow into large cracks; grain may loosen, and boards cup and warp and pull away from fasteners. The roughened surface changes color, gathers dirt and mildew, and may become unsightly; the wood loses its surface coherence and becomes friable—splinters and fragments can come off. All these effects, brought about by a combination of light, water, and heat, are comprehended in one word: *weathering*.

The deleterious effect of wood weathering has been ascribed to a complex set of reactions induced by a number of factors. The weathering factors responsible for changes in wood surfaces are solar radiation (UV, visible, and IR light), moisture (dew, rain, snow, and humidity), temperature, and oxygen. Of these factors, the photon energy in solar radiation is the most damaging component of the outdoor environment and initiates a wide variety of chemical changes at wood surfaces. Moreover, an additional weathering factor has arisen with the presence of atmospheric pollutants such as sulfur dioxide, nitrogen dioxide, and ozone in the presence or absence of UV light.

Anatomic Structure of Wood and Its Weatherability. The cell walls of wood are multilayered. They consist of the middle lamella, primary wall (P), and layers of the outer (S_1), middle (S_2), and inner (S_3) secondary walls. These layers differ from one another with respect to their structures, orientations and number of fibrils or fibers, as well as their chemical composition. The distribution of chemical constituents in the cell walls at the surfaces has a great influence on the weathering stability of wood. The chemical components across the cell wall are depicted in Figure 1. Cellulose, a linear, highly crystalline polymer of (1,4)-β-D-glucopyranose, is the major component of the cell wall (~45% of total dry weight), and is located mostly in the secondary wall. Hemicellulose (~20%) is an amorphous, polymeric carbohydrate having a slightly branched structure. Lignin, a three-dimensional network of polyphenols (~20–30%), is distributed throughout the cell wall but is highly deposited in the middle lamella region. These polymeric materials vary widely in their vulnerability to weathering. The variations in stability are caused primarily by differences in chemical structures, particularly in chromophoric functional groups. Metallic ions and other impurities may also promote deterioration by light (1, 7, 8).

Weathering Factors. MOISTURE. One of the principal causes of weathering is frequent exposure of the wood surface to rapid changes in moisture content (1). Rain or dew falling upon unprotected wood is quickly absorbed by capillary action on the surface layer of

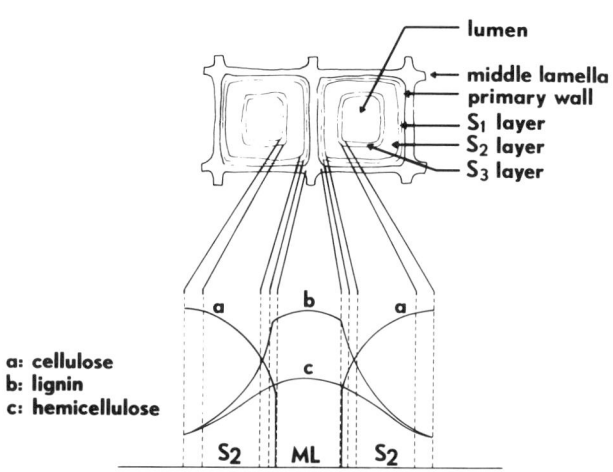

Figure 1. Chemical components across a wood cell wall.

the wood, followed by adsorption within wood cell walls. Water vapor is taken up directly by adsorption under increased relative humidities; consequently, the wood swells. Stresses are set up in the wood as it swells and shrinks due to moisture gradients between the surface and the interior. These induced stresses are greater the steeper the moisture gradient and are usually largest near the surface of the wood. Unbalanced stresses may result in warping and face checking (9–16).

LIGHT. The photochemical degradation of wood due to sunlight occurs fairly rapidly on the exposed wood surface (1, 8, 17). The initial color change of wood exposed to sunlight is a yellowing or browning that proceeds to an eventual graying. These color changes can be related to the decomposition of lignin in the surface wood cells and are strictly a surface phenomenon (17–20). These changes occur only to a depth of 0.05–2.5 mm (see section entitled "Penetration of Light and Wood Surface Deterioration") and are a result of sunlight, particularly UV light, which initiates photodegradation. Photodegradation by UV light induces changes in chemical composition, particularly in the lignin, and subsequent color changes (7, 8, 21–26).

The two most important elements of weathering—sunlight and water—tend to operate at different times. Exposed wood can be irradiated after having been wet by rain or when surface moisture content is high from overnight high humidity or dew. Time of wetness, therefore, is important in relating climatic conditions to exterior degradation. The action of the combined elements can follow different degradation paths, with irradiation accelerating the effect of water or the converse.

OTHER FACTORS. Heat may not be as critical a factor as UV light or water, but as the temperature increases, the rate of photochemical and oxidative reactions increases (1). Visible light may also contribute to the breakdown of wood during weathering (27, 28). A loss in strength was associated with light-induced depolymerization of lignin and cell wall constituents and to the subsequent breakdown of wood microstructure. The decisive factor in wood weathering in the summer is the intensity of solar radiation, and in the winter the increased amount of SO_2 in the surrounding air is the main weathering factor (central Europe exposure) (29).

Freezing and thawing of absorbed water can also contribute to wood checking. Abrasion or mechanical action, such as wind, sand, and dirt, can be an important factor in the rate of surface degradation and removal of wood. Small particles such as sand can become lodged in surface checks and, through swelling and shrinking, weaken fibers in contact with the particles. Solid particles in combination with wind can have a sandblasting effect (1, 8).

Penetration of Light and Wood Surface Deterioration.
Although the weathering of wood materials depends on many environmental factors, there is mounting evidence that only a relatively narrow band of the electromagnetic spectrum, i.e., the UV-light portion of sunlight, is responsible for the primary photooxidative degradation of wood.

The first law of photochemistry [the Grotthus–Drapper principle (30)] states that for a photochemical reaction to occur, some component of the system must first absorb light. The second law of photochemistry [the Stark–Einstein principle (31)] states that a molecule can only absorb one quantum of radiation. The absorbed energy causes the dissociation of bonds in the molecules of the wood constituents. This homolytic process produces free radicals as the primary photochemical products. This event, with or without the participation of oxygen and water, can lead to depolymerization and to formation of chromophoric groups such as carbonyls, carboxyls, quinones, peroxides, hydroperoxides, and conjugated double bonds.

Because light must be absorbed before a photochemical reaction can occur, the concentration, location, and nature of chromophores are highly significant in determining the rate of photooxidation of wood. Essentially, wood is an excellent light absorber. Although cellulose is not, it does absorb light strongly below 200 nm with indications of some absorption between 200 and 300 nm, and a tail of absorption extending to 400 nm (32, 33). Because of structural similarity, the UV absorption characteristics of hemicellulose resemble those of cellulose. Lignin and polyphenols absorb light strongly below 200 nm and have a strong peak at 280 nm with absorption down through the visible region (33). Extractives usually have the ability to absorb light between 300 and 400 nm (33, 34). As a consequence, most of the components in wood are obviously capable of absorbing enough visible and UV light to undergo photochemical reactions leading ultimately to discoloration and degradation.

Because of the wide range of chromophoric groups associated with its surface components, wood cannot easily be penetrated by light. Essentially, discoloration of wood by light is a superficial surface phenomenon. The dark brown surface layer of ponderosa pine and redwood that is affected by light extends only 0.5–2.5 mm into the wood (1, 17, 35). As weathering progresses, most woods change to a grayish color, but only to a depth of about 0.10–0.25 mm. Visible (400–750 nm) light as measured spectrophotometrically can penetrate into wood as far as 2540 μm (35). The gray wood surface layer was reported to be 125 μm thick; beneath the gray layer was a brown layer from 508 to 2540 μm thick. These color changes are a result of photochemical reactions that always involve free radicals.

The use of UV light transmission techniques to measure penetration of light through radial and tangential surfaces of different woods as a function of thickness has been reported (36). Electron spin resonance (ESR) techniques were used to monitor free radicals generated underneath different layers of wood. It was found that UV light cannot penetrate deeper than 75 μm; visible light, on the other hand, penetrates up to 200 μm into wood surfaces. Visible light of 400–700 nm is insufficient to cleave chemical bonds in any of the wood constituents (36) because the energy is less than 70 kcal/mol (33, 37). The brown color formed beneath a depth of 508–2540 μm could not be caused by light, as claimed by Browne and Simonsen (35). They suggested that the aromatic moieties of wood components at wood surfaces initially absorb UV light, and that an energy transfer process from molecule to molecule dissipates the excess energy.

The energy transfer processes between electronically excited groups at the outer layer of the wood surface and another group underneath the wood surface account for the photoinduced discoloration of wood underneath the surface, which absorbs practically no UV light. Furthermore, free radicals generated by light are high in energy and tend to undergo chain reactions to stabilize parent radicals. Consequently, new free radicals formed in this way may migrate deeper into wood to cause discoloration reactions.

Property Changes During Weathering

Chemical Changes. Over a century ago, Wiesner (5) reported that the intercellular substance of wood had been lost because of weathering and concluded that the remaining gray layer consists of "cells that, leached by atmospheric precipitation, have been robbed entirely or in large part of their infiltrated products so much that the remaining membranes consist of chemically pure or nearly chemically pure cellulose." Similar observations were reported by others (6, 38, 39).

The increase in cellulose content of the weathered wood surface was shown (40) and reported (19). Analytical data on white pine wood that had been weathered outdoors for 20 years was compiled. The results showed that weathering degraded and solubilized lignin. Cellulose appeared to be affected considerably less, except for the top surface layer of the wood. Similar results were obtained with various kinds of wood exposed on a test fence for 30 years. The top gray layer consistently exhibited very low lignin content. The brown layer immediately under the outer gray layer had a lignin content varying from that normally found for fresh unexposed wood by 40–60%. The interior wood layers only a few millimeters under the outer gray surface had a wood composition similar to that of normal, unweathered wood. Analysis of wood sugars from hydrolysis of a water extract of the weathered wood showed that xylan and araban were solubilized more rapidly than was glucosan. Glucose did not predominate in the hydrolyzed water extract during analysis, although glucose units do predominate in unaltered wood polysaccharides.

The UV-degradation process is initiated by the formation of free radicals and presumably begins with oxidation of phenolic hydroxyl (7, 8, 19, 24, 41). This degradation process results in a decrease in methoxyl and lignin content and an increase in acidity and carboxyl concentration of wood substance (see also References 24 and 25). These photochemical changes are enhanced more by moisture than by heat (41). The products of decomposition of weathered wood, in addition to gases and water, are mainly organic acids, vanillin, syringaldehyde, and higher molecular weight compounds, which are all leachable (19, 24). Chemical changes following artificial light irradiation of wood have also been reported by several authors (19, 21–23, 25, 26, 42–45).

Our conclusion is that absorption of UV light by lignin on the wood surface results in preferential lignin degradation. Most of the solubilized lignin degradation products are washed out by rain. Fibers high in cellulose content and whitish to gray in color remain on the wood surface and are resistant to UV degradation.

IR studies revealed that, during UV irradiation of wood, absorption due to carbonyl groups at 1720 cm^{-1} and 1735 cm^{-1} increased, whereas the absorption for lignin at 1265 cm^{-1} and 1510 cm^{-1} gradually decreased (Figure 2). The increment of carbonyl groups was the result of oxidation of cellulose and lignin. The reduction in the amount of lignin was due to its degradation by light.

A convenient measure of the change in carbonyl groups and lignin is given by the ratio of the IR absorbance bands of carbonyl groups and lignin to the absorption band at 895 cm^{-1}—an absorption band due to hydrogen located at the C-1 position, which is normally unchanged during photoirradiation. Results are shown in Table II. The change in lignin content can also be determined based on a calibration curve of lignin vs. absorption at 1510 cm^{-1}. Results of the change in lignin content at the photoirradiated wood surface are shown in Table III. These results show that carbonyl groups are generated, whereas lignin content is reduced, at the exposed wood surface. Moreover, surface washings of the photoirradiated wood exhibited increasing concentrations of water-soluble oxidation products, which can be detected by UV spectroscopy (Figure 3).

Wood exposed to the outdoors completely lost its absorption at 1265 cm^{-1} and 1510 cm^{-1}, due to the leaching of degraded lignin,

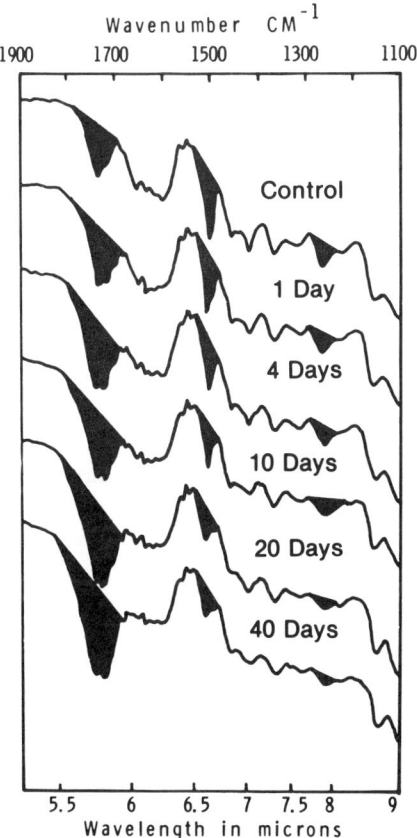

Figure 2. Change in IR spectra of UV-irradiated wood.

Table II. IR Absorbance of Wood Irradiated with UV Light

Irradiation Time (d)	Ratio of Optical Densities			
	1735:895	1720:895	1510:895	1265:895
0	1.349	1.209	1.751	0.733
1	1.701	1.649	1.636	0.636
4	1.871	1.866	1.260	0.524
10	2.164	2.730	1.206	0.448
20	2.581	2.658	1.100	0.420
40	2.954	3.031	0.969	0.373

Table III. Change of IR Absorbance and Lignin Content of Wood Irradiated with UV Light

Irradiation Time (d)	Absorbance at 1510 cm^{-1}	Lignin Content (%)
0	0.138	28.0
1	0.131	26.6
4	0.114	23.4
10	0.092	19.2
20	0.076	16.5
40	0.065	14.5

after 30 d of outdoor weathering (Figure 4). Absorption of carbonyl groups at 1720 cm^{-1} and 1735 cm^{-1} was also reduced. This observation shows that the oxidized chemical constituents at the wood surface, particularly lignin components, were moved away from the exposed surface by water. A study of an ionization difference curve of lignin in alkaline conditions revealed that water-soluble fractions of weathered wood exhibited characteristics of phenolic absorption. Electron spectroscopy for chemical analysis (ESCA) studies substantiated that oxidized surfaces have higher oxygen content than carbon

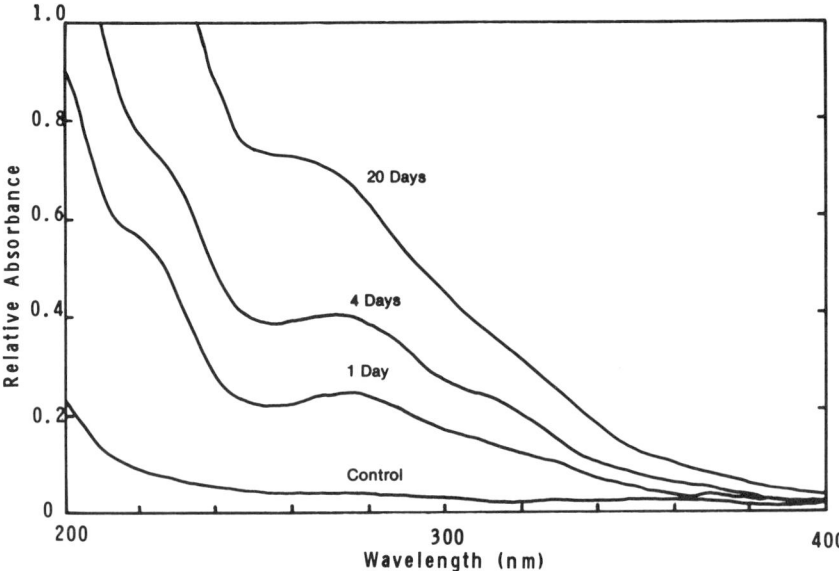

Figure 3. UV absorption spectra of water-soluble fraction of UV-irradiated wood.

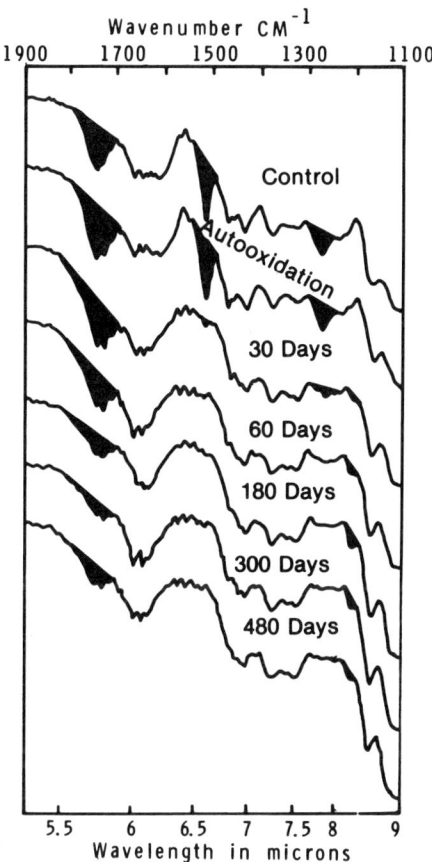

Figure 4. Change in IR spectra of outdoor exposed wood.

content, thus indicating that weathered wood surfaces are rich in cellulose with carbonyl groups, whereas lignin was degraded and leached away by water (46).

Color Changes. The color of wood exposed to the outdoors is affected very rapidly. Generally, all woods change toward a yellow to brown due to the chemical breakdown (photooxidation) of lignin and wood extractives (1, 5, 7, 17, 47a). This yellowing or browning occurs after only several months of exposure in sunny, warm climates (Figure 5). Woods rich in extractives may become bleached before the browning becomes observable.

When wood is exposed to the outdoors or in artificial UV light for a relatively short period, changes in brightness and color are readily observed. The decreases in brightness and color during 480 d of outdoor weathering are shown in Figures 6 and 7, respectively.

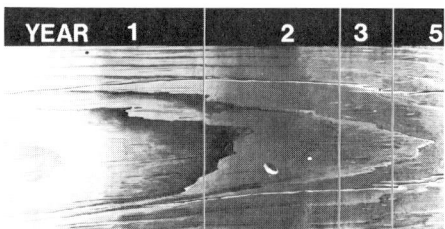

Figure 5. Artist's rendition of color changes and surface wood change during the outdoor weathering process of a typical softwood.

The change of color, ΔE, is based on CIELAB unit (47b). Some wood species, such as redwood, southern yellow pine, and Douglas-fir, lost their brightness significantly in the first month of exposure. These wood species, however, regained their brightness after 180 d of outdoor exposure. Beyond this weathering period, the brightness decreased again. Western redcedar gained in brightness for the first

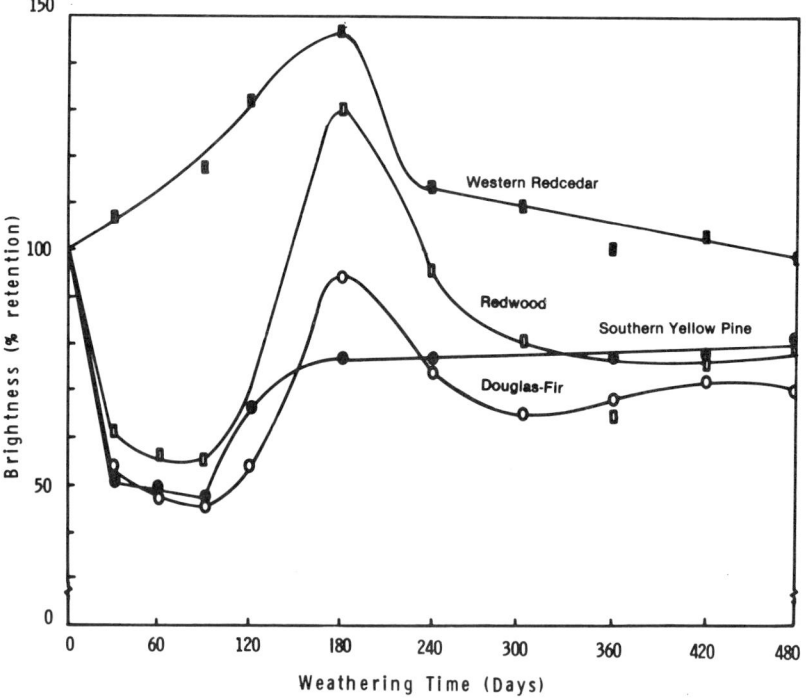

Figure 6. Decrease in brightness of outdoor weathered wood. Key: ■, western redcedar; □, redwood; ●, southern yellow pine; and ○, Douglas-fir.

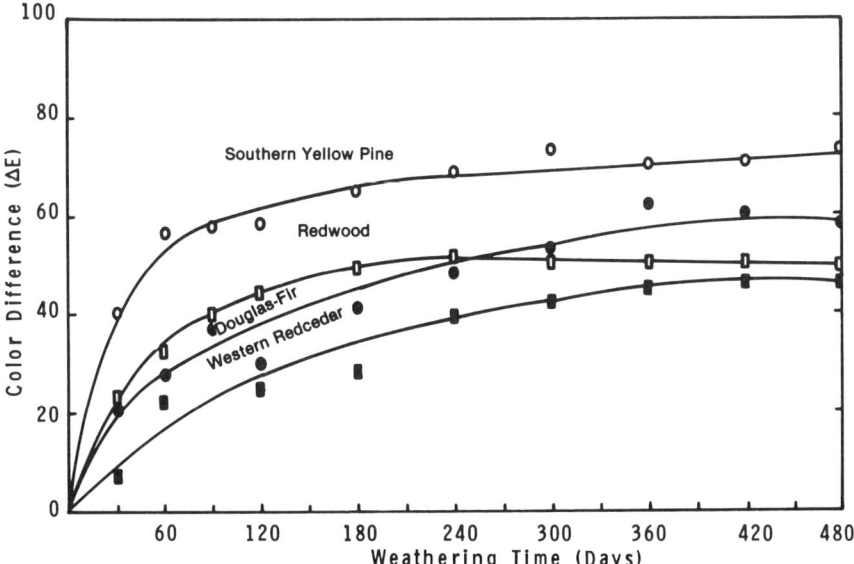

Figure 7. Change in color of outdoor weathered wood. Key: ○, *southern yellow pine;* □, *redwood;* ●, *Douglas-fir; and* ■, *western redcedar.*

180 d of outdoor exposure, followed by a decrease in brightness after 180 d of exposure.

In addition to the change in brightness, all wood species exposed to the outdoors changed in color from pale yellow to brown and to gray after 180 d of exposure. As shown in Figure 7, the significant discoloration took place between 90 and 120 d of exposure.

Changes in wood color reveal chemical changes in wood during weathering. Only those parts of the wood close to the exposed surface are affected (*see* section entitled "Penetration of Light and Wood Surface Deterioration"). As rain leaches the brown decomposition products of lignin, a silver-gray layer consisting of a disorderly arrangement of loosely matted fibers develops over the brown layer (*see* Chapter 5, Figure 18). The gray layer is composed chiefly of the more leach-resistant parts of the partially degraded wood cellulose. This surface color change to gray is observed when the wood is exposed to the sun in climates with little precipitation. However, another mechanism of surface graying of weathered wood—fungal action—usually predominates, particularly in the presence of moisture.

The discoloration (graying) of woods in the presence of moisture is practically always due to growth of fungi on the surface of the wood (*1, 41, 48–52*). The most frequently observed fungus species is *Aureobasidium pullulans* (*Pullularia pullulans*), which under favorable conditions grows not only on wood surfaces, but also on the surface of coatings and various organic and inorganic materials (53). This fungus is commonly referred to as mildew. The ecological requirements of this fungus and related fungi are modest, the most important condition for its growth being the sporadic supply of bulk water. The fungus is otherwise relatively resistant and adaptable.

A. pullulans grows on finished as well as unfinished or untreated softwood and hardwood surfaces (*11*). Discoloration of wood by mildew is more general than commonly believed. Fungal infection was the result of wetting the wood surface with water. Twenty European and non-European softwood and hardwood species of widely different density and mechanical strength properties were subjected to unprotected outdoor weathering of wood exposed in Switzerland facing south and inclined at 45° (*41*). Although behavior among the different species was at first distinctly different, this gradually changed, and photochemical and mechanical deterioration as well as intensity of attack by the blue stain fungi evened out. After only 1 year of weathering, all wood surfaces had a uniformly weathered and gray appearance.

Physical Changes. Weathering of the wood surface due to the combined action of light and water causes surface darkening and leads to formation of macroscopic to microscopic intercellular and intracellular cracks or checks. Strength of cell wall bonds is lost near the wood surface. As weathering continues, rainwater washes out degraded portions and further erosion takes place (Figure 8). Because of the different types of wood tissue on the surface, erosion and checking differ in intensity, and the wood surface becomes increasingly uneven. Hardwoods erode more slowly than do softwoods.

Browne (*54*) reports that the weathering process is so slow that "only 1/4 inch (6.4 mm) of thickness is lost in a century." However, a value of 1 mm/century has been reported for wood exposed in northern climates (*51*). An erosion value of 13 mm/century for western redcedar has been reported (*55*). This value was based on exposure data of 8 years of outdoor weathering at 90° facing south. Erosion data obtained on controlled accelerated weathering of redwood, Douglas-fir, Engelmann spruce, and ponderosa pine were used to estimate outdoor weathering. These data showed that these species would erode at a rate of approximately 6 mm/100 years (a value similar to Browne's) (*54*). Borgin (*56*) reported on erosion of wall cladding on stave churches in Norway and estimates that 10-mm-thick cladding had been reduced by 50% over a few hundred years of weathering. Jemison (*57*) found that ponderosa pine dowels of 5-mm diameter lost 7.8% of their weight after 10 years of exposure in full sunlight; dowels of 13-mm diameter lost 16.4%. Weight losses

Figure 8. Weathered surface of softwood after 15 years of exposure (in Madison, Wisconsin).

up to 10% were found (58) after heartwood samples of western redcedar, redwood, iroko, and teak had been weathered for 3 years. Surface profile was found to affect the erosion of wood only insignificantly (59).

The erosion rate of wood exposed to the outdoors has also been estimated from data obtained by controlled accelerated weathering of several woods (Table IV) (55). Specimens were exposed to a high-density xenon arc light in an accelerated weathering chamber. Exposure was cycles of 20 h of light followed by 4 h of distilled water spray. Erosion measurements were made microscopically (1, 55). The results show that the hard, dense hardwoods erode at a rate similar to that observed for the latewood of softwood species (estimated at 3 mm/century compared to 6 mm for earlywood of softwoods). Generally, the higher the density, the less the erosion rate. Lower density woods, such as basswood, erode at a higher rate than woods such as the oaks, but at a lower rate than the earlywood of softwoods.

Microscopic Changes. Microscopic changes accompany the gross physical change of wood during weathering. The first sign of deterioration in softwood surfaces is enlargement of apertures of bordered pits in radial walls of earlywood tracheids (60–62). Next, microchecks occur which enlarge principally as a result of contraction in cell walls. During weathering, the leaching and plasticizing effects of water apparently facilitate enlargement of the microchecks. Changes were more rapid for redwood than for Douglas-fir.

The scanning electron microscope was used to study the breakdown of the structure of wood due to weathering (56, 63–65). Old wood surfaces, both protected and exposed, were investigated. These studies revealed the slow deterioration and ultimate destruction of the middle lamella, the various layers of the cell wall, and the cohesive strength of wood tissue. Single individual fibers were remarkably stable and durable. The most stable part of the whole fiber seemed to be the microfibril. Various layers of the cell wall failed due to loss of cohesive structure between microfibrils and loss of adhesion between layers. All apertures or voids were enlarged, causing a weakening of the whole fiber structure. The destructive weathering process was limited to a thin surface layer of 2–3 mm. In very old, protected wood there was only a slight breakdown of certain elements at the ultrastructural level, and samples retained their normal macroscopic appearance and properties (65). As long as the main reinforcing structural elements, the microfibrils, remain intact, the major properties of wood do not undergo drastic changes.

Several publications describe the closely related observations of microscopic changes on artificial weathering (UV irradiation) of wood surfaces (45, 60, 62). Changes on the wood surface after accelerated artificial weathering were observed (9) that were very similar to those found for natural outdoor weathering. These changes include the formation of longitudinal checks between adjacent walls of neighboring elements that apparently occur in or close to the middle la-

Table IV. Erosion of Wood Surfaces After Accelerated Weathering

Species	Specific Gravity (g/cm^3)	Erosion After Exposure to Light (μm)			
		600 (h)	1200 (h)	1800 (h)	2400 (h)
Hardwoods					
White oak	0.641	65	105	135	180
Red oak	0.566	75	135	150	200
Maple					
hard	0.572	95	175	200	240
soft	0.450	85	160	195	250
Basswood	0.370	130	195	320	385
Yellow poplar	0.449	115	170	260	305
Birch, yellow	0.555	100	200	245	300
Softwoods					
Southern pine					
Sapwood	0.558				
Earlywood	0.30^1	95	190	325	410
Latewood	0.70^1	20	25	55	75
Western redcedar					
Heartwood	0.291				
Earlywood	—	145	380	515	615
Latewood	—	20	75	110	145
Sapwood	0.272				
Earlywood	—	200	395	495	655
Latewood	—	105	175	255	335
Redwood					
Heartwood	0.302				
Earlywood	—	100	225	375	510
Latewood	—	60	75	120	155
Sapwood	0.324				
Earlywood	—	160	375	520	650
Latewood	—	65	100	125	150
Douglas-fir					
Heartwood	0.437				
Earlywood	—	85	240	340	455
Latewood	—	50	100	130	155
Sapwood	0.392				
Earlywood	—	115	215	305	460
Latewood	—	65	100	105	135

NOTE: Values are for hardwoods and represent heartwood latewood erosion; earlywood erosion was only slightly greater.
[1] Estimated values

mella, longitudinal checks in element walls, and diagonal checks through pits that probably follow the fibril angle of the S_2 layer.

The pattern of breakdown of surface wood cells and cells adjacent to the surface was studied (66, 67) in radiata pine sapwood exposed outdoors to the weather for 4.5 years. The pattern of breakdown was characterized by a progressive deterioration of cells toward the surface. Evidence of deterioration was found 10–12 cells from the surface. The nature of the deterioration was twofold; initial loss of histochemical staining properties of lignin followed by progressive thinning of the cell walls. The thinning of the tracheid walls occurred centrifugally, the inner secondary wall appearing to be lost first.

Deterioration of wood surfaces after exposure to artificial UV light was observed after wood was exposed for only 500 h (68). Photodegradative effects on transverse, radial, and tangential surfaces of a typical southern yellow pine specimen are described in the following sections.

TRANSVERSE SECTION. The transverse section of southern yellow pine is normally quite simple and homogeneous. Its axial system is essentially composed of wood tracheids with only a relatively small number of parenchyma cells. An SEM micrograph of a transverse southern pine surface before exposure is shown in Figure 9.

A microtomed transverse wood face was exposed to UV light for 500 h. Surface deterioration of the exposed wood surface was observed readily from the SEM micrograph (Figure 10). The cell walls were separated at the middle lamella zone. In the extreme case, the secondary wall almost collapsed. Roughening of the surfaces could

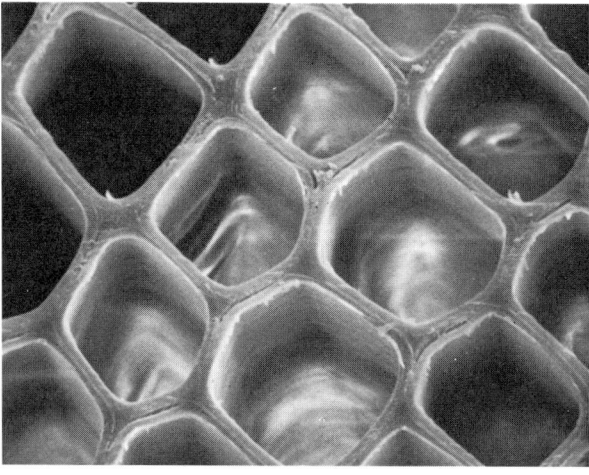

Figure 9. Cross section of southern yellow pine (700 ×).

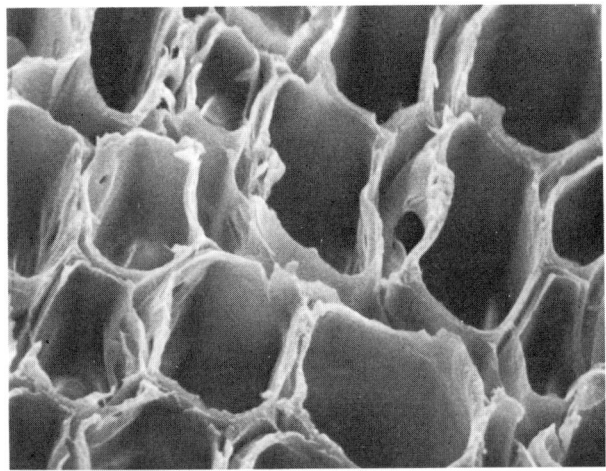

Figure 10. Cross section of southern yellow pine exposed to UV light for 500 h (700 ×).

be observed visually. Surface deterioration further developed when specimens were exposed for a total of 1000 h (Figure 11). Bordered pits located at the tracheid walls were totally destroyed. The color of the exposed wood changed from pale yellow to light brown and then dark brown after 500 and 1000 h of UV light exposure, respectively.

RADIAL SECTION. Bordered pits in southern yellow pine could

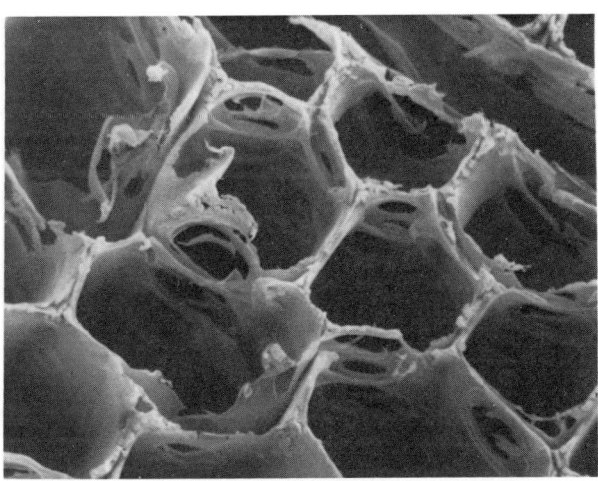

Figure 11. Cross section of southern yellow pine exposed to UV light for 1000 h (700 ×).

be observed at radial walls in both earlywood and latewood. Generally, bordered pits located in the earlywood were larger and more numerous than those in the latewood. Typical SEM micrographs for half-bordered pits and bordered pits at radial walls before UV exposure are shown in Figures 12 and 13.

The first perceptible change in the anatomical structure of the radial section of southern yellow pine upon exposure appears to take place at the pits. After 500 h of UV exposure, half-bordered pits were damaged. Bordered pits also interacted with light, but to a lesser extent (Figure 14). The bordered pits could still be recognized. In addition, checking and void formation in radial walls occasionally could be seen from the exposed specimen. After 1000 h of exposure, however, severe deterioration of the bordered pits was observed. The SEM micrograph (Figure 15) shows that the apertures of bordered pits were enlarged to the limit of the pit chambers. The pit domes were destroyed completely. At the extreme, the deterioration also spread over the radial surface of the tracheid wall. Complete degradation of these cell walls would probably take place at a longer exposure time. Disappearance of bordered pits has also been observed for redwood exposed to UV light (60, 62).

TANGENTIAL SECTION. Bordered pits were rarely found in the tangential surfaces observed. SEM studies revealed that diagonal microchecks passing through bordered pits in tracheid cell walls were the most conspicuous anatomical change at the tangential section upon UV exposure. The narrow microchecks were oriented diago-

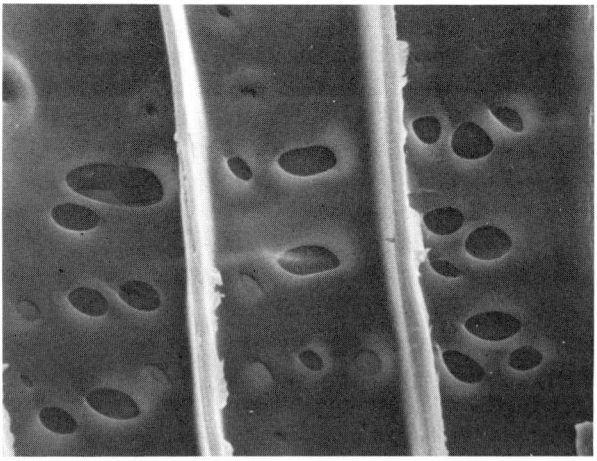

Figure 12. Half-bordered pit structures of southern yellow pine on radial section (700 ×).

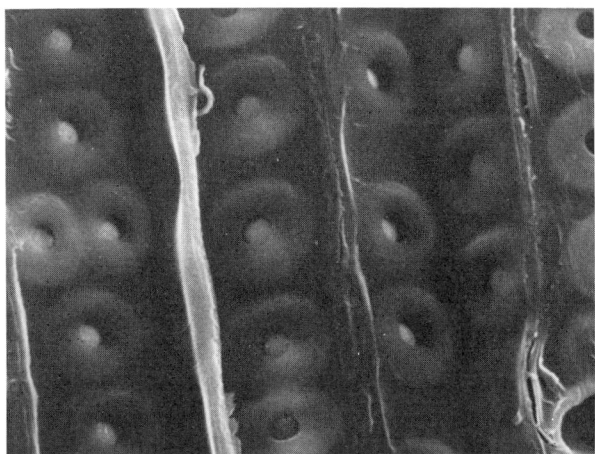

Figure 13. Bordered pit structures of southern yellow pine on radial section (700 ×).

nally to the axis of the cell wall, thus indicating that microchecks occur at the fibril angles of the S_2 cell wall (Figures 16 and 17). Similar observations have been reported (60). The common appearance of the diagonal microchecks during UV exposure was suggested to be the result of local concentrations of tensile stress at right angles to the fibril direction of the S_2 layer. Relatively wider diagonal checks

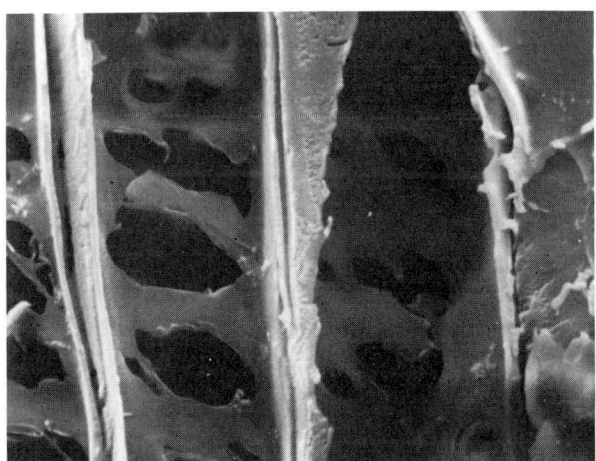

Figure 14. Deterioration of half-bordered pits and cell wall of southern yellow pine at radial section after exposure to UV light for 500 h (700 ×).

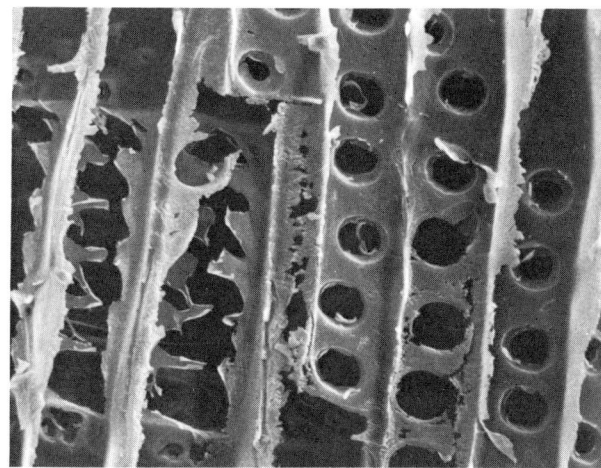

Figure 15. Deterioration of bordered pits and cell wall of southern yellow pine at radial section after exposure to UV light for 1000 h (700 ×).

were observed in the tangential section of tracheid walls of latewood (Figure 17).

Weathering of Wood-Based Materials

The weathering process described thus far has been for solid wood. The introduction of another variable, the adhesive, in the weathering of wood-based materials such as plywood and particle board creates additional complications. Wood substance is still ex-

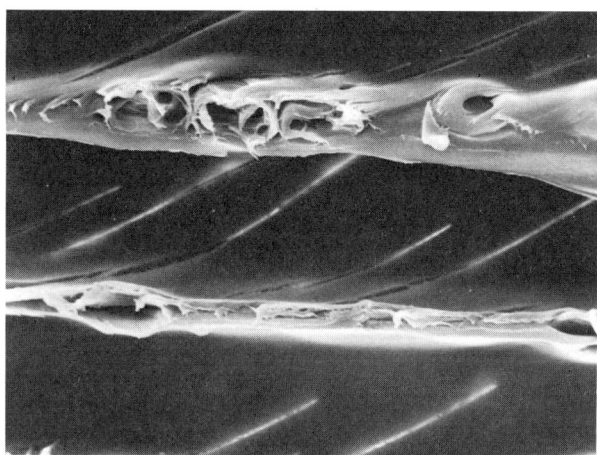

Figure 16. Microchecks of cell wall of southern yellow pine at tangential section (earlywood) after exposure to UV light for 500 h (700 ×).

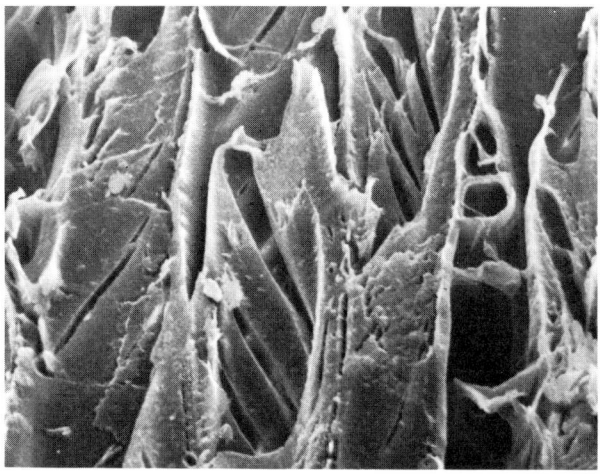

Figure 17. Microchecks of cell wall of southern yellow pine at tangential section (latewood) after exposure to UV light for 500 h (550 ×).

posed to the elements in these reconstituted products and deteriorates in a manner similar to that for solid wood. The wood–adhesive bond is the new element in exposure (1, 69, 70).

Plywood. The weathering of plywood is related directly to the quality of the veneer exposed and to the adhesives used. Because of its tendency to check, most exterior plywood is protected with a finish or with overlay material. Such plywood weathers and performs similarly to solid wood (1, 2, 71).

The swelling and shrinking that results from periodic wetting and drying plays an important role in weathering by forming checks that expose more wood surface area to weathering. In plywood the checks may expose the glueline to weathering, particularly as they become enlarged by the weathering process (72).

Plywood undergoes many visible changes in appearance during the weathering process. These changes can be described according to the following sequence (72):

1. Large checks that normally originate with lathe checks are first formed. These become wider as weathering along their borders progresses.
2. Microchecks are formed on the surface during the early stages of weathering.
3. The microchecks become deeper, wider, and more and more numerous until they actually separate individual cells and bundles of cells.

4. Particles of degraded wood—cells, cell bundles, and degraded materials—are removed through leaching, volatilization, and mechanical actions; the surface becomes roughened and cratered with a pitted appearance.
5. Lower density areas (earlywood) generally erode more quickly than higher density areas (latewood), thus giving a raised-grain appearance which becomes more and more pronounced as weathering progresses (similar to solid wood).

Because the earlywood of a given softwood species usually weathers away much more quickly than the latewood of that species, the grain pattern of the face ply becomes important in determining the rate at which weathering proceeds to the glueline. Figure 18 illustrates cross-sectional views of weathered plywood at the stage where the exposed, easily weathered earlywood on the face veneer has been eroded away to leave the denser latewood bands exposed. Four different grain patterns have been selected to illustrate their effect on erosion rate. Apparently the glueline can be exposed rapidly when face veneers are taken from either fast-growth trees (Figure 18C) or when they possess a vertical-grain pattern (Figure 18D). When these conditions occur, weathering can proceed directly to the

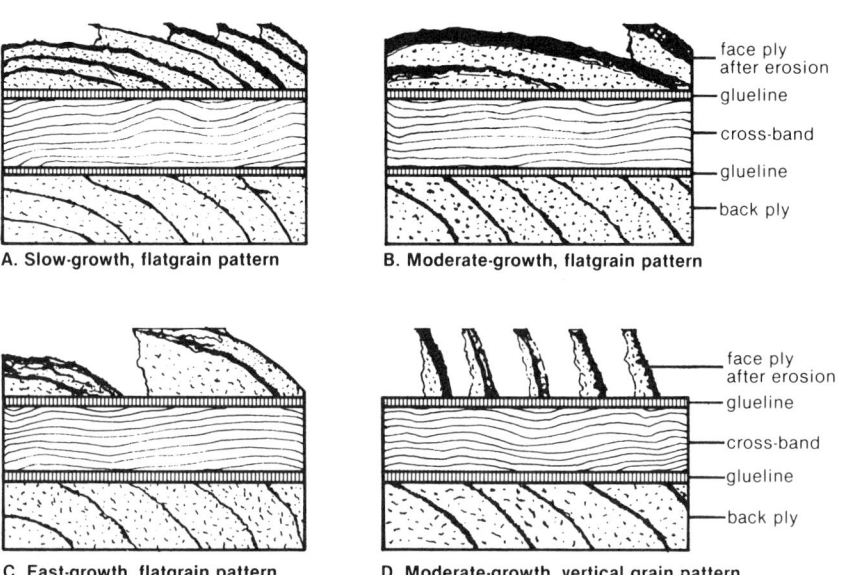

Figure 18. *Cross-sectional views of plywood illustrating the effects of weathering on face veneers with selected grain patterns.*

glueline through a path of easily eroded earlywood. The thinner the face veneer, the more probable that a situation such as that depicted in Figure 18C will occur.

Latewood also erodes away through weathering. Its erosion rate for most softwood species is slow (Table IV). Eventually, however, the face veneer of unprotected plywood will erode away, regardless of the grain pattern or wood species.

Reconstituted Panel Products. As in the case of plywood, durability of reconstituted panel products such as hardboard and particle board (waferboard, flakeboard, chipboard, oriented strand board) in outdoor weathering depends very much on wood species and on the amount and nature of resin (binder, adhesive) used in preparing the board (*1*, *73*). Hardboard is generally never exposed directly to the weather. However, it is not unusual for particle boards to be exposed to the outdoors where their outer layers are subjected to greater degradation than the inner layers. As long as the outer cover layers are intact, inner layers are protected from the elements of weathering. When outer layers of exposed boards deteriorate and loosen, and shrinking and swelling of the inner layers result from changes in moisture content. Accelerated deterioration of inner layers generally results, cohesion is lost, and boards may fail under mechanical loads (*1*). Only 1 or 2 years of weathering can cause significant strength loss and increased swelling (*74*). Deterioration of particle board during outdoor weathering takes place because of the combined effects of springback from compression set, deterioration of resin, and differential shrinkage of adjacent wood particles during moisture content change. Phenolic resins appear to give the best overall performance. Additional related studies have reported on the effect of natural outdoor and artificial accelerated weathering on durability and strength properties of particle board and related materials (*1*).

Weathering of Chemically Modified Woods

The chemical modification of wood can play a very important role in controlling the natural weathering process. Researching the effects of chemical modification of wood on weatherability and elucidating the mechanism(s) of UV degradation of modified woods have been undertaken (*75*, *76*). Chemical modification of wood cell walls with butyl isocyanate or butylene oxide, lumen-filling modification with methyl methacrylate, and combined cell wall modification and lumen-filling modifications were compared to unmodified southern pine. Physical, microscopic, and chemical changes occurring on the wood surfaces after UV irradiation in controlled accelerated weathering environments were evaluated for earlywood and latewood. Both UV light and UV light–water combinations of exposure were included in the studies.

The earlywood and latewood of southern pine chemically modified with butyl isocyanate or butylene oxide were not resistant to the degradative effects of UV light. Surface deterioration, color changes, and small weight losses occurred during accelerated weathering (UV light and water spray). Accelerated weathering produced little surface erosion until water washed away degraded wood elements. Degradation and loss of latewood during accelerated weathering was much less than that found for earlywood. This was characteristic of unmodified wood as well. Latewood erosion was greater for butylene oxide-modified wood than all others. Weight loss increased markedly as lignin degradation products were washed away by water, and chemical modification did not reduce this weight loss. Increasing the dimensional stability of the wood and blocking lignin phenolic hydroxyl groups apparently was not enough to stop the extreme degradative effects of UV light in the weathering process. UV absorbers or screens chemically bound may be necessary to protect the exposed wood surfaces.

Lumen-filling modification with methyl methacrylate polymer reduced the extent of erosion. The erosion rate of earlywood and latewood and wood substance loss during accelerated weathering was reduced significantly when compared to chemically modified or unmodified wood. In UV-light exposure, even with water spray action, degradation was minimal. The methyl methacrylate polymer, polymerized *in situ* within the wood structure, probably reduced water uptake and retarded subsequent leaching of wood degradation products. The polymer can be regarded as a gluelike material holding the surface wood fibers in place even though the natural glue (native lignin) had been degraded on the wood surface by the action of the UV light. As the methacrylate polymer holds the cellulose-rich fibers on the wood surface, the fibers may act as partial screens to protect the underlying wood substance.

Although chemical modification with butyl isocyanate or butylene oxide was not successful in controlling UV light degradation of wood, a combination of either of these chemical modifications with methyl methacrylate lumen-fill treatment resulted in a modified wood that had good resistance to accelerated weathering. The combination of the lumen-filling polymer and the cell wall-modifying chemical treatments provided a dimensional stabilization that significantly increased weatherability. Weight losses for these combined chemical treatments were at least 50% less than those of the chemically modified specimens, and wood erosion and erosion rates were low.

Chemical Aspects of Weathering Reactions

Sunlight, especially a small portion of UV light, is the principal instigator of weathering reactions. The immediate consequence of the interaction of wood with light is the generation of free radicals at the exposed surface (7, 19). As these labile free radicals terminate and stabilize, chromophoric and auxochromic groups are formed and discoloration and deterioration occur.

Wood does not contain any intrinsic free radicals (77). However, wood is a good light absorber. It interacts readily with electromagnetic radiation with wavelengths equal to or shorter than visible light and various types of free radicals are generated. They can be detected by electron spin resonance (ESR) spectroscopy (77b). Typical ESR signals of free radicals originating from wood irradiated with different light sources, i.e., fluorescent light, sunlight, and UV light, are shown in Figure 19. The shorter wavelength and greater light energy of UV light generate the highest amount of free radical concentration on the wood surface. This is followed by sunlight and fluorescent light when wood is irradiated under identical conditions. Regardless of the light source, the free radicals formed rapidly interact with oxygen molecules to generate thermal and light sensitive hydroperoxide via a hydroperoxide radical intermediate. This has an adverse effect on wood stabilization against weathering (78). The hydroperoxide impurities generated at wood surfaces can be determined by spectrophotometric techniques using iodometric and triphenylphosphine methods (79, 80).

Free Radical Reactions in Cellulose and Hemicellulose. The light sensitivity of cellulose has been recognized for nearly a century. In 1883, Witz showed that the photodegradation of cellulose is chemical in nature (81). Free radical intermediates are produced in cellulose during photodegradation reactions, and most of them have been identified (7). The photodegradation rate of cellulose and hemicellulose depends markedly on the intensity and energy distribution of the light. The formation of free radicals is a sign of initiative degradation of the polymer.

Pure cellulose is not influenced in vacuo by the irradiation of light longer than 340 nm, and cellulose degradation by light is confined to a narrow band of the electromagnetic spectrum. However, in the presence of air (mainly oxygen), cellulose degradation may take place at a slow rate when exposed to light of wavelength longer than 340 nm. When cellulose is subjected to sunlight, the glycosidic linkages are cleaved which causes a loss of strength and of degree of polymerization. The formation of free radicals located due to the chain scission at the C-1 and C-4 positions can be detected by ESR spectrophotometry. Discoloration and formation of hydroperoxide on exposed surfaces can be recognized easily.

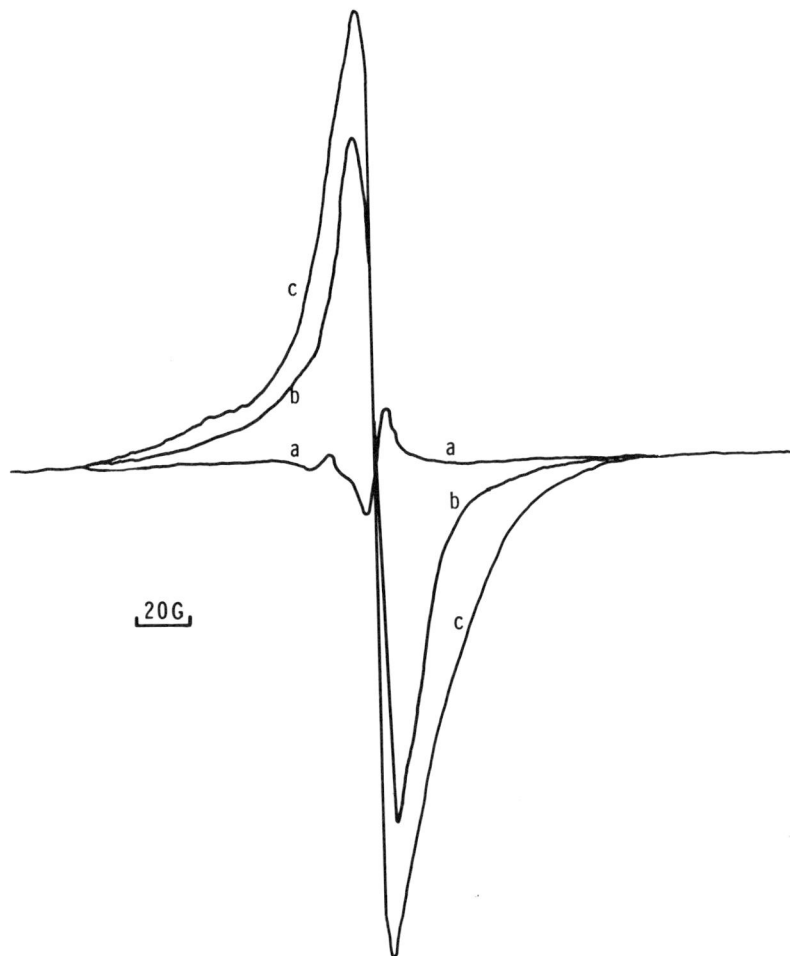

Figure 19. Electron spin resonance (ESR) signals of wood free radicals originating from wood irradiated with different light sources at 77 K for 60 min. Key: a, fluorescent light; b, sunlight; and c, UV light.

When cellulose is exposed to light of wavelength longer than 280 nm, in addition to chain scission, dehydrogenation takes place, preferentially at the C-1 and C-5 positions. Dehydroxymethylation due to the cleavage of the C-5–C-6 side chains of cellulose is observed when cellulose is exposed to light longer than 254 nm (82). The formation of carbon radicals, alkoxy radicals, formyl radicals, and hydrogen atoms in cellulose irradiated with various light sources can be detected by ESR. The degree of degradation with different light sources can be evaluated by the change of viscosity, the loss of degree of polymerization, and weight loss.

In general, alkoxy radicals generated in cellulose are stable as compared to carbon radicals. The carbon radicals readily undergo secondary termination reactions. Carbon radicals in vacuo have an affinity for recombination and hydrogen abstraction to stabilize themselves in the presence of oxygen, and they are transformed rapidly into hydroperoxide radicals to build up hydroperoxide. This rapid oxygenation reaction is further accelerated when excited oxygen is presented (83).

Although cellulose is not sensitive to UV light of wavelengths longer than 340 nm, the presence of metal ions, particularly ferric ions, dyes, and many sensitizers, promotes free radical formation even when cellulose is exposed to light longer than 340 nm (84). In addition to wavelengths, other factors that have significant effect on free radical formation and degradation rate are oxygen and sensitizers, humidity and wetness (85), and morphology (86a).

Free Radical Reactions in Lignin. The conventional lignin model gives a broad picture of the reactive groups available in native lignin that make it an excellent light absorber. Lignin has an absorption peak at 280 nm with its tail extending to over 400 nm (Figure 20). The reactive groups available in lignin consist of ethers of various types, primary and secondary hydroxyl groups, carbonyl groups, and carboxyl groups. There also exist a number of aromatic and phenolic sites and activated locations capable of interacting with light to initiate free radical chain reactions. Because of the complexity of the lignin structure, identifying the free radical sites formed is extremely

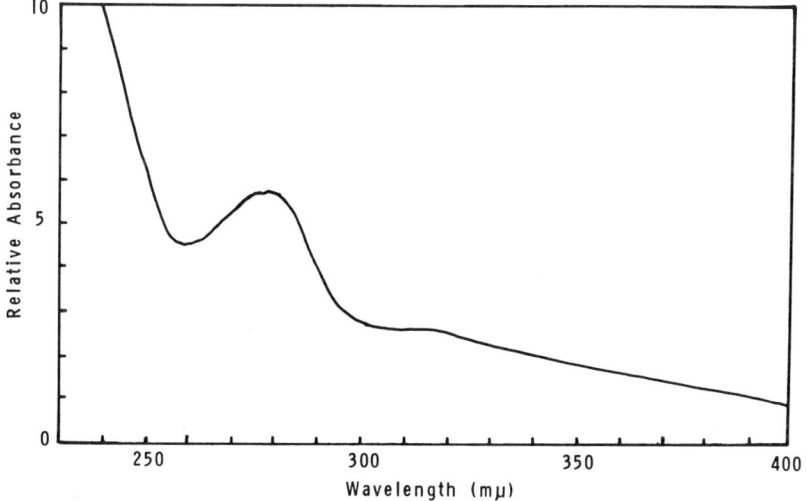

Figure 20. UV absorption curve for lignin.

difficult. However, with careful selection of model compounds, detailed study of photoinduced free radicals has been possible (7).

Several facts on photochemical reactions have been elicited. They are summarized as follows:

1. Lignin is degraded easily by light of wavelength shorter than 350 nm. Significant color buildup or formation of chromophoric groups is recognized.
2. Lignin is not degraded by light longer than 350 nm, but photobleaching or whitening of lignin can be observed when it is exposed to light longer than 400 nm.
3. Reduction of methoxy content of lignin occurs.
4. Phenoxy radicals are produced readily from phenolic hydroxy groups.
5. Carbon–carbon bonds adjacent to α-carbonyl groups are photodissociated via the Norrish Type I reaction (86b).
6. The Norrish Type I reaction does not occur efficiently in those compounds with ether bonds adjacent to the α-carbonyl group. Photodissociation takes place at the ether bond.
7. Compounds bearing benzoyl alcohol groups are not susceptible to photodissociation except when photosensitizers are present.
8. α-Carbonyl groups function as photosensitizers in the photodegradation of lignin (7).

Because of the phenolic hydroxy groups and ether bonds in lignin, the phenoxy radicals are the major intermediate formed in photoirradiated lignin. Although phenoxy radicals are rather stable intermediates, they are capable of being excited by light, or reacting with oxygen to induce demethylation of the guaiacyl unit of lignin to produce o-quinonoid structures. Leary suggested that o-quinone is the end product of the reaction (87). Consequently, quinonoid moieties formed in lignin are apparently the major chromophoric groups contributing to the discoloration of lignin and wood materials.

Free Radical Characteristics and Reactions in Weathered Wood. Wood, wood fiber components, and isolated lignin contain certain amounts of free radicals that are detectable by ESR spectroscopy (88, 89). Unexposed green wood with 69% moisture content (in dark and in vacuo) was found (77a) to contain no free radicals. A trace amount of free radicals may be produced in the presence of oxygen, and most of these free radicals are generated in wood during mechanical preparation (90) as well as in wood exposed to electromagnetic irradiation. ESR studies revealed that wood interacts readily with sunlight, fluorescent light, and artificial UV light to produce free radicals, either in the presence of air or in vacuo (Figures 21 and 22). Higher amounts of free radicals were generated in vacuo than in air for all light sources at 77 K. Oxygen is a mandatory element to activate wood surfaces for promoting free radical formation when fluorescent light is used at ambient temperature. For all systems, free radicals generated in vacuo have a relatively long lifetime compared to those generated in the presence of air. Addition of oxygen to wood treated in vacuo promotes free radical formation; peroxy radicals are formed readily at the wood surface. The peroxy radical also seeks to complete its unsatisfied valence, which it may do by abstracting a proton from a nearby molecule to form a hydroperoxide. The hydroperoxide is relatively unstable toward heat and light, and is usually transformed into a new chromophoric group, such as a carbonyl or carboxylic group.

Effect of Water and Moisture on the Formation and Stability of Free Radicals. Water is considered to be a critical element in wood's weatherability. Because water is a polar liquid it readily penetrates and swells the wood cell walls. Water molecules may interact with free radicals generated by light. In order to study the influence

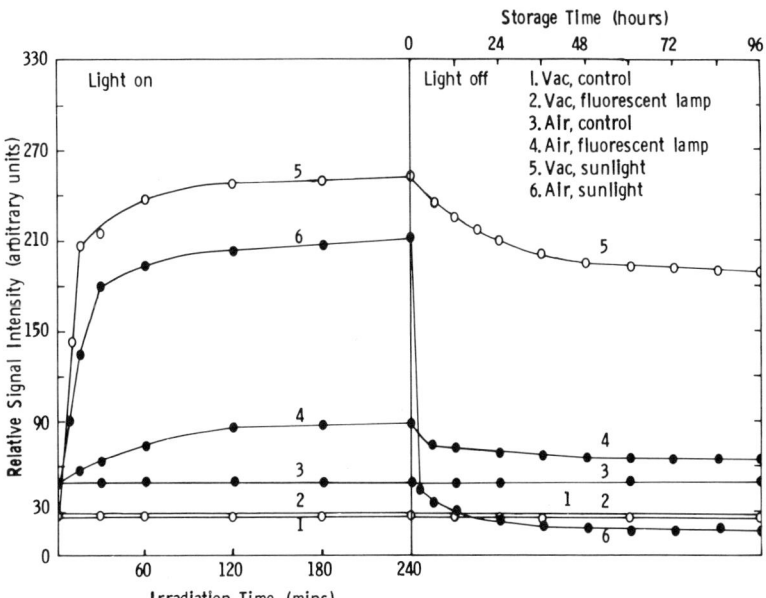

Figure 21. ESR signal intensity (recorded at 77 K) of wood as a function of irradiation time and storage time at ambient temperature. Key: 1, vac, control; 2, vac, fluorescent lamp; 3, air, control; 4, air, fluorescent lamp; 5, vac, sunlight; and 6, air, sunlight.

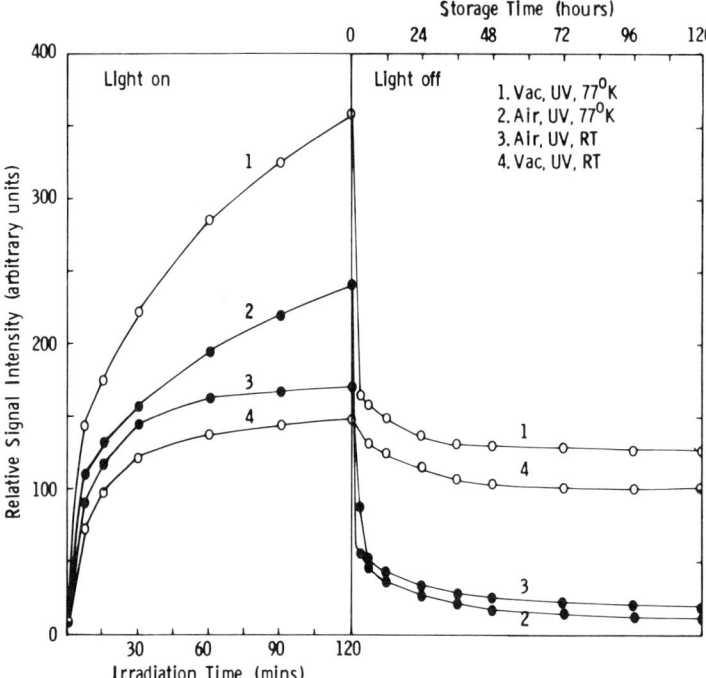

Figure 22. ESR signal intensity (recorded at 77 K) of wood as a function of UV irradiation time and storage time at ambient temperature. Key: 1, vac, UV, 77 K; 2, air, UV, 77 K; 3, air, UV, room temperature; and 4, vac, UV, room temperature.

of water molecules upon the reactivity of free radicals, wood with different moisture contents was prepared and exposed to fluorescent light. Different levels of ESR intensity (91), which are directly proportional to free radical concentration, were obtained from the wood. The ESR intensity (either in vacuo or air) initially increased as the moisture content increased from 0 to 3.2%, and reached a peak at 6.3%. At 15.9% moisture content, a significant decrease in intensity was observed. At 31.4% moisture content, only a weak signal was detected (Figure 23). From the stereotopochemistry point of view, the principal role of water is to facilitate light penetration into the accessible regions and to open up the nonaccessible regions for light penetration. Thus, more free radicals are generated in these regions. The excess water molecules present probably trap free radicals to form wood free radical–water complexes (91).

Participation of Singlet Oxygen in the Weathering Process. In addition to sunlight and water, oxygen molecules are among the most ubiquitous in nature. They play a unique role in many photophysical and photochemical processes. We have explained that oxygen is an important element to promote free radical formation, and possibly that peroxide impurity is formed due to the interaction of free radicals and oxygen molecules. However, the rate of oxidation of most polymers is usually very small without radiation at ambient temperature. The acceleration of the reaction rate by electromagnetic energy may be due to the generation of excited oxygen species. Considerable

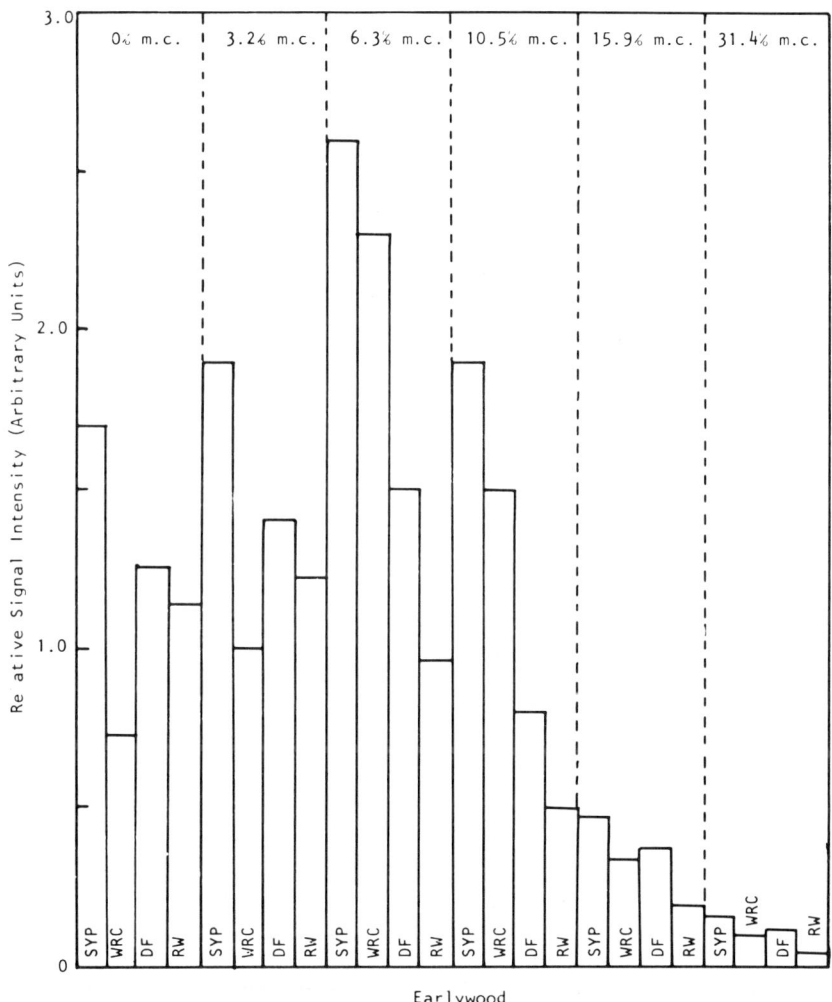

Figure 23. Comparison of ESR relative signal intensities (carried out in a vacuum) of free radicals in earlywood with different moisture contents. Key: SYP, southern yellow pine; WRC, western redcedar; DF, Douglas-fir; and RW, redwood.

evidence exists that many photooxidation reactions involve the low-lying singlet state of oxygen ($^1\Delta_g$ and $^1\Sigma_g{}^+$) as intermediates (92).

Wood is a polymer blend containing cellulose, hemicellulose, lignin, and extractives. These wood components contain internal chemical entities such as carbonyl, carboxyl, aldehyde, phenolic hydroxyl, and unsaturated double bonds, and external entities such as wax, fat, and metal ions. The absorption of light energy by these components may bring them to an excited triplet state that transfers the energy to triplet ground state oxygen molecules to create singlet oxygen (7, 8, 78, 83). The participation of singlet oxygen in the photooxidation of wood was evidenced by using singlet oxygen generators and singlet oxygen quenchers during irradiation. Iodometry studies revealed that hydroperoxide was formed in wood photoirradiated in the presence of oxygen. The formation rate of hydroperoxide at the wood surface increased when methylene blue and rose bengal solutions were added to the wood prior to irradiation (Figure 24). Peroxide radicals involved in the interim were detected by an ESR spectrophotometer, i.e., an asymmetric singlet signal of peroxy radicals with the average g-value of 2.021 ($g_\perp$ = 2.034; $g_\parallel$ = 2.007) was detected. On the other hand, when singlet quenchers, such as triethylamine and 1,4-diazabicyclo[2.2.2]octane (DABCO) were used in the identical experiment conditions, the hydroperoxide content was reduced in some cases, even in the presence of rose bengal (Figure

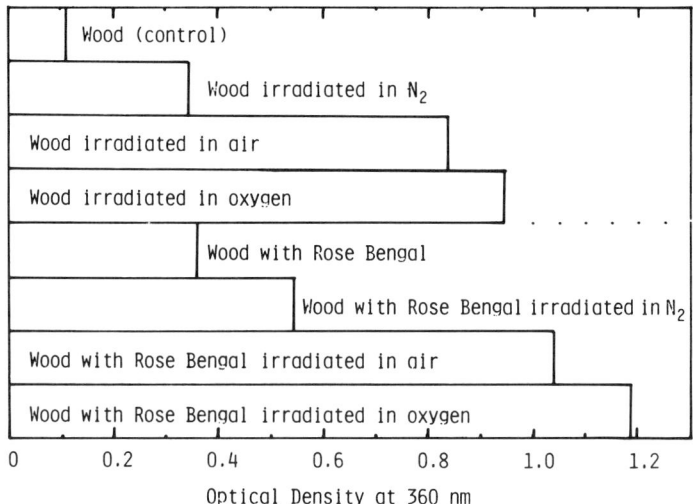

Figure 24. Effect of oxygen and rose bengal on the rate of peroxide formation in wood photoirradiated for 24 h.

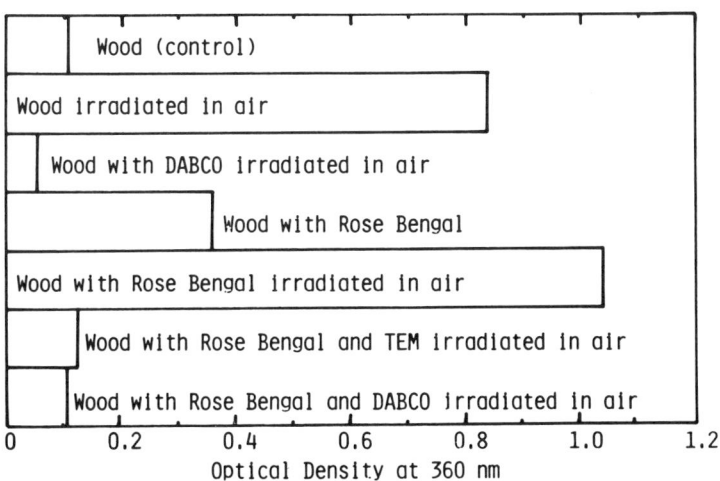

Figure 25. Inhibiting effect of DABCO and triethylamine on peroxide formation in wood photoirradiated for 24 h.

25). This evidence supports the theory that a singlet oxygen is formed during photoirradiation and that it interacts rapidly with free radicals of wood to produce hydroperoxides. Due to the instability against heat and light, the hydroperoxides decompose rapidly under ambient conditions to create chromophoric groups, such as carbonyl and carboxyl groups. These groups contribute to the discoloration of wood surfaces.

Summary of the Chemical Aspects of Weathering. The deterioration of wood materials upon weathering involves a very complex reaction sequence. Penetration of UV light into wood does not traverse deeper than 75 μm. Nonetheless, wood surface reactions initiated or accelerated by light can be observed by discoloration, loss of brightness, and change in surface texture after artificial UV light irradiation or long-term solar irradiation.

Free radical species are generated in wood readily by light. These radicals rapidly interact with oxygen to produce hydroperoxide impurities that are decomposed easily to produce chromophoric groups such as carbonyl and carboxyl groups. The use of singlet oxygen generators, such as rose bengal and methylene blue, as well as singlet oxygen quenchers, such as 1,4-diazabicyclo[2.2.2]octane and triethylamine, suggests the participation of singlet oxygen as an effective intermediate in photooxidation reactions at the wood surface. The presence of water in wood also influences the rate of free radical formation. When moisture content in wood is increased from 0 to 6.3%, more free radicals are formed; beyond this stage, the rate of radical decay increases. IR studies reveal that carbonyl groups are generated in cellulose and lignin. Water-soluble fractions of degraded wood exhibit characteristics of phenolic absorptions due to the loss of lignin. ESCA studies show that oxidized wood surfaces contain higher oxygen contents than unexposed wood surfaces. A general mechanistic scheme, able to account for the weathering, or more commonly photooxidation, of wood surfaces, is illustrated in Figure 26. Free radical chain reactions in the presence of oxygen and light are responsible for the discoloration and deterioration of wood surfaces.

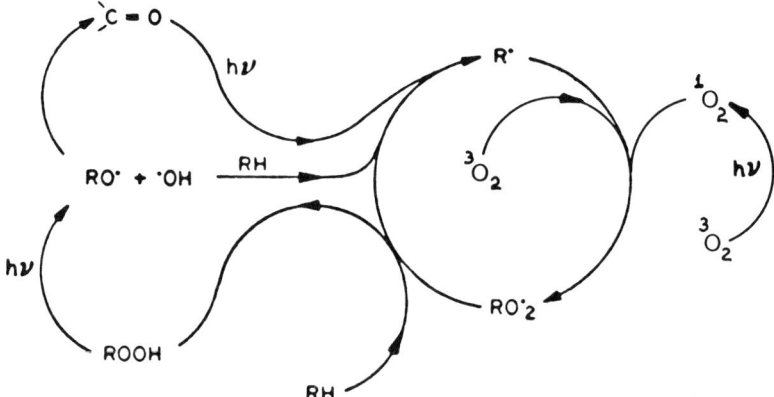

Figure 26. Mechanisms of photodegradation of wood.

Protection Against Weathering

Paint and other coatings on wood (finishes) used indoors can protect and last for many decades without refinishing (93, 94). Indoor finishes are relatively unaffected by wood properties. The durability of finishes on wood exposed outdoors to the natural weathering process, however, depends first on the wood itself. Wood properties that are important in finishing are moisture content; density and texture; resin and oil content; width and orientation of growth rings; and defects such as knots, reaction wood, and fungi-infected (diseased) wood (95). Other contributing factors are the nature and quality of the finish used, application techniques, pretreatments, the time between refinishings, the extent to which the surfaces are sheltered from the weather, and climatic and local weather conditions (96–99).

The primary function of any wood finish is to protect the wood surface from the natural weathering elements (sunlight and water) and help maintain appearance (2). Where appearance does not matter, wood can be left unfinished to weather naturally, and such wood will often protect the structure for an extended time (1, 2, 54). Different finishes give varying degrees of protection to the weather (96, 97).

Surface treatment provides protection against light and water and will be affected by the weather resistance of the bonding agents of the finish (drying oils, synthetic resins, latexes, etc.). These bonding agents are subject to photodegradation to some degree. The mechanism of failure of paints and other finishes has been described (93, 100), and will not be discussed further here. Wood exposed to the outdoors is protected from the effects of weather by various finishes, by construction practices, and by design factors (2, 11, 96, 97, 101, 102).

Two basic types of finishes (or treatments) are used to protect wood surfaces during outdoor weathering: those that form a film, layer, or coating on the wood surface, and those that penetrate the wood surface to leave no distinct layer or coating. Film-forming materials include paints of all description, varnishes, lacquers, and also overlays bonded to the wood surface. Penetrating finishes include preservatives, water repellents, pigmented semitransparent stains, and chemical treatments.

Film-Forming Finishes. PAINTS. Of all the finishes, paints most protect wood against erosion by weathering and offer the widest selection of colors. A nonporous paint film will retard penetration of moisture, thereby reducing problems of paint discoloration by wood extractives, paint peeling and checking, and warping of the wood (98, 99). Proper pigments will essentially eliminate photodegradation of the wood surface. Paint, however, is not a preservative; it will not prevent decay if conditions are favorable for fungal growth. The durability of paint coatings on exterior wood is affected by variables in the wood surface, by moisture, and by type of paint.

Paints are commonly divided into the oil-base or solvent-borne systems, and latex or waterborne systems (93, 103). Oil-base (or alkyd) paints are essentially a suspension of inorganic pigments in an oleoresinous vehicle that binds the pigment particles and the bonding agent to the wood surface. Latex paints are suspensions of inorganic pigments and various latex resins in water. Acrylic latex resins are particularly durable, versatile materials for finishing wood and wood-related materials.

VARNISHES. The most natural appearance for wood is obtained by use of clear varnishes or lacquers. Other treatments either change wood color or cover it up completely. Unfortunately, clear varnish finishes used on wood exposed to sun and rain require frequent maintenance to retain a satisfactory appearance. Durability of varnish on exposed wood is limited and many initial coats are necessary for reasonable performance. Maintenance of the varnish surface must be carried out as soon as signs of breakdown occur. This may be as little as 1 year in severe exposures. Lacquers and shellacs are usually not suitable as exterior clear finishes for wood because of water sensitivity and ease of cracking or checking of their rather brittle films.

The addition of colorless UV light absorbers to clear finishes has found only moderate success to help retain the natural color and original surface structure of wood (*1, 18, 104–108*). Opaque pigments found in paints and stains generally provide the most effective and long-lasting protection against light (*1, 42, 109–111*). Even when using relatively durable, clear, synthetic resin varnishes, the weatherproof qualities of the wood–varnish system are still limited because UV light penetrates the transparent varnish film and gradually degrades the wood under it (*21, 112, 113*). Eventually, the varnish begins to flake and crack off, taking with it fibers of the wood that have been degraded photochemically (*51, 60, 114*).

Studies for predicting the durability of clear finishes for wood from basic properties have been reported (*115, 116*) and the relations between composition, water absorption, water vapor permeability, tensile strength, and elongation reviewed. Single and multiple regressions were used to establish the relative importance of the different properties in determining durability ratings on outdoor exposure. The durability of clear phenolics and alkyds can be predicted from water absorption and permeability properties. Tensile strength and mechanical properties are of less importance.

Penetrating Finishes. WATER REPELLENTS. A large proportion of the damage done to exterior woodwork (paint defects, deformations, decay, leakage, etc.) is a direct result of moisture changes in the wood and subsequent dimensional instability (*1, 11, 117–123*). This is discussed in detail in the section on the effect of water. Water generally enters wood through open cracks, unprotected end-grain surfaces, and defects in surface treatments. Although the negative effects of such problems can be avoided, or at least reduced, by proper design or correct choice of materials, it is extremely difficult to stop checks or cracks from appearing where woodwork is subjected to harsh long-term exposure. Even a quality coating often loses its protective qualities because it cannot tolerate the stresses and strains of shrinkage and swelling, especially around joints. Eventually, the coating gives way.

Because of these problems, woodwork exposed to the outdoors should be given a protective treatment that is both water-repellent and water-resistant to decay fungi. Such treatment could be used either as the finish itself or prior to the final finish. Materials developed for such purposes are termed *water-repellent preservatives*

(WRP). Generally, they are comprised of a resin (10–20%), a solvent, a wax (as the water repellent), and a preservative [generally pentachlorophenol, but others such as bis(tri-n-butyltin) oxide, copper naphthenate, zinc naphthenate, and copper 8-quinolinolate are used] (1, 11, 118–120).

A great store of information has been accumulated on the effectiveness of WRPs in protecting exterior wood (1, 117–119, 121, 122, 124–129). The treatments can be applied by vacuum impregnation, by immersion (which is preferred), by brush, or by spray application. They improve the performance of many finishes applied over them and add greatly to the durability of exposed wood. Even chipboard or particleboard, which is very susceptible to moisture, can be protected quite effectively against the effects of outdoor exposure by using a WRP pretreatment followed by a diffusion-resistant coating (126, 130–133).

The WRP treatments give wood the ability to repel water, thus denying stain and decay fungi the moisture they need to live. Wood surfaces that remain free of mildew fungi retain an attractive "natural" appearance. A WRP reduces water damage to the wood and helps protect applied paint from blistering, peeling, and cracking, which often occurs when excessive water penetrates wood.

A WRP usually contains a fungicide as preservative that kills any surface mildew living on the wood. In many medium- to low-hazard decay situations, preservative may not be needed for successful performance of the water repellent (122). Therefore, restriction of water from wood is of prime importance in improving the durability of exposed wood (117).

STAINS. When pigments are added to water repellent or WRP solutions or to similar penetrating transparent wood finishes, the mixture is classified as a *pigmented penetrating stain* (1, 2, 124, 134). Addition of pigment provides color and greatly increases the durability of the finish because UV light is blocked. These semitransparent stains permit much of the wood grain to show through; they penetrate the wood to a degree without forming a discrete, continuous layer. Therefore, they will not blister or peel even if excessive moisture enters the wood. The durability of any stain system is a function of pigment content, resin content, preservative, water repellent, and quantity of material applied to the wood surface. Their performance during outdoor exposure has received a great deal of attention (1, 69, 71, 124, 134–137).

Penetrating stains are suitable for both smooth- and rough-textured surfaces; however, their performance improves markedly if applied to roughsawn, weathered, or rough-textured wood (1, 69, 99, 124, 134, 137, 138) because more material can be applied to such surfaces. They are especially effective on lumber and plywood that does not hold paint well, such as flat-grained and weathered surfaces, or dense species. Penetrating stains can be used effectively to finish such exterior surfaces as siding, trim, exposed decking, and fences. Stains can be prepared from both solvent-base resin systems and latex systems; however, latex systems do not penetrate the wood surface. Commercial finishes known as heavy-bodied, solid-color, or opaque stains are also available, but these products are essentially similar to paint because of their film-forming characteristics. Such stains do find wide success when applied on textured surfaces and panel products such as hardboard. They can be oil-based or latex-based (*see also* "Natural Wood Finishes").

PRESERVATIVES. Although not classified generally as wood finishes, preservatives in wood do protect against weathering and decay, and a great quantity of preservative-treated wood is exposed without any additional finish (2). There are three main types of preservative: (1) the preservative oils (e.g., coal-tar creosote), (2) the organic solvent solutions (e.g., pentachlorophenol), and (3) waterborne salts (e.g., chromated copper arsenate) (2). These preservatives can be applied in several ways, but pressure treatment gives the greatest protection against decay. Greater preservative content of pressure-treated wood generally results in greater resistance to weathering and improved surface durability. The chromium-containing preservatives also protect against UV degradation (*1, 138, 139*).

SURFACE TREATMENTS. Certain inorganic chemicals (especially hexavalent chromium compounds), when applied as dilute aqueous solutions to wood surfaces, provide the following benefits (*70, 135, 139–141*): (1) retarding degradation of wood surfaces by UV radiation; (2) improving durability of UV-transparent polymer coatings; (3) improving durability of paints and stains; (4) providing a degree of dimensional stabilization to wood surfaces; (5) providing fungal resistance to wood surface and to coatings on the surface; (6) serving without further treatment as natural finishes for wood; and (7) fixing water-soluble extractives in woods such as redwood and cedar, thereby minimizing subsequent staining of applied latex paints.

The most successful treatments investigated were those containing chromium trioxide (chromic acid, chromic anhydride), copper chromate (mixtures of soluble copper salts and soluble chromates), or ammoniacal solutions of these chemicals. Successful results have been obtained by using zinc-containing compounds (*142*). Sell et al. (*143*) described surface treatments with chromium–copper–boron salts. Field weathering tests, leaching tests, and electron-probe microanalysis showed that this treatment was resistant to leaching and weathering.

The coating durability of organolead-treated southern pine in exterior exposure was reported (144). Significant improvement in the durability of a vinyl–acrylic latex and an alkyd paint on treated wood was demonstrated. Improved durability appeared to be independent of the type or concentration of the organolead-treated compounds.

The role of chromium in the treatment of wood has been described in detail (145–148). Wood waterproofing, lignin cross-linking, kinetics of reaction, and kinetic behavior have been studied.

Williams and Feist (149) described the application of electron ESCA techniques to evaluate wood and cellulose surfaces that had been modified by aqueous chromium trioxide treatment. ESCA data showed at least 80% Cr(VI) to Cr(III) reduction on all substrates. Leaching experiments confirmed this reduction to a highly water-insoluble or "fixed" chromium complex on both wood and on filter paper (cellulose). Similar oxidation products were observed with wood and filter paper. These experiments indicate that chromium–cellulose and chromium–lignin interactions are involved in the mechanism of chromium(VI) stabilization of wood surfaces.

When wood was treated with 0.1% chromic acid solution, protective effects could be recognized on transverse surfaces after 500 h of UV light irradiation (Figure 27) (68). Although some longitudinal microditches in the middle lamella zone were observed with the chromic acid-treated wood, the deterioration of cell walls was less, and the microditches were narrower than those of the untreated wood. The degree of protection was directly proportional to the con-

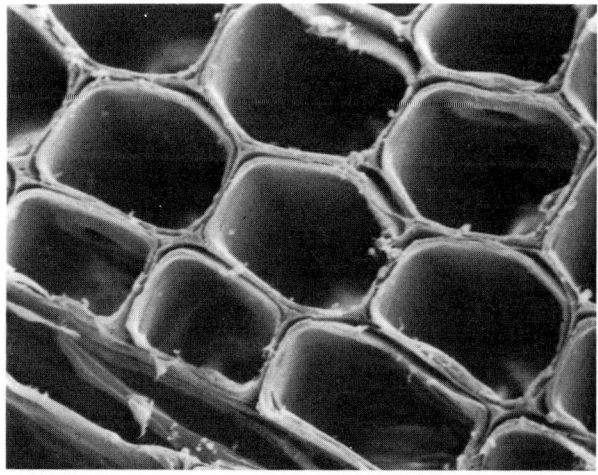

Figure 27. Cross section of southern yellow pine treated with 0.1% chromic acid solution after exposure to UV light for 500 h (700 ×).

centration of chromic acid used in the treatment. When woods were treated with 5 and 10% chromic acid solution, most of the cell walls were protected (Figure 28 and 29). The cell walls appeared to be very resistant to photodeterioration; only small voids created at the cell corners were observed after 1000 h of exposure. For the most part, the middle lamella region was preserved.

Normally ferric chloride is a strong oxidative agent for cellulosic textiles (150); however, a photoprotective effect was observed with wood treated with ferric chloride solution similar to those treated with chromic acid. Similar photoprotective effects of ferric chloride were observed with thermomechanical pulp fibers (151).

Protective effects of chromic acid and ferric chloride treatments on surface degradation also were observed on radial surfaces. The preservation of both simple and bordered pits in woods treated by these inorganic salts was observed. At a 10% chromic acid treatment concentration, the structure of the pits retained most of the original shape after 1000 h of UV irradiation. The diagonal microchecks passing through the bordered pits in radial walls of tracheids, however, can still be observed (Figure 30).

Both treated and untreated tangential wood surfaces were quite resistant to photodegradation. Frequently observed diagonal microchecks in untreated woods were minimal in the chromic acid- and ferric chloride-treated specimens.

These findings show changes in wood microstructure during photodegradation that lead to cell wall separation at the middle la-

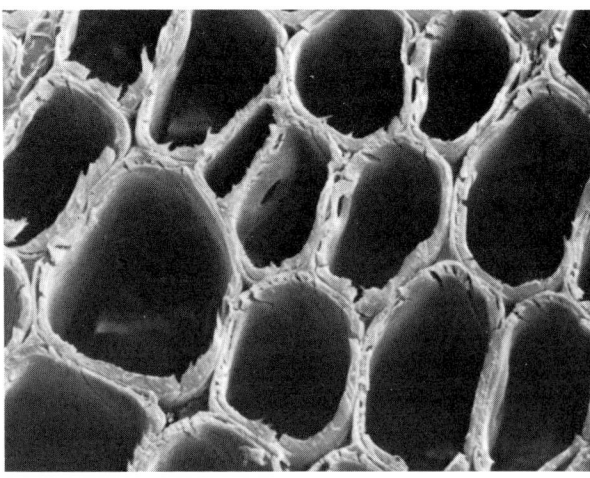

Figure 28. Cross section of southern yellow pine treated with 5% chromic acid solution after exposure to UV light for 500 h (700 ×).

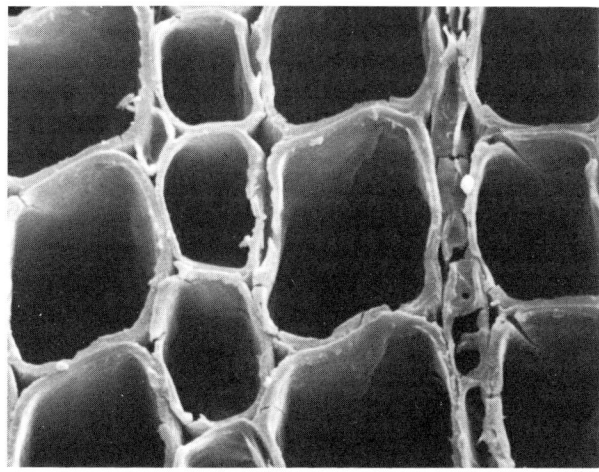

Figure 29. Cross section of southern yellow pine treated with 10% chromic acid solution after exposure to UV light for 500 h (700 ×).

mella region and to damage of half-bordered and bordered pits on radial surfaces. The degradative effect of UV light can be minimized by treating wood surfaces with aqueous solutions of chromic acid and ferric chloride.

Natural Wood Finishes

Some wood finishes are often applied as so-called natural finishes for wood. Each finish system offers various advantages and disadvan-

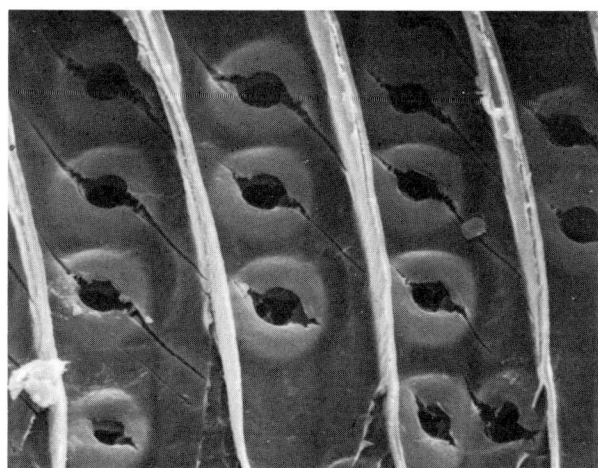

Figure 30. Microchecks of cell wall of southern yellow pine treated with 10% chromic acid solution after exposure to UV light for 1000 h (700 ×).

tages for this use. These systems can be classified as film-forming or penetrating finishes. The penetrating finishes can be subdivided further into transparent, semitransparent, and waterborne salts as described earlier. The simplest natural finish, as supplied by nature, is the weathering process.

Film-Forming. Varnishes are the primary transparent film-forming materials used for natural wood finishes, and they greatly enhance the natural beauty of wood. Varnishes lack exterior permanence unless protected from direct exposure to sunlight. Varnish finishes on wood exposed to the outdoors without protection will generally require refinishing every 1–2 years.

Penetrating. The penetrating finishes are the second broad classification of natural wood finishes. These finishes do not form a film on the wood surface.

TRANSPARENT. Water-repellent preservatives (WRP) are the most important of the transparent penetrating natural finish systems. Treating wood surfaces with WRP will protect wood exposed to the outdoors with little change in appearance. A clean, golden-tan color can be achieved with most wood species. The treatment reduces warping and cracking, prevents water staining at edges and ends of wood siding, and helps control mildew growth. The first application of WRP may protect exposed wood surfaces for only 1–2 years, but subsequent reapplications may last 2–4 years because the weathered boards absorb more of the finish.

SEMITRANSPARENT. The semitransparent oil-based stains are the second of the penetrating natural wood finishes. These stain finishes provide a less natural appearance because they contain pigment that partially hides the original grain and color of the wood. They are generally much more durable than are varnishes or WRPs, and they provide more protection against weathering. With these stain systems, weathering is slowed by retarding the alternate wetting and drying of wood and the presence of pigments on the wood surface minimizes the degrading effects of sunlight. The amount of pigment in the semitransparent stains varies considerably, and different degrees of protection against UV degradation and masking of the appearance of the original wood surface can be achieved.

Latex stains are also described as semitransparent. These pigmented finishes are generally nonpenetrating and retain the surface texture of the wood but often obliterate the natural wood color.

WATERBORNE SALTS. Waterborne inorganic salts are a special group of penetrating finishes. These surface treatments result in a finish similar to the semitransparent penetrating finishes because they change the color of the wood and leave a surface deposit of material similar to the pigment found in the semitransparent stains.

Opaque Stains. Solid color or opaque stains are another classification of finishes sometimes inaccurately described as natural wood finishes. These finishes are high in pigment content and completely mask the color and figure of the wood. Surface texture is retained and these finishes yield a flat appearance. They do protect wood against UV degradation, but tend to perform more like paints in that they do not penetrate the wood surface to any degree.

Protection of Wood-Based Materials

Wood-based panel products for exterior applications often require special finishes and special finishing practices (1, 69, 71, 74, 96). Pretreatments and edge treatments are often required.

Protection of exterior reconstituted panel products (in particular particle board) by surface coatings, treatments and overlays, and subsequent weathering performance, has received considerable attention (1, 11, 74, 133, 153–155, 157–159). Painted and overlaid boards are much more durable than unfinished boards. Addition of wax as a water repellent benefits performance of the boards, but does not protect against water vapor (1, 11, 74).

Outdoor weathering experiments with different surface treatments show that the most important prerequisites for long-term protection are the quality of the finish and the protection of board edges against moisture (1, 11). Complementary construction measures are necessary for good performance, and generally, waterproofing and sealing of the wood surface is imperative. Above a minimum value, the gradient of water vapor diffusion of the coating is only of secondary importance. The visual state of the board and thickness swelling are considered the best criteria for evaluating the protective effect of surface treatments. Surface stability is one of the most important criteria in retaining finish and maintaining satisfactory protection and appearance.

Studies on exterior finishes on plywood and plywood composite panels include coating formulations and accelerated testing (1, 69, 71, 136, 158–162).

Wood–Coating Interactions

Many studies contribute to an overall understanding of wood–coating interactions (1, 13–15, 19, 60, 106). Some studies involve adhesives that interact similarly to coatings. Schneider (163) reviewed much of the early work (before 1970) relating to wood–coating interactions. He concluded that even though there was some clear evidence about the nature of wood–coating interactions, much remained in the realm of speculation and untried theory. Subsequent studies by Schneider (164–168) using electron microscopy, fluorescence microscopy, and pyrolysis gas–liquid chromatography techniques have dealt with coating penetration in wood.

Scanning electron microscopes and energy dispersion X-ray analysis techniques have been used (169) to examine the distribution and location of the components of water-based preservatives applied to wood. Subsequently, the selective penetration of anions into the cell wall with the cations remaining in the cell lumen has been shown; iron oxide pigments were completely deposited at the wood surface.

Summary and Future Considerations

All wood materials are sensitive to outdoor weathering. Wood exposed to the outdoors without protection undergoes: photodegradation by UV light; leaching, hydrolysis, and swelling by water; and discoloration and degradation by staining and decay microorganisms. Unfinished wood surfaces exposed to weather change color, are roughened by photodegradation and surface checking, and they erode. Although physical as well as chemical changes occur because of weathering, these changes affect only the surface of the exposed wood.

UV light cannot penetrate deeper than 75 μm, and visible light no deeper than 200 μm into wood surfaces. There is a rapid loss of brightness and color change when wood is exposed to UV light or to sunlight outdoors. SEM studies show that most of the cell walls on exposed transverse surfaces are separated at the middle lamella region. Half-bordered and bordered pits on exposed radial surfaces are degraded by UV light severely.

Free radicals are generated at the wood surface during irradiation. The rate of free radical formation is enhanced when moisture content increases from 0 to 6.3%. Electron spin resonance and UV studies on the behavior of free radicals generated and their interactions with oxygen molecules to form hydroperoxides revealed that free radicals and singlet oxygen play important roles in discoloration and deterioration reactions of wood surfaces.

Wood exposed to the weather can be protected by paints, stains, and similar materials. Paints provide the most protection to exposed wood surfaces because they are generally opaque to the degradative effects of UV light and protect wood to varying degrees against water. Paint performance may vary greatly on different woods. Pigmented stains also provide durable finishes for wood exposed to the outdoors. Pretreatments such as water-repellent preservatives or certain inorganic chemicals (chromium compounds) can improve the performance of finishes significantly when applied over treated woods.

Many aspects of wood weathering are not understood completely. A complete understanding of the mechanisms involved in outdoor weathering would allow the development of new pretreatments and finishes that would greatly enhance durability. The ever-changing wood substrate, with previously unused species and new adhesive–wood combinations introduced at increasing frequency, poses particular challenges to modern wood finishes. A detailed study of the various interactions that affect the performance of wood-derived materials is needed to provide suitable protection for these products when they are used outdoors.

Newer techniques and tools for the study of wood surfaces such as Fourier transform IR spectroscopy, electron spectroscopy for chemical analysis, and electron spin resonance spectroscopy will be able to provide a great deal of insight into the weathering process for both finished or unfinished wood substrates. Use of these techniques will allow in-depth study of treatment of wood surface interactions and the importance of these interactions in ultimate performance.

Understanding the role of chemical modification of wood and wood surface in controlling the outdoor weathering process is significant to the future use of wood exposed to the outdoors. The role of modification will become larger as greater demands are placed on the newer wood-based products. The future of such modification lies in end-product property enhancement. Permanently bonded chemicals that provide UV stabilization, color control, water resistance, and dimensional stability could enhance outdoor performance greatly.

Acknowledgments

The authors are indebted to the expert technical assistance of S.-T. Chang and T. DeVilbiss of Virginia Polytechnic Institute and State University, and to P. G. Sotos of the Forest Products Laboratory, Madison, Wisconsin.

Literature Cited

1. Feist, W. C. in "Structural Use of Wood in Adverse Environments"; Meyer, R. W.; Kellogg, R. M., Eds.; Van Nostrand Reinhold Co.: New York, 1982; pp. 156–78.
2. U.S. Forest Products Laboratory, "Wood Handbook: Wood as an Engineering Material". USDA Agric. Handb. No. 74, Rev., Washington, D.C., 1974.
3. Stalker, I. N. *Chem. Ind.* **1971**, *50*, 1427–31.
4. Berzelius, J. *Lehrb. Chem.* **1827**, *3*, 603.
5. Wiesner, J. *Sitzungsber. Akad. Wiss. Wien* **1864**, *49*, 61–94.
6. Schramm, W. H. *Jahresber. Vereinigun Angew. Bot.* **1906**, *3*, 116–53.
7. Hon, D. N.-S. in "Developments in Polymer Degradation"; Grassie, N., Ed.; Applied Science Publishers: London, 1981; Chap. 8.
8. Hon, D. N.-S. in "Proc. 2nd Intl. Conf. on Environ. Degrad. of Eng. Mat."; Virginia Poly. Inst. and St. Univ., Blacksburg, Va., 1981; pp. 519–29.

9. Coupe, C.; Watson, R. W. *Proc. Ann. Conv. Br. Wood-Preserv. Assoc.* **1967**, 37–49.
10. Borgin, K. *For. S. Afr.* **1968**, *9*, 81–93.
11. Sell, J. *Holz Roh- Werkst.* **1975**, *33*(9), 336–40.
12. Stamm, A. J. *Am. Paint J.* **1963**, *48*(25), 72–88.
13. Stamm, A. J. "Wood and Cellulose Science"; The Ronald Press Co.: New York, 1964.
14. Stamm, A. J. *Off. Dig., Fed. Soc. Paint Technol.* **1965**, *37*(485), 654–69.
15. Stamm, A. J. *Off. Dig., Fed. Soc. Paint Technol.* **1965**, *37*(485), 707–19.
16. Stamm, A. J.; Loughborough, W. K. *Trans. ASME* **1942**, *64*, 379–85.
17. U.S., For. Serv. Res. Note FPL 0135, 1975.
18. Schlumbom, F. *Holz-Zentralbl.* **1963**, *89*(110), 153–56.
19. Kalnins, M. A. U.S., For. Serv. Res. Pap. FPL 57 (Part II).
20. Desai, R. L. *Pulp Pap. Mag. Can.* **1968**, *69*, 53–61.
21. Sandermann, W.; Schlumbom, F. *Holz Roh- Werkst.* **1962**, *20*(7), 245–52.
22. Sandermann, W.; Schlumbom, F. *Holz Roh- Werkst.* **1962**, *20*(8), 285–91.
23. Webb, D. A.; Sullivan, J. D. *For. Prod. J.* **1964**, *14*(11), 531–534.
24. Leary, G. J. *Tappi* **1967**, *50*(1), 17–19.
25. Leary, G. J. *Tappi* **1968**, *51*(6), 257–60.
26. Kleinert, Th. N. *Holzforsch. Holzverwert.* **1970**, *22*(2), 21–24.
27. Derbyshire, H.; Miller, E. R. *Holz Roh- Werkst.* **1981**, *39*(8), 341–50.
28. Miller, E. R.; Derbyshire, H. "Proc. on the 2nd Intl. Conf. on Durab. of Build. Mat. and Components"; Natl. Bur. Stand.: Gaithersburg, Md., 1981.
29. Raczkowski, J. *Holz Roh- Werkst.* **1980**, *38*(6), 231–34.
30. Ranby, B.; Rabek, J. F. "Photodegradation, Photooxidation and Photostabilization of Polymers"; John Wiley and Sons: New York, 1975; p. 6.
31. Calvert, J. G.; Pitts, Jr., J. N. "Photochemistry"; John Wiley and Sons: New York, 1966; p. 19.
32. Hon, N.-S. *J. Polym. Sci., Polym. Chem. Ed.* **1975**, *13*, 1347.
33. Hon, D. N.-S. in "Preservation of Paper and Textiles of Historic and Artistic Value"; Williams, J. C., Ed., ADVANCES IN CHEMISTRY SERIES No. 193, ACS: Washington, D.C., 1981; Chapter 10.
34. Launer, H. F.; Wilson, W. K. *J. Res. Natl. Bur. Stand.* **1943**, *30*, 55.
35. Browne, F. L.; Simonsen, H. C. *For. Prod. J.* **1957**, *7*(10), 308–14.
36. Hon, D. N.-S.; Ifju, G. *Wood Sci.* **1978**, *11*(2), 118–27.
37. Hon, N.-S. *J. Macromol. Sci., Chem.* **1976**, *A10*(6), 1175–92.
38. Wislicenus, H. Z. *Angew. Chem.* **1910**, *23*, 1441–46.
39. Möbius, M. *Ber. Dtsch. Bot. Ges.* **1924**, *42*, 341–44.
40. Browne, F. L. unpublished report, U.S., For. Serv., Forest Products Laboratory, Madison, Wis., 1957.
41. Sell, J.; Leukens, U. *Holz Roh- Werkst.* **1971**, *29*(1), 23–31.
42. Ashton, H. E. *Can. Build. Dig.* **1970**, CBD 122 Nat'l. Res. Council, Canada, Ottawa, Can.
43. Stout, A. W. *Off. Dig., Fed. Soc. Paint Technol.* **1958**, *30*(407), 1423–26.
44. Futo, L. P. *Holz Roh- Werkst.* **1976**, *34*(1), 31–36.
45. Futo, L. P. *Holz Roh- Werkst.* **1976**, *34*(2), 49–54.
46. DeVilbiss, T.; Hon, D. N. -S; Feist, W. C., Paper presented at the 56th ACS Colloid and Surface Sci. Symp., Blacksburg, Va, 1982.
47a. Frey-Wyssling, A. *Schweiz. Z. Forstwes.* **1950**, *101*, 278–282.
47b. Judd, D. B. "Color in Business, Science, and Industry"; Wiley and Sons, Inc.: New York, 1952.
48. Duncan, C. G. *Off. Dig., Fed. Soc. Paint Technol.* **1963**, *35*(465), 1003–12.
49. Sell, J. *Holz Roh- Werkst.* **1968**, *26*(6), 215–22.
50. Sell, J.; Wälchli, O. *Mater. Org.* **1969**, *4*(2), 81–87.
51. Kühne, H.; Leukens, U; Sell, J; Wälchli, O. *Holz Roh- Werkst.* **1970**, *28*(6), 223–29.

52. Schmidt, E. L.; Franch, D. W. *For. Prod. J.* **1976**, *26*(7), 34–37.
53. Brand, B. G.; Kemp, H. T. "Mildew Defacement of Organic Coatings"; The Paint Res. Inst., Fed. of Soc. for Paint Tech.: Philadelphia, Pa, 1973.
54. Browne, F. L. *South. Lumberman* **1960**, *201*(2513), 141–43.
55. Feist, W. C.; Mraz, E. A. *For. Prod. J.* **1978**, *28*(3), 38–43.
56. Borgin, K. *Archit. Builder* **1969**, June-July.
57. Jemison, G. M. *J. For.* **1937**, *35*(5), 460–62.
58. Arndt, U.; Willeitner, H. *Holz Roh- Werkst.* **1969**, *27*(5), 179–88.
59. Böttcher, P. *Holz Roh- Werkst.* **1977**, *35*(7), 247–51.
60. Miniutti, V. P. *U.S., For. Serv. Res. Pap. FPL* 74, 1967.
61. Miniutti, V. P. *Microscope* **1970**, *18*(1), 61–72.
62. Miniutti, V. P. *J. Paint Technol.* **1973**, *45*(577), 27–33.
63. Borgin, K. *J. Microsc. (Oxford)* **1970**, *92*(1), 47–55.
64. Borgin, K. *J. Inst. Wood Sci.* **1971**, *5*(4), 26–30.
65. Borgin, K.; Parameswaran, N.; Liese, W. *Wood Sci. Technol.* **1975**, *9*(2), 87–98.
66. Bamber, R. K. XVI IUFRO World Congress, Oslo, Norway, 1976, Swedish Wood Preserv. Inst. Rpt. No. 124.
67. Bamber, R. K.; Summerville, R. *J. Inst. Wood Sci.* **1981**, *9*(2), 84–87.
68. Chang, S.-T; Hon, D. N.-S; Feist, W. C. *Wood Fiber*, **1982**, *14*(2), 104–17.
69. Feist, W. C. *J. Coat. Technol.* **1982**, *54*(686), 43–50.
70. Feist, W. C. in "Wood Technology: Chemical Aspects," Goldstein, I. S., Ed., ACS SYMPOSIUM SERIES No. 43; ACS: Washington, D.C., 1977; pp. 294–300.
71. Am. Plywood Assoc., 303 Plywood Siding Guide, APA Pamphlet E300, 1981, Tacoma, Wash.
72. Emery, J. A. Am. Ply. Assoc., Tacoma, Wash., personal communication, 1982.
73. Clad, W.; Schmidt-Hellerau, C. *Holz-Zentralbl.* **1976**, *102*(40), 543.
74. Geimer, R. L.; Heebink, B. G.; Hefty, F. V. *U.S. For. Serv. Res. Pap. FPL* 212, 1973.
75. Rowell, R. M.; Feist, W. C.; Ellis, W. D. *Wood Sci.* **1981**, *13*(4), 202–8.
76. Feist, W. C.; Rowell, R. M. in "Graft Copolymerization of Lignocellulosic Fibers," Hon, D. N.-S., Ed; ACS SYMPOSIUM SERIES No. 187; ACS: Washington, D.C., 1982; pp. 349–70.
77a. Hon, D. N.-S.; Ifju, G.; Feist, W. C. *Wood Fiber* **1980**, *12*(2), 121–30.
77b. Ranby, B; Rabek, J. F. "ESR Spectroscopy in Polymer Research"; Springer-Verlag: Berlin, Heidelberg, New York, 1977.
78. Hon, D. N.-S.; Chang, S. T.; Feist, W. C. *Wood Sci. Technol.* **1982**, *16*(3), 193–201.
79. Banerjee, D. K.; Budke, C. C. *Anal. Chem.* **1964**, *36*(4), 792–96.
80. Stein, R. A.; Slawson, V. *Anal. Chem.* **1963**, *35*(8), 1008–10.
81. Witz, G. *Bull. Soc. Ind. Rouen* 1893, *11*, 188.
82. Hon, N.-S. *J. Polym. Sci., Polym. Chem. Ed.* **1976**, *14*(10), 2497–512.
83. Hon, D. N.-S. *J. Polym. Sci., Polym. Chem. Ed.* **1979**, *17*(2), 441–54.
84. Hon, N.-S. *J. Polym. Sci., Polym. Chem. Ed.* **1975**, *13*(8), 1933–41.
85. Hon, N.-S. *J. Polym. Sci., Polym. Chem. Ed.* **1975**, *13*(4), 955–59.
86a. Hon, N.-S. *J. Polym. Sci., Polym. Chem. Ed.* **1976**, *14*(10), 2513–25.
86b. McKellar, J. F.; Allen, N. S. "Photochemistry of Man-Made Polymers"; Applied Scientific Publishers Ltd.: London, 1979.
87. Leary, G. J. *Nature (London)* **1968**, *217*, 672–73.
88. Ludwig, C. H.; Sarkanen, K. V. "Lignin"; Wiley-Interscience: New York, 1971; p. 326.
89. Steelink, C. *Recent Adv. Phytochem.* **1972**, *4*, 239–71.
90. Hon, D. N.-S.; Glasser, W. G. *Tappi* **1979**, *62*(10), 107–10.
91. Hon, D. N.-S.; Feist, W. C. *Wood Sci.* **1981**, *14*(1), 41–48.
92. Kaplan, M. L.; Trozzolo, A. M. in "Singlet Oxygen"; Wasserman, H. H.; Murray, R. W., Eds.; Academic Press: New York, 1979; Chap. 11.

93. Banov, A. "Paints and Coatings Handbook"; Structures Publ. Co.: Farmington, Mich., 1973.
94. Browne, F. L. *U.S., For. Serv., For. Prod. Lab. Rep.* R1053, 1948.
95. Browne, F. L. *U.S., For. Serv., For. Prod. Lab., Misc. Publ.* 629, 1962.
96. Cassens, D. L.; Feist, W. C. N. Cent. Reg. Ext. Publ. No. 132, Coop. Ext. Serv., 1980; Purdue University: West Lafayette, Ind.
97. Cassens, D. L.; Feist, W. C. N. Cent. Reg. Ext. Publ. No. 133, Coop. Ext. Serv., 1980; Purdue University: West Lafayette, Ind.
98. Cassens, D. L.; Feist, W. C. N. Cent. Reg. Ext. Publ. No. 134, Coop. Ext. Serv., 1980; Purdue University: West Lafayette, Ind.
99. Cassens, D. L.; Feist, W. C. N. Cent. Reg. Ext. Publ. No. 135, Coop. Ext. Serv., 1980; Purdue University: West Lafayette, Ind.
100. Hamburg, H. R.; Morgans, W. M. "Hess' Paint Film Defects: Their Causes and Cures", 3rd ed.; Chapman and Hall: London, 1979.
101. Anderson, L. O. *U.S., For. Serv., Inf. Bull.* 311, 1972.
102. U.S. Forest Products Lab., *U.S., For. Serv. Home and Garden Bull.* 203, 1975.
103. Bufkin, B. G.; Wildman, G. C. *For. Prod. J.* **1980**, *30*(10), 37–44.
104. Estrada, N. *For. Prod. J.* **1958**, *8*(2), 66–72.
105. Ashton, H. E. *Can. Paint Finish.* **1974**, *48*(2), 13–16.
106. Tarkow, H.; Southerland, C. F.; Seborg, R. M. *U.S., For. Serv. Res. Pap.* FPL 57, Part I, 1966.
107. Rothstein, E. C. *J. Paint Technol.* **1967**, *39*(513), 621–28.
108. Ljulijka, B. *Holz Roh- Werkst.* **1971**, *29*(6), 224–31.
109. Ashton, H. E. *J. Paint Technol.* **1967**, *39*(507), 212–24.
110. Feist, W. C.; Mraz, E. A. *U.S., For. Serv. Res. Pap.* FPL 366, 1980.
111. Janotta, O. *Holzforsch. Holzverwert.* **1975**, *27*(4), 82–87.
112. Sanderman, W.; Puth, M. *Farbe & Lack* **1965**, *71*(1), 13–26.
113. Hill, R. R. *Timber Rev.* **1974**, *158*(33), 49–51.
114. Zicherman, J. B.; Thomas, R. J. *J. Paint Technol.* **1972**, *44*(570), 88–94.
115. Ashton, H. E. *J. Coat. Technol.* **1979**, *51*(653), 41–52.
116. Ashton, H. E. *J. Coat. Technol.* **1980**, *52*(663), 63–71.
117. Verrall, A. F. *U.S., For. Serv. Tech. Bull.* 1334, 1965.
118. Feist, W. C.; Mraz, E. A. *U.S., For. Serv. Res. Note.* FPL 0124, 1978.
119. Banks, W. B. *Br. Wood Preserv. Assoc. Ann.* **1971**, 129–47.
120. Johansson, S. For. Prod. Lab., STFI Series A No. 467, 1977, Sweden.
121. Sell, J. *Holz Roh- Werkst.* **1977**, *35*(1), 75–78.
122. Feist, W. C.; Mraz, E. A. *For. Prod. J.* **1978**, *28*(5), 31–35.
123. Banks, W. B.; Voulgaridis, E. V. *Br. Wood Preserv. Assoc. Ann.* **1980**, 43–53.
124. Feist, W. C.; Mraz, E. A. *U.S., For. Serv., Res. Pap.* FPL 366, 1980.
125. Banks, W. B. *Wood Sci. Technol.* **1973**, *7*(4), 271–84.
126. Meierhofer, U. A.; Sell, J. *For. Prod. J.* **1977**, *27*(9), 24–27.
127. Voulgaridis, E. V.; Banks, W. B. *Int. J. Wood Preserv.* **1979**, *1*(2), 75–80.
128. Voulgaridis, E. V.; Banks, W. B. *J. Inst. Wood Sci.* **1981**, *9*(2), 72–83.
129. Sell, J. *Holz Roh- Werkst.* **1982**, *40*(6), 225–32.
130. Sell, J.; Meierhofer, U. A. *Holz Roh- Werkst.* **1974**, *32*(10), 390–96.
131. Sell, J.; Krebs, U. *Holz Roh- Werkst.* **1975**, *33*(6), 215–21.
132. Sell, J. *Holz Roh- Werkst.* **1978**, *36*(5), 193–98.
133. Sell, J.; Sommerer, S.; Meierhofer, U. A. *Holz Roh- Werkst.* **1979**, *37*(10), 373–78.
134. Black, J. M.; Laughnan, D. F.; Mraz, E. A. *U.S., For. Serv. Res. Note* FPL 046, 1979.
135. Feist, W. C.; Mraz, E. A.; Black, J. M. *For. Prod. J.* **1977**, *27*(1), 13–16.
136. Emery, J. A. Am. Plywood Assoc. Res. Rep. No. 140, 1980, Tacoma, WA.
137. Feist, W. C.; Mraz, E. A. *For. Prod. J.* **1980**, *30*(5), 43–46.
138. Feist, W. C. *U.S., For. Serv. Res. Pap.* FPL 339, 1979.
139. Black, J. M.; Mraz, E. A. *U.S., For. Serv. Res. Pap.* FPL 232, 1974.

140. Feist, W. C. *For. Prod. J.* **1977**, *27*(5), 50–54.
141. Feist, W. C.; Ellis, W. D. *Wood Sci.* **1978**, *11*(2), 76–81.
142. Desai, R. L.; Clarke, M. R. *Can. For. Ind.* **1972**, *92*(12), 47–49.
143. Sell, J.; Muster, W. J.; Wälchli, O. *Holz Roh- Werkst.* **1974**, *32*(2), 45–51.
144. Barnes, H. M. *J. Coat. Technol.* **1979**, *51*(651), 43–45.
145. Pizzi, A. *J. Appl. Polym. Sci.* **1980**, *25*(11), 2547–53.
146. Pizzi, A.; Mortert, D. *Holzforsch. Holzverwert.* **1980**, *32*(6), 150–2.
147. Pizzi, A. *J. Polym. Sci., Polym. Chem. Ed.* **1981**, *19*(12), 3093–121.
148. Kubel, H.; Pizzi, A. *J. Wood Chem. Technol.* **1981**, *1*(1), 75–92.
149. Williams, R. S.; Feist, W. C. paper presented at the 56th ASC Colloid and Surface Sci. Symp., 1982, Blacksburg, Va.
150. Hon, N.-S. *J. Appl. Polym. Sci.* **1975**, *19*(10), 2789–97.
151. Yusoff, M. M.; Hon, D. N.-S. 1980, unpublished results.
152. Lehmann, W. F. "Proc. 2nd Wash. State Univ. Particleboard Symp.," Pullman, Wash., 1968, pp. 275–306.
153. Meierhofer, U. A.; Sell, J. *Holz Roh- Werkst.* **1975**, *33*(12), 443–50.
154. Meierhofer, U. A.; Sell, J. *For. Prod. J.* **1977**, *27*(9), 24–27.
155. Sell, J.; Sommers, S.; Meierhofer, U. A. *Holz Roh- Werkst.* **1979**, *37*(10), 373–8.
156. Deppe, H. J.; Schmidt, K. *Holz Roh- Werkst.* **1979**, *37*(8), 287–94.
157. Deppe, H. J.; Schmidt, K. *Holz Roh- Werkst.* **1981**, *39*(4), 139–48.
158. Gunn, D. J. *J. Oil Colour Chem. Assoc.* **1980**, *63*(1), 28–33.
159. Boxall, J. *J. Oil Colour Chem. Assoc.* **1981**, *64*(4), 135–9.
160. Boxall, J.; Worley, W. *J. Oil. Colour Chem. Assoc.* **1981**, *64*(2), 75–82.
161. Emery, J. A. *J. Coat. Technol.* **1981**, *53*(673), 61–67.
162. Emery, J. A. Am. Plywood Assoc. Rep. No. PT 81–9, 1981, Tacoma, WA.
163. Schneider, M. H. *J. Paint Technol.* **1972**, *44*(564), 108–10.
164. Schneider, M. H.; Cote, W. A., Jr. *J. Paint Technol.* **1967**, *39*(511), 465–71.
165. Schneider, M. H. *J. Paint Technol.* **1970**, *42*(547), 457–60.
166. Schneider, M. H. *J. Oil Colour Chem. Assoc.* **1979**, *62*(11), 441–4.
167. Schneider, M. H. *Wood Sci. Technol.* **1980**, *14*(2), 107–14.
168. Schneider, M. H. *J. Coat. Technol.* **1980**, *52*(65), 64–67.
169. Desai, R. L.; Cote, W. A. *J. Coat. Technol.* **1976**, *48*(614), 33–37.

RECEIVED for review May 19, 1983. ACCEPTED August 8, 1983.

DEGRADATION CHEMISTRY

Biological Decomposition of Solid Wood

T. KENT KIRK

U.S. Department of Agriculture, Forest Service, Forest Products Laboratory, Madison, WI 53705

ELLIS B. COWLING

School of Forest Resources, North Carolina State University, Raleigh, NC 27605

Decomposition of wood is an important part of the carbon cycle of nature. Decomposition is caused by fungi, insects, and marine borers that use the wood as food or shelter, or both. Lignin in wood provides a physical barrier to enzymatic decomposition of cellulose and hemicelluloses. This barrier is breached mechanically by insects and marine borers, biochemically by white- and soft-rot fungi, and possibly by small nonenzyme catalysts in the case of brown-rot fungi. Cellulose is degraded by endo- and exo-glucanases and β-glucosidases, hemicelluloses by endo-glycanases and glycosidases, lignin by nonspecific enzymes, and perhaps by nonenzymatic, oxidative agents. Rapid strength loss occurs with all decay fungi, but especially with brown-rot fungi. Strength loss due to insect attack is roughly proportional to the amount of wood removed. Fungal decomposition of wood can be prevented by keeping it below its fiber-saturation moisture content (approximately 27% of its dry weight) and by using the heartwood of naturally durable woods (species) or preservative-treated wood. Useful application of wood-decomposing fungi is limited currently to production of edible mushrooms. Potential applications include biological pulping, pretreatment for enzymatic conversion of wood to sugars, and waste treatment. Many aspects of wood biodecomposition have not been researched adequately.

Millions of tons of wood are produced every year in the forests of the world. Observation, however, tells us that the sum-total of wood upon the surface of the earth remains fairly constant from year to year and from century

This chapter not subject to U.S. copyright.
Published 1984, American Chemical Society

> to century. We must, therefore, conclude that there are destructive agencies at work by which millions of tons of wood are destroyed annually. Regarded in this light the problem of what these destructive agencies are, and how they act, becomes of general scientific and economic interest.
>
> A. H. R. Buller,
preeminent mycologist,
1906 (1)

BULLER RECOGNIZED THAT WOOD IS THE MOST ABUNDANT ORGANIC MATERIAL on earth. He also recognized that the trees that form wood through photosynthetic processes, and the fungi and other "destructive agencies" that destroy wood through respiratory processes are engaged in a never-ending cycle of biosynthetic and biodecompositional forces. These relationships are shown by the following simplified reaction for the predominant part of the carbon cycle of the earth:

Photosynthesis by trees and
other plants

$$6n CO_2 + 5n H_2O + 677{,}000n \text{ calories} \rightleftarrows (C_6H_{10}O_5)_n + 6n O_2 \quad (1)$$

Respiration by fungi and
other destructive organisms

Life as we know it would stagnate for lack of atmospheric carbon dioxide in about 20 years if wood destruction were to cease while photosynthesis continued unabated (2). Thus, the biological decomposition of wood is both a great blessing and a serious limitation to the usefulness of wood.

Man has probably always recognized that wood disintegrates on the forest floor and in other moist environments. But it was only a century ago, in 1878, that wood decay was recognized as a biological process. The pioneering German forest pathologist Robert Hartig was the first to prove that fungi are the cause rather than the product of wood decay (3). Forty years later Oshima (4) demonstrated that some insects digest the structural polymers of wood.

This chapter presents an overview of the biological decomposition of wood. It begins with a brief description of the major types of wood destruction and their causal agents, and it continues with a description of the progressive changes that take place in wood as it is decomposed. Special emphasis is given to the chemistry and biochemistry involved. The chapter ends with a brief treatment of how wood in use can be protected from decomposition and some beneficial uses of wood-decomposing organisms.

We have used the terms *decomposition* and *degradation* to refer to the conversion of one or more of the structural polymers of wood to simpler molecules. *Degradation* can also be used to mean *deterioration*, i.e., to decrease the value of wood for some use; we have used only *deterioration* in this narrower sense. *Decay* and *rot* refer to the fungal decomposition of wood.

Hundreds of research papers on wood biodecomposition have been published since the pioneering works of Hartig (3) and Buller (1). We aim to introduce the reader to the principles that those investigations have developed. With this in mind we have categorized and generalized, but we hope that we have not oversimplified. The biodecomposition processes are quite complex biologically, chemically, and biochemically. We have pointed to some of the unknowns in a field that retains many. In Table I we have listed a number of reviews that can be consulted for details.

Susceptibility and Resistance

When wood is laid down by the cambium of a living tree, two major types of wood cells are formed—thick-walled fiber cells that make wood strong and thin-walled parenchyma cells in which reserve foods are stored. Wood fiber cells die a few days or weeks after they are formed and lose their cytoplasmic contents as soon as they become functional in water transport. Thus, mature wood fiber cells consist almost entirely of cell wall polymers—cellulose, hemicelluloses, and lignin. For this reason, wood fiber cells can be degraded only by organisms that have the ability to decompose these structurally complex high-polymeric materials.

Table I. Some Reviews on the Biological Decomposition of Wood

Subject	References
General, historical	5–10
Insects, marine borers	6, 11–14
Chemical, physical changes during decomposition	5, 6, 15–18
Cellulose and hemicellulose decomposition	19–27
Lignin decomposition	28–33
Control of biodecomposition	Chapter 11 this book, 34–40
Uses and potential uses of biocomposing organisms	26, 41, 42

By contrast, wood-storage cells remain alive for many years and only lose their cytoplasmic contents when sapwood is transformed into heartwood. The sugars, starch, amino acids, and proteins in the wood-storage cells make sapwood highly susceptible to invasion by a large number of fungi and bacteria that can use the reserve food materials but cannot attack the more complicated cell wall polymers.

The heartwood of certain species of trees is moderately to highly resistant to decomposition even by organisms that can degrade the cell wall polymers. Highly resistant species include trees such as cypress, various cedars, and osage-orange; moderately resistant species include white oak, Douglas-fir, and certain pines. Resistance is caused by phenols, terpenes, alkaloids, and other substances that are deposited in heartwood and are toxic to wood-destroying fungi, bacteria, insects, and marine borers (43). Because these toxic substances do not occur in sapwood, the dead sapwood of all tree species is highly susceptible to biological decomposition. It should be noted that the living sapwood in trees is resistant to decay by virtue of active defense mechanisms; the heartwood is actually more susceptible than the living sapwood.

Although numerous fungi and some insects can cause decomposition of the dead heartwood tissues inside living trees, these organisms rarely continue to cause decomposition of timber products after the trees are harvested. Despite much speculation about why this is so, no scientific explanation of this phenomenon has been provided to date (44).

Types of Wood Deterioration

Table II summarizes the major types of wood deterioration and the causal organisms. The following discussion deals with deterioration without cell wall decomposition; the bulk of the chapter deals with deterioration with decomposition of cell walls.

Deterioration Without Decomposition. When fresh-cut lumber or veneer is properly air-seasoned, the stored food materials in the sapwood are soon depleted by the respiratory processes of the wood parenchyma cells themselves. If drying is delayed, however, the fresh-cut wood can be invaded by so-called sap-stain fungi and algae, or by bacteria and molds that develop over the surface or penetrate deep into the sapwood by growing through the ray cells from one wood storage cell to another. These organisms use the contents of the wood storage cells as food and thus do not affect the strength of wood seriously; they primarily discolor the wood or alter its permeability.

When fresh-cut wood is kiln-dried immediately, the living cells of the sapwood are killed by the heat and the reserve foods are

Table II. Types of Biological Deterioration of Wood and the Organisms Responsible

Type of Deterioration	Organism(s)
Deterioration without decomposition	
Loss of stored food reserves	Living wood cells in sapwood
Mechanical boring, pecking, cutting	Insects, birds, mammals
Stains	Fungi
Surface discoloration	Fungi, algae
Pit membrane destruction	Bacteria, fungi
Decomposition of structural polymers	
Mechanobiochemical	Insects, marine borers
Biochemical (decays)	Fungi

retained in the wood storage cells. If kiln-dried wood becomes wet again, these stored foods can again become substrates for growth of discoloring fungi and bacteria.

When fresh-cut logs are converted quickly into large piles of chips, the living cells of the sapwood, together with the fungi and bacteria mentioned above, rapidly convert the stored food reserves into carbon dioxide (CO_2), water, and heat (see Reaction 1 for respiration) (45). If this metabolic heat is not dissipated, the pile becomes hot, and under conditions of very poor ventilation can lead to spontaneous combustion (46). For all of these reasons, fresh-cut sapwood must be considered to be alive and, therefore, must be handled as a perishable raw material.

The most common discoloring organisms are of two general types—surface molds with colored spores, and algae that grow on wood surfaces; and fungi with dark-colored hyphae that discolor the wood interior as they penetrate deep into sapwood. *Aspergillus* spp. and *Penicillium* spp. are among the most common surface molds. Discolorations caused by these fungi usually are so superficial that they can be removed by brushing, planing, or sanding. *Ceratocystis* spp. are among the most common deep-penetrating sapstain fungi (47). These discolorations usually cannot be removed even by vigorous bleaching chemicals.

Bacillus polymyxa (Prazmowski) Macè and certain other bacteria (48), as well as stain fungi and some molds such as *Trichoderma viride* Pers. ex Fr. (49,50), degrade the pectin membranes in the bordered pits between wood cells. This degradation greatly increases the permeability of the wood to water and organic solvents. Increased permeability is a problem in wood finishing but can be a help in the penetration of pulping chemicals and preservatives into sapwood.

Mechanical disintegration of wood is caused by many species of insects, birds, and some mammals. In some cases this disintegration can be quite serious.

Deterioration with Decomposition. The susceptibility of the wood cell wall polymers to biological decomposition is determined largely by their accessibility to enzymes and other metabolites produced by wood-destroying fungi or, in the case of certain insects and marine borers, by microorganisms that live in the digestive tracts of these animals. Direct physical contact between the enzymes or other metabolites and the wood cell wall polymers is prerequisite to hydrolytic or oxidative degradation. Because the cellulose, hemicelluloses, and lignin are all water-insoluble polymers and are deposited in wood cell walls in intimate physical admixture with each other, this necessary physical contact can be achieved only by diffusion of the enzymes or other metabolites into this complex matrix or by fine grinding of the wood prior to digestion.

The crucial structural component of wood governing wood's biological decomposibility is lignin. In wood, the cellulose microfibrils are coated or overlayered by hemicelluloses which in turn are under a lignin sheath (51). The lignin is covalently bonded to, and to some extent physically intermixed with, the hemicelluloses; the bonds between lignin and hemicelluloses are probably infrequent (52). Whatever the exact relationship between the hemicelluloses and lignin, the lignin physically prevents enzyme access to both the hemicelluloses and cellulose. Digestibility of solid wood and other intact lignified tissues (lignocelluloses) is largely a function of lignin content (Figure 1).

Three biological mechanisms have evolved for overcoming the lignin barrier: (1) insects and marine borers physically disrupt the barrier by grinding the wood very finely; (2) some microorganisms, primarily higher fungi, decompose lignin and thus expose the polysaccharides; and (3) certain other higher fungi apparently secrete nonenzymatic cellulose-depolymerizing agents that are small enough to penetrate the lignin sheath. Mechanism 1 permits *mechano-biochemical decomposition* of solid wood; Mechanisms 2 and 3 permit *biochemical decomposition*. We will discuss each mechanism in turn.

Mechanobiochemical Decomposition

Circumventing the lignin barrier to enzymatic digestion of the polysaccharides occurs when wood is finely ground. Below a certain wood particle size, the polysaccharides (celluloses and hemicellu-

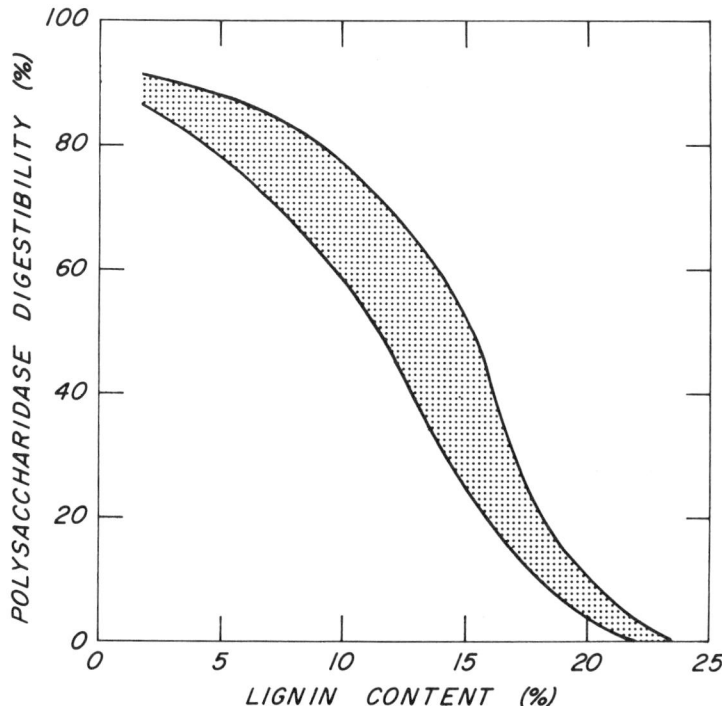

Figure 1. Digestibility of wood by mixtures of cellulases and hemicellulases is largely a function of lignin content (adapted from Reference 53).

loses) should become maximally digestible by enzymes. This size probably varies somewhat with lignin content and perhaps distribution, and hence wood species. Maximum digestibility has been achieved with some woods (sweetgum, red oak, aspen) by vibratory ball milling, but with other woods (red alder, conifers) this milling technique has other effects that adversely affect digestibility (54). The effect of particle size reduction on enzyme accessibility deserves further study with other milling procedures. Virtanen and coworkers (55) first demonstrated the effect of reducing particle size on digestibility by showing that cellulolytic bacteria unable to degrade intact wood are able to utilize fine sander dust. Pew (56) later demonstrated that fine milling makes wood susceptible to mixtures of cellulases and hemicellulases.

The effect of fine grinding had already been demonstrated by many insects and marine borers whose mouth parts, augmented in some cases by internal grinding organs, reduce wood to a digestible particle size. Cellulases and hemicellulases in the guts digest the exposed polysaccharides; excreta are enriched in lignin. Digestion in many insects is incomplete, reflecting failure to grind the wood finely enough, absence of a full enzyme complement, insufficient residence time, or other factors. Certain wood-boring insects, including ambrosia beetles, lyctus beetles, carpenter ants, and carpenter bees do not digest the structural polymers of wood. Wood passes through the gut of the lyctus beetles, but only the easily digested nonstructural materials—primarily starch in parenchyma cells—are removed (6, 11, 12). The ambrosia beetles, bees, and ants do not ingest wood.

Some insects, such as the Indian longhorn beetle, *Stromatium barbatum* Fabricius (57), and the common marine borers *Limnoria tripunctata* Menzies (an isopod) (58) and *Bankia setacea* Tryon (a mollusk) (59), are thought to have endogenous cellulases (and perhaps other polysaccharide hydrolases). Termites and most of the other wood-digesting insects rely on polysaccharolytic microbes in their guts. The beetle *Stromatium barbatum* (57) and the marine borer *Bankia setacea* (60) are thought to utilize both their own cellulases and those of gut microbes.

Wood-decomposing insects and the marine borers commonly digest only the cellulose and hemicelluloses. Reports of extensive decomposition of lignin (6, 60) deserve further study with modern techniques. Limited decomposition of lignin, however, probably does occur in some insects. One of the most convincing reports is that conversion of ^{14}C lignin to $^{14}CO_2$ in the gut of the termite *Nasutitermes exitiosus* (Hill) was demonstrated (61). Because anaerobic decomposition of lignin has not been observed (62, 63) it is probable that some oxygen is present in the guts of *Nasutitermes* (61) and in those of certain other insects that reportedly digest lignin.

Table III lists representative wood-decomposing insects and marine borers and summarizes some features of their action on wood. Figure 2 shows the extensive damage caused by some of these wood-decomposing animals. Only a few detailed studies of the chemistry and biochemistry of wood degradation by the insects and marine borers have been made. Because the animals derive nourishment from the structural polymers of wood, the subject is of practical significance and deserves more research attention.

Biochemical Decomposition: The Wood Decays

Types of Decay. As shown in Figure 3, wood decay fungi can be divided into three classes based on the type of decay they cause: white, brown, and soft rots. The white rots can be further separated into pocket rots and uniform rots (Figure 3). In North America, the white and brown rots are caused by about 1700 species of wood-decaying fungi in the class Basidiomycetes; over 90% of these cause the white rot type of decay (67). Early reports of non-basidiomycete white rots (e.g., decay by *Xylaria* spp.) may be incorrect (26). White- and brown-rot fungi are closely related, which makes the evolutionary basis for their very different effects on wood most intriguing.

Soft rots are caused by fungi in the classes Ascomycetes and Fungi imperfecti. Most are normally found in the soil or in marine environments (68). The list of soft-rot fungi is long and growing longer. A streptomycete has been shown to cause at least limited soft rot in beech (69); it is likely that other soil bacteria will be shown to cause soft rot. We have observed that certain litter-decomposing basidiomycetes, such as *Collybia butyracea*, cause a slow decomposition that resembles soft rot macroscopically.

Macroscopic and microscopic characteristics of white, brown, and soft rots are summarized in Table IV.

Progressive Changes in Chemical Composition. The chemical composition of wood begins to change as soon as it is colonized by the hyphae of wood-destroying fungi. Initial invasion of the ray cells and vessels (70) is primarily at the expense of soluble sugars, starch, and other carbohydrates. The hyphae then penetrate and become established in virtually every cell of the wood in which environmental and other factors are favorable. This attack results in a progressive depletion of all three structural polymers during white and soft rots, and in the polysaccharides during brown rot (Figure 4). Depletion of lignin roughly parallels loss of polysaccharides in white rot, lags behind polysaccharide depletion in soft rot, and is insignificant in brown rot. A slight initial increase in lignin is frequently seen during brown rot (*see* Figure 4). This increase is probably due to partial oxygenation of the lignin polymer (31, 72).

Some variation in the relative rates of cellulose vs. hemicellulose depletion is observed among various fungi and woods. Marked selectivity for the hemicelluloses has been achieved through genetic manipulation of several white-rot fungi (73) and occurs naturally in some species, e.g., *Polyporus pargamenus* Fr. (74). Selective removal of cellulose is not observed, a consequence of its location within the lignin–hemicellulose sheath. Selective removal of lignin does not occur, apparently because polysaccharides provide energy necessary for lignin decomposition (28, 75–77).

Progressive Changes in Strength Properties. Decay of wood has profound effects on strength properties. Of the various measures of wood strength (78), toughness and the related property of work to maximum load are most sensitive to decay. Toughness, which is also called *impact bending strength,* is relatively easy to measure and has been the most widely studied.

All three types of decay cause losses in toughness and related properties that far exceed their losses in weight. Wood decayed to

Table III. Decomposition of Wood by Some Insects and Marine Borers

Classification (example)	Description of Damage	Notes and References
Insects		
Coleoptera, Anobiidae [*Xestobium rufovillosum* (De G.), "deathwatch beetle"]	Round, powder-filled galleries (<3 mm in diameter); wood becomes riddled (Figure 2)	"Powderpost" beetles; damage done primarily by larvae; limited digestion of cellulose and hemicelluloses (6)
Coleoptera, Cerambycidae [*Hyalotrupes bajulus* (L.), "old house borer"]	Oval-shaped galleries winding irregularly through the wood, galleries powder-filled (6–10 mm in diameter) (Figure 2)	Longhorn beetles; larvae ("round-headed borers") do damage; limited digestion of cellulose and hemicelluloses (11)
[*Stromatium barbatum* (Fabricius) "Indian longhorn beetle"]	As above	Larvae do the damage; extensive digestion of cellulose and hemicelluloses; lignin decomposition reported (57)

Isoptera, Kalotermitidae (*Kalotermes flavicollis* Fabr.)	Galleries, chambers within wood; honeycombing; wood structure thoroughly destroyed (Figure 2)	"Drywood" termites; polysaccharides are extensively digested; lignin may be partly decomposed (*6, 64*)
Isoptera, Rhinotermitidae (*Reticulitermes* spp.)	First attack is on less dense part of wood, so that earlywood is removed preferentially; wood structure is thoroughly destroyed (Figure 2)	"Subterranean" termites; polysaccharides are extensively digested; lignin at least slightly altered (65)
Hymenoptera, Siricidae (*Sirex phantoma*)	Small (4–6 mm in diameter) round galleries tightly packed with frass	"Wood wasps"; some digestion of cellulose (6)
Marine Borers		
Adapedonta, Teredinidae (bivalve mollusks) (*Bankia, Teredo* spp.)	Small entry holes on wood surface; interior riddled with much larger bore holes (Figure 2)	Digest cellulose extensively, and probably hemicelluloses, but not lignin; *Bankia* secretes own cellulase, and also has cellulolytic gut microbes (59)
Isopoda, Limnoridae (*Limnoria tripunctata* Menzies)	Maze of tunnels from surface inward, resulting in progressive erosion (Figure 2)	Preferentially digest cellulose (and probably hemicelluloses); lignin unaltered (66); thought to secrete own cellulases; gut is sterile (58)

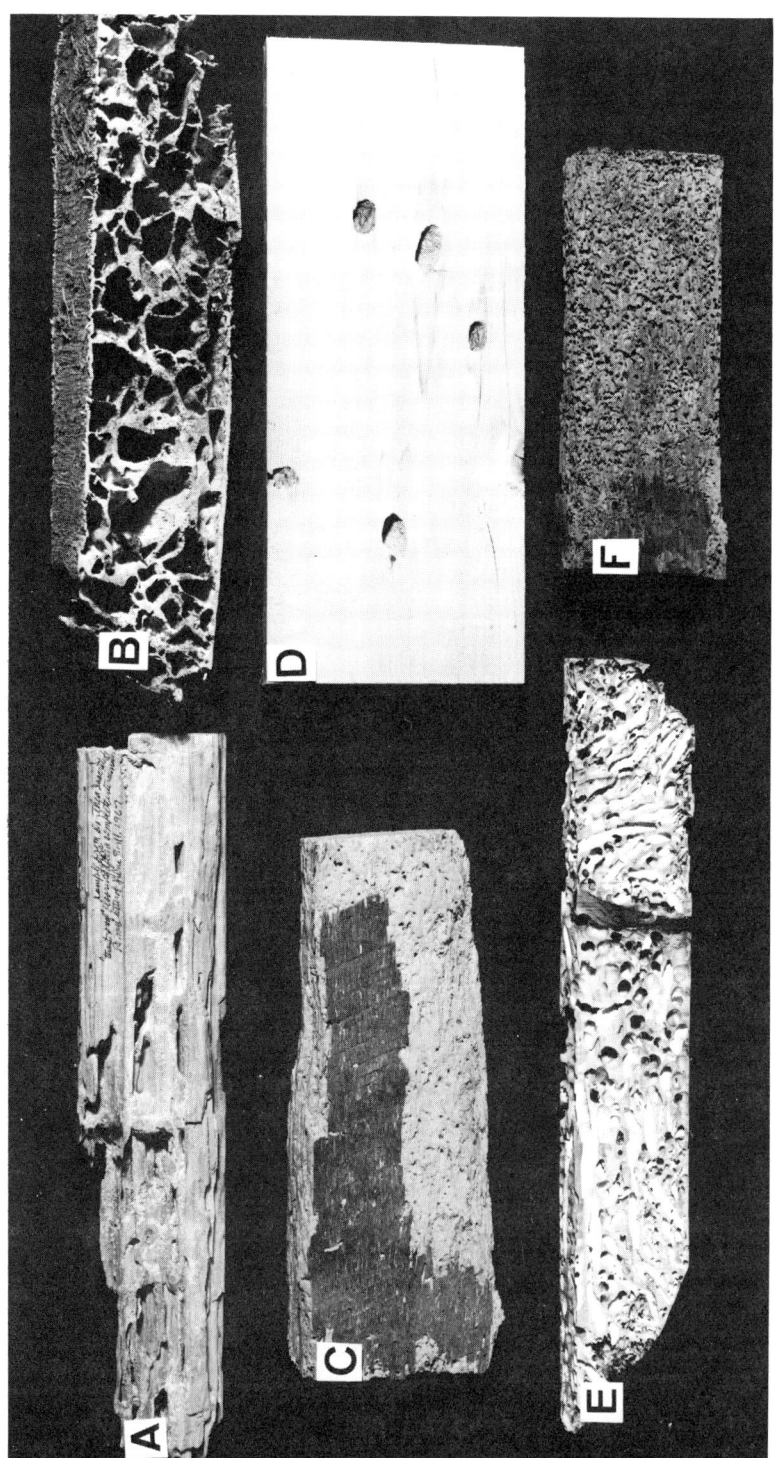

Figure 2. Wood attacked by some insects and marine borers. Key: A, oak flooring by subterranean termites; B, cross section of albizzia board by drywood termites (chamber walls within board are unattacked wood); C, cuangare by anobiid beetle; D, white pine by a longhorn beetle; E, yellow pine by a bivalve marine borer (Teredo sp.); and F, Douglas-fir by an isopod marine borer (Limnoria sp.). Board A is 27 cm long.

Figure 3. Wood partially decomposed by fungi. Key: A, uniform-white rot in northern red oak: decay has permeated sapwood (broad, tangential-face), but is only slowly penetrating into heartwood (narrow, radial-face); B, white oak heartwood decayed by a white-pocket rot fungus; C, western redcedar decayed by a brown-rot fungus (white mycelium is visible on surface); and D, yellow pine decayed by a soft-rot fungus. Board A is approximately 20 cm long.

Table IV. Macroscopic and Microscopic Features of White, Soft, and Brown Rots of Wood

Type of Decay (representative causal organisms)	Macroscopic Features[a]	Microscopic Features[b]
White rot [*Coriolus versicolor* (L. ex Fr.) Quel., *Phanerochaete chrysosporium* Burds.]	Usually not discernible in early stages, wood later becomes bleached or discolored; mottling or flecking is common; some species form white pockets in wood; dark zone lines often observed in nature; wood retains size and shape	Hyphae colonize lumens of cells, penetrate first from cell to cell via pits, later directly via bore holes; progressive decay observed as thinning of cell walls; (in "pocket rots," decay is limited to the developing pockets)
Soft rot (*Chaetomium globosum* Kunze, *Paecilomyces* spp., *Allescheria*	Only water-soaked wood attacked; decay even in early stages characterized by softening and	Hyphae grow longitudinally within the secondary walls; characteristic catenate, spindle-, or dia-

terrestris Apinis)	minute checking of surfaces; dull gray to brown color. Decay progresses gradually from the surface inward	mond-shaped cavities, oriented with cellulose microfibrils
Brown rot [*Gloeophyllum trabeum* (Pers. ex Fr.) Murr., *Poria placenta* (Fr.) Cke., *Merulius lacrymans* (Wulf.) Fr.]	Often not discernible during early stages, but wood rapidly becomes brash; later, wood becomes discolored, finally brown and soft; drying causes extensive checking across the grain, giving a cubical pattern; mycelial fans may be present on surfaces or within the checks	Hyphae colonize lumens of cells, penetrate from cell to cell via pits and often directly via bore holes. As decay progresses, cell walls may shrink and collapse

[a] *See* Figure 3 *and* References 7 and 8.
[b] *See* Reference 70.

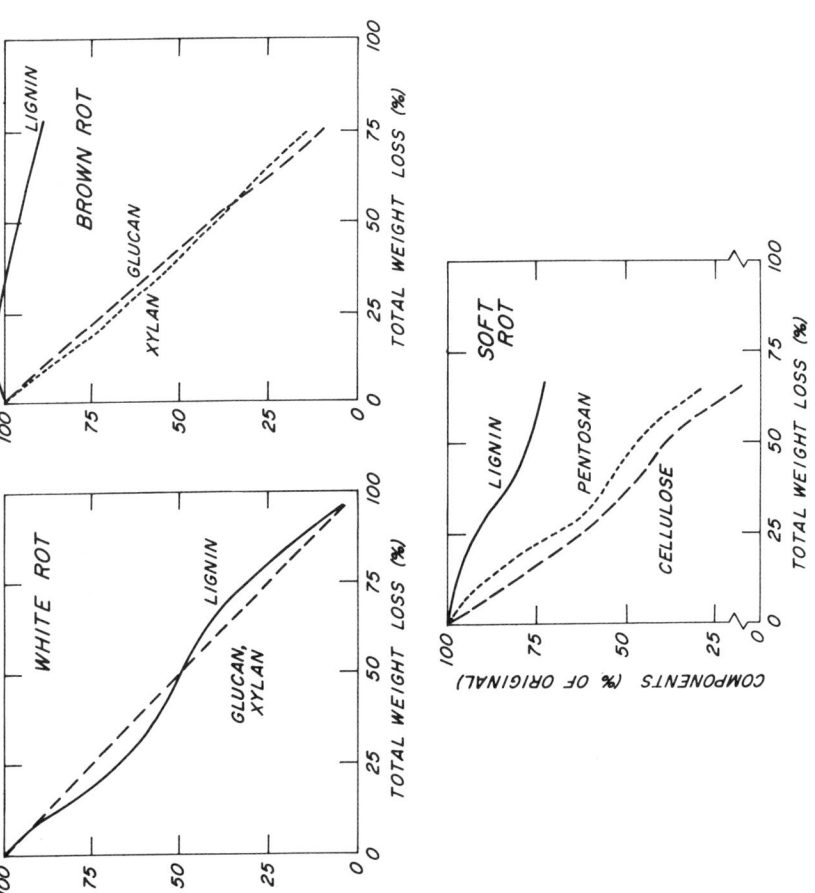

Figure 4. Cellulose (glucan) and hemicelluloses (xylan, pentosan) are depleted during decay by white-, brown-, and soft-rot fungi; lignin is decomposed efficiently only by white-rot fungi. Key: white rot, sweetgum by Coriolus versicolor (L. ex Fr.) Quel. (15); brown rot, sweetgum by Poria placenta (Fr.) Cke. (15); and soft rot, beech by Chaetomium globosum Kunze (71).

weight losses of less than 3% by white-, soft-, or brown-rot fungi frequently has lost over 50% of its strength measured as toughness (18, 79–85). Microscopic observations reveal that this loss is not due to a localized destruction of part of the test piece, but rather to a general decomposition.

It is not known why toughness is so sensitive to decay by all three groups of fungi. Campbell (5) concluded that the reduction is due to a "shortening of the cellulose chain molecules in wood." This is reasonable for brown rot, but unlikely for white rot or soft rot. Cowling (15) demonstrated that the brown-rot fungus *Poria monticola* Murr. [now *Poria placenta* (Fr.) Cke.] causes a sharp reduction in the degree of polymerization (DP) of cellulose during decay of sweetgum wood (Figure 5). In contrast, the white-rot fungus *Polyporus versicolor* L. ex Fr. [now *Coriolus versicolor* (L. ex Fr.) Quel.] only gradually reduces cellulose DP (Figure 5). Levi and Preston (71) showed that the residual cellulose in beech wood is also only gradually decreased in DP during decay by the soft-rot fungus *Chaetomium globosum* Kunze. Thus the basis for toughness loss in white and soft rots lies elsewhere and deserves further research.

The effect of brown-rot fungi on other wood strength properties is much more pronounced than that of white-rot fungi (*see* Reference 18), and reflects cellulose depolymerization. Interestingly, wood species vary substantially in strength loss at a given weight loss by brown-rot fungi (18). The basis for this variation is not known and deserves additional research

Mechanisms of Wood Decay. Microscopy indicates that the enzymes or other agents that degrade the wood cell wall polymers diffuse away from the hyphae, and that they are secreted by the laterals as well as by the growing tips (70, 86, 87). Gelatinous sheaths encase both hyphae and substrate when brown- or white-rot fungi attack isolated cellulose (88). The nature and actual importance of such sheaths in the decay of solid wood need further investigation. The sheaths might provide an optimum environment for enzyme action and might also help prevent loss of enzymes with their precious nitrogen. Because the nitrogen content of wood is quite low, this important element is limiting in the nutrition of wood-decomposing organisms. Its importance in wood decay has been discussed (89).

Studies of the biochemical mechanism of wood decay have, with very few exceptions, been conducted with isolated cellulose, hemicelluloses or lignin, or with appropriate model compounds, rather than with solid wood. The few studies with enzymes produced during growth on or in wood have revealed no peculiarities that would bring into question the results of studies with the isolated components.

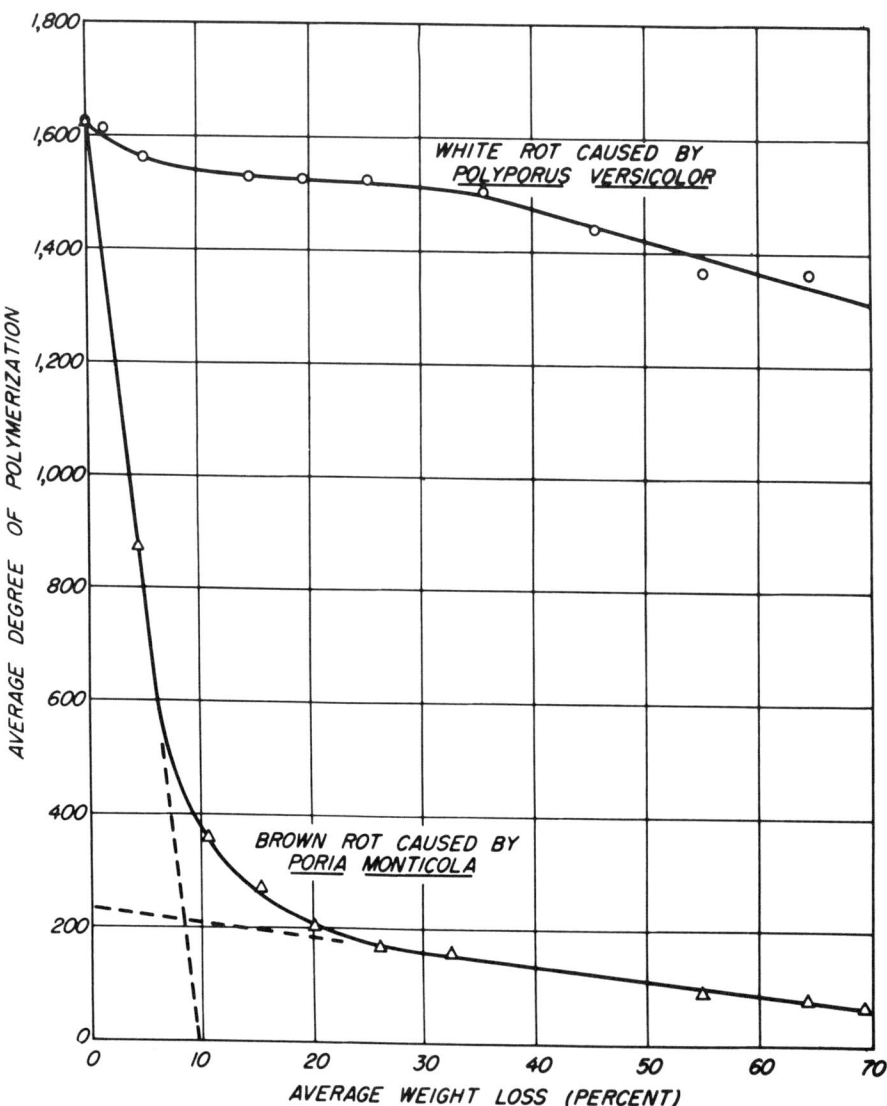

Figure 5. Cellulose in wood is depolymerized early during brown rot, but only gradually during white rot. The data are for sweetgum sapwood decayed by the brown-rot fungus Poria placenta *(Fr.) Cke. (formerly P. monticola Murr.) and for the white-rot fungus* Coriolus versicolor *(L. ex Fr.) Quel. (formerly* Polyporus versicolor *L. ex Fr.) (15).*

Isolated cellulose was not degraded by the brown-rot fungus *Poria placenta* unless the fungus was in contact with whole wood (90). Further study indicated, however, that the wood could be replaced with simple sugars, i.e., that nutrients in the wood were simply serving as starter carbon/energy sources (90).

In the following discussion, we have assumed that wood decay is due to the concerted action of the individual enzyme systems responsible for cellulose, hemicellulose, and lignin decomposition.

CELLULOSE DECOMPOSITION. Decomposition of crystalline cellulose by white-rot fungi and various soil fungi (including the soft-rot fungi), results from the synergistic action of three types of hydrolytic enzymes: *endo*-1,4-β-glucanases, *exo*-1,4-β-glucanases, and β-glucosidases. Presumably, bacterial and animal cellulases are similar. Brown-rot fungi also decompose crystalline cellulose, but most possess no *exo*-glucanase; their unique mechanism of cellulose degradation is discussed later.

The mechanism of cellulose degradation by fungi has been the subject of extensive research (19–21, 23–25), primarily with the mold *Trichoderma viride* and the white-rot fungus *Sporotrichum pulverulentum* Novo. (now *Phanerochaete chrysosporium* Burds.). Results indicate that the *endo*-1,4-β-glucanases (C_x enzymes) act randomly over the exposed surfaces of cellulose microfibrils. Nonreducing termini generated by this action are then hydrolyzed by *exo*-1,4-β-glucanases (cellobiohydrolases, C_1 enzymes), releasing cellobiose. Cellobiose may be cleaved by a β-glucosidase to yield glucose, or in white rot and perhaps other fungi it may be oxidized to cellobionic acid and then cleaved. The *endo*- and *exo*-glucanases act synergistically, perhaps as a loose complex (91). Generally, the *endo*-glucanases and probably the *exo*-glucanases are repressed by high concentrations of monosaccharides (23). Both types of glucanases have molecular weights ranging up to about 75,000 (92, 93), whereas the β-glucosidases are considerably larger (94).

In addition, oxidizing enzymes probably are involved in cellulose decomposition by certain white-rot fungi. *Phanerochaete chrysosporium* possesses cellobiose oxidase, which converts cellobiose to cellobiono-δ-lactone, with O_2 serving as electron acceptor (23). This enzyme is responsible for the more rapid hydrolysis of cellulose under aerobic conditions than under anaerobic conditions, presumably because it removes cellobiose and prevents the transglycosylation reactions and the inhibition of *endo*-glucanase activity that occurs when cellobiose accumulates (23, 73). Similar oxidizing activity is not found in the species of Fungi Imperfecti examined (91). Another enzyme, cellobiose:quinone oxidoreductase, has the same action, but requires quinones as electron acceptors (23). A glucose oxidase also has been implicated in the overall process (23); it oxidizes glucose to gluconolactone with O_2 or quinones as electron acceptors. Presumably, these oxidizing activities regulate the amounts of glucose and cellobiose, ultimately coordinating the rates of cellulose hydrolysis and the metabolism of end products. Phenols, which are intermediates in lignin decomposition, also affect the amounts of endo-glucanase activity in white-rot fungi (23). Figure 6 summarizes the interconversions and regulatory interactions in cellulose hydrolysis by white-rot fungi (23).

Cellulose decomposition by brown-rot fungi is an unusual process. In the early stages of their decay of wood, and in contrast to decay by white-rot fungi, the cellulose is severely depolymerized (see Figure 5). This discovery (15) explained the highly destructive effects of these fungi on wood, as described in the section entitled "Progressive Changes in Strength Properties" (page 463), but it presented an enigma because only a very small percentage of the cellulose in wood is accessible to cellulolytic enzymes. *Enzymatic* degradation must cause a gradual loss in cellulose integrity, as seen in white rot (15).

Because the depolymerizing agent of the brown-rot fungi completely penetrates the crystalline microfibrils, only very small molecules can be responsible (15). In discussing this, Cowling and Brown (92) noted that G. Halliwell (95) had described experiments on the depolymerization of cellulose under physiological conditions with Fenton's reagent (H_2O_2 and ferrous salts). Subsequent studies demonstrated that brown-rot fungi secrete H_2O_2 and that wood contains enough iron for a possible involvement of an $Fe-H_2O_2$ system in cellulose depolymerization (96). Cellulose is in fact oxidized during attack by the brown-rot fungus *Poria placenta* (90). Oxalic acid, which is secreted by brown-rot fungi, can reduce the Fe^{3+} normally present in wood to Fe^{2+}, the active form in Fenton's reagent (97, 98). Figure 7 shows the proposed mechanism for the depolymerization of cellulose. Nicholas et al. (personal communication) demonstrated degradation of ^{14}C-cellulose by *Gloeophyllum trabeum* (Pers. ex Fr.) Murr. through a membrane with a nominal molecular weight limit of 1000.

This initial oxidative depolymerization of cellulose evidently opens up the wood cell wall structure so that cellulolytic and hemicellulolytic enzymes can reach their substrates despite the presence of lignin. Solubility of wood in 1% NaOH increases markedly on brown-rot attack (15) and reflects cellulose depolymerization and the opening up of the wood structure.

Interestingly, enzyme preparations from most brown-rot fungi possess *endo-* but not *exo*-1,4-β-glucanase activity (99–101). These fungi differ from other cellulolytic organisms, too, in that *endo*-glucanase production is not repressed by monosaccharides (102). Still another unusual feature is a large enzyme complex that hydrolyzes carboxymethyl cellulose, xylans, glucomannans, and various glycosides (103).

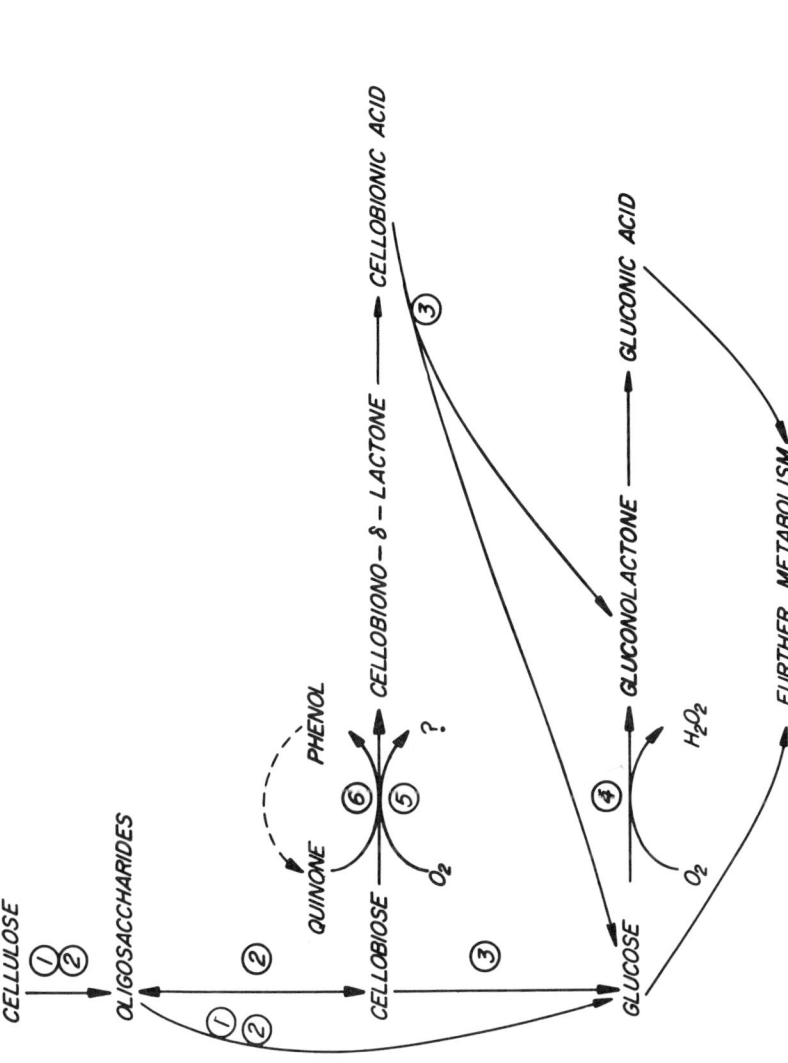

Figure 6. Both hydrolytic and oxidative enzymes participate in cellulose decomposition by the white-rot fungus Phanerochaete chrysosporium Burds. Hydrolytic enzymes: 1, endo-1,4-β-glucanases; 2, exo-1,4-β-glucanase; and 3, β-glucosidase. Oxidative enzymes: 4, glucose oxidase; 5, cellobiose oxidase; and 6, cellobiose:quinone oxidoreductase. Adapted from Reference 23.

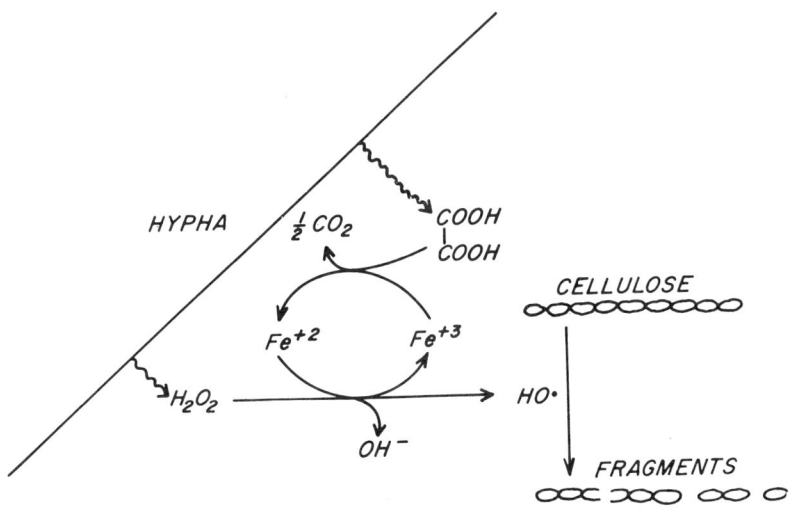

Figure 7. Depolymerization of cellulose in wood by brown-rot fungi might involve Fe^{2+} and H_2O_2 (92, 96, 97).

Although white-rot fungi also secrete H_2O_2 (104), they have not been found to depolymerize cellulose oxidatively. One reason might be that they possess oxalate decarboxylase, which decomposes oxalate, whereas brown-rot fungi apparently do not (98). This problem deserves further investigation.

The cellulolytic system of soft-rot fungi has received relatively little attention. It is probably similar to that of the closely related Ascomycetes and Fungi Imperfecti (19–25, 27), which apparently differs little from that of white-rot fungi, except perhaps in the absence of oxidizing enzymes.

HEMICELLULOSE DECOMPOSITION. Dekker and Richards have reviewed the microbial hemicellulases (22). Wood-rotting fungi produce enzymes capable of hydrolyzing a variety of β-(1 → 4)-linked glycan (mannan and xylan) substrates, as well as various glycosides (94, 101, 105–8). Judging from the relatively little research done, corresponding enzymes from the three groups of decay organisms appear to be similar, and are also similar to those from other microbes (22, 106). To our knowledge, the enzymes responsible for *complete* depolymerization of a wood hemicellulose have not been described in any microbe. This is another subject that needs additional study. *endo*-Glycanases from white-, brown-, and soft-rot fungi apparently all act randomly and produce dimeric and higher oligomeric products (22, 108–11). Uronic acid-substituted oligosaccharides are produced from glucuronoxylan substrates (107–9, 111) but enzymes catalyzing hydrolysis of the xylose–uronic acid linkage have not been reported. A mannanase purified from a brown-rot fungus hydrolyzed both mannose–(1 → 4)-glucose and glucose–(1 → 4)-mannose linkages (110). Glycosidases active on hemicellulose-derived disaccharides also are produced by the wood-rotting fungi (94), but *exo*-hydrolases degrading hemicellulose oligosaccharides have not been reported in these organisms.

The enzyme complex with multiple glycan and glycoside hydrolase activity in the brown-rot fungus *Poria placenta*, mentioned above (103), has a molecular weight of 1.85×10^5; other hemicellulases (*endo*-glycanases) of wood-rotting fungi have molecular weights of 3×10^4–6×10^4 (94).

Information regarding regulation of the synthesis of the hemicellulases is somewhat contradictory (22). Regulation in wood-decomposing fungi, however, has not been the subject of the detailed study that it deserves. Multiple hemicellulase activity is found in culture filtrates of both white- and brown-rot fungi after growth on a variety of substrates (107). Hemicellulase production by the two types of fungi is different, however, in that brown-rot fungi exhibit good hemicellulase (as well as cellulase) activities during growth on simple sugars, whereas white-rot fungi do not (107, 112). Production of the enzymes on simple sugars might be induced in response to hyphal wall constituents following substrate depletion, as is an *endo*-xylanase in the white-rot fungus *Stereum sanguinolentum* Fr. (94). Hemicellulases in the soft-rot fungus *Chaetomium globosum* are induced specifically by their substrates (108).

LIGNIN DECOMPOSITION. Research on the fungal decomposition of lignin has accelerated greatly in recent years. Some of the reactions comprising degradation have been elucidated, and the unusual biochemical and physiological features are being described. Several reviews (28–33) provide specific literature references.

Research with the white-rot fungi has shown that the process is oxidative, that the ligninolytic system is nonspecific, that its rate-limiting component is not induced by lignin, and that depolymerization may not be an obligatory first step. Lignin degradation, therefore, is distinct from cellulose and hemicellulose degradation; indeed, it differs from the biodegradation of all other biopolymers studied. Prominent reactions of lignin polymer degradation are oxidations and oxidative cleavages in the propyl side chains, demethylations of methoxyl groups, and even cleavages in aromatic rings. The chemistry of degradation is being investigated both by characterization of the partially degraded polymer and of low molecular weight degradation products, and through studies of the metabolism of low molecular weight dimeric model compounds representing substructures in the polymer. With the model compounds, the specific reactions that comprise degradation are rapidly being described; for example, Scheme 1 shows the fate of the nonphenolic diarylpropane-type of substructure model in cultures of the white-rot fungus *Phanerochaete chrysosporium* (113–15). Models representing other important substructures have been studied similarly.

Many past studies on the enzymes of lignin decomposition focused on phenol-oxidizing enzymes, such as laccase and peroxidase, produced by white-rot fungi. It is unlikely, however, that this type of activity is important in structural decomposition (33), although the enzymes may have some other role in lignin decomposition (28).

Because the lignin polymer is attacked by extracellular nonspecific oxidizing agents, it is possible that enzymes may not be involved directly, just as cellulose apparently is nonenzymatically oxidized by brown-rot fungi. Hall (116) suggested that diffusible species such as superoxide (O_2^-), derived from molecular oxygen, may participate in lignin decomposition. Research has since indicated that activated oxygen species apparently are commonly produced by the lignin-degrading fungi. Singlet oxygen, an excited state of O_2 reportedly in-

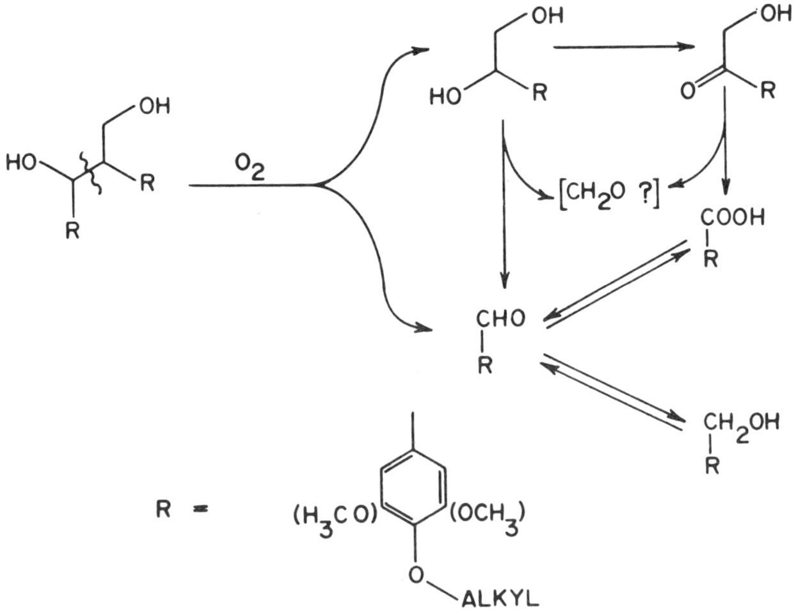

Scheme 1. Model compounds are being used to elucidate the specific reactions of lignin decomposition by white-rot fungi. The fate of a nonphenolic diarylpropane-type of model in ligninolytic cultures of Phanerochaete chrysosporium Burds. is shown (113–15).

volved in other biological reactions, was implicated in lignin biodegradation (117), but was later shown not to be involved (118). Hydroxyl radical (·OH) has now been implicated (119–21), but not proven to be involved. Whether it is or not, good evidence has been gained for participation by its immediate precursor, H_2O_2 (119, 121, 144). Hydroxyl radical is formed by metal reduction of H_2O_2 (see Figure 7).

Findings in this very active research area indicate that enzymes are directly involved. An extracellular enzyme in ligninolytic cultures of *P. chrysosporium* has been discovered (122). It requires H_2O_2 for activity and catalyzes the cleavage of nonphenolic model compounds of both the diarylpropane and the arylglycerol-β-aryl ether types. In the former type of model, the degradative reaction is the initial cleavage shown in Scheme 1. The new enzyme catalyzes the partial depolymerization of lignin. It is not the same as earlier reported "lignin-degrading enzymes" (123–26).

Further metabolism of low molecular weight products of the initial degradation of lignin, however it occurs, is probably via classical modes. Vanillic acid, a prominent product of the fungal degradation of lignin (127), is degraded by substrate-inducible enzymes in *P. chrysosporium* (128, 129).

Although brown-rot fungi are poor degraders of lignin (see Figure 4) they do apparently possess the basic machinery. The main effect they have on lignin is demethylation of arylmethoxyl groups (130), although oxidative changes occur, including some cleavage of aromatic rings (72). Indeed, limited oxidation of aromatic and propyl side chain carbon as well as methoxyl carbon to CO_2 has been demonstrated (131, 132). Extensive depolymerization apparently does not occur (133, 134), and it seems unlikely that the limited degradation of lignin by brown rot fungi is sufficient to open up the wood structure to polysaccharidases. The nonenzymatic oxidative depolymerization of cellulose, discussed in the section entitled "Cellulose Decomposition" (page 473), is evidently what opens up the structure.

Analyses of soft-rotted wood have revealed limited depletion of lignin (71, 135). That these fungi can oxidize lignin to CO_2 was shown by using ^{14}C-lignins (132). Rates did not approach those seen with white-rot fungi, but optimization studies have not yet been conducted. Clearly, the wood-decomposing machinery of the soft-rot fungi has received too little attention.

Control and Uses of Wood-Decomposing Organisms

Control. When used properly, wood will retain its strength and other desirable properties for many centuries. When wood is used improperly, however, it can be decomposed by various organisms, as discussed in this chapter.

The cardinal rules for proper use of wood can be stated simply: keep wood dry, and, if you can't keep it dry, use naturally durable or preservative-treated wood. The first rule is based on a simple biological principle: liquid water is needed in wood cells to provide a medium for diffusion of the enzymes or other metabolites by which wood-decomposing organisms digest the wood substance. If there is no liquid water present inside the wood cells, there will be no medium for diffusion, and therefore no biological decomposition except for certain insects of relatively minor importance. Thus, as long as wood is kept below its fiber-saturation point (about 27% of its dry weight), it will never decay.

A few brown-rot fungi, notably *Serpula incrassata* (Berk. and Curt.) Dank. and *Merulius lacrymans* (Wulf.) Fr. have the unusual ability to conduct liquid water from moist soil or other sources of moisture into dry wood (36). Similarly, subterranean termites can attack very dry wood and, if they have access to water, can transport it through the tubes they construct between moist soil and wood. But chemical soil treatments around wood buildings can prevent attack by even these organisms. Construction practices that thwart both fungal decay and insect attack have been described (36, 38).

The second cardinal rule is based on an equally simple biological principle: some chemicals inhibit living organisms. When wood must be used in moist environments, use of naturally durable or preservative-treated timber will provide long-lasting protection against biological decomposition. Although the heartwood of many tree species is naturally durable (36, 43, 136, 137), sapwood of all tree species is highly susceptible to decomposition. Most construction timbers of temperate regions are sapwood and require preservative treatment for use in moist environments.

The chemicals that have found wide use and acceptance as wood preservatives, primarily creosote, pentachlorophenol, chromated copper arsenate, and ammoniacal copper arsenate, are broad-spectrum pesticides. To achieve greater specificity, advantage could be taken of the unique physiological features of the causal organisms. One such feature is digestion of wood. Treatments that make wood a nonsubstrate (Chapter 4), or chemicals that interfere with the synthesis, secretion, or activity of the wood-decomposing enzymes can be envisioned. A chemical that prevents the synthesis of chitin, which is an essential component of both insects and fungi, is a goal of current research (138). Many other possibilities for more specific interference with the growth or activity of the wood-decomposing organisms could result from a better understanding of the physiology and biochemistry of the decomposition processes and of the organisms responsible.

Uses and Potential Uses (26, 139). Although wood partly degraded by fungi or attacked by insects sometimes has increased aesthetic appeal for use in decorative paneling, wooden bowls, or other household items, a decrease in attractiveness and usefulness is more often the case. Thus biological decomposition usually leads to a decrease rather than an increase in the value of wood.

However, deliberate conversion of solid wood by the agents of biodecomposition is done commercially and has considerable additional potential. The world's second most important commercial mushroom, comprising 20% of world sales (140–42), is shiitake (*Lentinus edodes*) (Figure 8). Cultivated in Asia, primarily on oak logs, shiitake is a major food in Japan and that nation's largest agricultural export. Several other commercial mushrooms are also grown on solid wood (Table V). These mushrooms have much potential in the West where they are currently of minor importance.

Most, and perhaps all, of these edible mushroom-forming fungi cause the white-rot type of wood decomposition. *Lentinus edodes* and some of the others increase the ruminant digestibility of wood—sometimes to over 60% (26)—because they remove the lignin and hemicelluloses before the cellulose. Thus, they have potential for direct conversion of wastewood into feed for ruminants. Although the residue from shiitake production is sometimes fed to cattle in Japan, a process aimed at converting wood to cattle feed by this fungus or other fungi apparently has not been developed.

The possibility of using brown-rot fungi to open up the wood structure for ruminants or for conversion via enzymatic hydrolysis or direct fermentation has received virtually no research attention.

Another potential use of white-rot fungi that has received some research attention is in biomechanical pulping. Treatment of wood to a weight loss of less than 3% can lower the energy requirements for subsequent mechanical pulping by more than 20% (139). Mechanical pulps make up a growing share—now about 10%—of U.S. pulp production. Energy consumption is high and makes any treatment attractive that decreases energy demand. We are struck by the similarity in the rapid loss in toughness (*see* the "Progressive Changes in Strength Properties" section) and the rapid decrease in energy requirements for mechanical pulping during white rot. The physical–chemical basis for this decrease should be investigated; it might actually have little to do with lignin degradation, which has been assumed to be the basis (139). Biomechanical pulping deserves additional investigation.

Production of microbial chemical products during growth on

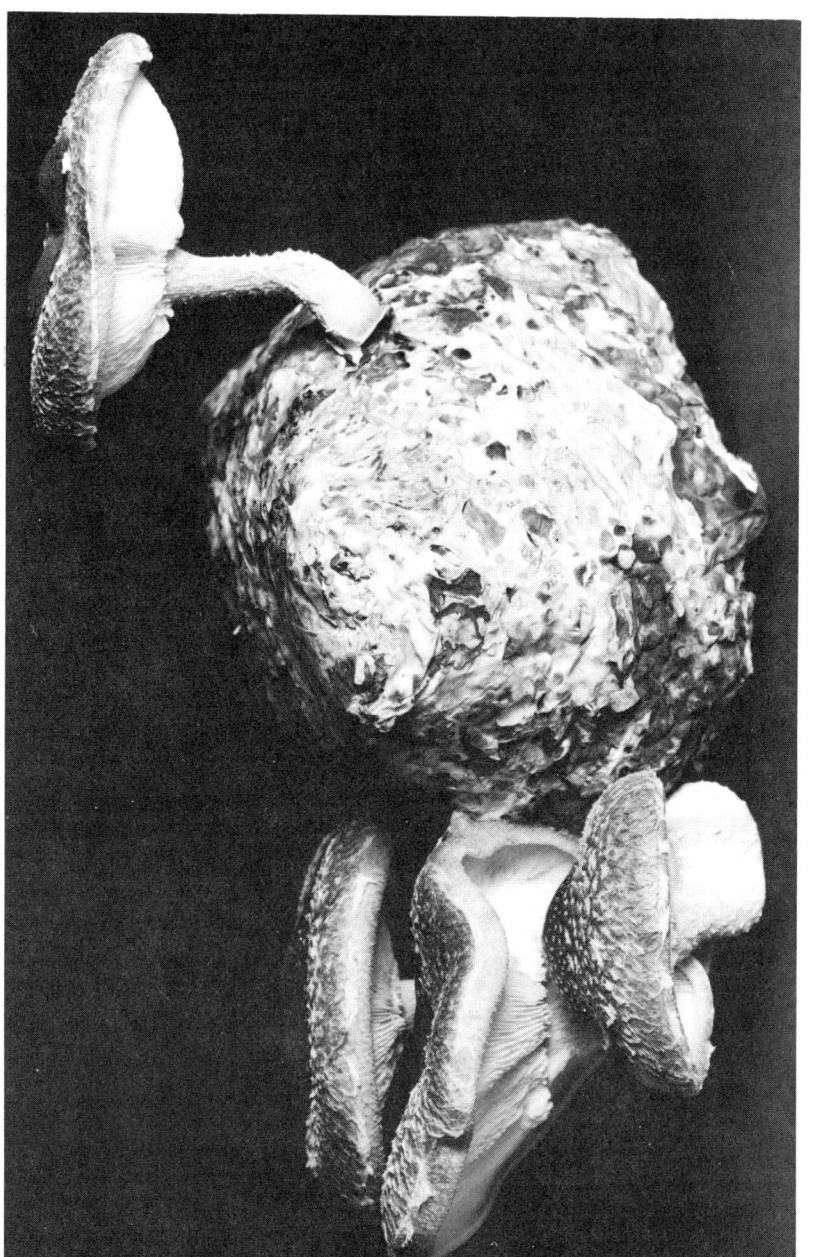

Figure 8. Shiitake (Lentinus edodes (Berk.) Sing.), the Japanese forest mushroom, is grown commercially on oak logs in Asia but can be grown on shredded wood-bark mixtures as shown (average cap diameter here is approximately 10 cm). The world's second most important commercial mushroom, shiitake is Japan's largest agricultural export (142).

(Courtesy of G. F. Leatham, unpublished data.)

Table V. Some Commercial Mushrooms Cultivated on Solid Wood (*140, 143*)

Species	Substrates
Auricularia spp.	Hardwood logs
Flamulina velutipes (Fr.) Sing.	Sawdust
Pholiota nameko (T. Ito) S. Ito et Imai	Hardwood logs
Pleurotus spp.	Hardwood logs, sawdust
Tremella fuciformis Berk.	Hardwood logs
Lentinus edodes (Berk.) Sing.	Hardwood logs, sawdust

NOTE: Adapted from References 140 and 143.

solid wood, although possible, seems unlikely to become important, because of the availability of more practical substrates.

Altering the properties of wood components for particular uses is another possible use of wood-decomposing microbes. As an example, in their attack on solid wood, brown-rot fungi leave a lignin residue that is enriched in phenolic hydroxyl groups (*72, 130*); such lignin might serve well in phenolic adhesives.

One additional potential application of wood-decomposing fungi is in waste treatment. Although not directed at solid wood, such applications may be possible simply because the fungi have evolved the capacity to degrade such a complex solid substrate. The potential use of the lignin-decomposing system of white-rot fungi to decolorize lignin-derived wastes from pulp bleaching has been investigated with promising results (*139*). The lack of specificity of the oxidative ligninolytic system suggests a broader applicability than just to wood-derived wastes.

Acknowledgments

We thank T. L. Amburgey, Forest Products Utilization Laboratory, Mississippi State University, Mississippi State, Mississippi, and G. R. Esenther, W. E. Eslyn, T. L. Highley, B. R. Johnson, and G. F. Leatham, all of the U.S. Forest Products Laboratory, Madison, Wisconsin, for critically reviewing portions of the manuscript. Samples used in Figures 2 and 3 were provided by Esenther and Eslyn. Leatham provided Figure 8. The expert clerical assistance of Jane D. Kohlman is gratefully acknowledged.

Literature Cited

1. Buller, A. H. R. *J. Econ. Biol.* **1906**, *1*, 101.
2. Rabinowitch, E. I. "Photosynthesis"; Interscience: New York, 1945; Vol. I, 599.

3. Hartig, R. "Die Zerzetzungserscheinungen des Holzes der Nadelholzbaüme und der Eiche"; Springer: Berlin, 1878; 151 p.
4. Oshima, M. *Philipp. J. Sci.* **1919**, *15*, 319.
5. Campbell, W. G. in "Wood Chemistry"; Wise, L. E.; Jahn, E. C., Eds.; Reinhold: New York, 1952; Vol. 2, pp. 1061–116.
6. Seifert, K. *Holzforschung* **1962**, *16*, 78.
7. Cartwright, K. St. G.; Findlay, W. P. K. "Decay of Timber and Its Prevention"; Her Majesty's Stationery Office: London, 1958; 332 p.
8. Duncan, C. G. *U.S., For. Serv., For. Prod. Lab. Rep. No. 2173* **1960**, 28.
9. Dickinson, C. H.; Pugh, G. J. F. "Biology of Plant Litter Decomposition"; Academic Press: New York, 1974; 2 vols.
10. "Wood Deterioration and Its Prevention by Preservative Treatments"; Nicholas, D. D., Ed.; Syracuse Univ. Press: New York, 1973; Vol. 1, 380 p.
11. Moore, H. B. "Wood-Inhabiting Insects in Houses: Their Identification, Biology, Prevention, and Control"; *Report, Interagency Agreement IAA-27-75, U.S. Dep. Agric., For. Serv., and U.S. Dep. Housing Urban Dev.* Superintendent of Documents: Washington, D.C., 1979; 133 p.
12. Coulson, R. N.; Lund, A. E. "Wood Deterioration and Its Prevention by Preservative Treatments"; Nicholas, D. D., Ed.; Syracuse University Press: New York, 1973; Vol. 1, p. 277–305.
13. Hochman, H. "Wood Deterioration and Its Prevention by Preservative Treatments"; Nicholas, D. D., Ed.; Syracuse University Press, New York, 1973; Vol. 1, p. 247–75.
14. Clapp, W. F.; Kenk, R. *Off. Nav. Res., Dep. Navy, Rep. ACR–74*, 1135 p. (U.S. Govt. Printing Off. No. 1963-0-679485), 1963.
15. Cowling, E. B. *U.S., Dep. Agric., Tech. Bull.* **1961**, *1258*, 79.
16. Kirk, T. K. "Wood Deterioration and Its Prevention by Preservative Treatments"; Nicholas, D. D., Ed.; Syracuse University Press, New York, 1973; Vol. 1, p. 149–81.
17. Seifert, K. *Holz Roh- Werkst.* **1963**, *21*, 85.
18. Kennedy, R. W. *For. Prod. J.* **1958**, *8*, 308.
19. "Cellulases and Their Applications"; Hajny, G. J.; Reese, E. T., Eds.; ACS ADVANCES IN CHEMISTRY SERIES No. 95, ACS: Washington, D.C., 1969; 470 p.
20. "Cellulose as a Chemical and Energy Resource"; Wilke, C. R., Ed.; Biotechnol. Bioeng. Symp. 5, Wiley-Interscience: New York, 1975; 361 p.
21. "Enzymatic Conversion of Cellulosic Materials: Technology and Applications"; Gaden, E. L., Jr.; Mandels, M. H.; Reese, E. T.; Spano, L. A., Eds.; Biotechnol. Bioeng. Symp. 6, Wiley-Interscience: New York, 1976; 319 p.
22. Dekker, R. H.; Richards, G. N. *Adv. Carbohydr. Chem. Biochem.* **1976**, *32*, 277.
23. Eriksson, K.-E. *Biotechnol. Bioeng.* **1978**, *20*, 317.
24. "Hydrolysis of Cellulose: Mechanisms of Enzymatic and Acid Catalysis," Brown, R. D., Jr.; Jurasek, L., Eds.; ACS ADVANCES IN CHEMISTRY SERIES No. 181, ACS: Washington, D.C., 1979; 400 p.
25. Ryu, D. D.; Mandels, M. *Enzyme Microb. Technol.* **1980**, *2*, 91.
26. Kirk, T. K. in "The Filamentous Fungi"; Smith, J. E.; Berry, D. R.; Kristiansen, B., Eds.; Edward Arnold: London, 1983; Vol. IV, p. 266–95.
27. "Bioconversion and Biochemical Engineering: Symposium 2"; Ghose, T. K., Ed.; Indian Inst. Technol.: Delhi, India, 1981; 2 vols.
28. Ander, P.; Eriksson, K.-E. in "Progress in Industrial Microbiology"; Bull, M. T., Ed.; Elsevier: Amsterdam, 1978; p. 1–58.
29. "Lignin Biodegradation: Microbiology, Chemistry, and Potential Applications"; Kirk, T. K.; Chang, H-m.; Higuchi, T., Eds.; CRC Press: Boca Raton, Fla., 1980; 2 vols.

30. Amer, G. I.; Drew, S. W. in *Annu. Rep. Ferment. Processes* Perlman, D., Ed.; Academic Press: New York, 1980; p. 67.
31. Crawford, R. L. "Lignin Biodegradation and Transformation"; Wiley: New York, 1981; 154 p.
32. Zeikus, J. G. *Adv. Microb. Ecol.* **1981**, 5, 211.
33. Kirk, T. K. in "Biochemistry of Microbial Degradation"; Gibson, D. T., Ed.; Marcel Dekker: New York, 1984; p. 399.
34. Nicholas, D. D., Ed. "Preservatives and Preservative Systems"; Syracuse Univ. Press: Syracuse, New York, 1973; Vol. II, 402 p.
35. Scheffer, T. C.; Verrall, A. F. *U.S., For. Serv., Res. Pap. FPL 190* 1973; 56 p.
36. Verrall, A. F.; Amburgey, T. L. *Report, Interagency Agreement IAA-25-75, U.S. Dep. Agric., For. Serv., and U.S. Dep. Housing Urban Dev.*, Superintendent of Documents: Washington, D.C., 1979; 148 p.
37. Nicholas, D. D.; Cockcroft, R. "Wood Preservation in the USA"; *Inf. No. 288*, Swedish Board for Technical Development: Stockholm, 1982; 129 p.
38. Amburgey, T. L. in "Evaluation, Maintenance and Upgrading of Wood Structures. A Guide and Commentary"; Am. Soc. Civil Eng., New York, 1982; p 338.
39. Verrall, A. F. *U.S., Dep. Agric., Tech. Bull.*, **1968**, *1385*, 27.
40. Findlay, W. P. K.; Savory, J. G. *For. Prod. Res., London, Bull. No. 1* (6th ed.), 1960; 35 p.
41. Kirk, T. K.; Chang, H-m. *Enzyme Microb. Technol.* **1981**, 3, 189.
42. Zadražil, F. *Eur. J. Appl. Microbiol.* **1980**, 9, 243.
43. Scheffer, T. C.; Cowling, E. B. *Annu. Rev. Phytopathol.* **1966**, 4, 147.
44. Highley, T. L.; Kirk, T. K. *Phytopathology* **1979**, 69, 1151.
45. Feist, W. C.; Springer, E. L.; Hajny, G. J. *Tappi* **1971**, 54, 1295.
46. Cowling, E. B.; Hafley, W. L.; Weiner, J. *Tappi* **1974**, 57, 120.
47. Scheffer, T. C.; Lindgren, R. M. *U.S., Dep. Agric., Tech. Bull.* **1940**, *714*, 123.
48. Johnson, B. *Wood Fiber* **1979**, 11, 10.
49. Lindgren, R. M.; Scheffer, T. C. *Proc.—Annu. Meet. Am. Wood-Preserv. Assoc.* **1939**, 35, 325.
50. Lindgren, R. M. *Proc.—Annu. Meet. Am. Wood-Preserv. Assoc.* **1952**, 48, 158.
51. Kerr, A. J.; Goring, D. A. I. *Cellul. Chem. Technol.* **1975**, 9, 563.
52. Obst, J. R. *Tappi* **1982**, 65, 109.
53. Baker, A. J. *J. Anim. Sci.* **1973**, 35, 768.
54. Millett, M. A.; Effland, M. J.; Caulfield, D. F. in "Hydrolysis of Cellulose: Mechanisms of Enzymatic and Acid Catalysis," Brown, R. D., Jr., Jurasek, L., Eds.; ACS ADVANCES IN CHEMISTRY SERIES No. 181; ACS: Washington, D.C., 1979; 71.
55. Virtanen, A. I. *Nature (London)* **1946**, 158, 795.
56. Pew, J. C.; Weyna, P. *Tappi* **1962**, 45, 247.
57. Misra, S. C.; Singh, P. *Mater. Org.* **1978**, 13, 59.
58. Seeter, T. D.; Boyle, P. J.; Cundell, A. M.; Mitchell, R. *Mar. Biol. (Berlin)* **1978**, 45, 329.
59. Crosby, N. D.; Reid, R. G. B. *Can. J. Zool.* **1971**, 49, 627.
60. Seifert, K.; Becker, G. *Holzforschung* **1965**, 19, 105.
61. Butler, J. H. A.; Buckerfield, J. C. *Soil Biol. Biochem.* **1979**, 11, 507.
62. Hackett, W. F.; Connors, W. J.; Kirk, T. K.; and Zeikus, J. G. *Appl. Environ. Microbiol.* **1977**, 33, 43.
63. Zeikus, J. G.; Wellstein, A. L.; Kirk, T. K. *FEMS Microbiol. Lett.* **1982**, 15, 193.
64. Leopold, B. *Sven. Papperstidn.* **1952**, 55, 784.
65. Esenther, G. R.; Kirk, T. K. *Ann. Entomol. Soc. Am.* **1974**, 67, 989.
66. Seifert, K. *Holz Roh-Werkst.* **1964**, 22, 209.

67. Gilbertson, R. L. *Mycologia* **1980**, *72*, 1.
68. Leightley, L. E. *Bot. Mar.* **1980**, *23*, 387.
69. Baecker, A. A. W.; King, B. *Biodeter. Mater., Proc. Fourth Int. Biodeterior. Symp.* Berlin, **1978**, 53.
70. Wilcox, W. W. *Bot. Rev.* **1970**, *36*, 1.
71. Levi, M. P.; Preston, R. D. *Holzforschung* **1965**, *19*, 183.
72. Kirk, T. K. *Holzforschung* **1975**, *29*, 99.
73. Eriksson, K.-E. *Pure Appl. Chem.* **1981**, *53*, 33.
74. Heuser, E.; Shema, B. F.; Shockley, W.; Appling, J. W.; McCoy, J. F. *Arch. Biochem.* **1949**, *21*, 343.
75. Kirk, T. K.; Connors, W. J.; Zeikus, J. G. *Appl. Environ. Microbiol.* **1976**, *32*, 192.
76. Jeffries, T. W.; Choi, S.; Kirk, T. K. *Appl. Environ. Microbiol.* **1981**, *42*, 290.
77. Yang, H. H.; Effland, M. J.; Kirk, T. K. *Biotechnol. Bioeng.* **1980**, *22*, 65.
78. Forest Products Laboratory, Forest Service "Wood Handbook: Wood as an Engineering Material"; Agric. Handb. 72, 1974.
79. Scheffer, T. C. *U.S., Dep. Agric., Tech. Bull.* **1936**, *527*, 45.
80. Armstrong, F. H.; Savory, J. G. *Holzforschung* **1959**, *13*, 84.
81. Cartwright, K. St. G.; Findlay, W. P. K.; Chaplin, C. J. *G. Br. For. Prod. Res., Bull.* **1931**, *11*, 18.
82. Zycha, H. *Holz Roh-Werkst.* **1964**, *22*, 4.
83. Richards, D. B. *J. For.* **1954**, *52*, 260.
84. Pechmann, H. von; Schaile, O. *Forstwiss. Centralbl.* **1950**, *69*, 441.
85. Richards, C. A.; Chidester, M. S. *Proc.—Annu. Meet. Am. Wood-Preserv. Assoc.* **1940**, *36*, 24.
86. Schmid, R.; Liese, W. *Arch. Mikrobiol.* **1964**, *47*, 260.
87. Eriksson, K.-E.; Grünewald, A.; Nilsson, T.; Vallander, L. *Holzforschung* **1980**, *34*, 207.
88. Highley, T. L.; Murmanis, L. L.; Palmer, J. G. *Holzforschung* **1983**, *37*, 179.
89. Cowling, E. B.; Merrill, W. *Can. J. Bot.* **1966**, *44*, 1539.
90. Highley, T. L. *Mater. Org.* **1977**, *12*, 25.
91. Wood, T. M.; McCrae, S. I. in Ref. 24, 1979; p. 181.
92. Cowling, E. B.; Brown, W. in Ref. 19, 1969; p. 152.
93. Cowling, E. B.; Kirk, T. K. in Ref. 21, 1976; p. 95.
94. Ahlgren, E.; Eriksson, K.-E. *Acta Chem. Scand.* **1967**, *21*, 1193.
95. Halliwell, G. *Biochem. J.* **1965**, *95*, 35.
96. Koenig, J. W. *Wood Fiber* **1974**, *6*, 66.
97. Schmidt, C. J.; Whitten, B. K.; Nicholas, D. D. *Proc.—Annu. Meet. Am. Wood-Preserv. Assoc.* **1981**, *77*, 157.
98. Shimazono, H. *J. Biochem. (Tokyo)* **1955**, *42*, 321.
99. Highley, T. L. *Appl. Environ. Microbiol.* **1980**, *40*, 1145.
100. Nilsson, T.; Ginns, J. *Mycologia* **1979**, *71*, 170.
101. King, N. *J. Biochem. J.* **1966**, *100*, 784.
102. Highley, T. L. *Wood Fiber* **1973**, *5*, 50.
103. Wolter, K. E.; Highley, T. L.; Evans, F. J. *Biochem. Biophys. Res. Commun.* **1980**, *97*, 1499.
104. Koenig, J. W. *Phytopathology* **1972**, *62*, 100.
105. Lyr, H. *Arch. Microbiol.* **1960**, *35*, 258.
106. Keilich, G.; Bailey, P. J.; Liese, W. *Wood Sci. Technol.* **1969**, *4*, 273.
107. Highley, T. L. *Mater. Org.* **1976**, *11*, 33.
108. Sørensen, H. *Physiol. Plant.* **1952**, *5*, 183.
109. King, N. J.; Fuller, D. B. *Biochem. J.* **1968**, *108*, 571.
110. Ishihara, M.; Shimizu, K. *Mokuzai Gakkaishi* **1980**, *26*, 811.
111. Ishihara, M.; Shimizu, K.; Ishihara, T. *Mokuzai Gakkaishi* **1978**, *24*, 108.
112. Eriksson, K.-E.; Goodell, B. *Can. J. Microbiol.* **1974**, *20*, 371.
113. Nakatsubo, F.; Reid, I. D.; Kirk, T. K. *Biochim. Biophys. Acta* **1982**, *719*, 284.

114. Enoki, A.; Gold, M. H. *Arch. Microbiol.* **1982**, *132*, 123.
115. Kirk, T. K.; Nakatsubo, F. *Biochem. Biophys. Acta* **1983**, *756*, 376.
116. Hall, P. *Enzyme Microb. Technol.* **1980**, *2*, 170.
117. Nakatsubo, F.; Reid, I. D.; Kirk, T. K. *Biochem. Biophys. Res. Commun.* **1981**, *102*, 484.
118. Kirk, T. K.; Nakatsubo, F.; Reid, I. D. *Biochem. Biophys. Res. Commun.* **1983**, *111*, 200.
119. Forney, L. J.; Reddy, C. A.; Tien, M.; Aust, S. D. *J. Biol. Chem.* **1982**, *257*, 11455.
120. Kelley, R. L.; Reddy, C. A. *Biochem. J.* **1982**, *206*, 423.
121. Kutsuki, H.; Gold, M. H. *Biochem. Biophys. Res. Commun.* **1982**, *109*, 320.
122. Tien, M.; Kirk, T. K. *Science (Washington, D.C.)* **1983**, *221*, 661.
123. Ishikawa, H.; Oki, T. *Mokuzai Gakkaishi* **1966**, *12*, 101.
124. Fukuzumi, T.; Takatuka, H.; Minami, K. *Arch. Biochem. Biophys.* **1969**, *129*, 396.
125. Oki, T.; Watanabe, H.; Ishikawa, H. *Mokuzai Gakkaishi* **1981**, *27*, 696.
126. Oki, T.; Senba, Y.; Ishikawa, H. *Mokuzai Gakkaishi* **1982**, *28*, 800.
127. Chen, C.-L.; Chang, H-m.; Kirk, T. K. *Wood Chem. Technol.* **1982**, *3*, 35.
128. Buswell, J.; Ander, P.; Pettersson, B.; Eriksson, K.-E. *FEBS Lett.* **1979**, *103*, 98.
129. Yajima, Y.; Enoki, A.; Mayfield, M. B.; Gold, M. H. *Arch. Microbiol.* **1979**, *123*, 319.
130. Kirk, T. K.; Adler, E. *Acta Chem. Scand.* **1970**, *24*, 3379.
131. Kirk, T. K.; Connors, W. J.; Bleam, R. D.; Hackett, W. F.; Zeikus, J. G. *Proc. Natl. Acad. Sci. U.S.A.* **1975**, *72*, 2515.
132. Haider, K.; Trojanowski, J. in Ref. 29, 1980; Vol. I, p. 111.
133. Brown, W.; Cowling, E. B.; Falkehag, S. I. *Sven. Papperstidn.* **1968**, *71*, 811.
134. Kirk, T. K.; Brown, W.; Cowling, E. B. *Biopolymers* **1969**, *7*, 135.
135. Eslyn, W. E.; Kirk, T. K.; Effland, M. J. *Phytopathology* **1975**, *65*, 473.
136. DeGroot, R. C.; Esenther, G. R. in "Structural Use of Wood in Adverse Environments"; Meyer, R. W.; Kellogg, R. M., Eds.; Van Nostrand Reinhold Co.: New York, 1982; p. 219.
137. Scheffer, T. C. "Wood Deterioration and Its Prevention by Preservative Treatments"; Nicholas, D. D., Ed.; Syracuse University Press, New York, 1973; Vol. 1, p. 31.
138. Johnson, B. D. *Int. Biodeterior. Bull.* **1982**, *18*, 37.
139. Eriksson, K.-E.; Kirk, T. K. in "Comprehensive Biotechnology"; Robinson, C., Ed.; Pergamon Press: Toronto, 1983; Vol. 3.
140. "The Biology and Cultivation of Edible Mushrooms"; Chang, S. T.; Hayes, W. A., Eds.; Academic Press: New York, 1978; 819 p.
141. Leatham, G. F. *For. Prod. J.* **1982**, *32*, 29.
142. Royse, D. J.; Schisler, L. D. *ISR, Interdiscip. Sci. Rev.* **1980**, *5*, 324.
143. Kurtzman, R. H., Jr. in "Annual Report on Fermentation Processes"; Perlman, D., Ed.; Academic Press: New York, 1979; Vol. 3, p. 305.
144. Faison, B. D.; Kirk, T. K. *Appl. Environ. Microbiol.* **1983**, *46*, 1140.

RECEIVED for review May 19, 1983. ACCEPTED July 7, 1983.

The Chemistry of Pyrolysis and Combustion

FRED SHAFIZADEH[1]

Wood Chemistry Laboratory, University of Montana, Missoula, MT 59812

Cellulosic materials decompose on heating or exposure to an ignition source by two alternative pathways. The first pathway, which dominates at temperatures below 300 °C, involves reduction in the degree of polymerization by bond scission; elimination of water; formation of free radicals, carbonyl, carboxyl, and hydroperoxide groups; evolution of CO and CO_2; and, finally, production of a highly reactive carbonaceous char. The second pathway, which takes over at temperatures above 300 °C, involves cleavage of molecules by transglycosylation, fission, and disproportionation reactions to provide a mixture of tarry anhydro sugars and lower molecular weight volatile products. Oxidation of the reactive char gives smoldering or glowing combustion, and oxidation of the combustible volatiles gives flaming combustion. Flaming combustion could be retarded by inorganic materials that suppress the formation of the combustible volatiles through dehydration and charring of the substrate. The smoldering combustion could be suppressed or enhanced by catalysts that affect the rates of oxidation of the char to CO (ΔH = 22.9 kcal/mol) and CO_2 (ΔH = -88.5 kcal/mol). The kinetics and mechanisms of the thermal decomposition, the rates of combustion and heat release, the composition of the pyrolysis products, and the formation and reactivity of char have been investigated extensively to provide a chemical description for combustion and fire prevention.

O<small>NE OF THE GREATEST ASSETS OF CELLULOSIC MATERIALS</small> is their compatibility with nature, including their combustibility and degradability which allow for constant turnover and regeneration of these

---

[1] Deceased

natural resources. A fundamental understanding of these properties and possible methods for controlling them is essential for protection and better utilization of these materials.

Combustion of wood involves a complex series of physical transformations and chemical reactions that are further complicated by the heterogeneity of the substrate. Wood, and cellulosic materials in general, do not burn directly; under the influence of sufficiently strong heat sources they decompose to a mixture of volatiles, tarry compositions, and highly reactive carbonaceous char. Gas-phase oxidation of the combustible volatiles and tarry products produces flaming combustion. Solid-phase oxidation of the remaining char produces glowing or smoldering combustion, depending on the rate of oxidation (*see* Figure 1).

The following discussion shows how the chemical composition, rate of formation, and heat of combustion of the pyrolysis products are affected by the variations in the composition of the substrate, the time and temperature profile, and the presence of inorganic additives or catalysts. The latter aspect, however, is discussed in more detail in Chapter 14. *Combustion* may be defined as complex interactions among fuel, energy, and the environment. Consequently, the combustion process is controlled not only by the above chemical factors, but also by the physical properties of the substrate and other prevailing conditions affecting the phenomena of heat and mass transport. Discussion of this phenomenon is beyond the scope of this chapter.

The general literature on this subject is rather confusing and controversial, mainly because of the variations in the composition of substrate ranging from different types of wood to different types of pulp and natural plant fibers. Superimposed on these variations are the effects of inorganics or ash content, the time and temperature profile, the ambient atmosphere, and the conditions of the heat and mass transport, which are seldom the same in different experiments. Therefore, for a fundamental understanding of the subject, it is essential to examine the individual components of the substrate rather than a poorly defined and variable aggregate. The properties of the aggregate, however (discussed later), are expected to correspond with the collective properties of its components. In this chapter, the pyrolysis or thermal degradation reactions of cellulose are described in detail. Cellulose is the major component of wood and other cellulosic materials as well as the major source of combustible volatiles that fuel the flaming combustion.

Formation of Volatile Products from Cellulose

The general pathways for pyrolysis of cellulose, leading to production of char as well as gaseous and volatile products, are shown

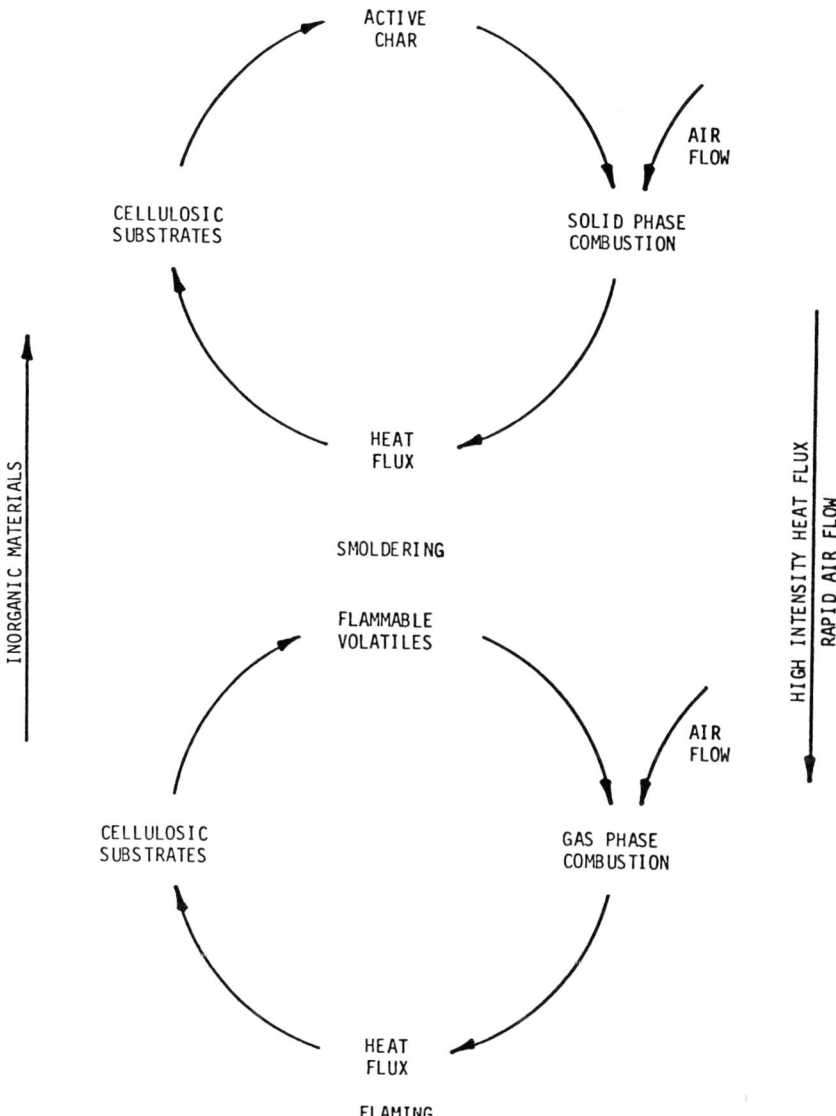

Figure 1. Graphic presentation of the flaming and smoldering combustion showing the respective roles of combustible volatiles and active char produced by pyrolysis under heat flux at different conditions.

in Scheme 1. The global kinetics for isothermal evolution of volatile pyrolysis products from purified cotton linter cellulose, in the temperature range of 257–310 °C, have been studied in air and nitrogen (*12*). The Arrhenius plot of the results based on first-order kinetics (shown in Figure 2) gave an activation energy of 17 kcal/mol for the

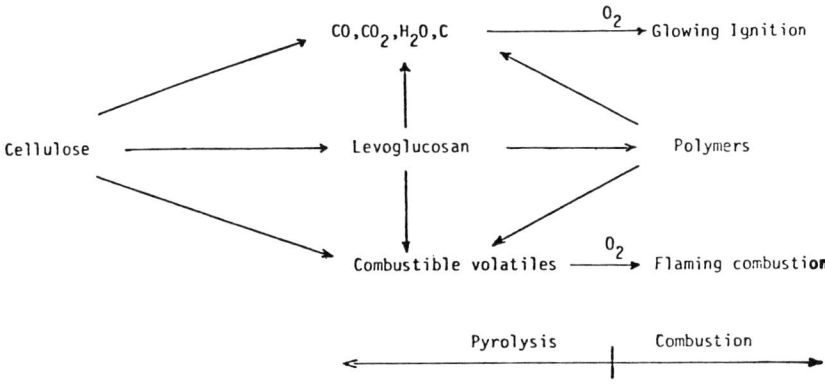

Scheme 1. *The pyrolysis and combustion of cellulose.*

weight loss due to the overall pyrolytic reaction of cellulose in air, and 37 kcal/mol in a nitrogen atmosphere. A transition around 300 °C (Figure 3) reflects the existence of two different pathways. The rate of pyrolysis determined by thermogravimetric analysis (TGA) under isothermal conditions (Figure 4) shows an initial period of acceleration that proceeds faster in air than in the inert atmosphere. As the pyrolysis temperature is increased, the initiation period and the difference between pyrolysis under nitrogen and air gradually diminish and disappear at 310 °C when pyrolysis by the second pathway takes over.

First Pathway. The reactions in the first pathway—which dominates at lower temperatures—involve reduction in the degree of polymerization by bond scission; appearance of free radicals; elimination of water; formation of carbonyl, carboxyl, and hydroperoxide groups (especially in air); evolution of CO and CO_2; and finally production of a charred residue. These reactions, which contribute to the overall rates of pyrolysis of cellulosic materials, have been investigated individually (2). Reduction in the degree of polymerization of cellulose on isothermal heating in air or nitrogen at a temperature within the range of 150–190 °C has been measured by the viscosity method (Figure 5). The resulting data have been correlated; the rates of bond scission are given in Table I and are used for calculating the Arrhenius plot shown in Figure 6. These calculations give an activation energy of 21 kcal/mol for bond scission in air and 27 kcal/mol in nitrogen. This indicates that at low temperatures a larger number of bonds are broken in air than in nitrogen.

The rates of production of CO and CO_2 at 170 °C in air and nitrogen (Figure 7) indicate that the rates for evolution of these gases are much faster in air than in nitrogen and that these rates accelerate

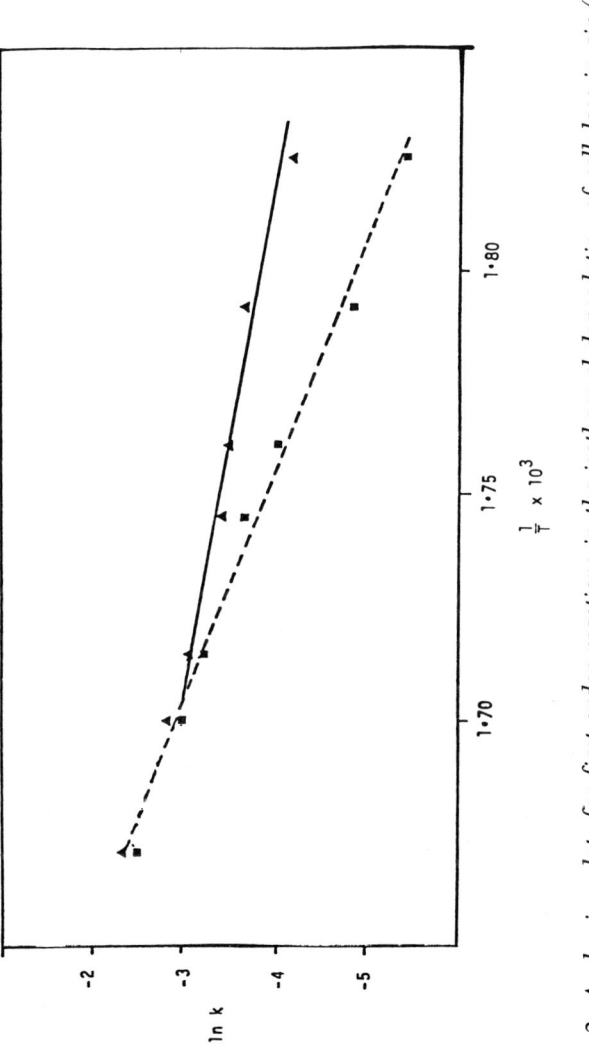

Figure 2. Arrhenius plots for first-order reactions in the isothermal degradation of cellulose in air (■) and nitrogen (▲).

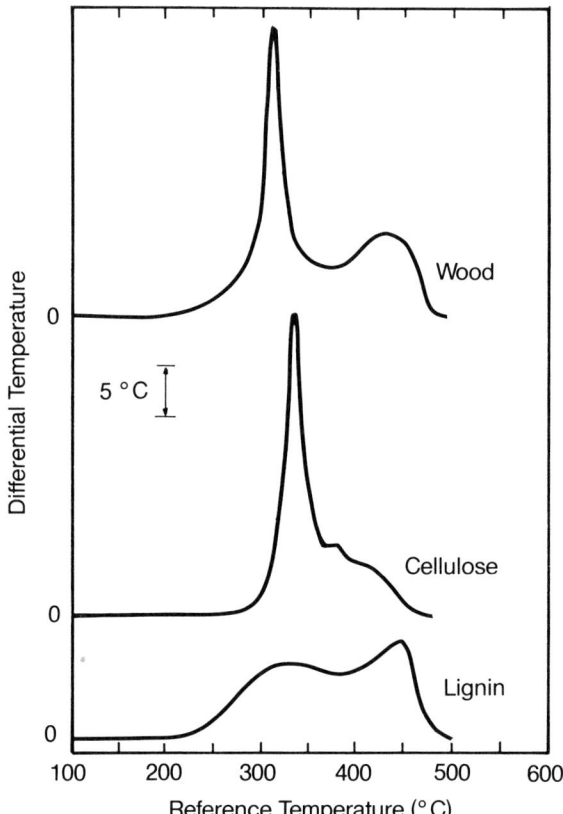

Figure 3. Differential thermal analysis of untreated wood, cellulose, and lignin run in an oxygen atmosphere.

on continued heating. Comparison of the initial linear rates for the evolution of these gases with the rates of bond scission obtained for depolymerization at 170 °C (Table II) shows that the rate of bond scission in air approximately equals the rate of production of CO_2 plus CO in moles per glucose unit. In nitrogen, however, the rate of bond scission is greater than the rates of CO and CO_2 evolution combined.

It is assumed that CO_2 and CO are formed by decarboxylation and decarbonylation, respectively. The significance of the former reactions was determined by measuring the net rate of accumulation of carboxyl and carbonyl groups in cellulose upon heating in air at 190 °C. The results shown in Figure 8 indicate an almost linear rate of formation on heating for 50 h. On heating for longer periods, the rate of accumulation of carboxyl groups falls off, and the rate of accumulation of carbonyl groups is increased. There was also a very

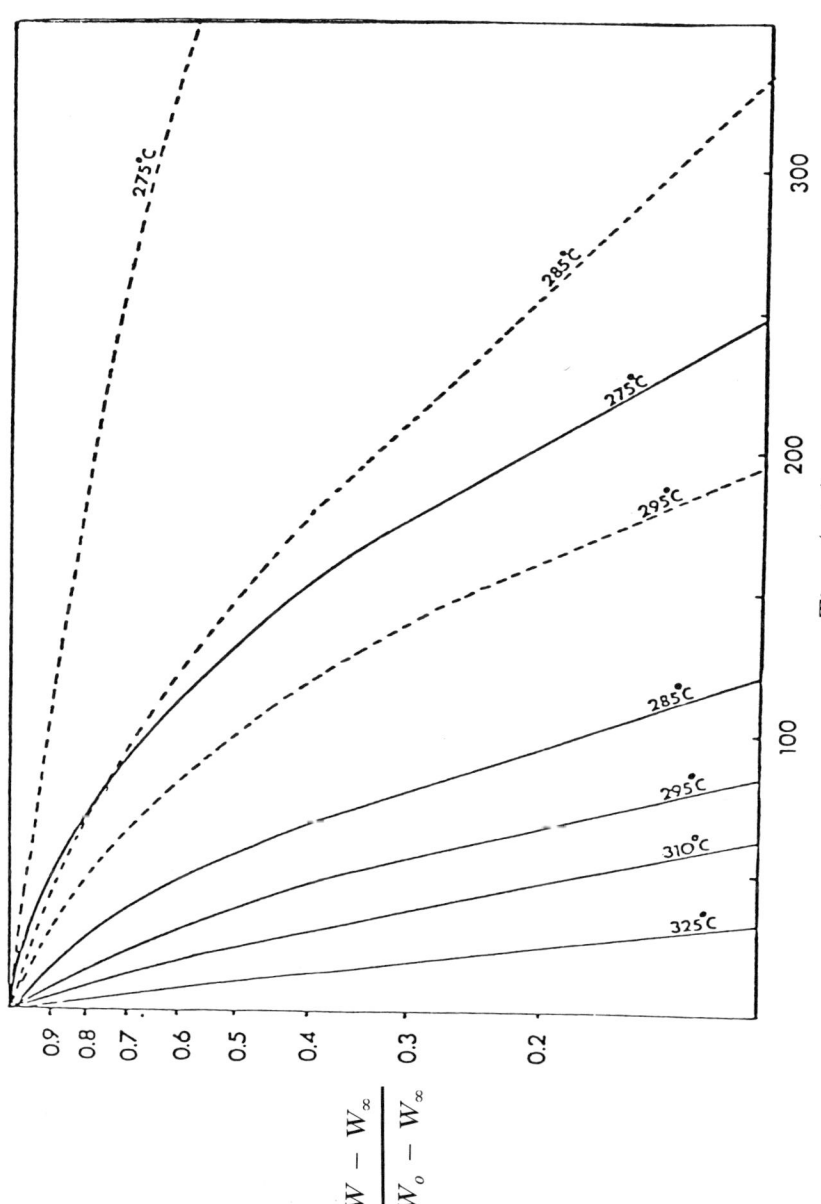

Figure 4. First-order plot for the residual cellulose weight (normalized) vs. time. Plots at 310 and 325 °C for air (—) and nitrogen (---) are similar.

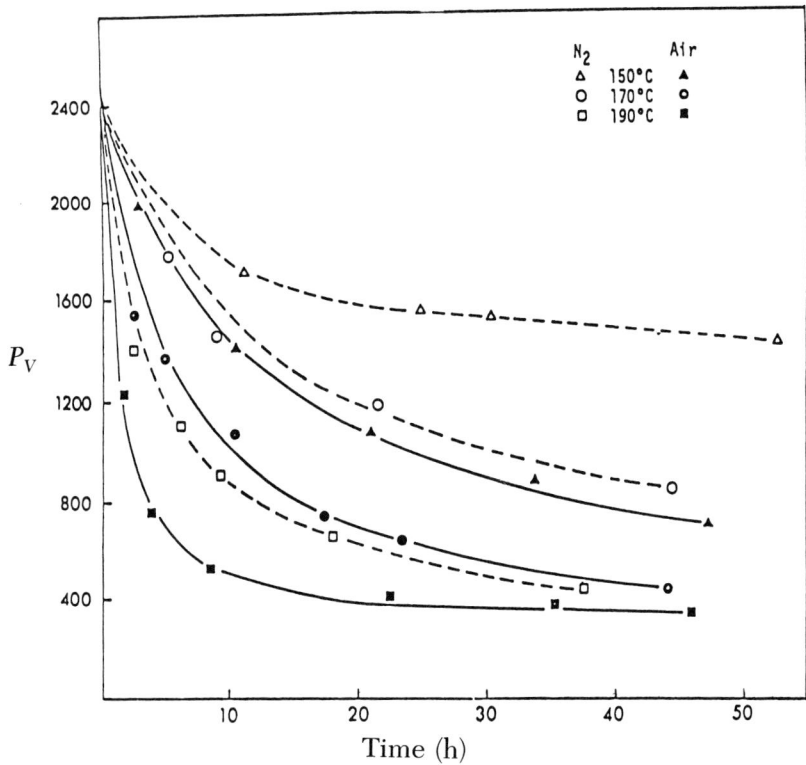

Figure 5. Viscosity average degree of polymerization (P_v) of cellulose heated in air or nitrogen at 150, 170, and 190 °C.

Table I. Rate Constants for the Depolymerization of Cellulose in Air and Nitrogen

Temperature (°C)	Conditions	$k_0 \times 10^7$ (mol/162 g min)[a]
150	N_2	1.1
	Air	6.0
160	N_2	2.8
	Air	8.1
170	N_2	4.4
	Air	15.0
180	N_2	9.8
	Air	29.8
190	N_2	17.0
	Air	48.9

[a] 162 g represents 1 mol of monomer unit.

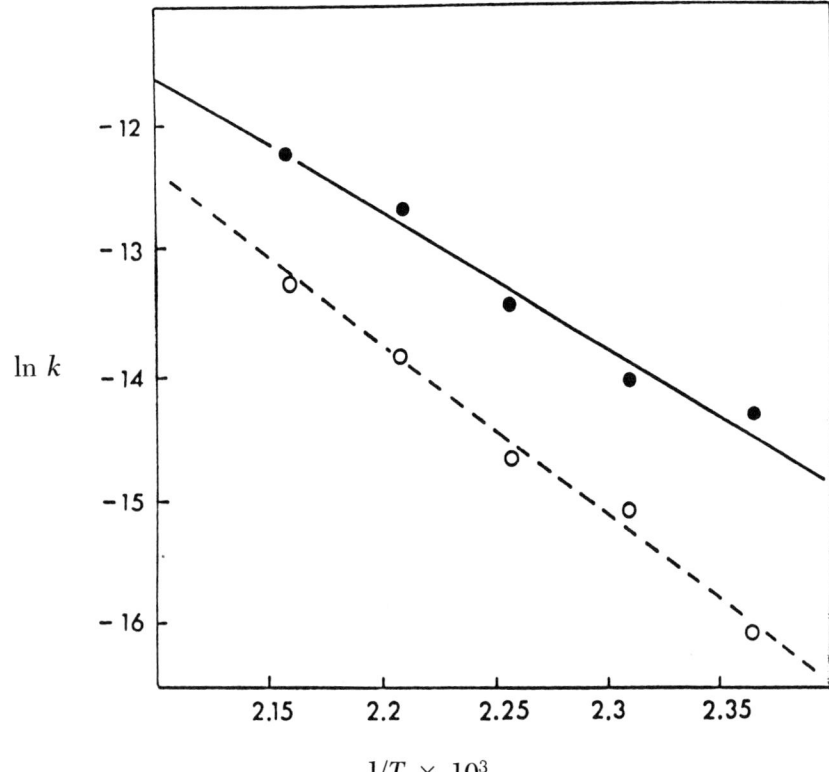

Figure 6. Arrhenius plot for the rate of bond scission in air (●) and nitrogen (○).

small increase in the number of these functions on heating in nitrogen, which could be formed by dehydration and rearrangement, as shown for model compounds (3, 4). The extent of the decarboxylation at the lower pyrolysis temperatures was determined with samples of carboxylcellulose having a low degree of substitution, with carboxyl groups at C-1, C-2, C-3, and C-6. The results generally were not conclusive, although the sample oxidized at C-2 and C-3 showed a definite reduction in carboxyl-group content.

The thermal degradation of cellulose may also involve a free radical mechanism. It was difficult to observe these radicals, but it was possible to monitor the formation of hydroperoxide groups on heating cellulose in air. The hydroperoxide functions are formed and decomposed simultaneously, and their concentration rapidly climbs until a steady state is reached. The decomposition of the hydroperoxide function appeared to follow first-order kinetics with a rate con-

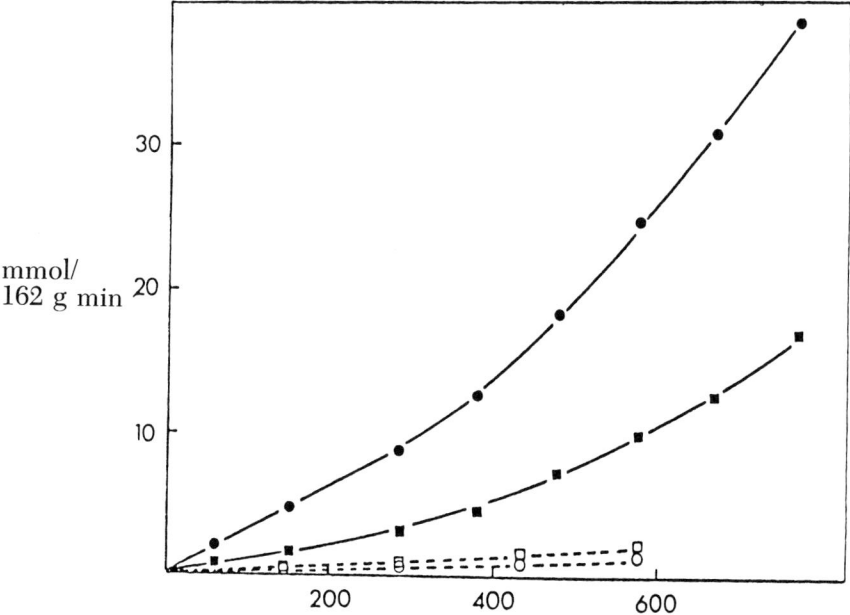

Figure 7. Yields of CO and CO_2 from heating cellulose at 170 °C. Key: ○, CO_2 in N_2; □, CO in N_2; ●, CO_2 in air; and ■, CO in air.

stant of 2.5×10^{-2} min^{-1} at 170 °C. From the steady-state concentration of 3.0×10^{-5} mol/164 g min, the rate of hydroperoxide decomposition is, therefore, 7.5×10^{-7} mol/162 g min. When compared with the initial rate of bond scission in air of 1.5×10^{-6} mol/162 g min at 170 °C (Table I), it is apparent that hydroperoxide formation could make a significant contribution to bond scission.

These considerations reveal that three stages are involved in the low temperature pathway of cellulose: initiation of pyrolysis, propagation, and product formation. The initiation period apparently in-

Table II. Initial Rates of Glycosidic Bond Scission and of CO and CO_2 Formation

Reaction	Rate × 10⁵ in N_2 (mol/162 g h)	Rate × 10⁵ in Air (mol/162 g h)
Bond scission[a]	2.7	9.0
CO evolution[b]	0.6	6.4
CO_2 evolution[b]	0.4	2.1

NOTE: Values are for reaction at 170 °C.
[a] Calculated from the rate constants in Table I.
[b] Calculated from the initial linear portion of plots in Figure 8.

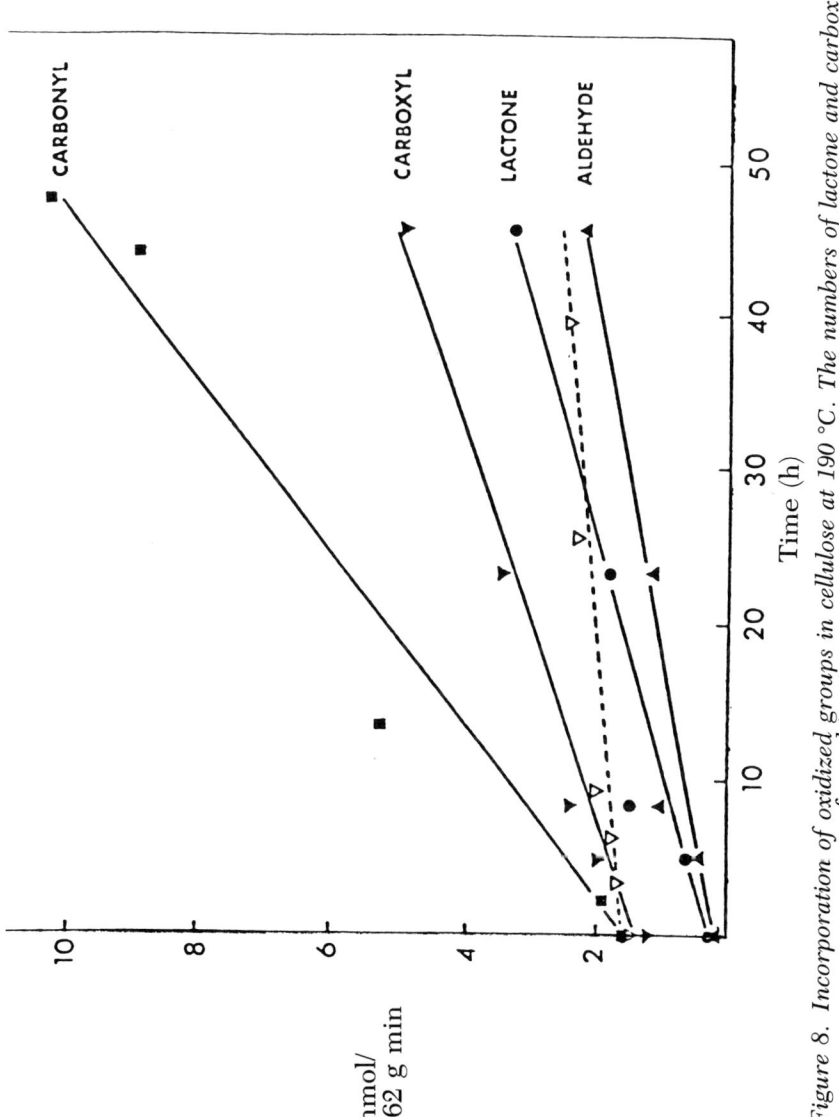

Figure 8. Incorporation of oxidized groups in cellulose at 190 °C. The numbers of lactone and carboxyl groups formed in nitrogen at 190 °C are represented by --▽--.

volves the formation of free radicals facilitated by the presence of oxygen or inorganic impurities. Subsequent reactions of the free radicals could lead to bond scission, oxidation, and decomposition of the molecule, to produce char, water, CO, and CO_2. Scheme 2 shows the initiation, propagation, and decomposition reactions involved in the thermal decomposition of cellulose by this pathway in air. In an inert atmosphere, a lactone could be formed by rearrangement and decomposed by dehydration and decarboxylation (3, 4).

Second Pathway. At temperatures of approximately 300 °C, the second pathway gradually takes over and dominates. The primary reaction in this pathway involves depolymerization by transglycosylation. This reaction takes place when the molecule has gained sufficient flexibility (activation) and produces levoglucosan (1,6-anhydro-β-D-glucopyranose), its furanose isomer (1,6-anhydro-β-D-glucofuranose) and randomly linked oligosaccharides as shown in Scheme 3 (5). The intermolecular and intramolecular transglycosylations shown in this scheme are accompanied by dehydration, followed by fission and disproportionation reactions in the gas phase, and further decomposition and condensation of the solid phase to produce a mixture of gases and volatile products and a "stable" carbonaceous char (3, 6).

On raising the temperature, the tar-forming reactions accelerate

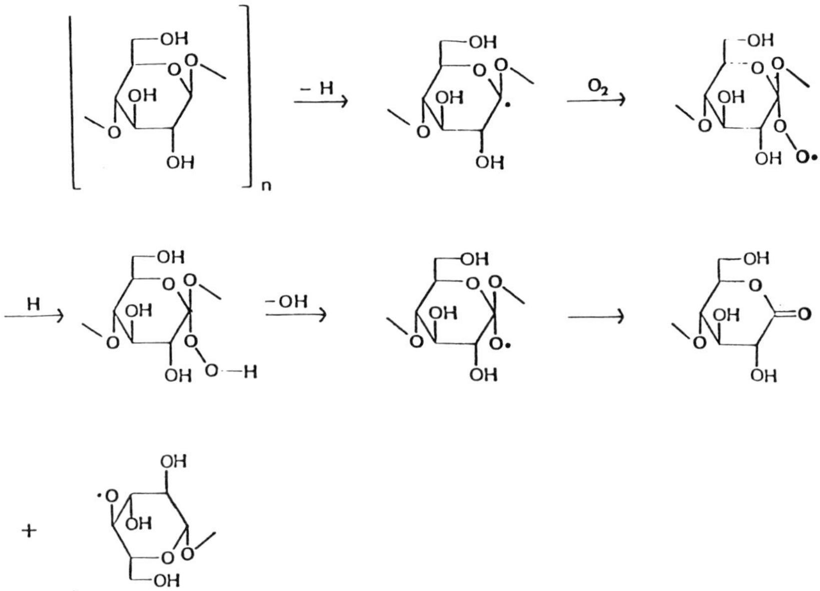

Scheme 2. Possible mechanism of formation and decomposition of cellulose hydroperoxide formed thermally in air.

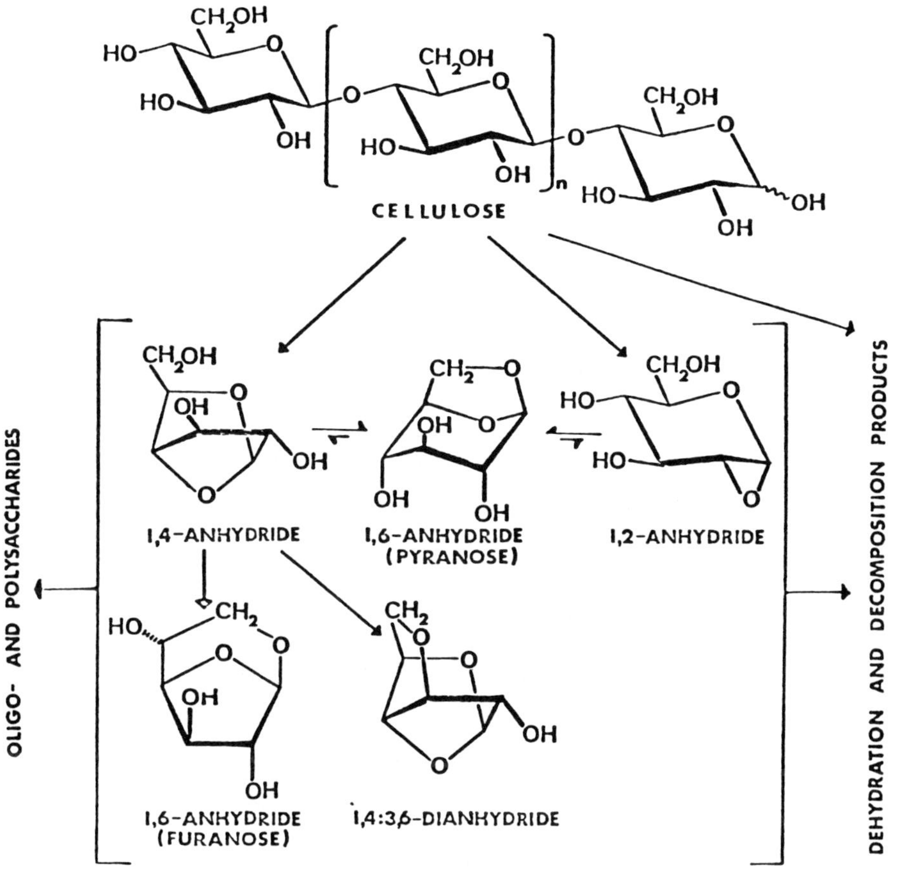

Scheme 3. *Pyrolysis of cellulose to anhydro sugars.*

rapidly and overshadow the production of char and gases. Figure 9 shows the production of diminishing amounts of char and increasing amounts of tar (containing anhydro sugar derivatives that could be hydrolyzed to reducing sugar) as the oven temperature is raised from 300 to 500 °C (7). Evaporation of levoglucosan and the volatile pyrolysis products is highly endothermic. Thus, the increased oven temperature could raise the rate of heat transfer but not necessarily the temperature of the ablating substrate which is cooled by the heat of evaporation, especially under vacuum. As shown in Figure 10 (8), the oven temperature is reached when the rapid evaporation is over. In other words, at the higher temperatures, the pyrolysis process may be controlled by the rate of heat transfer rather than the kinetics of the chemical reaction.

Material transport presents another major obstacle to the inves-

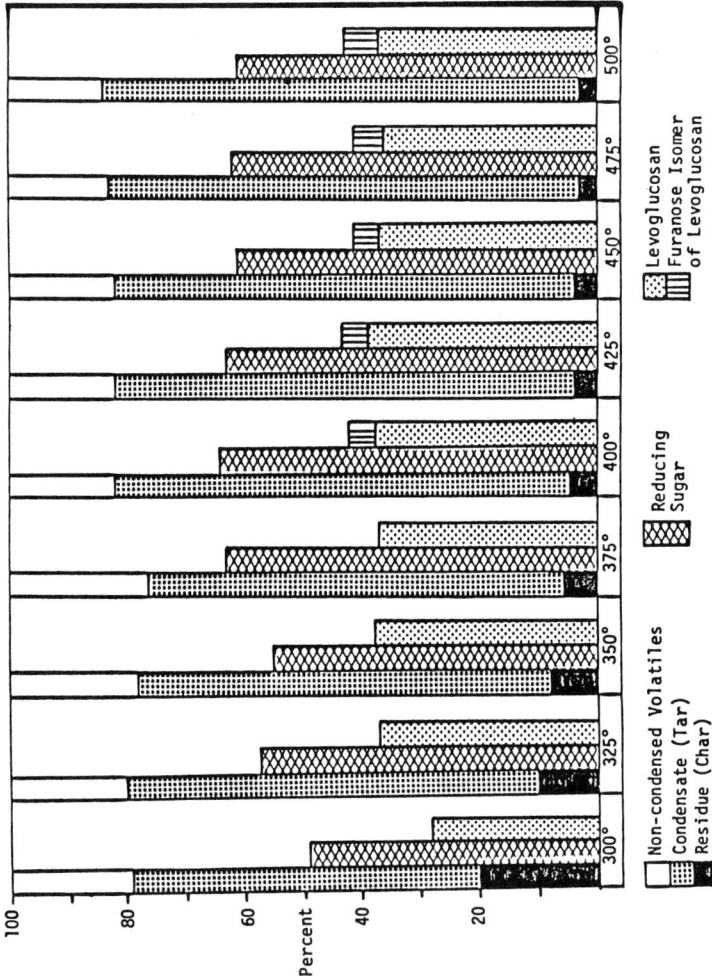

Figure 9. Products from pyrolysis of cellulose powder under vacuum.

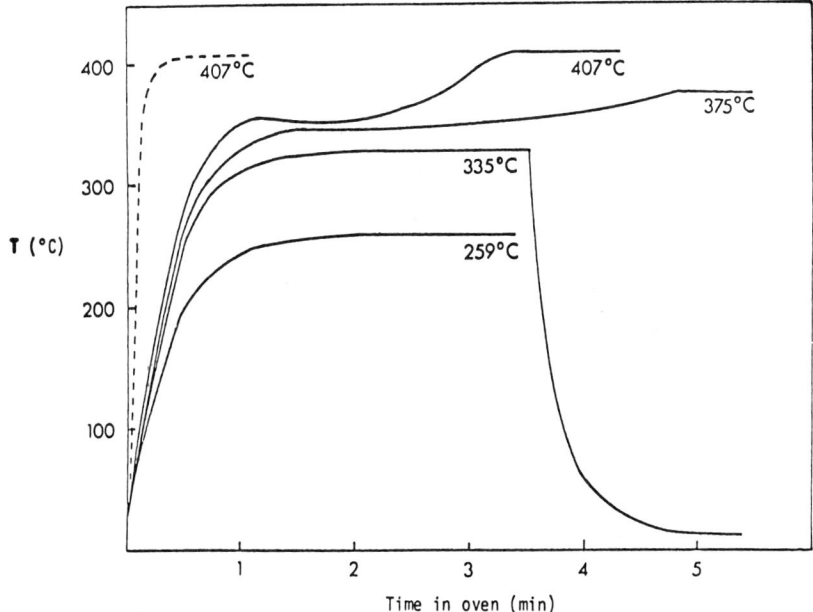

Figure 10. Temperature of pyrolysis as measured by a thermocouple in cellulose sample. Final temperatures, T_f, are indicated. The cooling curve follows removal of a sample from the oven at 335 °C. (Dotted line represents the temperature measured in an empty boat.)

tigation of chemical kinetics because if the products of primary reactions are not removed, they can undergo further decomposition reactions. Table III shows the difference between the yield of different pyrolysis products in vacuum, which removes the primary volatile products, and in nitrogen at atmospheric pressure, which allows for

Table III. Analysis of the Pyrolysis Products of Cellulose

Product	Atmospheric Pressure	1.5 mm Hg	1.5 mm Hg, 5% $SbCl_3$
Char	34.2	17.8	25.8
Tar	19.1	55.8	32.5
levoglucosan	3.57	28.1	6.68
1,6-anhydro-β-D-glucofuranose	0.38	5.6	0.91
D-glucose	trace	trace	2.68
hydrolyzable materials	6.08	20.9	11.8

NOTE: Values (in percent) are for pyrolysis at 300 °C under N_2 and the percentages are based on the original amount of cellulose.

more decomposition of the anhydro sugars (5). Table III also shows the effect of inorganic catalysts in changing the nature of the reactions and products. In view of these considerations, the chemical kinetics of cellulose pyrolysis have been investigated within the limited temperature range of 260–340 °C and under vacuum in order to obtain chemically meaningful data. Under these conditions, the chemical kinetics of cellulose pyrolysis (Scheme 4) could be represented by the three kinetic models shown in Equations 1–3 (8). In these models it is assumed that the initiation reactions lead to the formation of an active cellulose, which subsequently decomposes by two competitive first-order reactions—one yielding anhydro sugars (transglycosylation products) and the other char and a gaseous fraction. Figure 11 shows a comparison between predicted and experimental results.

Dehydration Reactions. Detailed analysis of the pyrolysis tar as discussed previously (Figure 12 and Scheme 3) shows the presence of levoglucosan, its furanose isomer (1,6-anhydro-β-D-glucofuranose) and their transglycosylation products as the main components. In addition to these compounds, the pyrolyzate contains minor amounts of a variety of products formed from dehydration of the glucose units. The dehydration products detected include 3-deoxy-D-erythrohexosulose, 5-hydroxymethyl-2-furaldehyde, 2-furaldehyde (furfural), other furan derivatives, levoglucosenone (1,6-anhydro-3,4-dideoxy-β-D-glycerohex-3-enopyranos-2-ulose), 1,5-anhydro-4-deoxy-D-hex-1-ene-3-ulose, and other pyran derivatives. The dehydration products are important as intermediate compounds in char formation.

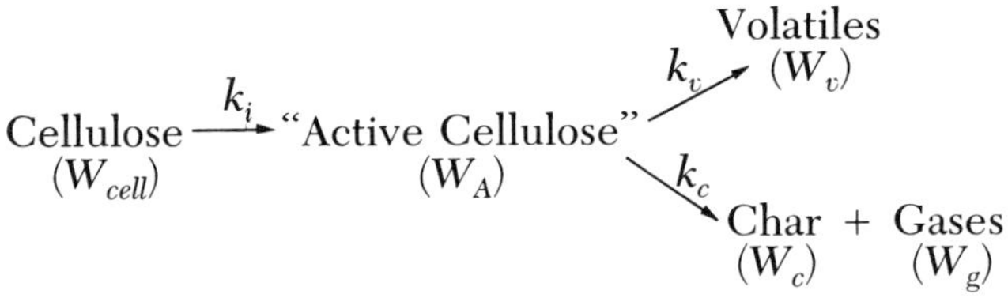

$$\text{Cellulose} \xrightarrow{k_i} \text{"Active Cellulose"} \begin{array}{c} \xrightarrow{k_v} \text{Volatiles } (W_v) \\ \xrightarrow{k_c} \text{Char + Gases} \\ (W_c) \quad (W_g) \end{array}$$
$$(W_{cell}) \qquad (W_A)$$

Scheme 4.

$$\frac{-d(W_{cell})}{dt} = k_i[W_{cell}] \tag{1}$$

$$\frac{d(W_A)}{dt} = k_i[W_{cell}] - (k_v + k_c)[W_A] \tag{2}$$

$$\frac{d(W_c)}{dt} = 0.35 k_c[W_A] \tag{3}$$

where

$k_i = 1.7 \times 10^{21} e^{-(58,000/RT)}$ min^{-1}, $k_v = 1.9 \times 10^{16} e^{-(47,300/RT)}$ min^{-1}, and $k_c = 7.9 \times 10^{11} e^{-(36,000/RT)}$ min^{-1}.

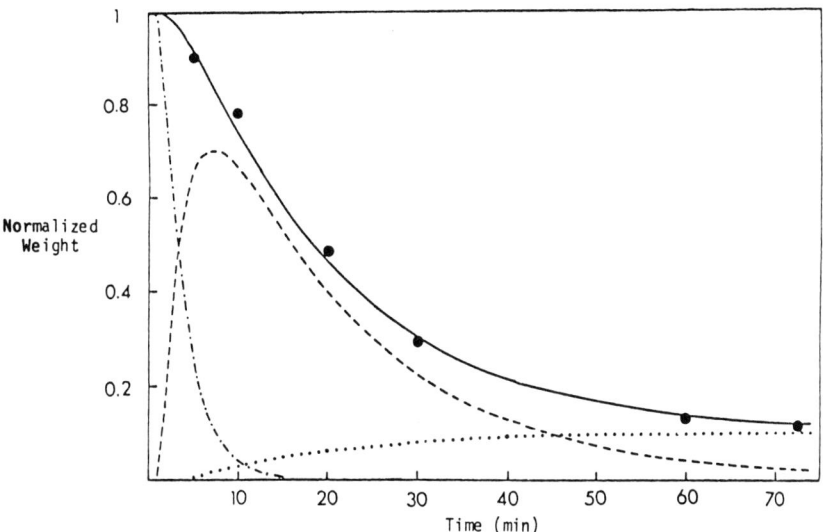

Figure 11. Comparison of experimentally measured residual weight to that predicted from the reaction model at 312 °C. Key to predicted residual weights: —, predicted; -·-, W_{cell}; ---, W_A; and ···, W_c. The experimental residual weight is shown by ●.

As in aqueous reactions, the dehydration reactions are strongly catalyzed by the presence of acidic reagents. Gas liquid chromatography (GLC) analysis (Figure 13) has shown that acid-catalyzed pyrolysis of cellulose at about 350 °C produces a pyrolyzate containing levoglucosenone (instead of levoglucosan) as the major component, and 1,4:3,6-dianhydro-α-D-glucopyranose, 2-furaldehyde, and 5-hydroxymethyl-2-furaldehyde as minor components (9). Levoglucosenone, formed by dehydration reactions shown in Scheme 5, can be separated by fractional distillation, and is a highly reactive compound that can be obtained by pyrolysis of waste paper treated with mineral acids. It could be converted readily to a variety of vinyl, substituted, and addition compounds. Some of these reactions have been investigated already, but the industrial applications of this compound still remain to be explored.

Fission and Disproportionation Reactions. On further heating, fission—or fragmentation—of the sugar units at higher temperatures accompanied by dehydration, disproportionation, decarboxylation, and decarbonylation provides a variety of carbonyl, carboxyl, and olefinic compounds, as well as water, CO_2, CO, and char (*3*, *10*, *11*). The analytical results from these products are closely similar to those obtained from pyrolysis of levoglucosan and wood (Table IV). These compounds may be divided into three categories.

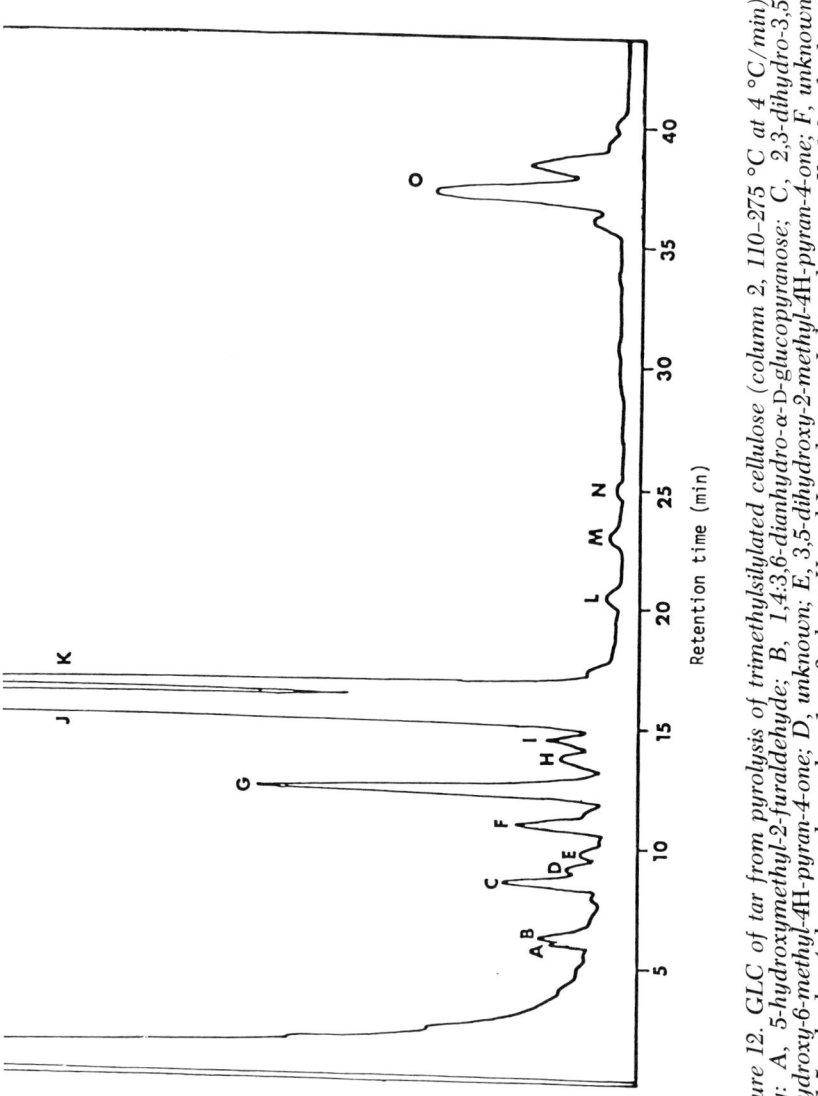

Figure 12. GLC of tar from pyrolysis of trimethylsilylated cellulose (column 2, 110–275 °C at 4 °C/min). Key: A, 5-hydroxymethyl-2-furaldehyde; B, 1,4:3,6-dianhydro-α-D-glucopyranose; C, 2,3-dihydro-3,5-dihydroxy-6-methyl-4H-pyran-4-one; D, unknown; E, 3,5-dihydroxy-2-methyl-4H-pyran-4-one; F, unknown; G, 1,5-anhydro-4-deoxy-D-glycerohex-1-en-3-ulose; H and I, unknown; J, levoglucosan; K, 1,6-anhydro-β-D-glucofuranose; L and M, β-D-glucose; N, 3-deoxy-D-erythro-hexosulose; and O, O-D-glucosyllevoglucosans.

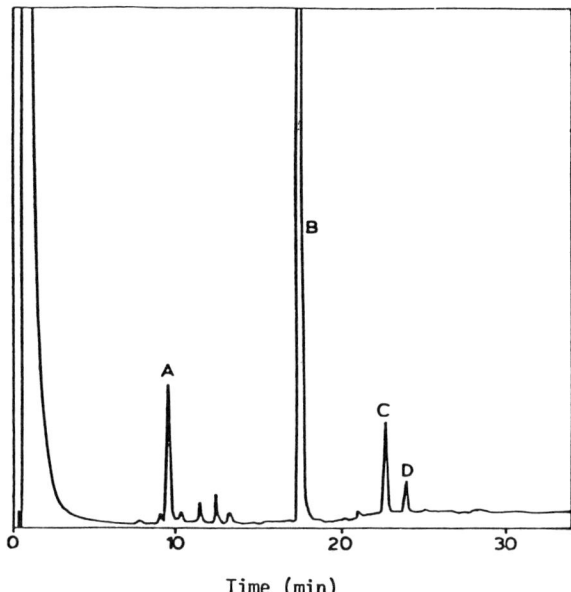

Figure 13. GLC analysis of the pyrolyzate from cellulose + 2% H_4PO_4 at 350 °C. Key: A, furfural; B, levoglucosenone; C, 1,4:3,6-dianhydro-α-D-glucopyranose; and D, 5-hydroxymethyl-2-furaldehyde.

The first category includes furan compounds, water, and char—all of which are the expected products of the better understood acid-catalyzed dehydration reactions of carbohydrates under aqueous acid conditions (1). The second category includes glyoxal, acetaldehyde, and other low molecular weight carboxyl compounds, which are sim-

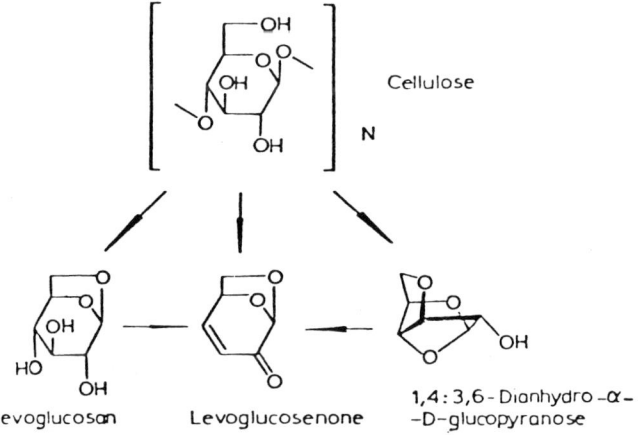

Scheme 5. Formation of levoglucosenone by dehydration reactions.

Table IV. Pyrolysis Products of Wood and Treated Wood

Product	Neat[a]	+5% $ZnCl_2$[a]
Acetaldehyde	2.3	4.4
Furan	1.6	7.9
Acetone } Propionaldehyde	1.5	0.9
Propenal	3.2	0.9
Methanol	2.1	2.7
2-Methylfuran	—[b]	—[b]
2,3-Butanedione	2.0	1.0
1-Hydroxy-2-propanone	2.1	trace
Glyoxal	2.2	trace
Acetic acid	6.7	5.4
2-Furaldehyde	1.1	5.2
Formic acid	0.9	0.5
5-Methyl-2-furaldehyde	0.7	0.9
2-Furfuryl alcohol	0.5	trace
Carbon dioxide	12	6
Water	18	18
Char	15	24
Balance (tar)	28	22

NOTE: Values (in percent) are for pyrolysis at 600 °C.
[a] Yield based on the weight of the sample
[b] Not clearly identifiable for wood

ilar to the alkaline-catalyzed fission products of the sugar molecule formed through the reverse aldol condensation mechanism. The products formed through these pathways are randomized further by disproportionation, decarboxylation, and decarbonylation reactions to provide a third category characterized by their pyrolytic reactions, especially at the elevated temperatures (3, 6, 11).

Formation and Properties of Char

Char Formation. The production of volatiles leaves a solid residue that is neither intact substrate nor pure carbon, but a different material at various stages of charring and carbonization. The intermediate chars are characterized by the functional groups present (including aromatic and olefinic structures); a high concentration of free spins trapped in a rigid structure or stabilized by aromatic and olefinic structures; a large surface area; and a high degree of reactivity. All of these depend on progression of the secondary reaction in the solid phase. Development of these structures and functionalities can be investigated by several chemical and physical methods and can be related to the pyrolysis conditions, particularly to the heating and to the presence of inorganic catalysts, as well as the reactivity of the products.

Development of aromaticity can be investigated by permanganate oxidation of the char which gives benzenepolycarboxylic acids derived from the aromatic nuclei present (see Scheme 6) (*12*). These compounds can be methylated by diazomethane and analyzed by GLC. These products, as shown in Figure 14, are grouped by the number of the carboxylic acid substituents and abbreviated as B2C–B6C. Analysis of the products indicates the concentration and degree of substitution or condensation of the aromatic units. However, because the permanganate oxidation leaves only the central aromatic rings, the resulting data represent only a minimum amount of the aromatic structures present in the char.

The yields, elemental composition, and empirical formulas of chars obtained by isothermal heating of cellulose samples for 5 min within the temperature range of 300–500 °C are given in Table V (*12*). The total aromatic carbon content of these chars and the aromatic carbon content of the char are given in Figure 15. These data show that the reduction in the char yield, or the increase in the weight loss on increasing the heat treatment temperature (HTT), is accompanied by the increased aromatization of the char as measured by the aromatic carbon content of the benzenepolycarboxylic acids (Figure 16). Furthermore, the increased aromaticity of the char is

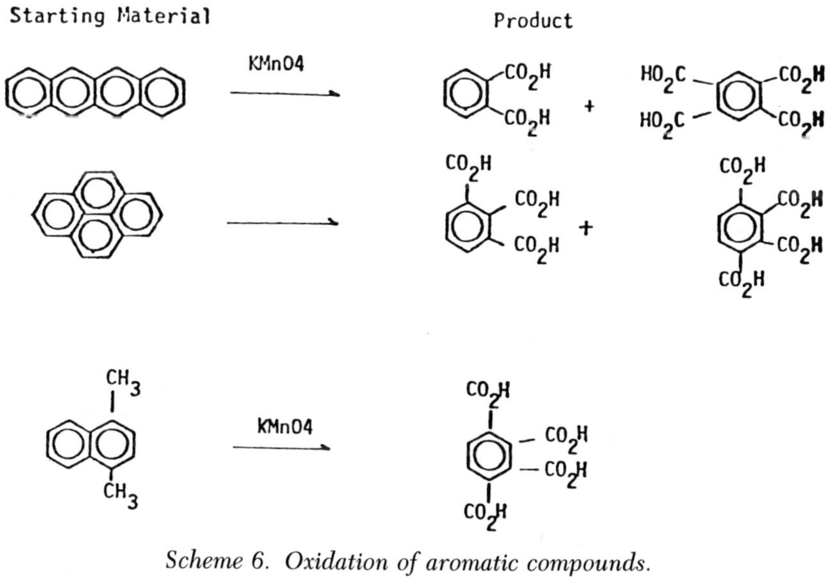

Scheme 6. Oxidation of aromatic compounds.

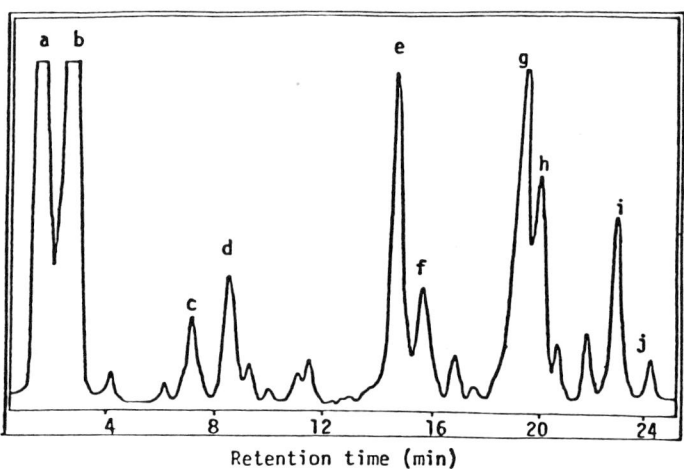

Figure 14. Typical gas chromatogram of oxidation product. Key: a,b, solvent; c, benzenedicarboxylic acid; d, benzenemonohydroxydicarboxylic acid; e, benzenetricarboxylic acid; f, unknown; g, benzenetetracarboxylic acid; h, unknown; i, benzenepentacarboxylic acid; j, benzenehexacarboxylic acid.

Table V. Elemental Composition of Starting Material and Its Char Prepared by the Isothermal Pyrolysis for 5 Min at the Temperatures Noted

Material	Temperature (°C)	Char Yield (wt%)	Composition			Empirical Formula (ref to C_6)
			C	H	O^a	
Cellulose	no treatment	—	42.8	6.5	50.7	$C_6H_{11}O_{5.3}$
	325	63.3	47.9	6.0	46.1	$C_6H_9O_{4.3}$
	350	33.1	61.3	4.8	33.9	$C_6H_{5.6}O_{2.5}$
	400	16.7	73.5	4.6	21.9	$C_6H_{4.5}O_{1.3}$
	450	10.5	78.8	4.3	16.9	$C_6H_{3.9}O_{1.0}$
	500	8.7	80.4	3.6	16.1	$C_6H_{3.2}O_{0.9}$
Wood	no treatment	—	46.4	6.4	47.2	$C_6H_{9.9}O_{4.6}$
	400	24.9	73.2	4.6	22.2	$C_6H_{4.5}O_{1.4}$
Lignin	no treatment	—	64.4	5.6	24.8^b	$C_6H_{6.5}O_{2.0}$
	400	73.3	72.7	5.0	22.3^c	$C_6H_{5.0}O_{1.3}$

NOTE: Char prepared by isothermal pyrolysis for 5 min at the temperatures noted.
^{a} By difference
^{b} Contains 1.2% sulfur
^{c} Contains 0.6% sulfur

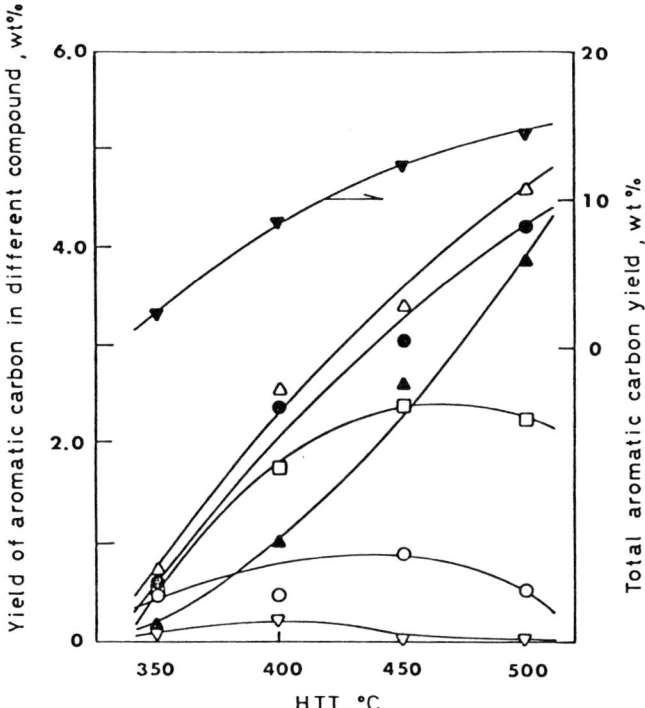

Figure 15. Aromatic carbon content of the char from benzenedicarboxylic acid (○), benzenetricarboxylic acid (□), benzenetetracarboxylic acid (△), benzenepentacarboxylic acid (●), benzenehexacarboxylic acid (▲), and the total yield of hydroxybenzenecarboxylic acid (▽).

accompanied by a reduction of the hydrogen-to-carbon ratio (H/C) (Figure 17). The yields of aromatic carbon contents can be calculated on the basis of the carbon content of the original cellulose rather than the carbon content of the char (Figure 18), as a function of HTT. In Figure 18 the yield of aromatic carbons approaches 2.5/100 carbons of the original cellulose at 400 °C HTT and then levels off. The aromatization, as indicated by drastic reduction of the H/C ratio (Figure 19) and the increased formation of B5C and B6C (Figure 17), continues at HTTs above 400 °C. Above 400 °C, the number of aromatic clusters that are oxidized to polycarboxylic acids remains constant, but the aromatization continues through condensation and the growth of the individual clusters, which results in lower H/C ratios. Furthermore, the weight loss in the "stable" char at temperatures above 400 °C, although relatively small, must take place by elimination of aliphatic substituents and must be accompanied by some dehydrogenation and condensation or fusion of the products. These

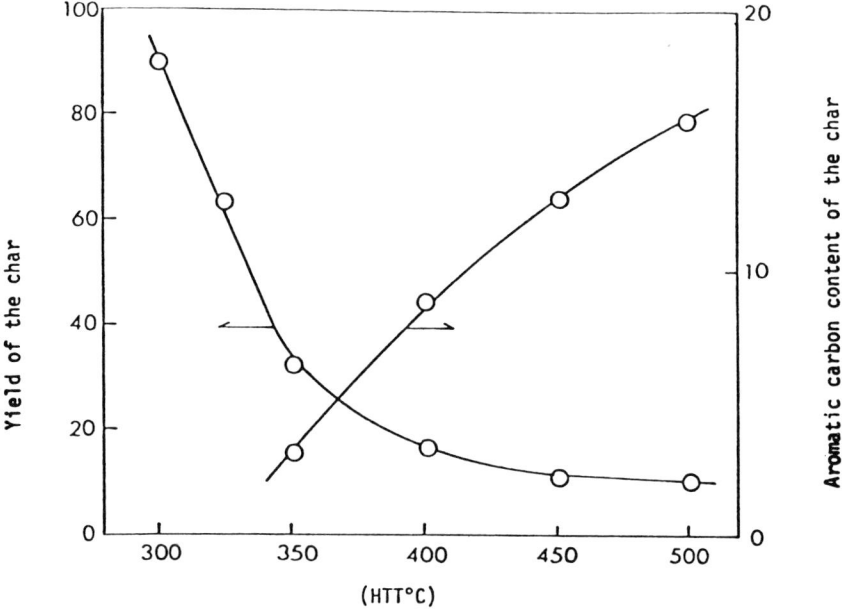

Figure 16. Changes in the yield and aromatic content of the char as a function of HTT.

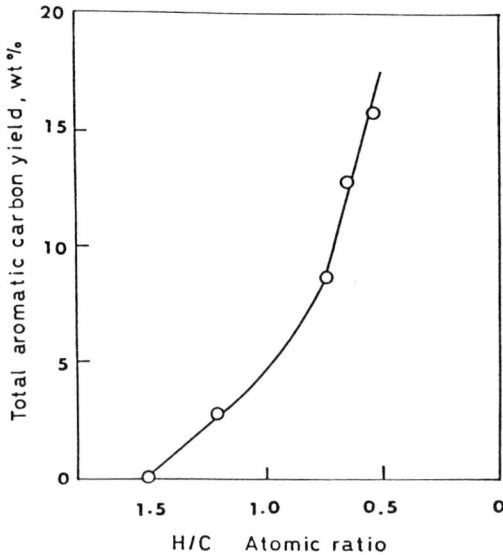

Figure 17. Relation between total aromatic carbon and H/C ratio of char.

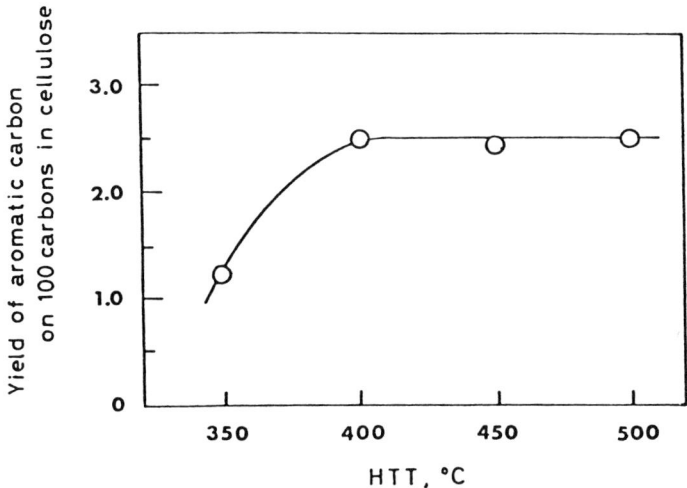

Figure 18. Number of aromatic carbons in 100 carbons of original cellulose.

processes apparently take place through homolytic cleavage and condensation of the resulting free radicals, evident from the measurement of the trapped free-spin concentration in char that is affected by HTT and other variables.

Further information about the structure and functionality of chars produced within the temperature range of 325–500 °C has been obtained by cross-polarized magic-angle spinning (CPMAS) ^{13}C-NMR spectroscopy, which allows quantitative investigation of the carbon skeleton in solid and by Fourier transform IR (FTIR) spectroscopy, which detects the functional groups. The data obtained from these studies are summarized in Figures 19 and 20 and Table VI (13). As can be seen in Figure 19, the CPMAS ^{13}C-NMR spectrum of untreated cellulose has a sharp peak corresponding to the carbon atoms of the glucose units. On further heating, new peaks appear that are associated with aliphatic (0–60 ppm), olefinic (100–110 ppm), carboxylic and ester (160–180 ppm), and carbonyl carbons (190–220 ppm). When the sample is heated at 325 °C, about 37% weight loss occurs (Table V) and the IR spectrum (Figure 20) shows new bands at 1600 and 1700 cm^{-1}, indicating the formation of olefinic and carbonyl functionalities, respectively. The functionalities are due to dehydration and rearrangement in the glycosylic structure. At the same time, small broad peaks are found in both sides of the glycosylic carbon region in the NMR spectrum. However, these peaks are not quantified easily. At 350 °C, the weight loss increases to 67% (Table V) and the NMR spectrum then shows new distinct resonances at 14

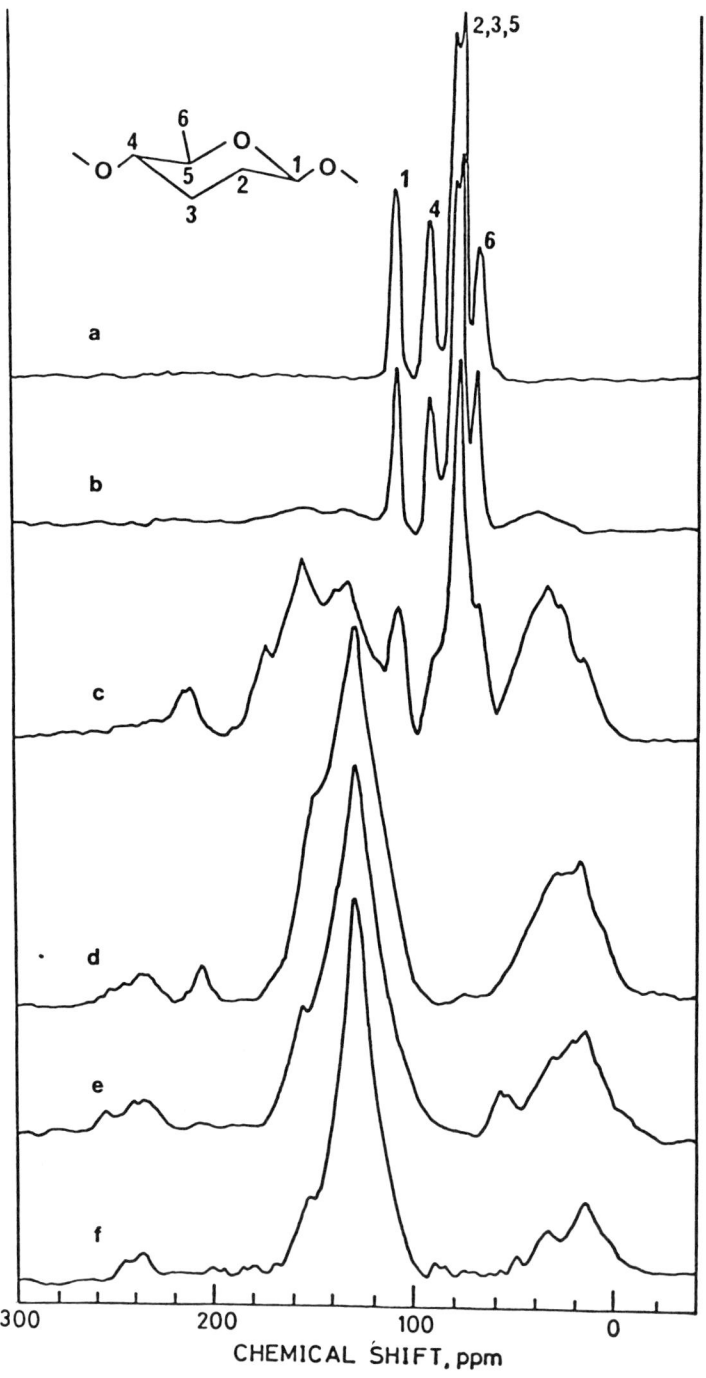

Figure 19. CPMAS ^{13}C-NMR spectra of cellulose chars prepared for 5 min heating at different temperatures (Table V). Key: a, no treatment; b, 325 °C; c, 350 °C; d, 400 °C; e, 450 °C, and f, 500 °C. The small peak located in the 240-ppm region is a spinning sideband.

Table VI. Effect of Char Preparation Temperature on Distribution of Carbon Atoms in Various Functional Groups

Functional Group	Chemical Shift Region (ppm)	Distribution (%) at Char Preparation Temperature (°C)					
		325	350	400	450	500	
Paraffinic							
CH_3	0–30	$4^a (3)^b$	9 (4)	14 (4)	10 (2)	6 (1)	
Others	30–60		15 (7)	13 (4)	11 (2)	6 (1)	
Subtotal		4 (3)	24 (11)	27 (8)	21 (4)	12 (2)	
Glycosylic	60–110	86 (59)	32 (15)	<1 (0.3)	≃0	—	
Aromatic							
C_6H_5-H, C_6H_5-C	110–150	3 (2)	23 (11)	56 (16)	66 (12)	79 (12)	
C_6H_5-O	150–170		13 (6)	13 (4)	11 (2)	9 (1)	
Subtotal		3 (2)	36 (17)	69 (20)	77 (14)	88 (13)	
Oxygen Functional Group							
–COOH, –COOR	170–190	8 (15)	5 (2)	1 (0.3)	1 (0.2)	<1 (0.2)	
>C=O, –CHO	190–220		5 (1.4)	2 (0.2)	1 (0.2)	—	
Subtotal		8 (15)	10 (3.4)	3 (1.3)	2 (0.4)	1 (0.2)	
Total		100 (69)	100 (46.4)	100 (29.6)	100 (18.4)	101 (15.2)	

[a] Based on the carbon atoms in char
[b] Based on the carbon atoms in original cellulose

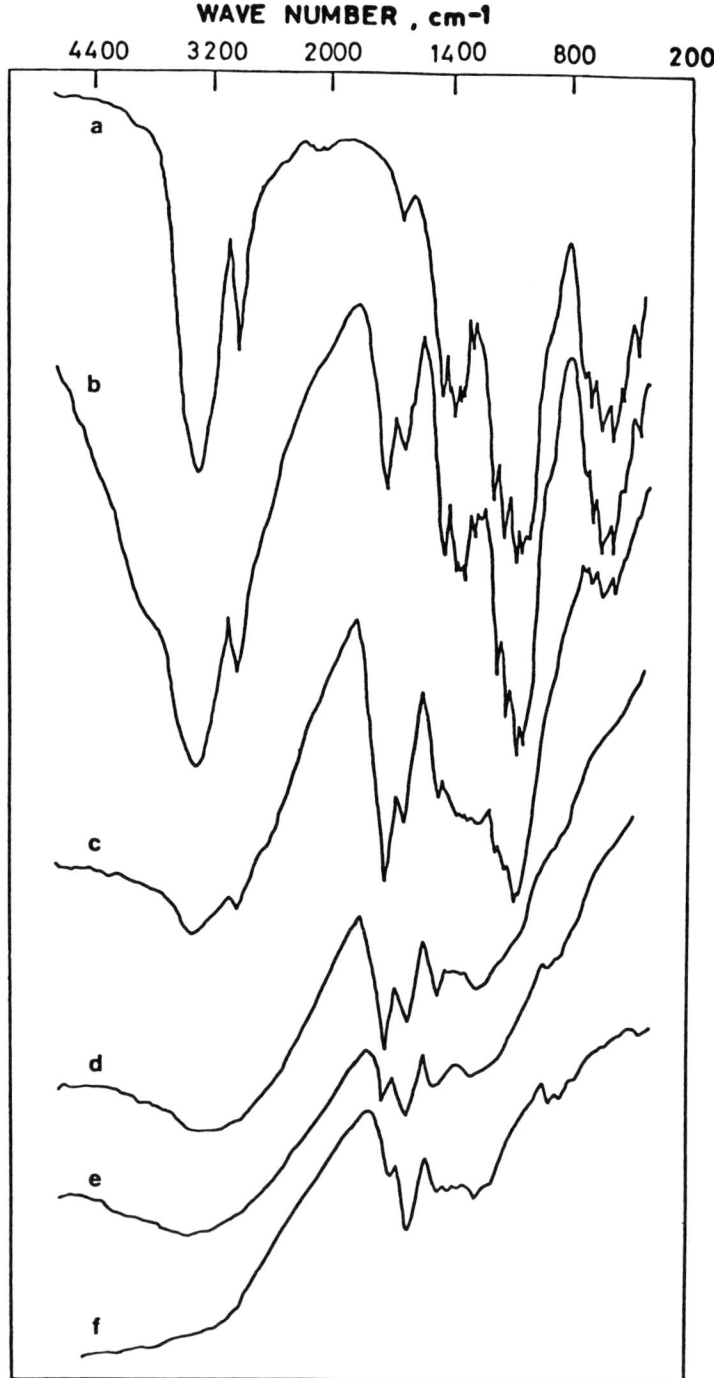

Figure 20. FTIR spectra of cellulose chars prepared for 5 min heating at different temperatures. Key is the same as in Figure 23.

ppm for methyl carbon, 34 ppm for paraffinic carbon without methyl, 132 ppm for olefinic and aromatic carbon adjacent to hydrogen or another carbon, 154 ppm for olefinic and aromatic carbon adjacent to oxygen, 173 ppm for $-COOH$, $-COOR$, and 211 ppm for $C=O$ and $-CHO$ carbons. Aromatic formation probably is initiated at this temperature because of the appearance of NMR peaks at 132 and 154 ppm. These observations are consistent with the results obtained by permanganate oxidation, with which aromatic carbon contents of 0% for 325 °C and 2.8% for 350 °C are obtained (12). At the pyrolysis temperature of 400 °C, the glycosylic structure is no longer found in either the IR or NMR spectrum, as indicated by the disappearance of the 1220–900-cm^{-1} and 60–110-ppm bands, respectively. The NMR spectrum tends to gather into two main regions, paraffinic and aromatic. Resonances of 110–170 ppm are assigned to aromatic carbons. However, this region corresponds to olefinic carbons that are very likely to exist at 350 °C but unlikely to survive above 400 °C when polycyclic structures are formed (12). The char formed at 400 °C corresponded to only 17 wt% of the original material and was relatively stable because, on heating to still higher temperatures, the weight loss was relatively small (Table V). IR spectra of the chars above 400 °C were also similar to each other, except the transmittance of the 1600-cm^{-1} band becomes greater than that of the 1700-cm^{-1} band at temperatures above 450 °C. These observations suggest the increase and predominance of aromatic structures. The extent of the aromatization is demonstrated by the NMR data, which show that the intensity of the aromatic signal increases with increased char preparation temperature.

The relative yields of various carbon species in these chars are shown in Table VI. These data indicate that the glycosylic carbon disappears on heating up to 400 °C. At this temperature the char contains 69% aromatic and 27% paraffinic carbons which change to 88% and 12%, respectively, at 500 °C.

These data, in conjunction with previous studies (12) showed that stable char contains mainly condensed aromatic structure with intermittent paraffinic groups. This structure is formed by successive dehydration, rearrangement, loss of carboxyl, carbonyl, and paraffinic groups, formation of free radicals, and condensation of the carbon skeleton to polycyclic aromatic structures.

This investigation was extended to wood and lignin chars prepared at 400 °C to determine the effect of preexisting aromatic nuclei of lignin in the charring reactions. The permanganate oxidation analysis indicated that these chars, like cellulose chars, have considerably condensed or cross-linked aromatic structures, even at 400 °C. The NMR data also showed that the chars from similar cellulose, wood, and lignin had a similar aromatic carbon content of about 70%. The latter chars, however, showed distinct NMR peaks for the $CH_3-O-C_6H_5$ group, arising from the methoxyl groups in lignin from which the char yield from the lignin component of wood was estimated. These data indicated that preexisting aromatic nuclei in lignin do not increase the aromatic carbon content (aromaticity of the char) although they will increase the char yield.

Char Reactivity. Freshly prepared char is highly reactive and pyrophoric. This reactivity is closely related to smoldering combustion and involves the formation of reactive char by pyrolysis, chemisorption of oxygen on this product, evolution of CO and CO_2, and generation of new reactive sites (14, 15). This process, which is accompanied by evolution of incompletely oxidized volatile pyrolysis products at relatively low temperatures, is usually distinguished from the more rapid and incandescent combustion of the char at higher temperatures and in the presence of more oxygen, which is known as *glowing combustion*.

Figures 21 (14) and 22 show the weight increase and heat of reaction due to chemisorption of oxygen on fresh char determined by thermogravimetry (TG) and differential scanning calorimetry (DSC). In low-density fibrous cellulosic materials where the heat loss is restricted but oxygen can penetrate by diffusion, the heat flux from chemisorption could play a significant role in the ignition of the active

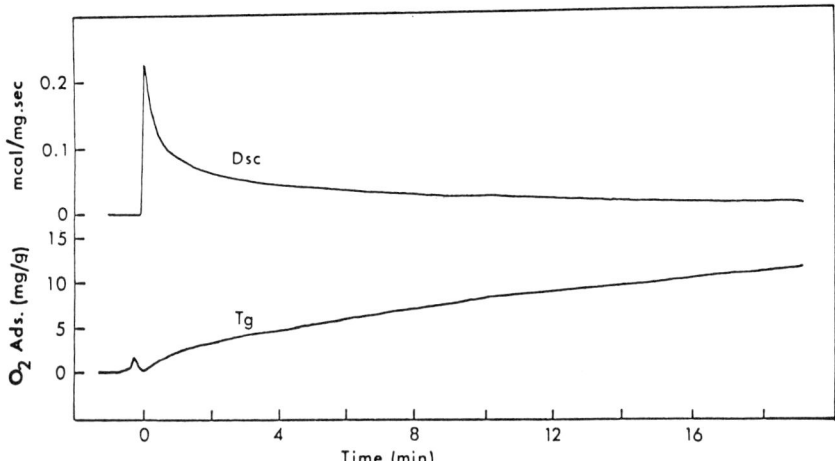

Figure 21. Differential scanning calorimetry and thermogravimetry of oxygen chemisorption on cellulose char at 118 °C. The analysis was carried out on 2.5-mg samples in aluminum pans using a Cahn R-100 electrobalance and a DuPont calorimeter cell attached to a DuPont model 990 thermal analyzer, and nitrogen and oxygen gas flows (60 mL/min, dried by passing through H_2SO_4) were rapidly interchangeable for DSC and TG.

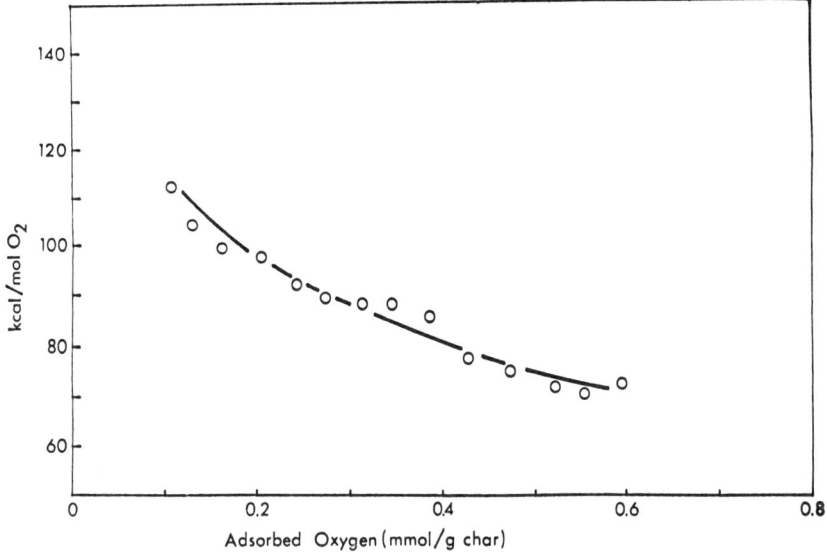

Figure 22. Differential heat of chemisorption, as a function of the amount of oxygen adsorbed on cellulose char at 118 °C.

char and initiation of the combustion. Figure 22 shows the differential heat of chemisorption at 118 °C, which is gradually reduced as the more reactive sites are occupied.

The chemisorption data determined gravimetrically were analyzed according to Elovich kinetics (15). These kinetics are based on the assumption that the rate of chemisorption (dg/dt) declines as the more reactive sites are quenched according to the equation:

$$dg/dt = a \exp(-bg)$$

which is integrated over time between the limits of $-t_o$ and t to give:

$$g = (1/b) \ln ab + (1/b) \ln (t + t_o)$$

where g is the amount of oxygen chemisorbed at time t and a and b are constants. The constant t_o allows for an initial "pre-Elovichian" period of chemisorption that is generally more rapid than that predicted by Elovich kinetics. The value of t_o was determined by an interaction method to give the best linearity of plots of g vs. $(t + t_o)$.

The activation energy for oxygen chemisorption on a typical char varied linearly from 13 to 22 kcal/mol with surface coverage of 0–2.5 mmol O_2/g char (Figure 23). These data show the significance of nascent or fresh unreacted char that is pyrophoric and becomes harder to ignite as the low activation energy sites are occupied and the heat of chemisorption (Figure 21) is reduced.

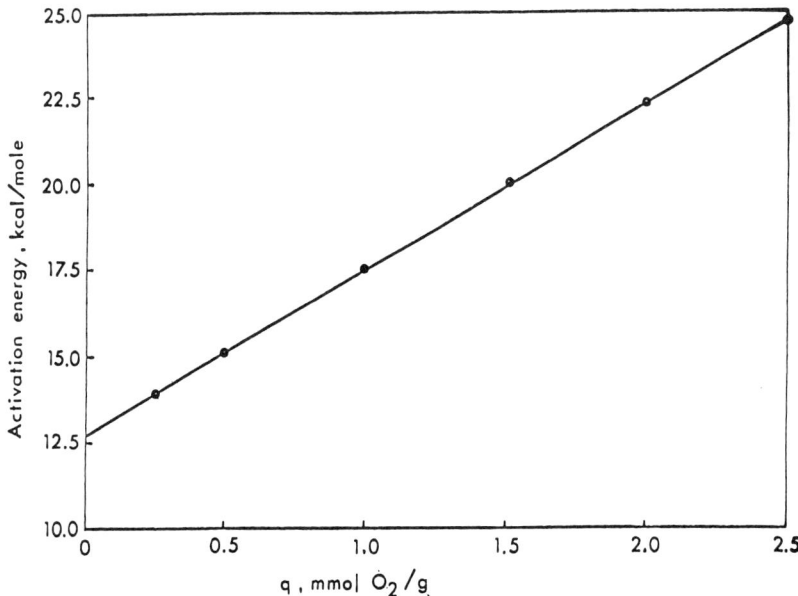

Figure 23. Variation of activation energy for oxygen chemisorption with surface coverage.

The reactivity of the char is related to a high concentration of trapped free radicals and a large surface area. Figure 24 shows the effect of temperature on the weight and free radical concentration of the chars produced by 1.5 min pyrolysis under nitrogen. The free radical concentration reaches a maximum for char produced at about 550 °C.

The surface areas of chars prepared from cellulose samples at different HTTs were determined by application of the Dubinin–Polany equation to CO_2 adsorption at room temperature and compared with the area occupied by surface oxides calculated from oxygen chemisorption at 230 °C. The results shown in Figure 25 indicate that cellulosic chars have large surface areas that vary according to the HTT, and peak at about 550 °C. The surface oxides formed by chemisorption occupy only a portion of the total surface area, and the chemisorption also shows a peak for chars formed at about 550 °C, corresponding to the temperature of smoldering combustion.

Combustion

Combustibility. The intensity of combustion may be expressed by the following general equation:

$$I_R = -\Delta H \frac{dw}{dt}$$

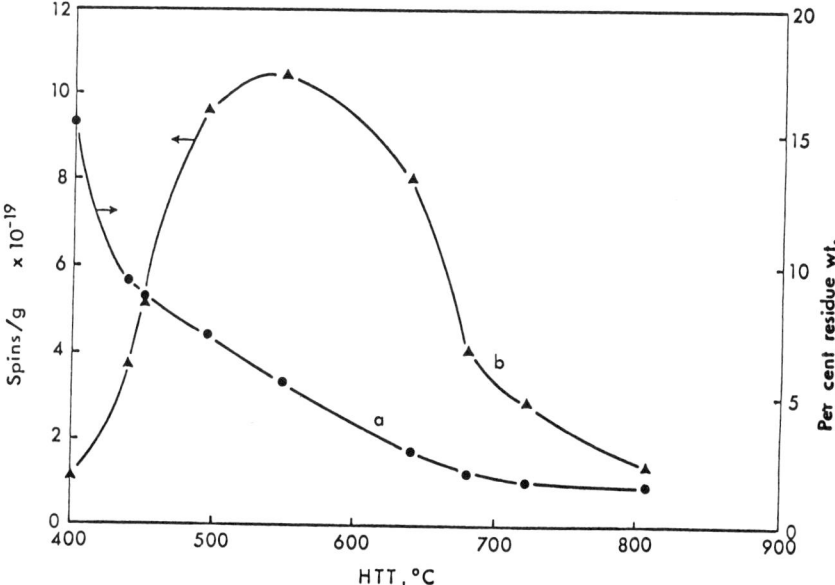

Figure 24. Comparison between char residue weight (Curve a) and free spin concentration (Curve b) as a function of char HTT.

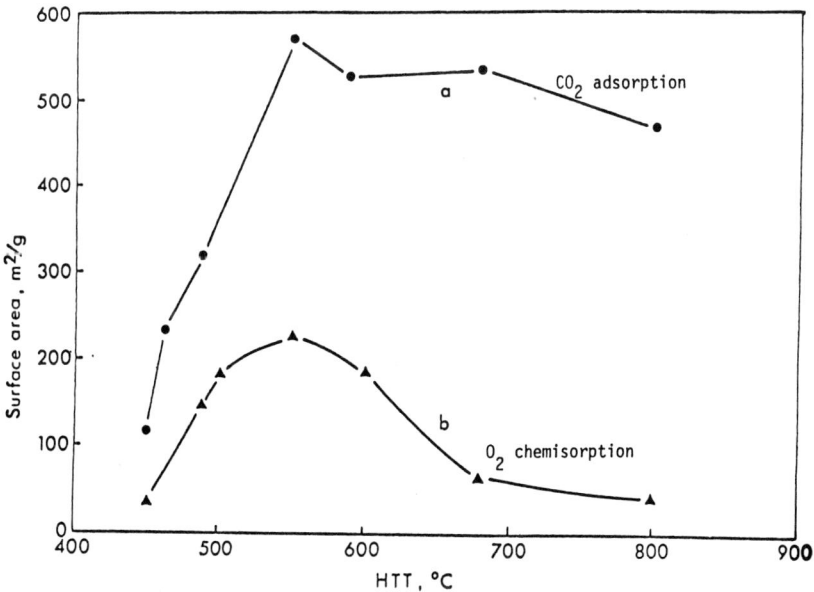

Figure 25. Comparison between total surface area calculated from CO_2 adsorption at 25 °C (Curve a) and that occupied by surface oxides calculated from oxygen chemisorption at 230 °C (Curve b).

where I_R is the reaction density, $-\Delta H$ is the heat of combustion, and dw/dt equals the rate of mass of fuel loss (16). The heats of combustion of different types of forest fuels and their components are given in Table VII (17). This table also shows the amount of char left at 400 °C on heating at the rate of 200 °C/min and the heat of combustion of the gaseous and carbonaceous products. Table VIII shows the distribution of the heat of combustion between the char and the volatile products.

For flaming combustion, the $-\Delta H$ and the rate of producton of the volatiles are important. The latter value could be determined by thermogravimetry and its derivative under simulated pyrolysis conditions, which indicate the progress and the rate of weight loss. Figure 26 shows the dynamic TG of cottonwood and its components and indicates that lignin contributes mainly to char, whereas cellulose and hemicelluloses form mainly the volatile pyrolysis products that are responsible for the flaming combustion. These data also indicate that wood shows the collective thermal properties of its components.

For quantitative determination of the rate and the amount of heat release, we have developed a method based on complete combustion of the volatiles on a platinum catalyst and measurement of amount of oxygen required (18). The required oxygen is generated by a reaction coulometer (Figure 27) and is related to the heat of combustion as shown in Figure 28. As discussed in Chapter 14, the amount of combustible, volatile pyrolysis products formed and the rate of heat release may be reduced drastically by the addition of flame retardants.

For smoldering combustion the driving force is measured by the heat and rate of combustion of the freshly formed and highly reactive pyrophoric chars that, on exposure to oxygen, could ignite at relatively low temperatures. The mechanisms of the oxidation reactions involved are not clear. However, it is believed that oxygen molecules are originally adsorbed on the active sites (C*) which may or may not be free radicals. As shown in Scheme 7 the adsorbed oxygen subsequently reacts to produce CO_2, CO, and new active sites. At 500 °C, these complex reactions could be summarized in the following reactions:

$$O_2 + C \rightarrow CO_2 \quad \Delta H = -88.5 \text{ kcal/mol}$$
$$O_2 + 2C \rightarrow 2CO \quad \Delta H = -22.9 \text{ kcal/mol}$$

These reactions show that incomplete oxidation of the char to CO produces about 25% of the total energy based on the available carbon and 50% of the total energy based on the available oxygen. Because the oxidation reactions are activated and exothermic, when

Table VII. Heat of Combustion of Natural Fuels and Their Pyrolysis Products as Char and Combustible Volatiles

Fuel			Char		Combustible Volatiles	
Source	Type	25 °C $\Delta H\ Comb$ (cal/g)	Yield (%)[a]	25 °C $\Delta H\ Comb$ (cal/g)	Yield (%)[a]	25 °C $\Delta H\ Comb$ (cal/g)
Cellulose	Filter paper	−4143	14.9	−7052	85.1	−3634
Douglas-fir lignin	Klason	−6371	59.0	−7416	41.0	−4867
Poplar wood (*Populus* ssp.)	Excelsior	−4618	21.7	−7124	78.3	−3923
Larch wood (*Larix occidentalis*)	Heartwood	−4650	26.7	−7169	73.3	−3732
Decomposed Douglas-fir (*Pseudotsuga menzeisii*)	Punky wood	−5120	41.8	−7044	58.2	−3738
Ponderosa pine (*Pinus ponderosa*)	Needles	−5145	37.0	−6588	63.0	−4298
Aspen (*Populus tremuloides*)	Foliage	−5034	37.8	−6344	62.2	−4238
Douglas-fir bark (*Pseudotsuga menzeisii*)	Outer (dead)	−5122	52.8	−5798	47.2	−4366
	Whole	−5708	47.1	−6406	52.9	−5087

[a] Heating rate 200 °C/min to 400 °C and held for 10 min

Table VIII. Distribution of the Heat of Combustion of Forest Fuels

| Fuel | | Char | Gas | Total |
Source	Type	(cal/g fuel)	(cal/g fuel)	(cal/g)
Cellulose	Filter paper	−1050	−3093	−4143
Douglas-fir lignin	Klason	−4375	−1995	−6370
Poplar wood				
(*Populus* spp.)	Excelsior	−1546	−3072	−4618
Larch wood				
(*Larix occidentalis*)	Heartwood	−1914	−2736	−4650
Decomposed Douglas-fir				
(*Pseudotsuga menzeisii*)	Punky wood	−2944	−2176	−5120
Ponderosa pine				
(*Pinus ponderosa*)	Needles	−2438	−2708	−5146
Aspen				
(*Populus tremuloides*)	Foliage	−2398	−2636	−5034
Douglas-fir bark				
(*Pseudotsuga menzeisii*)	Outer (dead)	−3061	−2061	−5122
	Whole	−3017	−2691	−5708

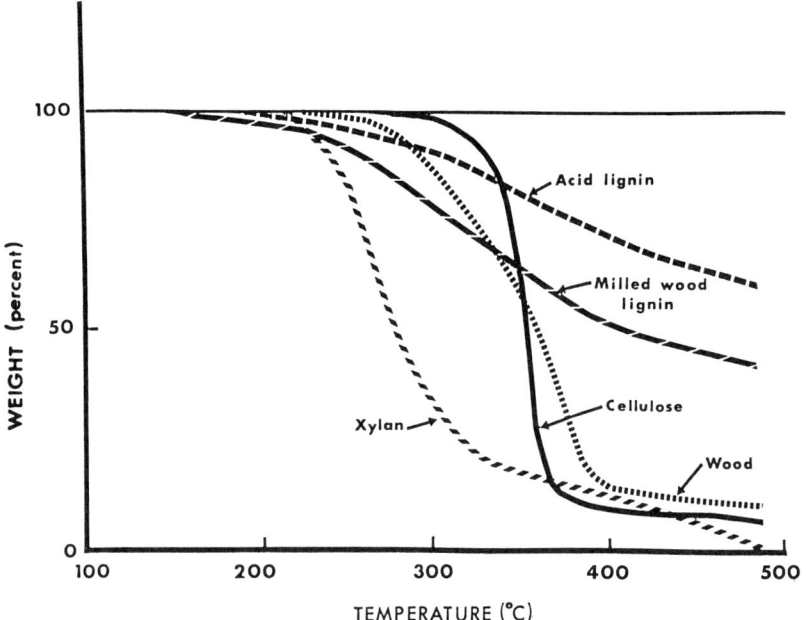

Figure 26. Thermogravimetry of cottonwood and its components.

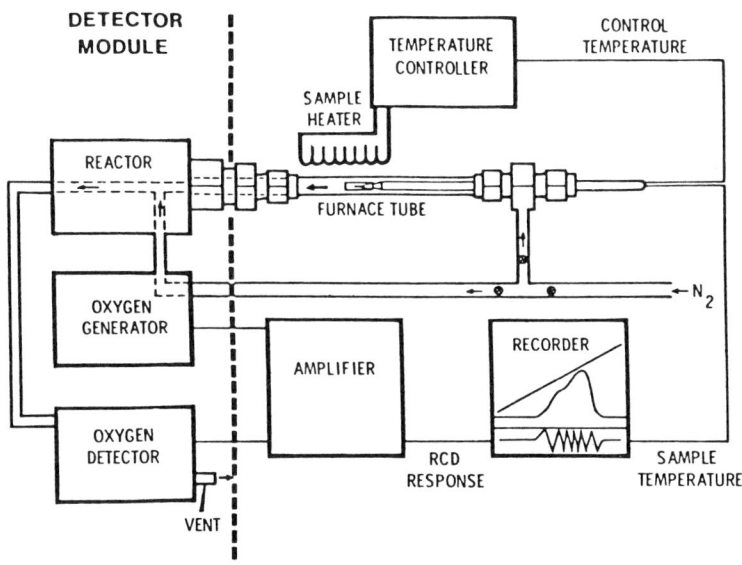

Figure 27. Block diagram of the thermal evolution analysis system coupled to a reaction coulometer detector.

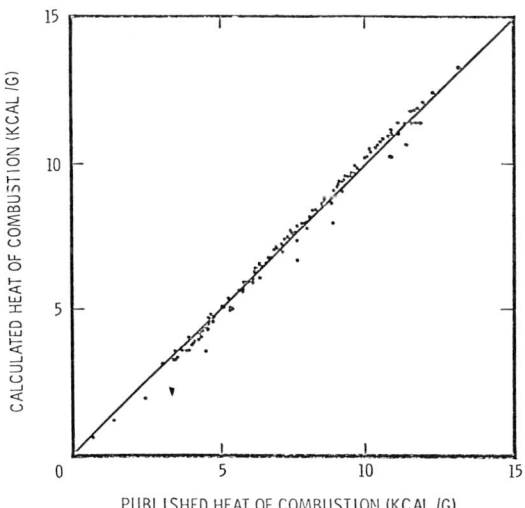

Figure 28. Correlation between the published heats of combustion and calculated values based on 3349 cal/g O_2 consumed.

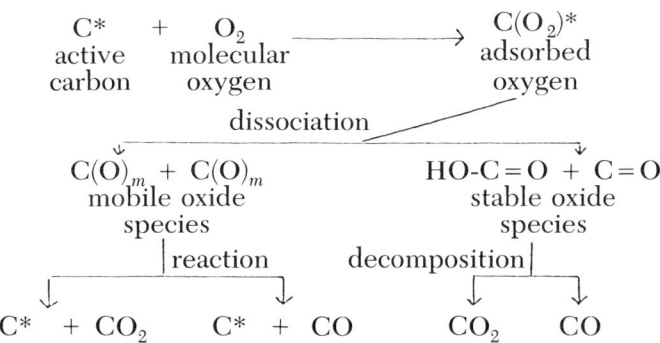

Scheme 7. *Oxygen adsorption, chemisorption, and oxidation of char.*

sufficient oxygen is available, the rate of reaction and the corresponding heat release could rapidly increase. This will raise the surface temperature to the level of glowing or incandescence, where much of the energy is lost to the surrounding environment by radiation.

The ratio of CO to CO_2 is influenced by various anions and cations. Anions such as phosphates and borates increase the ratio of CO to CO_2 and are known as glow retardants or smolder inhibitors. Cations such as sodium and potassium ions reduce the ratio of CO to CO_2 and promote the smoldering combustion. Addition of inorganic salts to cellulose leads to a wide range of reactivities, from enhancement to total inhibition of the smoldering combustion by variation in the rates of production of CO and CO_2 and the corresponding rates of heat release. The kinetic data presented in Tables IX and X show the effect of several additives on smoldering combustion ranging from enhancement by NaCl to inhibition by boric acid (*19, 20*). The catalytic effect of inorganic salt on the rate of production of CO and CO_2 explains the mechanism of smoldering inhibitors. The smoldering inhibitors should not be confused with the flame retardants that reduce the production of flammable volatiles by promoting the char production in the pyrolysis pathways discussed before, although ammonium phosphates and borax also function as flame retardants.

During flaming combustion, a relatively high rate of heat release generated by the combustion of volatile products (*see* Figure 1) provides the energy required for gasification of the substrate and the propagation of fire. When the intensity of the heat flux and the concentration of the combustible volatiles falls below the minimum level required for the propagation of flaming combustion, gradual oxidation of the reactive char initiates smoldering combustion. This smoldering process, which is accompanied by emission of noncombustible or unoxidized volatile products, is often observed with low density

Table IX. Arrhenius Parameters and Calculated Rate Constants for Production of CO and CO_2 from Chars Formed by Pyrolysis of Cotton Fabrics

	Arrhenius Parameters				Calculated Maximum Rate Constants					
	CO Production		CO_2 Production		500 °C			750 °C		
Additive	A (min^{-1})	E_a (kcal)	A (min^{-1})	E_a (kcal)	k_{CO} (min^{-1})	k_{CO_2} (min^{-1})	Rate of Heat Release ($kcal \cdot min^{-1} \cdot mol\ C^{-1}$)	k_{CO} (min^{-1})	k_{CO_2} (min^{-1})	Rate of Heat Release ($kcal \cdot min^{-1} \cdot mol\ C^{-1}$)
NaCl	66	10.1	194	11.1	0.092	0.141	−14.6	0.46	0.83	−83.5
$Na_2B_4O_7$	2810	16.9	245	12.4	0.047	0.076	−7.8	0.69	0.55	−64.3
Untreated	163	11.5	115	11.2	0.091	0.078	−9.0	0.57	0.47	−54.2
$(NH_4)_2HPO_4$	1526	16.0	39	11.2	0.046	0.027	−3.4	0.58	0.16	−27.3
H_3BO_3	108	11.4	33	10.3	0.065	0.040	−5.4	0.40	0.21	−27.5

NOTE: Cotton fabrics were pyrolyzed at 550 °C.

TABLE X. Effect of Prepyrolysis Additives on Rate of Char Oxidation and CO/CO_2 Ratio

Additive	Smoldering Behavior of Char	$d\dfrac{(CO + CO_2)}{dt\ max}$ at 450 °C $(mmol\ g\ char^{-1}\ min^{-1})$	CO/CO_2 Yield at 450 °C
—	smolders	5.7	1.01
HBO_3	none	4.6	1.69
$Na_2B_4O_7$	smolders	6.1	1.03
$(NH_4)_2HPO_4$	none	2.9	1.71
$(NH_4)_2SO_4$	none	3.9	1.43
$ZnCl_2$	none	5.1	0.90
$AlCl_3$	none	5.3	0.95
$NaCl$[a]	smolders	13.9	0.70
Na_2SO_4[a]	smolders	15.2	0.86

NOTE: Char oxidation performed at 450 °C in air.
[a] Ignition occurred initially in these samples in CO and CO_2 measurements. Ignition peak prevented ΔH determination by DSC.

and fibrous materials. With these materials the char can be oxidized gradually by the in-diffusing air. In the absence of substantial heat loss by radiation, convection, or conduction, the low rate of heat release is sufficient for further charring and propagation of the smoldering combustion. In contrast, glowing combustion refers to the oxidation of the residual char after flaming combustion has occurred.

This brief description of the combustion process serves to indicate how the reactions involved could be controlled by the prevailing conditions of heat and mass transfer and by the chemical and physical properties of the fuel, and especially, by the rate of pyrolysis and heat of combustion of the pyrolysis products.

Acknowledgments

I wish to acknowledge the contributions of the postdoctoral fellows, research associates, and graduate students who are named in the following list of references as my coauthors. I am also grateful to the National Bureau of Standards for supporting this program through a series of grants including Grant No. NB81NADA2066.

Literature Cited

1. Shafizadeh, F. *Adv. Carbohydr. Chem.* **1968**, *23*, 419.
2. Shafizadeh, F.; Bradbury, A. G. W. *J. Appl. Polym. Sci.* **1979**, *23*, 1431.
3. Shafizadeh, F. *Appl. Polym. Symp.* **1975**, *28*, 153.
4. Shafizadeh, F; Lai, Y. Z. *Carbohydr. Res.* **1975**, *42*, 39.
5. Shafizadeh, F.; Fu, Y. L. *Carbohydr. Res.* **1973**, *29*, 113.
6. Shafizadeh, F. *J. Anal. Appl. Pyrol.* **1982**, *3*, 283.
7. Shafizadeh, F.; Furneaux, R. H.; Cochran, T. G.; Scholl, J. P.; Sakai, Y. *J. Appl. Polym. Sci.* **1979**, *23*, 3525.
8. Bradbury, A. G. W.; Sakai, Y.; Shafizadeh, F. *J. Appl. Polym. Sci.* **1979**, *23*, 3271.

9. Shafizadeh, F.; Furneaux, R. H.; Stevenson, T. T. *Carbohydr. Res.* **1979**, *71*, 169.
10. Shafizadeh, F.; Lai, Y. Z.; McIntyre, C. R. *J. Appl. Polym. Sci.* **1978**, *22*, 1183.
11. Shafizadeh, F.; Lai, Y. Z. *J. Org. Chem.* **1972**, *37*, 278.
12. Shafizadeh, F.; Sekiguchi, Y. *Carbon* in press.
13. Sekiguchi, Y.; Frye, J. S.; Shafizadeh, F. *J. Appl. Polym. Sci.* **1983**, *28*, 3513.
14. Bradbury, A. G. W.; Shafizadeh, F. *Combust. Flame* **1980**, *37*, 85.
15. Bradbury, A. G. W.; Shafizadeh, F. *Carbon* **1980**, *18*, 109.
16. Shafizadeh, F. "Biomass Conversion Processes for Energy and Fuels"; Plenum: New York, 1981; p. 103.
17. Susott, R. A.; DeGroot, W. F.; Shafizadeh, F. *J. Fire Flammability* **1975**, *6*, 311.
18. Susott, R. A.; Shafizadeh, F.; Aanerud, T. W. *J. Fire Flammability* **1979**, *10*, 94.

RECEIVED for review May 23, 1983. ACCEPTED August 16, 1983.

14
Chemistry of Fire Retardancy

SUSAN L. LeVAN

U.S. Department of Agriculture, Forest Service, Forest Products Laboratory, Madison, WI 53705

Fire retardancy of wood involves a complex series of simultaneous chemical reactions, the products of which take part in subsequent reactions. Most fire retardants used for wood increase the dehydration reactions that occur during thermal degradation so that more char and fewer combustible volatiles are produced. The mechanism by which this happens depends on the particular fire retardant and the thermal–physical environment. This chapter presents a literature review of the investigations into the mechanisms, a discussion of test methods used for determining fire retardancy, the various formulations used to make wood fire retardant, and the research needs in the field of fire retardancy.

WOOD WAS FIRST TREATED FOR FIRE RETARDANCY in the first century A.D. when the Romans used solutions of alum and vinegar to protect their boats against fire. In 1820, Gay-Lussac advocated the use of ammonium phosphates and borax for treating cellulosic material. Many of the promising inorganic chemicals used today were identified between 1800 and 1870. Since then, the development of fire retardants for wood has accelerated. Commercially treated wood became available after the U.S. Navy (1895) specified its use in ship construction, and New York City (1899) required its use in buildings over twelve stories tall (*1*). Production reached over 65 million board feet in 1943, but by 1964 only 32 million board feet was treated annually (*1*).

Increased efforts to expand the use of wood products in institutional and commercial structures may require wood to be treated with fire retardants. Therefore, research on fire-retardant treatments for wood has accelerated.

Early Studies

One of the earliest studies on fire-retardant treatments for wood was conducted between 1930 and 1935 (Forest Products Laboratory).

This chapter not subject to U.S. copyright.
Published 1984, American Chemical Society

This study resulted in a series of reports on a comprehensive evaluation of fire-retardant treatments for wood (2–6). One hundred and thirty single chemicals or combinations of chemicals in the form of various salts were evaluated for flame-spread reduction, smoke, and corrosivity. Diammonium phosphate ranked first in reducing flame spread, followed by monoammonium phosphate, ammonium chloride, ammonium sulfate, borax, and zinc chloride. Zinc chloride, although excellent as a flame retardant, promoted smoke and glowing. Ammonium sulfate was the least expensive, but under certain environmental conditions it was corrosive to metals. None of the 130 compositions tested was considered ideal because of the adverse effects on some of the properties of wood. Several reviews of the subject are available and provide additional background material (1, 7–10).

Protection of Wood with Fire Retardants

Fire-retardant treatments for wood can be classified into two general classes: (1) those impregnated into the wood or incorporated into wood composite products, and (2) those applied as paint or surface coatings. Chemical impregnation has the greater use, primarily for new materials, whereas coatings have been limited primarily to materials in existing constructions. There are advantages and disadvantages to each class. Coatings are applied easily and they are economical. Chemical impregnation usually involves full-cell pressure treatment and can be costly. A coating is subject to abrasion or wear that can destroy the effectiveness of the fire retardant. Chemical impregnations deposit the fire retardant within the wood, so that if the surface is abraded, chemicals are still present. On-site application of surface coatings requires strict control of the amount applied to ensure correct loading levels for a particular flame-spread rating (11). Both coating and impregnation systems are based on the same chemical compounds, although the formulations for each vary.

Most of the chemicals used in fire-retardant formulations have a long history of use for this purpose, and most formulations are based on empirical investigations for best overall performance. These chemicals include the phosphates, some nitrogen compounds, some borates, silicates, and more recently, amino-resins. These compounds reduce the flame spread of wood but have diverse effects on strength, hygroscopicity, durability, machinability, toxicity, gluability, and paintability (1, 12, 13).

Test Methods

Knowledge of various test methods used to evaluate the effectiveness of fire retardants is necessary to understand the mechanisms of fire retardancy and formulations of fire retardants. Some of these tests are used by regulatory agencies to evaluate building materials and some are used for research and development work only. The commonly used test methods applicable to evaluate fire-retardant treatments include thermogravimetric analysis (TG); differential thermal analysis (DTA), and a similar technique, differential scanning calorimetry (DSC); 2-, 8-, and 25-ft tunnel flame-spread tests; and the oxygen index test. Other test methods are used to evaluate the effect of fire-retardant treatments on such related properties as smoke development, heat release rate, and toxicity.

Thermogravimetric Analysis (TG). TG involves weighing a sample while it is exposed to heat. The chief use of this technique has been to study the thermal decomposition of polymeric materials and to accumulate kinetic information about such decomposition. A sample is suspended on a sensitive balance that measures the weight (Figure 1) as it is exposed to a furnace. Air, nitrogen, or another gas flows around the sample to remove the pyrolysis or combustion prod-

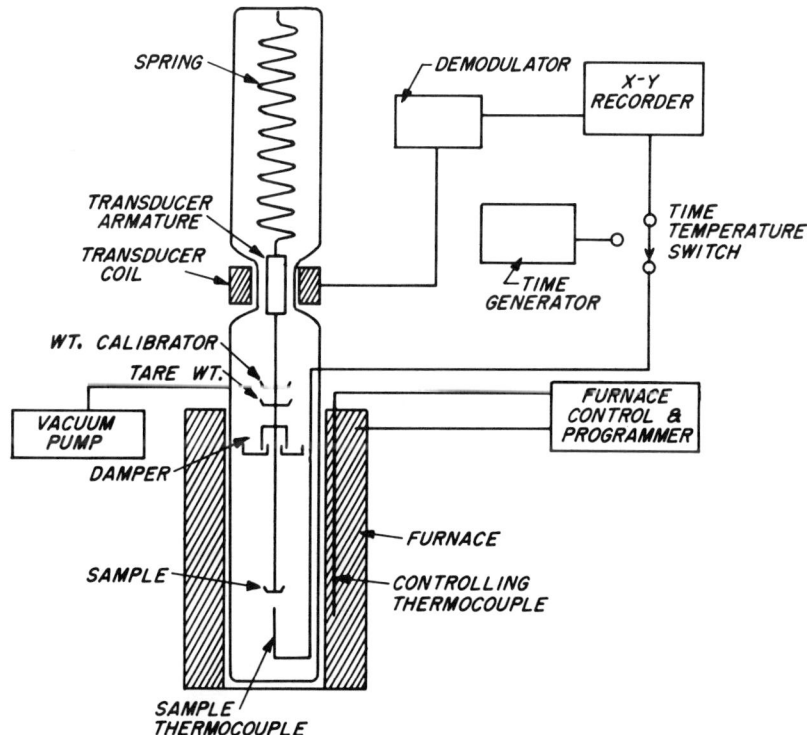

Figure 1. Schematic diagram of an early thermogravimetric analysis system.

ucts. Weight loss is recorded as a function of time and temperature. In isothermal TG, the change in weight of the sample is recorded as a function of time as the temperature remains constant. In dynamic or nonisothermal thermogravimetry, the change in weight is a function of both temperature and time as the temperature is raised at a given heating rate. With the use of a derivative computer, the rate of weight loss as a function of time and temperature can also be determined. The computer allows a more accurate determination of peak temperature transitions. This is referred to as derivative thermogravimetry (14).

Differential Thermal Analysis and Differential Scanning Calorimetry. DTA measures the amount of heat liberated or absorbed as a material moves from one physical transition state to another (i.e., melting, vaporization) or whenever it undergoes a chemical reaction. This heat is determined by measuring the temperature differences between a sample of the material and an inert reference. DTA can be used to measure heat capacity, to provide kinetic data, and to give information on transition temperatures. The test device consists of sample and reference pans exposed to the same heat source. Temperatures are measured by thermocouples embedded in the sample and reference pan. The temperature difference between sample and reference is recorded against time as the exposure temperature is increased at a linear rate. For calorimetry, the equipment is calibrated against standards at several temperatures. As in TG, air, nitrogen, or some other atmosphere flows through the sample cell to remove the resulting vapors (14).

DSC is similar to DTA except the actual differential heat flow is measured when the sample and reference temperature are equal. In DSC both the sample and reference are provided with individual heaters. If a temperature difference develops between the sample and reference because of exothermic or endothermic reactions in the sample, the power input is adjusted to remove this difference. Thus, the temperature of the sample holder is always kept the same as the reference.

The small sample size, rapid removal of pyrolysis or combustion products, and availability of huge excesses of reactant oxygen during thermal analysis can lead to erroneous interpretation of the material in terms of its performance in actual thermal situations. However, thermal analysis tests can provide basic information on the pyrolysis and combustion mechanism and can provide data on the relative performance of materials. This information should be supported by larger-scale fire tests.

Tunnel Flame-Spread Tests. The growth of a small fire in a building is influenced by the rate at which flames spread over the fire-exposed surfaces. Therefore, the combustibility and flame-spread characteristics of furnishings and the interior finish are important safety factors. Building standards designed to control fire growth often require certain flame-spread ratings for various parts of a building. For code regulations, flame-spread ratings are determined by a 25-ft tunnel test (Figure 2) which is an approved standard test method (15). For research and development work there are 2- and 8-ft tunnel tests. The 8-ft tunnel test (Figure 3) is also an approved standard (16). All tunnel tests measure the surface flame spread of a material although each differs in the method of the exposure. A specimen is exposed to an ignition source, and the rate at which the flames travel to the end of the specimen is measured. In the past, red oak flooring was used as a standard and was given a flame spread index of 100. Today, red oak flooring still has an index around 100, but is no longer used in the calculation of the ASTM E 84 flame-spread index.

The severity of the exposure and the time a specimen is exposed to the ignition source are the main differences between the tunnel test methods. The 25-ft tunnel test is the most severe exposure and the specimen is usually exposed for 10 min. An extended test of 30 min is performed on fire-retardant treated products. Materials that pass the extended test (have flame spread less than 25 with no evidence of glowing) qualify for a special "FR-S" rating. Because the 25-ft tunnel test is the most severe exposure it is used as the standard for building materials. The 2-ft tunnel test (17, 18) is the least severe. Because of the small specimen size required with this test, it is a valuable tool for development work on fire retardants. The 8-ft tunnel falls between the 2- and 25-ft tunnels in severity. It can be a valuable

Figure 2. *Underwriters Laboratory's 25-ft tunnel flame-spread test.*

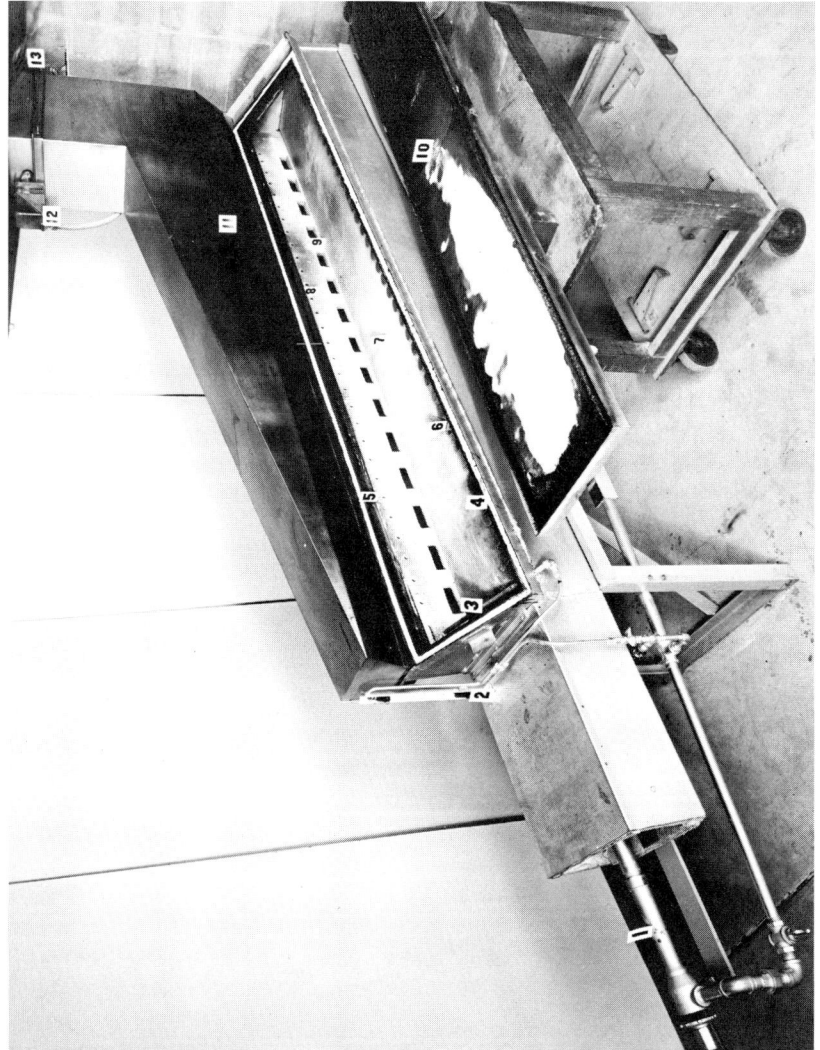

Figure 3. Forest Products Laboratory's 8-ft tunnel flame-spread test.

tool to determine the effectiveness of treatments as shown in Figure 4 (*19*).

Critical Oxygen Index Test. The oxygen index test measures the minimum concentration of oxygen in an oxygen–nitrogen mixture that will just support flaming combustion of a test specimen. This standard test method (*20*) provides the critical oxygen index required to maintain this flaming combustion under experimental conditions. Highly flammable materials have low oxygen index, less flammable materials have high values. The test was originally developed for plastics but it can be used for wood and, in particular, fire-retardant-treated wood (*21–23*). Figure 5 shows the effect of chemical retention levels on the oxygen index value for different chemical treatments on Douglas-fir.

One advantage of this test is the very small specimen size (*24*); another is that this method can be used to study the retardant mechanism in the gas phase which cannot be done with TG, DTA, or DSC because they only measure properties in the solid phase. A specimen treated with a chemical compound that acts as a gas-phase inhibitor should demonstrate the following: (1) show an increase in oxygen

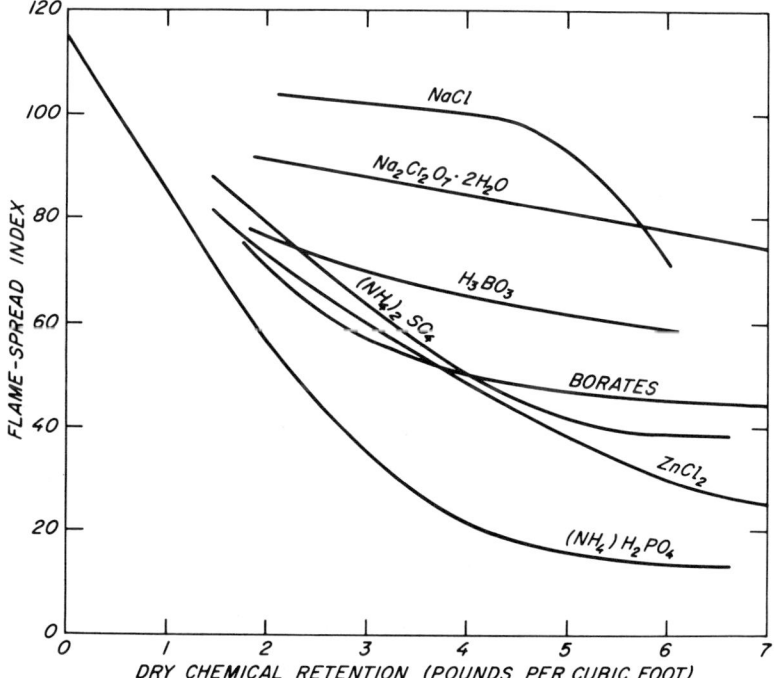

Figure 4. Effect of inorganic additives on flame-spread index, measured in 8-ft tunnel (19).

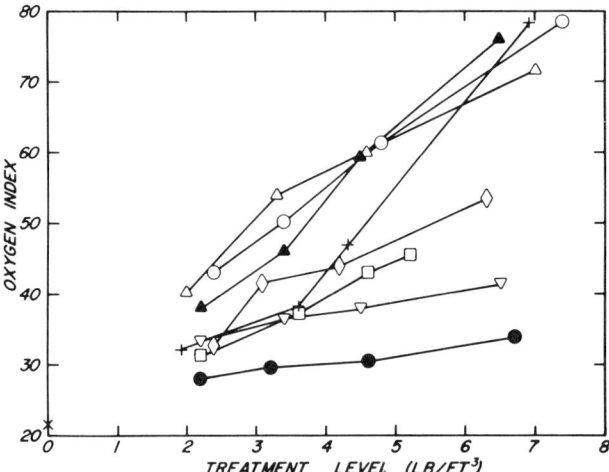

Figure 5. Effect of fire-retardant chemicals on oxygen index (21). Key: ▽, *ammonium sulfate;* +, *sodium tetraborate decahydrate;* ○, *disodium octaborate tetrahydrate;* □, *boric acid;* △, *monoammonium phosphate;* ▲, *ammonium polyphosphate (11-37-0);* ◇, *zinc chloride;* ●, *sodium dichromate;* ×, *untreated.*

index value when the sample is run in a regular oxygen–nitrogen atmosphere, and (2) show no change from the untreated specimen when the sample is run in N_2O_4. However, if the flame retardant acts in the solid phase, its effectiveness should not be affected by a change in oxidant. Therefore, discrimination is possible between vapor- and solid-phase activity (25, 26).

Test Methods for Related Properties. Test methods are available to evaluate such related physical properties of retardants as smoke production, heat release rate, and toxicity.

SMOKE PRODUCTION. Smoke production can be a critical problem in fire-retardant formulations. The 25-ft tunnel test uses a photoelectric cell to measure the amount of smoke evolved. The smoke density is measured continuously and is assigned a value relative to the behavior of red oak. The effect of fire retardants on smoke production varies depending on the chemical. Figure 6 demonstrates this effect as measured in the 8-ft tunnel; however, smoke values measured in various tunnel tests may not agree or correlate.

The National Bureau of Standards smoke density chamber (Figure 7) is a more recent technique used to evaluate smoke. This chamber can be used to measure adequately the smoke produced from untreated and fire-retardant-treated wood (27, 28). This method has three advantages over the tunnel method: (1) application to a variety of room situations, burning areas, and light-path lengths, (2)

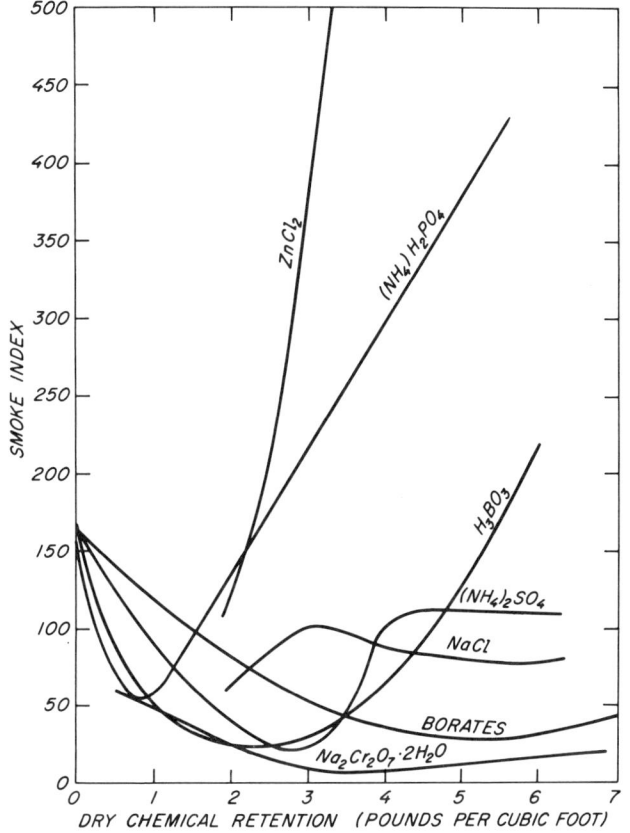

Figure 6. Effect of inorganic additives on smoke index measured in 8-ft tunnel (19).

control over exposure conditions, and (3) measurement of smoke production in flaming and nonflaming modes. This method is now a standard test procedure (29).

HEAT RELEASE RATE. Another measurement that is gaining acceptance as a tool for evaluation of fire-retardant treatments is the measurement of the heat release rate (30–35). The heat of combustion of wood varies depending on the species, resin content, moisture content, and other factors. The contribution to fire growth from wood depends on the total effect of these factors, along with the fire exposure and degree of combustion. Although the heat of combustion of a material never changes, fire retardants reduce the rate of heat release and extend the time at which the heat release begins to be measurable (31, 32, 35, 36). The rate of heat release for treated and

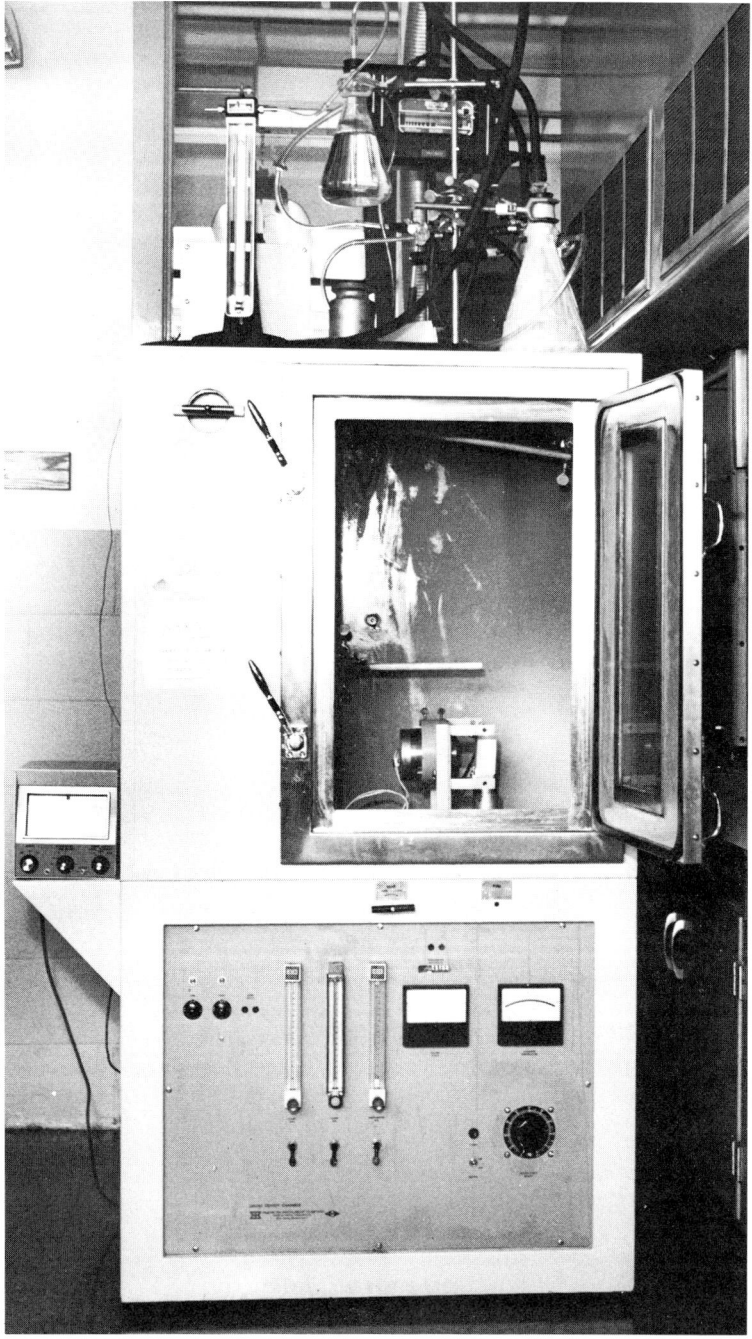

Figure 7. NBS smoke chamber is used to determine the obscured optical density from various materials.

untreated materials can be utilized in mathematical models to predict fire growth.

TOXICITY. Toxicity testing may become applicable to fire-retardant treatments because the toxicity of combustion products is becoming an important fire-performance characteristic. A large percentage of fire victims are not touched by flames but are overcome as a result of exposure to smoke, to toxic gases, and to depleted oxygen levels. The effect of fire retardants on the combustion products is, therefore, of increasing concern. A critical review by Kaplan et al. (37) discusses all the various proposed toxicity test methods and the advantages and disadvantages of each.

Mechanisms of Fire Retardancy

The burning of a solid is essentially a three-stage process consisting of a heating stage, a thermal degradation stage, and an ignition stage. Chapter 13 contains discussion of the chemistry of pyrolysis and combustion. Some of these concepts are reviewed here briefly in order to facilitate the discussion of fire-retardant mechanisms.

Chemistry of Burning. PYROLYSIS. The temperature of wood rises when it is exposed to an outside heat source. This rise in temperature is accompanied by the breaking of chemical bonds, which accelerates as the temperature increases further. In the absence of oxygen, this thermal degradation is called pyrolysis. As the wood undergoes thermal pyrolysis, volatile gases are produced and diffuse into the surrounding air.

VAPOR-PHASE COMBUSTION. If the volatile gases are mixed with air and heated by an external heat source to the ignition temperature, exothermic reactions known as combustion begin. The heat from these exothermic reactions in the vapor phase reradiates to the original material, thereby propagating the pyrolysis reactions in the solid phase. If the burning mixture accumulates enough heat to emit radiation in the visible spectrum, the phenomenon is known as flaming combustion and occurs in the vapor phase.

SMOLDERING AND GLOWING. In many materials, including wood and paper, a solid-phase combustion can also occur. This type of combustion is known as glowing or smoldering combustion. In this condensed phase, the activated char produced by the pyrolysis of the material combusts with oxygen to produce CO_2 and water vapor. Smoldering combustion usually is distinguished from glowing combustion in that combustion of the volatile pyrolysis products is not essential to propagate smoldering, and it may proceed in materials that have not undergone prior conversion to a char residue. Smoldering combustion usually is associated with materials that have a high surface-to-weight ratio, such as sawdust, upholstery material, and coal. The essential ingredient for smoldering combustion is low heat losses. Materials that smolder have good insulating properties and produce abundant char.

Theories of Fire Retardancy. Several theories have been proposed for the mechanism of fire retardants. Browne (38) made a complete literature search on these theories and summarized the research in an effort to understand the mechanisms involved in the inhibition of pyrolysis and combustion of wood. These mechanisms can be categorized under several theories.

1. *Barrier theories.* Fire-retardant chemicals prevent the escape of volatile products by forming a glassy barrier. This barrier also prevents oxygen from reaching the substrate and insulates the wood surface from high temperatures.
2. *Thermal theories.* Fire-retardant chemicals may increase the thermal conductivity of the wood to dissipate the heat from the surface faster than it is supplied by the igniting source, or they may affect chemical and physical changes so that heat is absorbed by the chemical, preventing the wood surface from igniting.
3. *Dilution by noncombustible gases theories.* Nonflammable gases released by the decomposition of the fire-retardant chemicals dilute the combustion gases formed by the pyrolysis of the wood and form a nonflammable gaseous mixture.
4. *Free radical trap theories.* Fire-retardant chemicals release free radical inhibitors at pyrolytic temperatures that interrupt the chain propagation mechanism of flammability.
5. *Increased char/reduced volatiles theories.* Fire-retardant chemicals lower the temperature at which pyrolysis occurs, directing the degradation pathway toward more char production and fewer volatiles.
6. *Reduced heat content of volatiles theories.* Fire-retardant chemicals lower the heat content of the combustible volatiles. This reduction in heat content always occurs when the amount of char is increased and the amount of volatiles is reduced. Therefore, Theories 5 and 6 function together, resulting in more char, fewer volatiles, and lower heat content of volatiles.

In most cases, a given fire retardant operates by several of these mechanisms, and much research has been done to determine the magnitude and role of each of these mechanisms in fire retardancy. The influence of the combining effect of several of these mechanisms is illustrated by phosphorus–nitrogen synergism. The theories involved in this synergistic system are discussed to demonstrate this interaction between mechanisms. Some mechanisms apply only to flaming combustion (vapor phase), some apply to both flaming and smoldering combustion, and others only apply to smoldering. The phase to which each theory pertains will be indicated, as well as a separate section on smoldering inhibition theories.

BARRIER THEORIES. A physical barrier can retard both smoldering combustion and flaming combustion by preventing the flammable products from escaping and by preventing oxygen from reaching the substrate. These barriers also insulate the combustible substrate from high temperatures. Common barriers include sodium silicates and coatings that *intumesce* (puff and form a cellular structure that remains attached to substrate). Intumescent systems swell and char on exposure to fire to form a carbonaceous foam and consist of several components. These components include a char-producing compound (polyhydric alcohol, carbohydrates, or epoxy resins), a blowing agent, a Lewis-acid dehydrating agent, and other optional components.

In the intumescent systems, the char-producing compound, such as polyol, will normally burn to produce CO_2 and water vapor and leave flammable tars as residues. However, the compound can esterify when it reacts with certain inorganic acids, usually phosphoric acid. The acid acts as a dehydrating agent and leads to increased amounts of char and reduced volatiles. Such char is produced at a lower temperature than the charring temperature of the wood substrate. *Blowing agents* decompose at characteristic temperatures and release gases that expand the char. Common blowing agents are dicyandiamide, melamine, urea, and guanidine (*39, 40*); they are selected on the basis of their decomposition temperature. Many blowing agents also act as the dehydrating agent. Other materials such as binders are added to the formulation to improve the toughness of the carbon foam.

Ingredients used in intumescent systems usually fulfill more than one function. Most compounds release some gas on heating, therefore they can be considered to be blowing agents. Many compounds produce some char.

THERMAL THEORIES. Researchers at Forest Products Laboratory impregnated wood with a metal alloy to determine whether change in thermal conductivity is a mechanism of fire retardants (*38*). The alloy was selected to melt at 105 °C. The treated and untreated specimens were subjected to a flame on one side and the temperature rise was recorded on the unexposed side. The rise of temperature was slower over the alloy-treated specimen than over the untreated specimen until the melt temperature of the alloy. Above this temperature the treated and untreated specimens then followed the same time–temperature regimes. The untreated specimen burst into flames and the treated specimens smoked and charred; however, all specimens did so at the same time and temperature. These observations could not be explained on the basis of changes in thermal conductivity alone (38).

Another thermal theory suggests that fire retardants cause chemical and physical changes so that heat is absorbed by the chemical to prevent the wood surface from igniting. This thermal absorption theory is based on chemicals that contain much water of crystallization.

Water will absorb its latent heat of vaporization from the pyrolysis reactions until all the water is vaporized. This serves to remove heat from the pyrolysis zone, thereby slowing down the pyrolysis reactions. This is demonstrated in the increased ignitability of very dry woods and forest fuels compared to woods and fuels with high moisture contents. However, Browne and Tang (41) and others (42–48) have demonstrated with TG and DTA that, after the water is lost, the pyrolysis of wood occurs and is independent of the past moisture content of the wood.

DILUTION OF NONCOMBUSTIBLE GASES THEORIES. Most of the evidence for this mechanism can be derived by considering the blowing agents in the intumescent systems discussed previously, or agents that release large amounts of water vapor. Agents such as dicyandiamide and urea release noncombustible gases at temperatures below the temperature at which the active pyrolysis begins. Borax compounds release water vapor in large quantities. The main difficulty with this theory has been that not enough noncombustible gas can be liberated to dilute the volatile gases. However, Browne (38) found that flammable gases account for only 23% of the total volume produced. Any reduction in this percentage would be beneficial because it increases the volume of combustible volatiles needed for ignition. Also the movement of gases away from the substrate may dilute the amount of oxygen near the boundary layer between the substrate and the vapor-phase reaction.

FREE RADICAL TRAP THEORIES. Combustion vapor-phase reactions have been studied using premixed gas flames such as methane. Considerable information concerning the mechanism of flame propagation has resulted from this work (40, 49, 50). Basically the process occurs predominantly by branching chain reactions among free radicals. The major chain branching reactions are

$$H \cdot + O_2 \rightarrow HO \cdot + O \cdot \tag{1}$$

$$O \cdot + H_2 \rightarrow HO \cdot + H \cdot \tag{2}$$

These two equations govern the exponential increase in free radical concentration; however, these postulations are based on premixed gas flames with excess oxygen available. Application of this theory to the combustion of solids must be treated with reservations because the combustion of wood proceeds in oxygen-deficient diffusion flames whose processes are a complex series of simultaneous reactions dependent on the material and the environment. Therefore, the exact role free radicals play in the combustion of wood is not known.

Certain fire retardants affect vapor-phase reactions by inhibiting the chain reactions in Reactions 1 and 2. Halogens such as bromine and chlorine are good free radical inhibitors and have been studied extensively in the plastics industry (40, 49, 50). Generally, large amounts of halogen are required (15–30% by weight) to attain a practical degree of fire retardance. The efficiency of the halogen decreases in the order Br > Cl > F. A mechanism for the inhibition of the chain branching reactions (using HBr as the halogen) is

$$H \cdot + HBr \rightarrow H_2 + Br \cdot \quad (3)$$
$$OH \cdot + HBr \rightarrow H_2O + Br \cdot \quad (4)$$

The hydrogen halide consumed in these reactions is regenerated to continue the inhibition. Although this proposed mechanism was based on experiments with premixed hydrocarbon flames, the same order of effectiveness exists with wood.

An alternate mechanism (Reactions 5–7) was suggested for halogen inhibition which involves recombination of oxygen atoms (50).

$$O \cdot + Br \cdot + M \rightarrow BrO \cdot + M^* \quad (5)$$
$$O \cdot + Br_2 \rightarrow BrO \cdot + Br \cdot \quad (6)$$
$$O \cdot + OBr \cdot \rightarrow Br \cdot + O_2 \quad (7)$$

Thus the inhibitive effect results from the removal of active oxygen atoms (O ·) from the vapor phase. Additional inhibition can result from removal of OH radicals in the chain-branching reactions:

$$BrO \cdot + \cdot OH \rightarrow HBr + O_2 \quad (8)$$
$$BrO \cdot + \cdot OH \rightarrow Br \cdot + HO_2 \quad (9)$$

Reactions 5–9 explain the lack of halogen inhibition in hydrocarbon–nitrous oxide flames where the hydrogen–oxygen chain is not required for oxidation (50). Some phosphorus compounds also have been found to inhibit flaming combustion by this mechanism (51).

INCREASED CHAR/REDUCED VOLATILES THEORIES. Most of the evidence relating to the mechanism of fire retardancy in the burning of wood indicates that retardants alter fuel production by increasing the amount of char and reducing the amount of volatile, combustible vapors. Many fire retardants for wood also lower the temperature at which active pyrolysis occurs.

Early studies involved treatment of wood specimens with fire-retardant chemicals, then subjecting the treated specimens to thermal analysis by TG. Browne and Tang (41) tested eight compounds, some of which were known to be effective fire retardants, and some of which were not. The TG results (Figures 8–10) indicate that all compounds increased the residual char weight of the material. Except for sodium tetraborate, the more effective the salt as a flame retardant the lower the temperature of active pyrolysis and the greater the amount of char. These results were confirmed through repeated experiments (45, 46).

Experiments were conducted on the pyrolysis products of wood samples to affirm that the increased amounts of char involved a decrease in the amount of combustible tars (52). The chemicals increased the yield of char, water, and noncondensable gases at the expense of the flammable tar fraction. These results confirmed that the increased amount of residual char in TG results was associated with the reduction of the combustible volatiles.

A possible chemical mechanism for the reduction of these combustible volatiles is that fire-retardant chemicals somehow inhibited the formation of levoglucosan (1,6-anhydroglucopyranose), a major volatile fraction obtained from the thermal degradation of cellulose (see Chapter 13). The results obtained from TG prompted many researchers to investigate this possible mechanism. The amount of levoglucosan produced by treated and untreated specimens of cellulose was analyzed and the results can be found in Table I (53). All the chemicals in Table I reduced the percentage of levoglucosan regardless of the relative effectiveness of the fire retardant as determined by the oxygen index test. Their findings include the effect of acidic, neutral, and basic additives on the levoglucosan yield (Table II). The acid treatment had the most pronounced effect on the breakdown. These results and the oxygen index results suggest that alkali and acid treatments impart flame retardancy to cellulose through different chemical mechanisms.

In degree of polymerization (DP) studies of borax treatments and ammonium dihydrogen orthophosphate (53), cellulose treated with the acid charred and depolymerized very rapidly. Its DP value decreased from 1110 to 650 after only 2 min of heating at 150 °C. Cellulose treated with borax showed a DP reduction from 1300 to 700 after 1 h of heat treatment at 150 °C. Both these compounds catalyzed the suppression of levoglucosan formation but they had different effects on the chain depolymerization reaction (53).

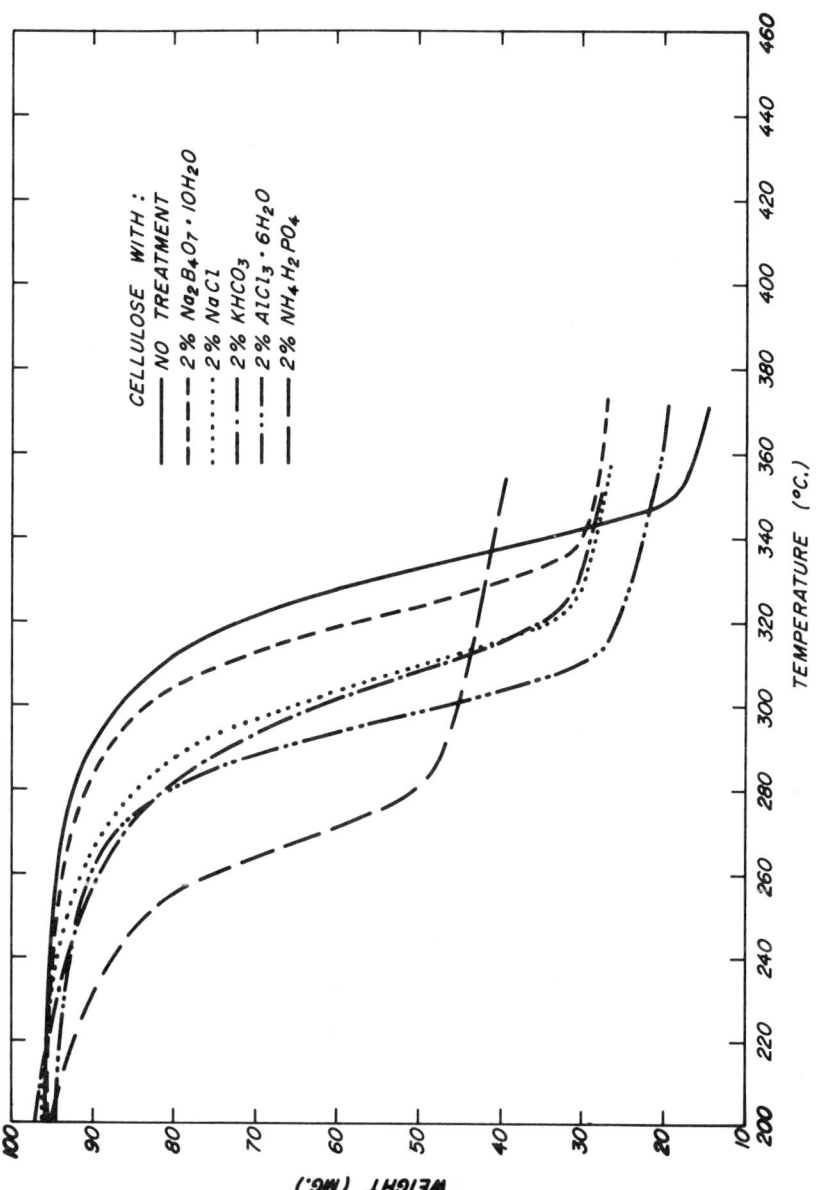

Figure 8. Thermogravimetric analysis of cellulose treated with various additives (46).

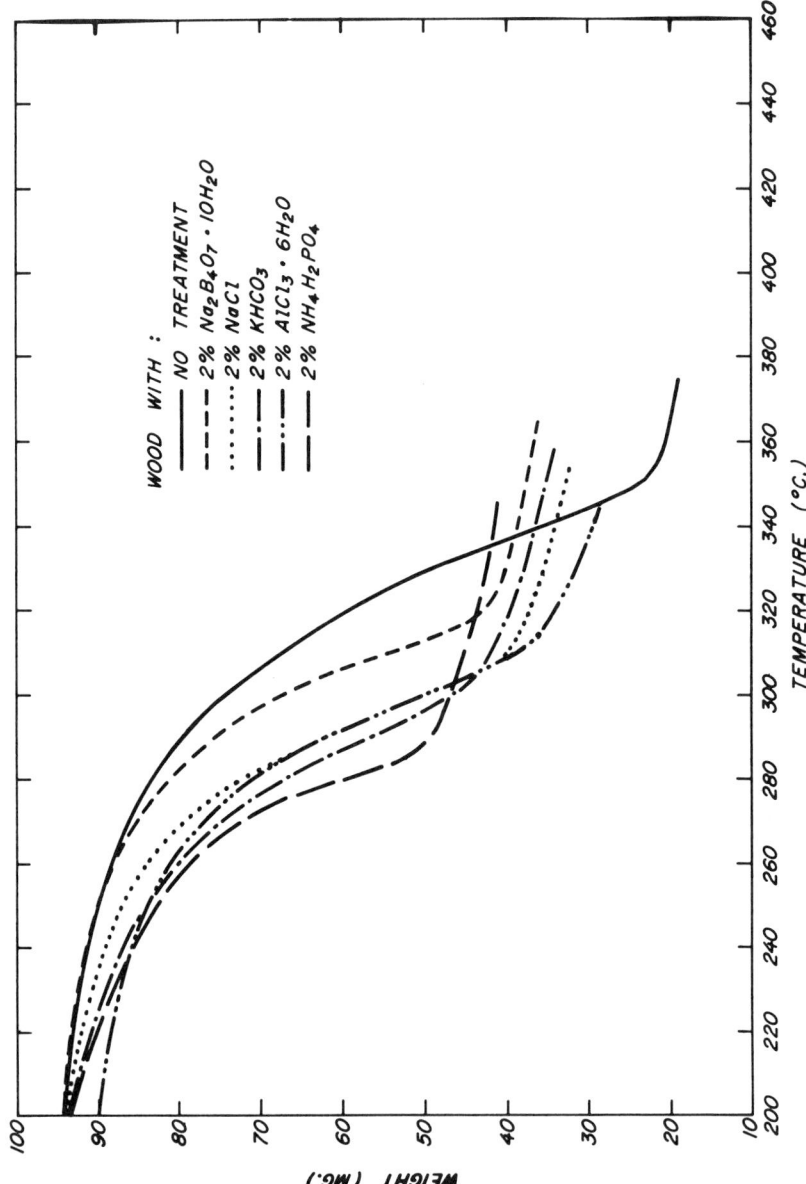

Figure 9. Thermogravimetric analysis of wood treated with various additives (46).

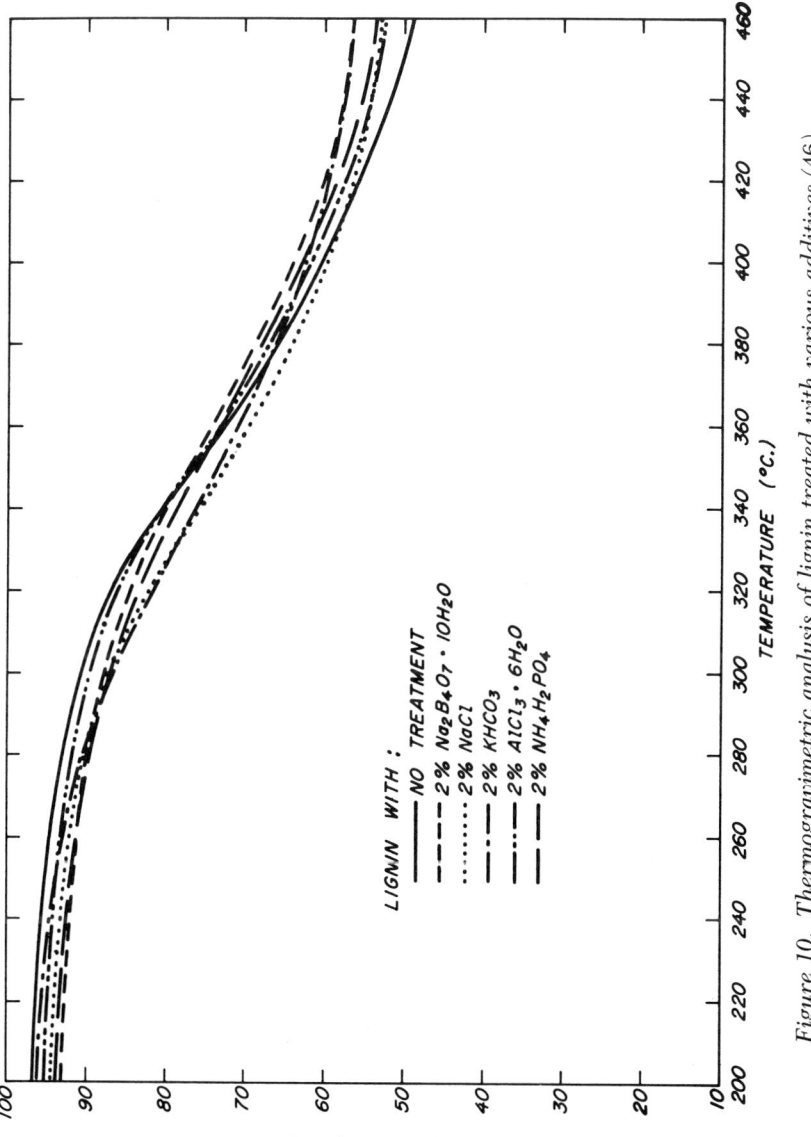

Figure 10. Thermogravimetric analysis of lignin treated with various additives (46).

Table I. Effect of Inorganic Additives on Oxygen Index and Levoglucosan Production

Treatment	Oxygen Index (%)	Levoglucosan Yield (%)
Untreated	17.3	10.1
KH_2PO_4	18.5	0.9
K_2HPO_4	18.6	0.2
$Na_2B_4O_7 \cdot 10H_2O$	19.3	<0.1
$ZnCl_2$	19.6	0.3
$NH_4H_2PO_4$	19.6	0.8
H_3PO_4	20.5	<0.1

(Adapted from Reference 53.)

The nature of uncatalyzed, acid-catalyzed, and alkali-catalyzed thermal decomposition of levoglucosan was studied (54–56). The data in Table III demonstrate that the same products are obtained from uncatalyzed and acid- or alkali-catalyzed reactions, although the quantities produced vary. The acid-catalyzed reaction produced the greatest amount of char. This agrees with the findings of Fung et al. (53).

Fung (57) and Halpern et al. (58) identified the pyrolysis product from cellulose treated with phosphoric acid as 1,6-anhydro-3,4-dideoxy-Δ^3-β-D-pyranosen-2-one (levoglucosenone). Halpern et al. (58) found the same compound when the cellulose was treated with monoammonium phosphate ($NH_4H_2PO_4$), sodium dihydrogen phosphate (NaH_2PO_4), or sodium hydrogen sulfate ($NaHSO_4$). They proposed the mechanism in Figure 11 for the acid-catalyzed reaction of 1,6-anhydro-β-D-glucopyranose (I) to 1,6-anhydro-3,4-dideoxy-Δ^3-β-D-pyranosen-2-one (IV), a combustible product. Work with the model compound glucovanillin revealed that Compound IV was produced

Table II. Effect of Acid, Alkaline, and Neutral Additives on Levoglucosan Production

Products	Levoglucosan (wt %)	Levoglucosan + 1% KBr	Levoglucosan + 1% K_2CO_3	Levoglucosan + 1% H_3PO_4
CO_2	0.12	—	0.21	4.05
CO	—	—	—	1.9
Levoglucosan (undecomposed)	86.05	73.30	63.90	2.28
Char residue	3.16	—	—	35.15
Total	97.38	95.55	80.41	80.58

(Adapted from Reference 53.)

Table III. Effect of Additives on Production of Various Pyrolysis Products

Compound	Untreated (%)	$ZnCl_2$ (%)	NaOH (%)
Acetaldehyde	1.1	0.3	7.3
Furan	1.0	1.3	1.6
Acrolein	1.7	<0.1	2.6
Methanol	0.3	0.4	0.7
2,3-Butanedione	0.5	0.8	1.6
2-Butenal	0.7	0.2	2.2
1-Hydroxy-2-propanone	0.8	<0.1	1.1
Glyoxal	1.4	<0.1	4.9
Acetic acid	1.7	0.7	1.5
2-Furaldehyde	0.9	3.0	0.4
5-Methyl-2-furaldehyde	0.1	0.3	—
CO_2	2.9	6.8	5.7
Water	8.7	20.1	14.1
Char	3.9	29.0	16.0
Balance	74.3	36.8	40.3

(Adapted from Reference 55.)

with phosphoric acid treatment (59). Compound I was found to decompose almost completely in the presence of phosphoric acid to give char, water, CO, CO_2, and just a small amount of Compound IV. The acid is believed to react with the cellulose to give Compound IV directly without going through the intermediate Compound I (Figure 11) (58). This mechanism of inhibiting formation of levoglucosan may only exist with phosphorus-type compounds. Some other mechanism may exist for nonphosphorus fire retardants.

Flame retardants may not only catalyze dehydration of the cellulose to more char and fewer volatiles but also enhance the condensation of the char to form cross-linked and thermally stable polycyclic aromatic structures (60). Cellulose was treated with various additives and then charred at 400 °C. The chars were then oxidized with permanganate (see Chapter 13) and the results are in Table IV. The char yield was slightly higher for the sodium chloride-treated sample (17.5%) and substantially more for the sample containing diammonium phosphate (28.9%), as compared to the yield from the untreated sample (15.3%). Furthermore, the increased char formation was accompanied by increased aromaticity, as measured by the amount of the aromatic carbon obtained from the char and the amount obtained from the original cellulose molecules (60).

The increased char formation is caused by the increased condensation and cross-linking of the carbon skeleton, in addition to any

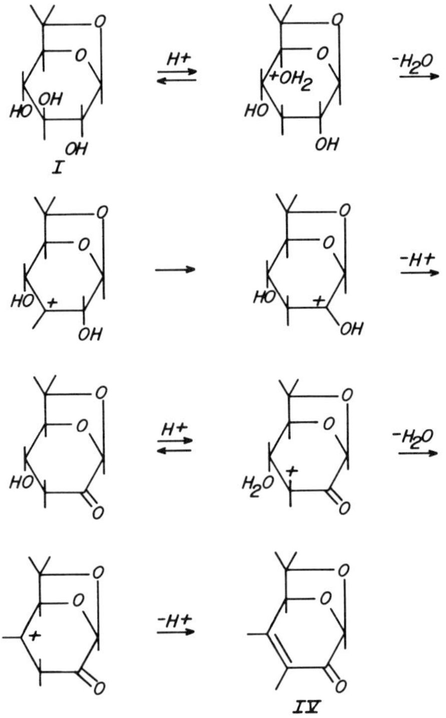

Figure 11. Proposed mechanism of levoglucosan inhibition (from Reference 58).

role that inorganic flame retardants play in dehydration of the glucose units in the cellulose molecule or in lowering the solid-phase combustion rate (60).

Nanassy (61) also examined the effects of fire retardants on the resulting char. He studied the effects of ammonium dihydrogen phosphate on the thermal diffusivity, thermal conductivity, and specific heat of treated and untreated Douglas-fir specimens that had been

Table IV. Effect of Inorganic Additives on Aromatic Char Formation

Additives	Char Yield (wt %)	Aromatic Carbon in Char (wt %)
None	15.3	13.7
NaCl	17.5	16.8
$(NH_4)_2HPO_4$	28.9	18.6

(Adapted from Reference 60.)

charred. The thermal diffusivity of treated specimens decreased from a value of 10.04 to 6.60 mm^2/s at a temperature of 100 °C. At the same temperature the thermal conductivity decreased only slightly, from 6.50 cW/m °C to 6.17 cW/m °C. The specific heat showed a large increase, from 9.1 dJ/g °C for untreated to 11.2 dJ/g °C for treated. The large decrease in diffusivity results in increasing the heat storage capacity of char as evidenced by the large increase in the specific heat.

REDUCED HEAT CONTENT OF VOLATILES THEORIES. Figure 9 shows that the inorganic additives (except for sodium tetraborate) lower the temperature at which active decomposition begins and this resulting decomposition leads to increased amounts of char and reduced amounts of volatiles. In the previous section, this increased amount of char and reduced amount of volatiles were attributed to the increased dehydration reactions, mainly of the cellulose component of wood. However, other competing reactions are also occurring such as decarbonylation, decomposition of simpler compounds, and condensation reactions. All these reactions compete with each other. As a result, shifts favoring one reaction over another also change the overall heat of reaction. Differential thermal analysis is used to determine these changes in heats of reactions and can help gain understanding about these competing reactions.

DTA of wood in helium (Figure 12) indicates two endothermic reactions followed by a feeble exothermic one. The first endothermic reaction, which peaks around 125 °C, is caused by evaporation of water and desorption of gases; the second, peaking between 200 and 325 °C, indicates depolymerization and volatilization (47). At 375 °C these endothermic reactions are replaced with a small exothermic one. When the wood samples are run in oxygen, these endothermic peaks are replaced with strong exothermic reactions, as evidenced in Figure 13 which has a 10-fold decrease in sensitivity compared to Figure 12. The first exotherm, around 310 °C for wood and 335 °C for cellulose, is attributed to the flaming of volatile products; the second exotherm, at 440 °C for wood and 445 °C for lignin, is attributed to glowing combustion of the residual char (47). These thermograms are qualitative, but they do indicate the temperatures where oxidation occurs.

DTA of inorganic fire retardants run in oxygen may shift the peak position temperature or the amount of heat released. Sodium tetraborate reduced the volatile products exotherm considerably, increased the glowing exotherm, and stimulated the appearance of a second glowing peak around 510 °C, as seen in Figure 14. Sodium chloride also reduced the first exotherm, increased the size of the second, but did not produce a second glowing exotherm as did the

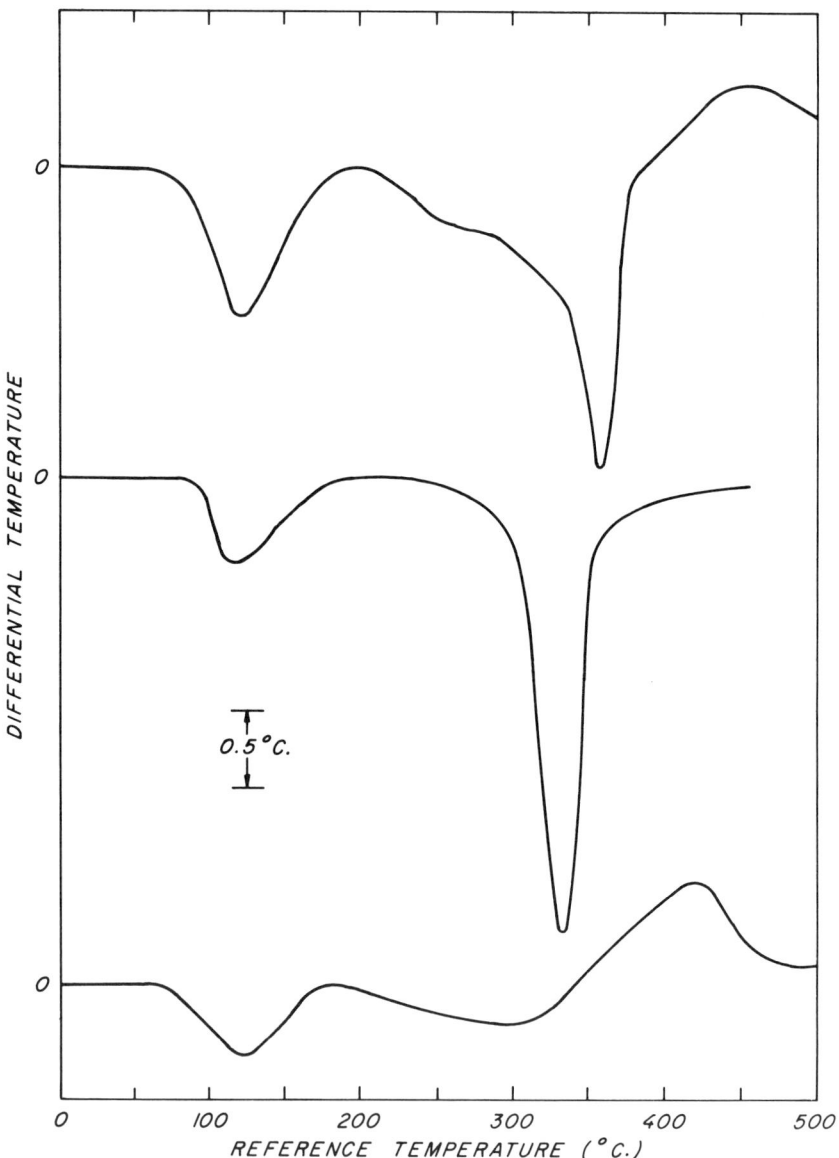

Figure 12. Differential thermal analysis of untreated wood (top), cellulose (middle), and lignin (bottom) run in helium atmosphere (47).

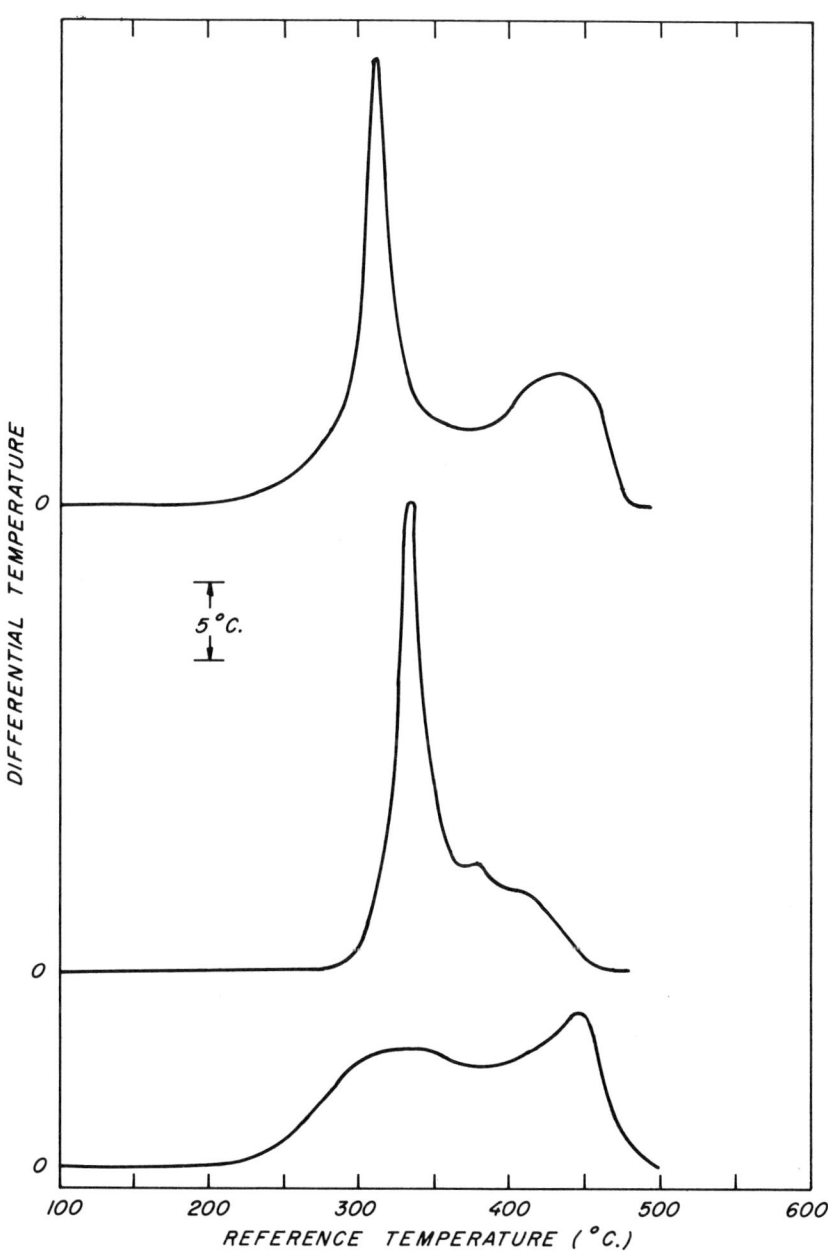

Figure 13. Differential thermal analysis of untreated wood (top), cellulose (middle), and lignin (bottom) run in oxygen atmosphere (47).

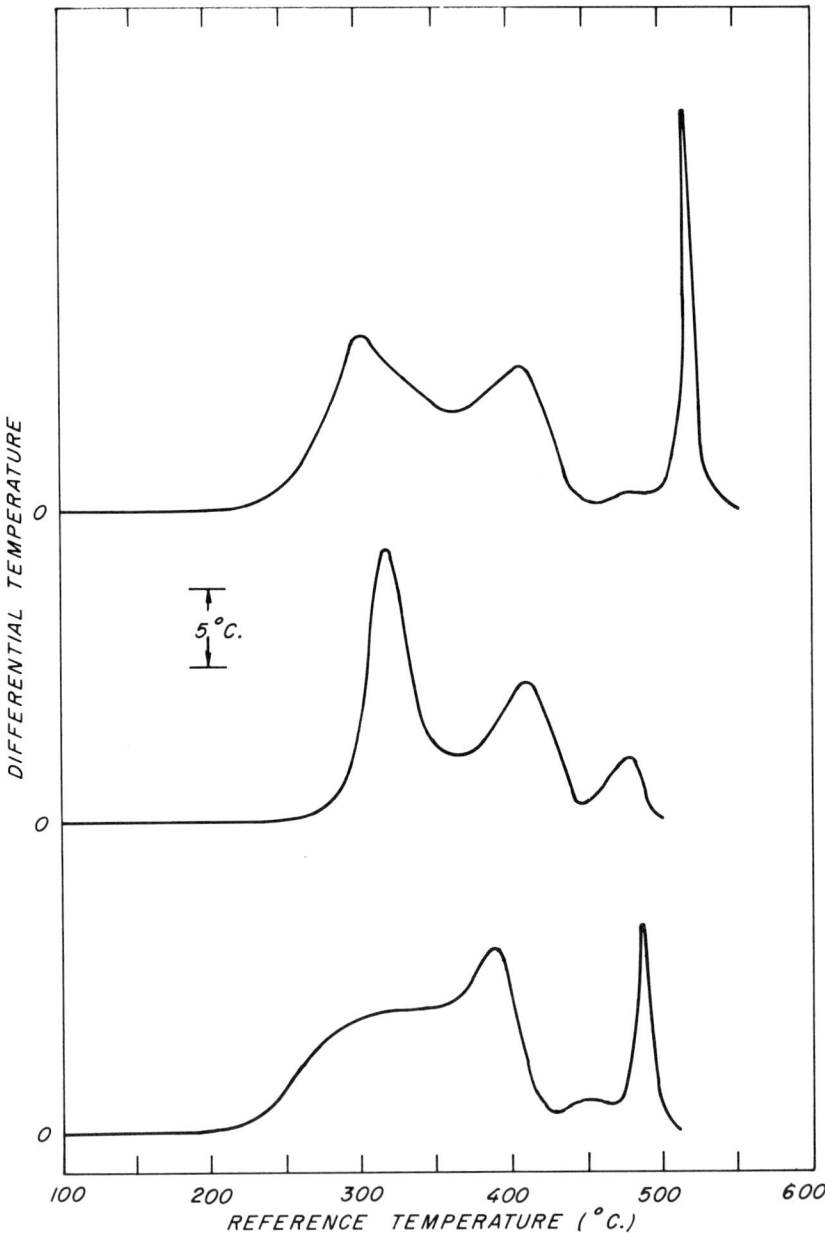

Figure 14. Differential thermal analysis of wood (top), cellulose (middle), and lignin (bottom) treated with 8% sodium tetraborate decahydrate run in oxygen atmosphere (47).

sodium tetraborate. Ammonium phosphate, in Figure 15, caused the most effective reduction in the height of the volatile products peak and also reduced the temperature at which this peak occurs. Also, ammonium phosphate almost eliminated the glowing exotherm. Table V lists the relative maximum heat intensities of the various inorganics and the temperatures at which these peaks occur (47). From this table, we can conclude that effective fire retardants reduced the heights of the volatile products exotherms but had different effects on the glowing peak. (The effect on the glowing peak will be discussed under the section on smoldering combustion.) Further information on the effects of fire-retardant additives on pyrolysis and combustion is contained in References 41, 42, 45–48, 56, 62, and 63.

The heats of combustion of the volatile pyrolysis products released at various stages of volatilization were determined from untreated and chemically treated ponderosa pine (64). Fire-retardant treatments reduced the average heat of combustion for the volatile pyrolysis products released at the early stage of pyrolysis below the value associated with untreated wood at comparable stages of volatilization. At 40% volatilization, untreated wood had released 29% of its volatile products' heat of combustion; treated wood had only released 10–19% of its total heat. Of all the chemicals tested, only NaCl, which is known to be an ineffective fire retardant, did not reduce the heat content. This reduction in heat content of the volatiles was confirmed by using thermal evolution analysis (TEA) (55).

The effectiveness of various compounds by TEA (Table VI) can be compared to the effectiveness determined by TG (Table VII) (56). Except for a few compounds, such as NaCl, $NaHSO_4$, and Na_2CO_3, the ranking of the effectiveness of various fire retardants by the two different methods agree—high effectiveness numbers by TG correspond to high effectiveness numbers by TEA and vice versa. This is as expected if we assume the mechanism for fire retardancy is to increase the amount of char produced and lower the amount of volatile combustible products.

A reaction coulometer has been used to determine the rate of heat release from these combustible volatiles (65). Table VIII shows these results on the effect of inorganic additives that were obtained by using reaction coulometry. The treated cellulose samples decomposed at lower temperatures and produced less heat than the untreated. Addition of 5% NaOH reduced the heat of combustion of cellulose volatiles at 500 °C to less than one-half of untreated (65).

PHOSPHORUS–NITROGEN SYNERGISM THEORIES. As mentioned previously, one role of phosphoric acid and phosphate compounds is to catalyze the dehydration reaction of wood to produce more char.

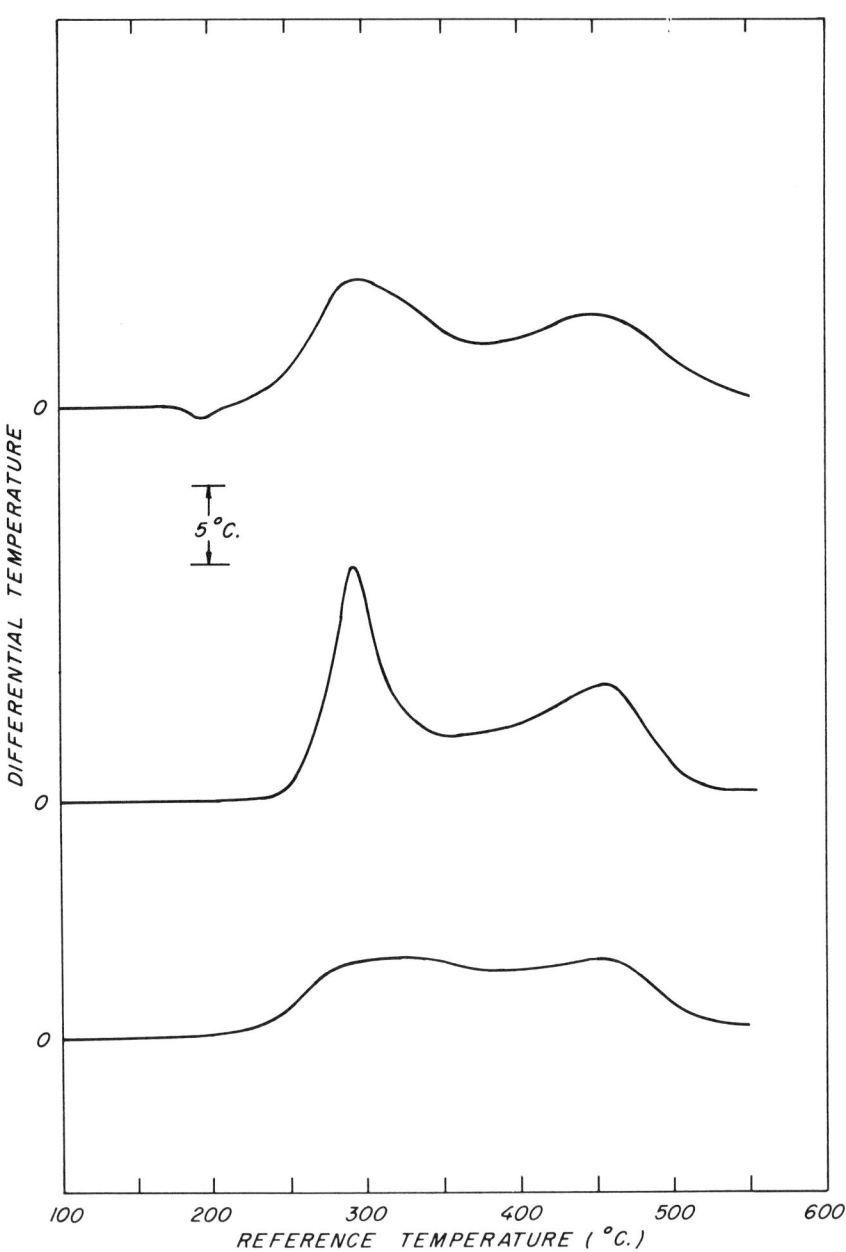

Figure 15. *Differential thermal analysis of wood (top), cellulose (middle), and lignin (bottom) treated with 8% monoammonium phosphate run in oxygen atmosphere (47).*

Table V. Relative Maximum Heat Intensity of Combustion by DTA for Wood, Cellulose, and Lignin

Sample Treatment[a]	Wood			Cellulose			Lignin		
		Solids			Solids			Solids	
	Volatiles	First Peak	Second Peak	Volatiles	First Peak	Second Peak	Volatiles	First Peak	Second Peak
8% aluminum chloride hexahydrate	310 480	425 235	— —	305 675	420 260	— —	330 465	425 810	— —
2% monoammonium phosphate	300 480	450 235	— —	315 635	450 275	— —	335 370	455 360	— —
8% monoammonium phosphate	295 340	450 240	— —	290 620	455 300	— —	330 290	455 270	— —
2% boric acid	300 650	440 190	— —	315 830	435 205	— —	310 465	430 540	— —
8% boric acid	315 505	440 240	— —	320 740	445 220	— —	320 415	435 390	— —
8% disodium phosphate	280 580	390 330	440 115	310 755	395 495	— —	295 505	575 600	440 105
8% ammonium sulfate	300 300	445 200	— —	285 460	450 260	— —	315 365	445 390	— —
8% ammonium pentaborate octahydrate	310 545	440 200	— —	324 770	440 215	— —	310 515	425 440	— —

NOTE: Values are for 50 mg wood, 50 mg cellulose, and 30 mg lignin in flowing oxygen. The temperatures at which the peak took place ±3 °C are the top lines for each entry. The heat intensity of combustion, I_c, in calories per gram-minute ±5%, at each peak are the lower lines for each entry.
[a] Chemical percentages
(Reproduced from Ref. 47.)

Table VI. Effect of Inorganic Additives on Thermal Evolution Analysis

Additive	Percent Carbon at 500 °C	Order of Effectiveness
H_3PO_4	13.9	1
$ZnCl_2$	23.4	2
NaOH	28.1	3
$(NH_4)_2HPO_4$	30.9	4
$NH_4H_2PO_4$	38.6	5
Na_2CO_3	39.5	6
$SbCl_3$ hydrate	45.0	7
H_3BO_3	55.0	8
$NaVO_3$	56.4	9
Na_3PO_4	59.1	10
$CaCl_2$	59.6	11
$SbCl_3$	60.6	12
$NaHCO_3$	63.9	13
NaCl	64.1	14
$Na_2WO_4 \cdot 2H_2O$	66.3	15
$(NH_4)_2SO_4$	66.5	16
$NaHSO_4$	68.7	17
$Na_2B_4O_7 \cdot 10H_2O$	69.9	18
$NaH_2PO_4 \cdot H_2O$	72.5	19
NH_4Cl	73.9	20
Untreated	76.2	21
$(NH_4)_2CO_3$	76.8	22

(Adapted from Reference 56.)

However, this reaction is always in competition with the other reactions that are taking place (i.e., decarbonylation, condensation, decomposition). The mechanism of a particular fire retardant is the summed effect of all simultaneous reactions. This summed effect is especially evident in the synergism of some compounds; the effect of two compounds together is greater than the summed effect of each individual one alone (9, 51, 71–73).

Phosphorus and nitrogen have displayed such a synergistic effect, and much work has been done to understand this. Although most of the work has been done on fabrics, the same synergistic effect between phosphorus and nitrogen appears in wood.

The interaction of phosphorus and nitrogen compounds produces a more effective catalyst for the dehydration because the combination leads to further increases in the char formation and greater phosphorus retention in the char (43, 71–73). This result may be caused by the cross-linking of the cellulose during pyrolysis through

Table VII. Effect of Inorganic Additives on Thermogravimetric Analysis

Additive	Percent Weight Loss at 500 °C	Order of Effectiveness
H_3PO_4	61	1
$(NH_4)_2HPO_4$	66	2, 3
$NH_4H_2PO_4$	66	2, 3
$ZnCl_2$	74	4
NaOH	79	5
H_3BO_3	81	6
NaCl	82	7
$CaCl_2$	83	8
$NaHSO_4$	84	9, 10
$SbCl_3$	84	9, 10
$NaVO_3$	85	11
$SbCl_3$ hydrate	86	12, 13
$(NH_4)_2SO_4$	86	12, 13
Na_2CO_3	87	14, 15
$NaH_2PO_4 \cdot H_2O$	87	14, 15
$Na_2B_4O_7 \cdot 10H_2O$	89	16
$Na_2WO_4 \cdot 2H_2O$	90	17
Na_3PO_4	91	18
$NaHCO_3$	92	19
NH_4Cl	93	20, 21, 22
$(NH_4)_2CO_3$	93	20, 21, 22
Untreated	93	20, 21, 22

(Adapted from Reference 56.)

Table VIII. Effect of Inorganic Additives on Heat Release as Determined by Reaction Coulometry

Additive	ΔH_1 (cal/g)			
	200 °C	300 °C	400 °C	500 °C
Untreated	12	69	3214	3364
$NaHCO_3$	9	175	2384	2565
Na_2CO_3	14	390	1842	2083
NaOH	45	543	1180	1441

(Adapted from Reference 65.)

ester formation with the dehydrating agents (71). Also the presence of amino groups causes retention of the phosphorus as a nonvolatile amino salt (71, 73, 74), in contrast to some phosphorus compounds that may decompose thermally and be released into the volatile phase (51). Another possibility is that the nitrogen compounds promote polycondensation of phosphoric acid to polyphosphoric acid (51). Polyphosphoric acid might also serve as a thermal and oxygen barrier because it forms a viscous fluid coating (75). Whatever the particular mechanism is, it is apparent that some other reactions are preceding the dehydration reaction in order to make it more effective.

SMOLDERING INHIBITION THEORIES. In Chapter 13, Shafizadeh describes the pyrolysis of cellulose by two pathways. The first pathway leads to abundant char that can promote glowing. Flaming is inhibited due to the lack of combustible volatiles. The second pathway leads to combustible levoglucosan and other tars that promote flaming; little char remains for glowing. Oxidation of the carbonaceous char promotes smoldering combustion which is a more localized and slower process than flaming combustion. This type of combustion proceeds as a moving front in the solid phase.

The low-intensity heat flux required for smoldering combustion is provided by the oxidation of the preceding char. Diffusion of oxygen into the char is the rate determining step in this process. The reactions occurring in smoldering compete with each other and, therefore, are influenced by the physical conditions such as material, density, temperature of exposure, radiation losses, and additives present. Shafizadeh in Chapter 13 and elsewhere (67, 68) provides a good explanation for smoldering behavior.

The ratio of CO to CO_2 produced in smoldering combustion is influenced by various additives (68–70). Phosphates and borates increase the $CO:CO_2$ ratio. Metal ions such as sodium and potassium reduce this ratio and promote smoldering combustion (68). Smolder promoters tend to be either monovalent metal cations or metals such as iron, lead, or chromium (70). The metal ions aid the dissociation of the adsorbed molecular oxygen, thereby promoting the smoldering process (60, 67–69). Ammonium phosphate and boric acid, which are known to inhibit smoldering or glowing, may interfere with the active sites thereby blocking the process.

Fire-Retardant Formulations

Many chemicals have been evaluated for their effectiveness as fire retardants. Today most fire retardants for wood are based on phosphorus, nitrogen, boron, aluminum trihydrate, and a few other compounds. Phosphorus and nitrogen are frequently used together because they behave synergistically; amino-resins are an example of such a combination. The chemicals discussed in this section may be either pressure impregnated into the wood or applied to the wood surface, depending on the particular formulation.

Most fire-retardant formulations are not resistant to leaching by water. Therefore, there have been increased efforts to develop leach-resistant chemicals that can be impregnated into wood products for use in exterior or high humidity applications. Some of the proposed leach-resistant systems include chemical combinations that form insoluble complexes, amino-resin systems, and monomers that polymerize in the wood.

Major Chemicals. PHOSPHORUS. Combination salts of the phosphates have been used for retarding wood since the time of Gay-Lussac. Monoammonium and diammonium phosphates have been the most effective. The efficiency of phosphorus compounds can be increased by the presence of certain nitrogen compounds that produce a synergistic effect. The advantage of such synergism is that increased flame-spread resistance can be achieved with lower chemical loading levels. The amino-resin systems are based on this synergistic effect.

Organophosphorus and polyphosphate compounds also have been used as fire retardants. In one study, ammonium polyphosphate was used at loading levels of 96 kg/m^3 to achieve a flame-spread index of 15 according to ASTM E 84 (12). This treatment produced low smoke yields; however, this treatment was corrosive to aluminum, slightly corrosive to mild steel, but not corrosive to brass (77). In a patent by Clermont (78), phosphorus pentoxide, dimethylformamide, and urea were used to produce fire-retardant paper or veneer. Other patents (79, 80) describe the reaction of ammonia with partial esters of polyphosphoric acid. All patents demonstrated some leach resistance of the phosphorus.

BORON. Boron compounds have been used to treat wood for fire retardancy. Borax and boric acid, the primary fire-retardant compounds, have low melting points and form glassy films on exposure to high temperature. Borax, also known as sodium tetraborate decahydrate, is available in other hydrated states. Sodium tetraborate pentahydrate can be used in place of the decahydrate at a weight ratio of 74 (pentahydrate) to 100 (decahydrate) (81).

The borax inhibits surface flame spread but also can promote smoldering or glowing. In contrast, boric acid reduces smoldering and glowing combustion but has little effect on flame spread (82). Therefore, these chemicals are used together. This combination of chemicals has some advantages over other inorganic salts. Strength tests indicate that the alkaline borate solutions produce a smaller reduction on modulus of rupture (MOR) than do the acid treatments (83). The borate solution is also less corrosive and less hygroscopic.

A form of the borax and boric acid solution frequently used to fire-retard wood products is called polybor and has a general formula of $Na_2B_8O_{13} \cdot 4H_2O$. When borax, $Na_2B_4O_7 \cdot 10H_2O$, is added to a saturated boric acid solution, the solubility increases. Polymerization of the polyborates removes boric acid and borate ions from the solution, thus permitting more boric acid or borax to dissolve. This resulting solution (which is near the Na_2O/B_2O_3 ratio of maximum solubility) is polybor. This material dissolves rapidly in water to form a supersaturated solution. The high solubility of this product is an asset for fire-retarding wood products (84).

Boron compounds have been used in several ways to achieve reduced flammability of wood products. Borax and boric acid can be incorporated into particle board chips before addition of a dicyandiamide, phosphoric acid, amino-resin system (85). They can also be used to produce a fire-retardant hardboard. Riem and Dwars (86) added water-soluble ammonium borate to wood fibers before the board was formed. A 6–7% boron content produced a hardboard that had a flame spread of 25 or less.

Boron compounds can be added in combination with other chemicals such as nitrogen and phosphorus. A solution containing sodium tripolyphosphate, boric acid, and ammonia provides a ready-to-use treatment on cellulose products such as plywood, fiberboard, and cardboard (87). The resulting products passed the British Standard 476, Section 6 (Fire Propagation test) Class 0 and Class I requirements of the British Standard Section 7 (Surface Spread of Flame).

Aluminum trihydrate also can be used in conjunction with boron compounds, because a synergistic effect between the boron and aluminum trihydrate exists (88). Hardboard, containing 28% aluminum trihydrate and 6% boron, can be produced and has a flame spread of 25 or less. The aluminum trihydrate is added to a slurry of water and wood fiber. The boron solution is added to the surface of the wetlap or as an impregnated solution in a secondary treatment (88).

ALUMINUM TRIHYDRATE. The utility of aluminum trihydrate as a flame retardant is based on its endothermic dehydration to aluminum oxide and water. In absorbing some of the heat of combustion and lowering the temperature of the substrate near the flame, the hydrate functions as a chemical heat sink. The water vapor provided by such action dilutes the gaseous reactants in the flame until all the water of crystallization is exhausted.

Aluminum trihydrate also can be used as the only fire-retardant ingredient in the production of fiberboard (89). However, other research indicates that it is more effective when used in combination with other chemicals (88, 90, 91). Hardboard and particle board can be produced by incorporating boron compounds (as mentioned previously) and amino-resin systems.

MISCELLANEOUS CHEMICALS. The possibilities for various combinations of the chemicals already discussed are endless. There have been some efforts with other chemicals that have not been studied as intensively as the phosphorus, boron, and aluminum compounds. Brominated lignin sulfonate and brominated Kraft lignin (92) reduced the char length of paper treated with this solution. Turnbo et al. (93) incorporated 1,1,2,4-tetrabromo-2-butene with an organic solvent in order to surface coat wood splints. The splints treated with the solution containing 80% of the 1,1,2,4-tetrabromo-2-butene had a limiting oxygen index of 42 compared to 21 for untreated according to ASTM D 2863 (93).

Oxalates also have been used as fire retardants for wood products (94). They behave like other inorganic salts. Specimens impregnated with potassium oxalate promoted degradation of wood components in the temperature range of 180–320 °C as well as retarding active decomposition during flaming combustion.

Leach-Resistant Chemicals. INSOLUBLE COMPLEXES. Leach-resistant fire retardants can be formed by reacting soluble salts with metal salts to form insoluble, metallic salt complexes. Sodium silicate reacted with calcium chloride formed an insoluble, hydrated calcium silicate (95). Application of a 20% diammonium phosphate solution, followed by a 20% magnesium sulfate solution, has been proposed as a ready-to-use treatment for wood roofs (96). This combination forms an insoluble magnesium ammonium phosphate and is recommended for roofs that are 5 years old or older. Test results indicate that this treatment provides increased flame-spread protection.

McCarthy et al. (97) tested a zinc, copper, chromium, arsenic, phosphorus preservative on fence posts. The addition of the zinc and phosphorus eliminated the afterglow problem caused by this treatment. However, incorporation of the phosphorus reduced the effectiveness of the decay resistance.

AMINO-RESINS. The most widely studied leach-resistant systems are the amino-resins. Goldstein and Dreher (98) first applied these systems as fire retardants. Basically, the amino-resin systems involve the combination of a nitrogen source (i.e., urea, melamine, guanidine, or dicyandiamide) with formaldehyde to produce a methylolated amine. The new product is then reacted with a phosphorus compound such as phosphoric acid. Because there is a synergistic effect between the phosphorus and the nitrogen, reduced loading levels can achieve the same level of fire retardancy as each compound alone (9, 99, 100). A composition containing 1.4% phosphorus and 0.4% nitrogen will give the same degree of fire-retardant effectiveness as one with 3.5% phosphorus alone.

In the past, the problems with the amino-resin systems were its limited pot life, leaching of the phosphorus, and excess formaldehyde emission. Recent research has addressed these problems.

Researchers at the Eastern Forest Products Laboratory in Canada have evaluated the urea and melamine amino-resin systems (9, 57, 99–110). Their work demonstrates that both systems show good leach resistance and reduced flame spread. The stability of these resins is controlled by the rate of methylolation of the urea, melamine, and dicyandiamide. The optimum mole ratio for stability of these solutions is 1:3:12:4 for urea or melamine, dicyandiamide, formaldehyde, and orthophosphoric acid. However, even at the optimum mole ratios, the pot life of the melamine system is less than that of the urea system. In both systems the nitrogen is fixed to a greater degree than the phosphorus. However, the degree of fixation of the phosphorus is greater with the melamine than with the urea. The melamine structure may promote formation of compounds with phosphoric acid that are less soluble than those from urea and dicyandiamide.

Another method to increase the stability of the solution, especially for transport purposes, is to use monomethylol dicyandiamide. This eliminates the need for adding formaldehyde and decreases the polymerization rate during transport. Solid monomethylol dicyandiamide was mixed with solid melamine (111, 112). This solid composition can then be shipped to treating facilities where it is mixed with water and then reacted with phosphoric acid. A similar modification allows dicyandiamide to react with formaldehyde at elevated temperatures until no free formaldehyde exists (111). The melamine is then added and the solution can be shipped. Both modifications increase the stability of the solution and eliminate the excess formaldehyde.

Another advantage of the amino-resin systems is their applicability to solid wood and wood-composite products. Cedar shingles were the first products treated with this type of fire-retardant system (99, 100, 113, 114). Commercially treated shingles available in the U.S. are based on these systems. Generally, these systems exhibit good durability to outdoor weathering when tested over extended periods (115–17).

The amino-resins are also suited for use on wood-composite products. In some cases the fire retardant can act as the binder for particle board (99, 100), the adhesive for plywood (99, 100), or a flame-retardant finish sealer for decorative plywood (*100, 118, 119*). In other cases, the amino resin is added as the fire retardant to the fiber finish used in making particle board (*120*) and hardboard (*121, 122*). All products demonstrate reduced flame spread. However, the amount of amino-resin incorporated governs the degree of flame-spread reduction. The amount incorporated involves a compromise among the properties of the board such as flame spread, strength, and dimensional stability.

OTHER METHODS. Other methods used to improve the leach resistance of fire retardants include many different techniques. Most involve incorporating a monomer into the wood, followed by a curing procedure. Most of the investigated monomers are organophosphorus compounds that can be used alone (*123–25*) or with other fire-retardant salts (*126–28*). Addition of combustible polymers (i.e., polymethyl methacrylate, polystyrene) result in higher values of percent burned than the control due to the additional amount of combustible material. The addition of fire-retardant salts reduces this value considerably, although not to the level of the fire-retardant salt alone. Other monomers investigated include tetrakis(hydroxymethyl)phosphonium chloride (THPC) (*115, 117, 129*), other salts of the tetrakis(hydroxymethyl)phosphonium group (*130*), and a cyclic sulfonium zwitterion monomer (*131*). Although several of these techniques may have possibilities they have not been researched thoroughly.

Of all the proposed leach-resistant formulations, only the amino-resin systems are used commercially. The high costs of many of the other proposed techniques limit their acceptability.

Future Research

Although much research has been done on fire retardants for wood, there are many areas where improvements are needed.

Leach-Resistant Compounds. Progress for improving the leachability of fire retardants has been made in the past decades. Several commercial treatments are available for exterior use. However, even these demonstrate some leaching of chemicals. Further work needs to be done to increase the leach resistance of these treatments without excessively increasing the cost. Improved leach resistance will be necessary to expand wood products into commercial and institutional buildings.

Improved Fire-Retardant Treatments for Panel Products. Fire-retardant treatment for panel products is also an area where research efforts need to be concentrated. Currently, there is only one commercially available fire-retardant-treated particle board that qualifies for use in commercial and institutional buildings. Expanding the wood products market to such buildings will increase the demand for this product. However, fire-retardant treatments for panel products still suffer from certain disadvantages, primarily reduction of physical properties because of the fire retardant. Further work needs to be done on the development of fire-retardant treatments that minimize these undesirable effects on the properties of treated wood. Alternate treating techniques would expand the range of fire-retardant treatments that could be used and also reduce the cost.

Effective Coating Systems. Research work in the area of intumescent coatings would benefit both solid and composite wood products. In some instances, the coating system is the more cost-effective treatment, particularly in cases of retrofitting a building. Further work needs to be done on improving the durability and effectiveness of coatings. Also, coatings are needed that are durable to exterior weathering, especially UV degradation. Intumescent coatings incorporated with adhesive binders have been suggested for use in panel products.

Reduced Smoke and Toxicity. The smoke and toxic products of combustion are a problem of growing concern. Until recently, this problem has been overlooked in developing fire retardants. Future formulations will not only have to limit flame spread, but also limit smoke and toxic combustion products. Addition of smoke suppressants to some formulations may improve some systems. Modification of systems may also be necessary to meet possible code restrictions. More research is necessary in this area to understand the mechanism of smoke production and accumulation.

Basic Mechanisms. Finally, further work is necessary on fundamental mechanisms of individual fire retardants. These mechanisms are a function of the particular chemicals involved and the environmental conditions of the fire exposure. There is a need to establish common methods and conditions for determining these mechanisms in order to compare different treatments. This would give us a better understanding of how these compounds work in action and would provide a more efficient approach for formulating fire-retardant systems than a trial and error approach. Correlations also need to be established between rapid precise thermal analysis methods and standard combustion tests. Retardant formulations could be evaluated initially on smaller (research and development size) samples. The more promising treatments could be tested for flame-spread index, heat release rate, and toxic smoke production.

Summary

The addition of fire retardants can reduce the flammability of wood; however, this may occur at the expense of related wood properties such as strength or increased smoke production. Therefore, fire retardants are formulated for best overall performance, including flame spread, smoke reduction, and reduced rates of heat release. The chemicals can be applied to wood products as either an impregnated solution or a coating. The application method depends on the formulation and the end-use of the product.

Mechanism. No single mechanism explains the action of all fire retardants, so they probably work through a combination of several mechanisms. The mechanisms of fire retardants in wood involve a complex series of simultaneous reactions whose products may affect subsequent reactions. Pyrolysis of cellulose involves dehydration, depolymerization, decarbonylation, decomposition of smaller compounds, condensation, and other reactions. These pyrolysis reactions occur both in the solid phase and vapor phase. Addition of fire retardants will alter the reactions; however, this alteration will depend on the additives, the material, and the thermal–physical environment. The presence of oxygen adds subsequent and competitive oxidation reactions to the above series. These oxidative reactions can take place in both the solid and vapor phases. Evidence indicates that most fire retardants reduce combustible volatiles production and limit combustion to the solid phase. The best retardants also inhibit solid-phase oxidation to effectively remove the fuel from the fire.

Lignin thermally decomposes to char and contributes little to flaming combustion. Most of the flaming combustion from wood is attributed to the hemicellulose and cellulose. However, lignin does support oxidation in the solid phase. Some fire retardants, such as phosphorus and boric acid, inhibit oxidation in the solid phase; other additives, such as sodium compounds, may promote it.

In addition to the chemical mechanisms of fire retardants, thermal or barrier-type mechanisms may be operative. Coatings may prevent oxygen from reaching the wood surface. Dilution of combustible gases by noncombustible gases and inhibition of flaming by free radicals can also be in effect. Therefore, fire retardancy of wood involves many complex reactions. The effectiveness of a particular fire retardant depends on the overall summation of these competitive and sequential reactions and the thermal and physical environment of the material.

Formulations. Fire-retardant formulations are numerous, although most of them are based on the inorganic salts, such as diammonium phosphates. Increased emphasis on improving the related wood properties associated with fire retardants has led to many interesting and creative formulations and processes.

Phosphorous compounds are the main chemicals used in most formulations. These compounds range from inexpensive ammonium phosphates to the more exotic ones such as phosphorous pentoxide and polyphosphoric acids. Improving the leach resistance of the phosphorous compounds is a major problem.

The borons are also effective and efficient fire retardants for wood. They are leachable but they do not reduce the strength or increase the hygroscopicity of the wood as some other compounds do. Little work has been done on the mechanism of action of the borons.

Other compounds such as aluminum trihydrate and silicate compounds have also been tried as fire retardants for wood. These compounds work best in combination with other chemicals, especially those in which the behavior is synergistic.

Future Research. Improvements in leach-resistant chemicals have been a primary concern over the past decade. Advances have been made in leach-resistant systems such as the amino-resin systems; however improvements still need to be made in leach-resistant compounds without increasing the cost. Other areas where research on fire retardants needs to be conducted are in coating systems, especially those that are durable to weathering and UV degradation; reduction of smoke and toxic products, improvements in fire-retardant treatments for panel products; and fundamental work on the mechanisms of particular formulations.

Literature Cited
1. Eickner, H. W. *J. Mater.* **1966**, *1*(3), 625–44.
2. Hunt, Geo. M.; Truax, T. R.; Harrison, C. A. *Proc.—Annu. Meet. Am. Wood-Preserv. Assoc.* **1930**, 130–64.
3. Hunt, Geo. M.; Truax, T. R.; Harrison, C. A. *Proc.—Annu. Meet. Am. Wood-Preserv. Assoc.* **1931**, 104–42.
4. Hunt, Geo. M.; Truax, T. R.; Harrison, C. A. *Proc.—Annu. Meet. Am. Wood-Preserv. Assoc.* **1932**, 71–96.
5. Truax, T. R.; Harrison, C. A.; Baechler, R. H. *Proc.—Annu. Meet. Am. Wood-Preserv. Assoc.* **1933**, 107–24.
6. Truax, T. R.; Harrison, C. A.; Baechler, R. H. *Proc.—Annu. Meet. Am. Wood-Preserv. Assoc.* **1935**, 231–48.
7. Lyons, John W. "The Chemistry and Uses of Fire Retardants"; Wiley-Interscience: New York, 1970; 108–42.
8. Goldstein, Irving S. "Wood Deterioration and Its Prevention by Preservative Treatment"; Nicholas, D. D., Ed.; Syracuse Univ. Press: Syracuse, NY, 1973; Vol. 1, 307–39.
9. Juneja, Subhash C. "Advances in Fire Retardants"; Bhatnagar, V. M., Ed.; Technomic Publ. Co. Inc.: Westport, Conn., 1973; Progress in Fire Retardancy Series, Vol. 3, Part 2, 31–53.
10. Bramhall, George "Wood Fire Behavior and Fire Retardant Treatment; A Review of the Literature"; Fang, J. B.; MacKay, G. D. M.; Bramhall, G., Eds.; Canadian Wood Council: Ottawa, Canada, 1966.
11. Baker, D. S. *Chem. Ind. (London)* **1977**, 74–79.
12. Holmes, C. A. "Wood Technology: Chemical Aspects"; Goldstein, Irving S., Ed.; ACS SYMPOSIUM SERIES No. 43; ACS: Washington, D.C., 1977; 82–106.

13. Brazier, J. D.; Laidlow, R. A. *BRE Information* IS 13/74, December 1974.
14. Slade, Philip E., Jr.; Jenkins, Lloyd T. "Techniques and Methods of Polymer Evaluation"; Marcel Dekker, Inc.: New York, 1966.
15. American Society of Testing and Materials. E 84–1979a, Philadelphia, Pa.
16. American Society of Testing and Materials. E 286–69 (reapproved 1975), Philadelphia, Pa.
17. Levy, Marshall M. *J. Cell. Plast.* **1967**, *3*(4), 168–73.
18. Vandersall, H. L. *J. Paint Technol.* **1967**, *39*(511), 494–500.
19. Eickner, H. W.; Schaffer, E. L. *Fire Technol.* **1967**, *3*(2), 90–104.
20. American Society of Testing and Materials. D 2863–1977, Philadelphia, Pa.
21. White, Robert H. *Wood Sci.* **1979**, *12*(2), 113–21.
22. Yoshimura, Mitsugu; Umemura, Kenzo *Mokuzai Gakkaishi* **1980**, *26*(3), 209–14.
23. Yoshimura, Mitsugu; Miwa, Akira *Mokuzai Gakkaishi* **1980**, *26*(4), 287–92.
24. Wharton, R. K. *J. Fire Flammability* **1981**, *12*, 236–39.
25. Fenimore, C. P.; Martin, F. J. *Combust. Flame* **1966**, *10*, 135–39.
26. Routley, A. F. Central Dockyard Laboratory, H. M. Dockyard, Portsmouth Great Britain, Report No. 5/73, 1973.
27. Brenden, J. J. *U.S., For. Serv., Res. Pap. FPL* 137, 1970.
28. Brenden, J. J. *U.S., For. Serv., Res. Pap. FPL* 249, 1975.
29. American Society of Testing and Materials, E662, 1979, Philadelphia, Pa.
30. Brenden, J. J. *U.S., For. Serv., Res. Pap. FPL* 217, 1973.
31. Brenden, J. J. *U.S., For. Serv., Res. Pap. FPL* 281, 1977.
32. Chamberlain, David A.; Brenden, John J. *Proc. Symp. Residential Fires and Wood Product Use* Society of Wood Science and Technol., Madison, Wis., 1980.
33. Smith, E. E. American Society for Testing and Materials, ASTM STP 502, Philadelphia, Pa., 1972, 119–34.
34. Smith, E. E. *J. Fire Flammability* **1977**, *8*, 309–23.
35. Chamberline, David L. "Rate of Heat Release—Tool for the Evaluation of the Fire Performance of Materials." Presented at joint meeting of General States and Western States Section of the Combustion Institute on Flammability and Burning Characteristics of Materials and Fuels, San Antonio, Tex., April 21–22, 1975.
36. Factory Mutual Research, Comparative fire tests of treated and untreated lumber for American Wood Preservers Institute. Serial No. 24679.01 (4840) Aug. 6, 1976.
37. Kaplan, Harold L.; Grand, Arthur F.; Hartzell, Gordon E. A critical review of the state-of-the-art of combustion toxicology. Southwest Research Institute, San Antonio, Tex., 1982.
38. Browne, F. L. U.S., For. Serv. No. 2136, For. Prod. Lab., Madison, Wis., 1958, reviewed and reaffirmed, 1963.
39. Kay, M.; Price, A. F. *J. Fire Retard. Chem.* **1979**, *6*, 69–91.
40. Hindersinn, Raymond R.; Witschard, Gilbert "Flame Retardancy of Polymeric Materials"; Kuryla, William C.; Papa, Anthony J., Ed.; Marcel Dekker, Inc.: New York, 1978; Vol. 4., 1–107.
41. Browne, F. L.; Tang, W. K. *U.S., For. Serv., Res. Pap. FPL* 6, 1963.
42. Egyed, O.; Simon, J. *J. Therm. Anal.* **1979**, *16*, 307–20.
43. Brady, T. P.; Langer, H. G. *Therm. Anal. (Proc. Int. Conf., 6th)* **1980**, *2*, 443–48.
44. Hirata, Toshimi; Abe, Hiroshi *Mokuzai Gakkaishi* **1973**, *19*(9), 451–59.
45. Browne, F. L.; Tang, W. K. *Fire Res. Abs. Rev.* **1962**, *4*(1 and 2), 76–91.
46. Tang, Walter K. *U.S., For. Serv., Res. Pap. FPL* 71, 1967.
47. Tang, Walter K.; Eickner, Herbert W. *U.S., For. Serv., Res. Pap. FPL* 82, 1968.

48. Nguyen, Tinh; Zavarin, Eugene; Barrall, Edward M., II *J. Macromol. Sci.—Rev. Macromol. Chem.* **1981**, *C21*(1), 1–60.
49. Hastie, John W. National Bureau of Standards special publication 411, Gaithersburg, Md., 1973.
50. Creitz, E. C. *J. Res. Natl. Bur. Stand.*, Sect. A **1970**, *74A*(4), 521–30.
51. Granzow, Albrecht *Acc. Chem. Res.* **1978**, *11*(5), 177–83.
52. Brenden, J. J. *U.S., For. Serv., Res. Pap. FPL* 80, 1967.
53. Fung, D. P. C.; Tsuchiya, Yoshio; Sumi, Kikuo *Wood Sci.* **1972**, *5*(1), 38–43.
54. Shafizadeh, F. *Adv. Carbohydr. Chem.* **1968**, *23*, 419–74.
55. Shafizadeh, F. *J. Polym. Sci. Part C* **1971**, *36*, 21–51.
56. Shafizadeh, Fred; Chin, Ping-Sen; DeGroot, William F. *J. Fire Flammability/Fire Retard. Chem. Suppl.* **1975**, *2*, 195–203.
57. Fung, D. P. C. *Wood Sci.* **1976**, *9*(1), 55–57.
58. Halpern, Yuval; Riffer, Richard; Broido, A. *J. Org. Chem.* **1973**, *38*(2), 204–9.
59. Domburg, G.; Rossinskaya, G.; Dobele, G. *Therm. Anal.* (*Proc. Int. Conf., 6th*) **1980**, *2*, 449–54.
60. Shafizadeh, F.; Sekiguchi, Y. *Carbon*, **1983**, 21, 511–16.
61. Nanassy, A. J. *Wood Sci.* **1978**, *11*(2), 111–17.
62. Hirata, Toshimi; Abe, Hiroshi *Mokuzai Gakkaishi* **1973**, *19*(11), 539–45.
63. Hirata, Toshimi; Abe, Hiroshi *Mokuzai Gakkaishi* **1973**, *19*(10), 483–92.
64. Browne, F. L.; Brenden, J. J. *U.S., For. Serv., Res. Pap. FPL* 19, 1964.
65. Susott, R. A.; Shafizadeh, F.; Aanerud, T. W. *J. Fire Flammability* **1979**, *10*(2), 94–103.
66. Shafizadeh, Fred; Bradbury, Allan G. W. *J. Therm. Insul.* **1979**, *2*(1), 141–52.
67. Bradbury, A. G. W.; Shafizadeh, F. *Combust. Flame* **1980**, *37*, 85–89.
68. Shafizadeh, Fred; Bradbury, Allan G. W.; DeGroot, William F.; Aanerud, Thomas W. *Ind. Eng. Chem. Prod. Res. Dev.* **1982**, *21*, 97–101.
69. Gann, R. G.; Earl, W. L.; Manka, M. J.; Miles, L. B. *Proc. 18th Int. Symp. Combustion* **1981**, 571–78.
70. McCarter, R. J. *Fire Mater.* **1981**, *5*(2), 66–72.
71. Hendrix, James E.; Drake, George L., Jr. *J. Appl. Polym. Sci.* **1972**, *16*, 257–74.
72. Gupta, Jagdish Chandra; Bhatnagar, K. L.; Bhatnagar, H. L. *Indian J. Text. Res.* **1979**, *4*, 40–42.
73. Bakos, D.; Kosik, M.; Antos, K.; Karolyova, M.; Vyskocil, I. *Fire Mater.* **1982**, *6*(1), 10–12.
74. Egyed, O.; Simon, J. *J. Therm. Anal.* **1979**, *16*, 321–27.
75. Kishore, K.; Mohandas, K. *Fire Mater.* **1982**, *6*(2), 54–57.
76. Satonaka, Seiichi "Advances in Fire Retardants" Bhatnagar, V. M., Ed.; Technomic Publ. Co. Inc.: Westport, Conn., 1973; Progress in Fire Retardancy Series, Vol. 3, Part 2, 74–98.
77. Eickner, H. W.; Stinson, J. M.; Jordan, J. E. *Proc.—Annu. Meet. Am. Wood-Preserv. Assoc.* **1969**, *70*, 95–102.
78. Clermont, Louis P. Canadian patent 934 105, 1973.
79. Brady, Thomas P.; Langer, Horst G. U.S. patent 4 283 501, 1981.
80. Langer, Horst G.; Brady, Thomas P. U.S. patent 4 287 131, 1981.
81. U.S. Borax Co., Technical data sheets 1C-1, 1C-2, 1982.
82. Day, M.; Wiles, D. M. *J. Consum. Prod. Flammability* **1978**, *5*, 113–22.
83. Middleton, J. C.; Draganov, S. M.; Winters, F. T., Jr. *For. Prod. J.* **1965**, *15*(12), 463–67.
84. Doonan, Daniel J.; Lower, Loren D. "Kirk-Othmer Encyclopedia of Chemical Technology"; John Wiley and Sons: New York, 1978; 3rd ed., Vol. 4, 67–110.
85. Surdyk, Lyle V. U.S. patent 3 874 990, 1975.

86. Riem, Roland H.; Dwars, Wilhelmus T. A. Canadian patent 872 192, 1971.
87. Salle, Nilo Eivar Great Britain patent 2 002 434A, 1977.
88. Moore, Gregory R.; Fischer, Craig A. U.S. patent 4 130 458, 1978.
89. Barnes, H. M.; Farrell, D. *For. Prod. J.* **1978**, *28*(6), 36–37.
90. Sobolev, Igor; Panusch, Erwin U.S. patent 4 008 214, 1977.
91. Sobolev, Igor; Panusch, Erwin U.S. patent 4 126 473, 1978.
92. Zeigerson, Esther; Bloch, Rudolf U.S. patent 3 962 208, 1976.
93. Turnbo, Roy G.; Walker, David G.; Rosen, Marvin U.S. patent 3 953 627, 1976.
94. Kosík, Martin; Grman, Dusan; Gasperík, Juray; Dolezal, Jozef; Mihálik, Peter *Drev. Vysk.* **1978**, *23*(3), 181–88.
95. Lilla, Allen G. U.S. patent 3 974 318, 1978.
96. Lipska, Anne E.; Amaro, Allen J. AD–A014 492 Stanford Research Institute, Final Report, April 1975.
97. McCarthy, D. F.; Seaman, W. G.; DaCosta, E. W. B.; Bezemer, Lynette D. *J. Inst. Wood Sci.* **1972**, *6*(1), 24–31.
98. Goldstein, Irving S.; Dreher, William A. U.S. patent 3 159 503, 1964.
99. Juneja, Subhash C. *For. Prod. J.* **1972**, *22*(6), 17–23.
100. Juneja, S. C.; Richardson, L. R. *For. Prod. J.* **1974**, *24*(5), 19–23.
101. Fung, D. P. C.; Doyle, E. E.; Juneja, S. C. Information Report OP–X–68, Dept. of the Environment, Canadian Forestry Service, Aug. 1973, Ottawa, Canada.
102. Fung, D. P. C.; Doyle, E. E.; Juneja, S. C. Report OPX103E, Study Number EFP–3–115, Eastern Forest Products Lab., Ottawa, Canada, 1974.
103. Fung, D. P. C.; Juneja, S. C.; Doyle, E. E. Report OPX157E, Study Number EFP–3–115, Eastern Forest Products Lab., Ottawa, Canada, 1976.
104. Juneja, Subhash C. U.S. patent 3 887 511, 1975.
105. Juneja, S. C.; Calve, L. *J. Fire Retard. Chem.* **1977**, *4*, 235–41.
106. Juneja, S. C.; Fung, D. P. C. *Wood Sci.* **1974**, *7*(2), 160–63.
107. Juneja, S. C.; Fung, D. P. C. Report OPX168E, Eastern Forest Products Laboratory, Ottawa, Canada, 1976.
108. Juneja, S. C.; Shields, J. K. *For. Prod. J.* **1973**, *23*(5), 47–49.
109. Juneja, S. C.; Richardson, L. R. *Proc. Int. Symp. Flammability Fire Retard.* **1976**, 167–74.
110. Juneja, S. C.; Richardson, L. R. Report OPX 185E, Eastern Forest Products Lab., Ottawa, Canada, 1977.
111. Oberley, William J. U.S. patent 3 986 881, 1976.
112. Oberley, William J. U.S. patent 4 010 296, 1977.
113. Bescher, Ralph H. *Fire J.* **1967**, *61*(5), 52–56.
114. King, F. W.; Juneja, S. C. *For. Prod. J.* **1974**, *24*(2), 18–23.
115. Holmes, C. A. *U.S., For. Serv., Res. Pap. FPL* 158, 1971.
116. Holmes, C. A. *U.S., For. Serv., Res. Pap. FPL* 194, 1973.
117. Holmes, C. A.; Knispel, R. O. *U.S., For. Serv., Res. Pap. FPL* 403, 1981.
118. Dolenko, A. J.; Clarke, M. R. *For. Prod. J.* **1973**, *23*(10), 22–27.
119. Yamagishi, Koichi; Ito, Hidetake; Kasai, Akira; Komazawa, Katsumi; Nunomura, Akiu *J. Hokkaido For. Prod. Res. Inst.* **1976**, 1–4.
120. Syska, Arthur D. *U.S., For. Serv., Res. Note FPL* 0201, 1969.
121. Myers, Gary C.; Holmes, Carlton A. *For. Prod. J.* **1975**, *25*(1), 20–28.
122. Myers, Gary C.; Holmes, Carlton A. *U.S., For. Serv., Res. Pap. FPL* 298, 1977.
123. Arni, P. C.; Jones, E. *J. Appl. Chem.* **1964**, *14*, 221–28.
124. Raff, R. A. V.; Adams, M. F. *Northwest Sci.* **1968**, *42*(1), 14–18.
125. Raff, R. A. V.; Herrick, I. W.; Adams, M. F. *For. Prod. J.* **1966**, *16*(2), 43–47.
126. Siau, John F.; Campos, Gerald S.; Meyer, John A. *Wood Sci.* **1975**, *8*(1), 375–83.

127. Siau, John F.; Meyer, John A.; Kulik, Roman S. *For. Prod. J.* **1972**, *22*(7), 31–36.
128. Kishore, K.; Mohandas, K.; Karuna Sagar, D. *Fire Mater.* **1980**, *4*(3), 115–18.
129. Handa, Takashi; Nagashima, Toshiaki; Takakashi, Yutaka; Ebihara, Naofumi *Proc. Jpn. Congr. Mater. Res.* **1978**, *22*, 306–12.
130. Elgal, Galoust M.; Perkins, Rita M.; Knoepfler, Nestor B. U.S. patent 4 246 031, 1981.
131. Brown, Frederick L. U.S. patent 4 049 849, 1977.

RECEIVED for review May 19, 1983. ACCEPTED August 22, 1983.

15
Degradation of Wood by Chemicals

IRVING S. GOLDSTEIN

Department of Wood and Paper Science, North Carolina State University, Raleigh, NC 27695–8005

> *The loss of its identity as wood is an inevitable consequence when wood is pulped or converted into chemicals. But wood also is exposed frequently to the action of chemicals during other types of processing and during ordinary use, yet it still retains its identity as solid wood. These initial processes of chemical attack and the accompanying changes in some important properties of the wood are described. Hydrolysis and oxidation are the most important degradative processes. The sites of initial hydrolytic attack are acetal linkages in carbohydrates and aryl ether linkages in lignin. Mechanisms for these hydrolytic and oxidative reactions are given. The principal consequences of chemical attack on the polymeric components of wood are depolymerization and solubilization.*

IN CONSIDERING THE EFFECT OF CHEMICALS ON WOOD, it is usual to focus on the end result of the chemical action. Such aspects as the fractionation of wood into its separate components, the conversion of the wood components into various low molecular weight or polymeric products, and the deterioration of wood by chemicals are some of the major categories into which wood chemistry and technology may be divided.

Although apparently diverse, the categories of solid wood processing, pulping, and conversion into chemicals do possess a common feature. Whatever the final product or products or the behavior of intermediates, all the processes begin with the interaction of a chemical reagent with the bonds and functional groups present in the wood. Because the structures of wood and its components are quite well defined, it is possible to relate these initial processes to specific reactions of reagents and reactive sites in the wood.

As these reactions proceed it is inevitable that changes in one or more important properties of the wood will occur. These changes are commonly called *degradation*, although the value of the resultant wood pulp or wood sugar may be greater than that of the starting material. Perhaps disorganization would be a better term, but degradation does apply to such properties as strength and color.

In this chapter the emphasis will be on the initial processes of chemical attack. During the early stages of any type of chemical processing the wood still retains its identity as solid wood. Thus, despite any changes in properties or degradation resulting from chemical action, these initial processes still fall within the framework of this book.

Sites of Initial Attack

The only chemical reactions that can affect the properties of wood are those involving chemical bonds or functional groups present in the wood. It will be helpful, therefore, to review the structure of the wood components with special attention to the most abundant features. Then, specific reactions at these sites can be considered.

Structure of Wood Components. As described in Chapter 2, wood consists of cellulose, hemicelluloses, lignin, and extractives. The first three are polymeric and are intimately associated with each other at the molecular level to form the cell wall. Carbohydrate content (cellulose and hemicelluloses) may reach 75% of the wood substance, so the reactions of carbohydrates are especially important. Although the extractives are extraneous materials, their presence can often influence reactions with the cell wall materials, and some wood properties also may be affected by reactions involving extractives.

CELLULOSE. Cellulose (Chapter 2, Figure 1) is a long chain polymer of β-D-glucose in the pyranose form. An important feature of the cellulose structure is the tendency for the individual cellulose chains to form bundles of crystalline order held together by hydrogen bonds between the hydroxyl groups of adjacent chains. This crystallinity influences reactivity by controlling the access of reagents or enzymes to functional groups and chemical bonds within the crystalline regions, and also by interfering with the changes in geometry required for the transition states of various reactions. The amorphous or less-ordered regions of the cellulose are not subject to these restrictions.

The hydroxyl groups in cellulose are of two types, a single primary hydroxyl on each anhydroglucose unit as well as two vicinal *trans* secondary hydroxyls. Next in abundance after the hydroxyl groups are the acetal linkages that form the pyranose rings and, by glycosidic bonds, connect the glucose rings in the cellulose chain. End groups are of much less importance in a high polymer such as cellulose. Aldehydic functionality is to be expected but, depending on the history of the sample, carboxyl groups are not unusual. The carbonyl content of the cellulose component may also vary with its history; this indicates that some reactions may take place along the chain.

HEMICELLULOSE. Hemicelluloses (Chapter 2, Figure 4) differ from cellulose in that they consist, for the most part, of pentose and hexose sugars other than glucose, are usually branched, and have much lower degrees of polymerization. They are not crystalline, so do not present the same barriers to accessibility as does cellulose.

Again hydroxyl groups are most abundant. However, primary hydroxyl groups may be absent as in the xylans. Acetal linkages and backbone end groups correspond to those in cellulose. In addition the hemicelluloses contain ester bonds in the form of acetyl groups. Uronic acid groups contribute both an acetal linkage and a carboxyl group. Methyl ethers are also present, but are of no concern in this discussion. Lignin–carbohydrate covalent bonding may also exist as esters, ethers, or acetals.

LIGNIN. Lignins (Chapter 2, Figure 6) are three-dimensional network polymers formed from phenylpropane units. The most common linkages between the phenylpropane units are shown in Figure 1 and their percentages are given in Table I (*1*). It is apparent that ether linkages are very important in lignin, with β-aryl ethers (A) most abundant and benzyl(α)-aryl ethers (C) and diphenyl ethers (G) significant. Lignins also contain 1–1.5 methoxyl groups per phenylpropane unit, and much smaller ratios of free phenolic hydroxyl (0.09–0.30), benzyl alcohol (0.15–0.20), and carbonyl (0.20) groups.

EXTRACTIVES. Extractives are the extraneous plant components that can be dissolved away from the insoluble cell wall material. They include many different kinds of chemicals. The reactive functional groups encountered include acids, aldehydes, alcohols, stilbenes, phenols, quinones, and esters.

Hydrolytic Reactions. Probably the most important reactions, with regard to their impact on the properties of wood, involve some type of hydrolysis. Both carbohydrates and lignin are affected, and hydrolysis is encountered in almost every kind of wood processing.

CARBOHYDRATES. The two functional groups subject to hydrolysis in wood polysaccharides are the ester and acetal linkages, of which the acetals are more important because of their greater abundance and their role in connecting the monosaccharides in the polymers. Hydrolysis may take place under both acidic and alkaline conditions. The extent and consequences of acid hydrolysis are greater.

Ester groups in wood polysaccharides are for the most part acetyl substituents on the hemicellulose components. In the presence of

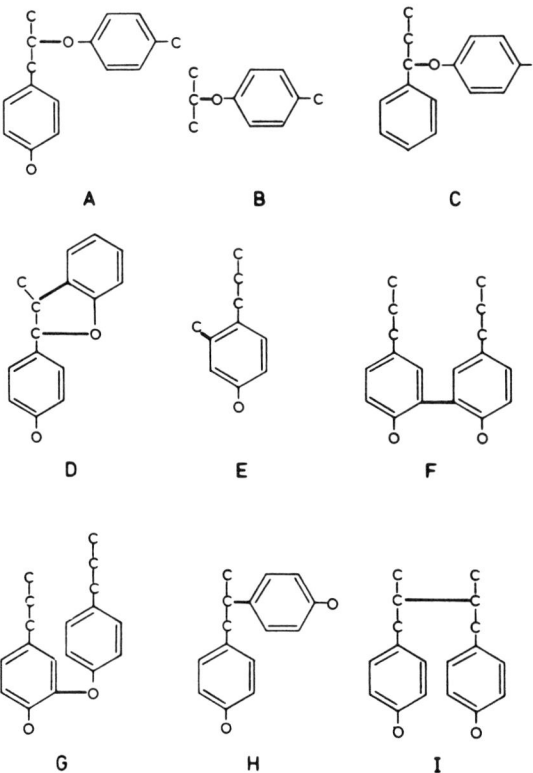

Figure 1. The most common linkages between phenylpropane units in lignins. (For proportions see Table I.) (Reproduced with permission from Ref. 18. Copyright 1981, Academic Press.)

alkali these are hydrolyzed to liberate free hydroxyl groups on the hemicellulose and acetate ions in the solution. Hydroxyl ions in solution are replaced by the acetate leading to a reduction in alkalinity. In conjunction with the free carboxyl groups present in the cell wall components and extractives this consumption of alkali can neutralize significant quantities of an alkaline reagent and possibly limit further alkaline hydrolysis.

Under acidic conditions, however, the liberation of free acetic acid during hydrolysis of the esters can increase the acidity and enhance further hydrolysis of not only additional ester groups, but acetal linkages and lignin bonds as well. A prime example is the so-called autohydrolysis reaction (2) in which acetic acid liberated by steam can cause complete hydrolysis of the hemicelluloses and convert the lignin into soluble fragments.

Acid hydrolysis of the acetal linkages in wood polysaccharides follows the normal hydrolysis of glycopyranosides with fission of the glycosyl oxygen bond between the rings. The mechanism involves protonation of this oxygen atom, followed by slow breakdown of the conjugate acid to the cyclic carbonium ion, which is then attacked rapidly by water. The reaction mechanism is depicted in Figure 2. Because the formation of the cyclic carbonium ion is the rate determining step, the hydrolysis rate is affected markedly by the crystallinity of the polysaccharide. The ion must assume a partially planar half-chair configuration, but the intermolecular hydrogen bonds in crystalline regions as in cellulose would tend to keep the pyranose ring in its puckered chair form and thus retard the formation of the planar intermediate required for hydrolysis (3). For this and other possible reasons the heterogeneous rate of cellulose hydrolysis is several orders of magnitude less than that of simple glycosides or noncrystalline polysaccharides (4). Thus, initial attack by acids involving acetal hydrolysis in cellulose would involve overwhelmingly the amorphous regions to the almost complete exclusion of the crystalline regions. In the noncrystalline hemicelluloses all acetals are susceptible to initial attack.

Under alkaline conditions hydrolysis of glycopyranosides is much slower and proceeds only at higher temperatures. The mechanism may involve intramolecular displacement of the exocyclic glycosidic oxygen with formation of a cyclic 1,6-anhydroglycopyranose (5).

Table I. Percentages of Different Types of Bonds in Lignins of Spruce and Birch

Bond Type[a]	Spruce (Picea abies)	Birch (Betula verrucosa)
Arylglycerol-β-aryl ether (A)	48	60
Glyceraldehyde-2-aryl ether (B)	2	2
Noncyclic benzyl(α)-aryl ether (C)	6–8	6–8
Phenylcoumaran (D)	9–12	6
2- or 6-Position condensed structures (E)	2.5–3	1.5–2.5
Biphenyl (F)	9.5–11	4.5
Diphenyl ether (G)	3.5–4	6.5
1,2-Diarylpropane-1,3-diol (H)	7	7
β,β-Linked structures (I)	2	3

[a] Letters A–I refer to Figure 1.
(Reproduced with permission from Ref. 1. Copyright 1977, Springer–Verlag.)

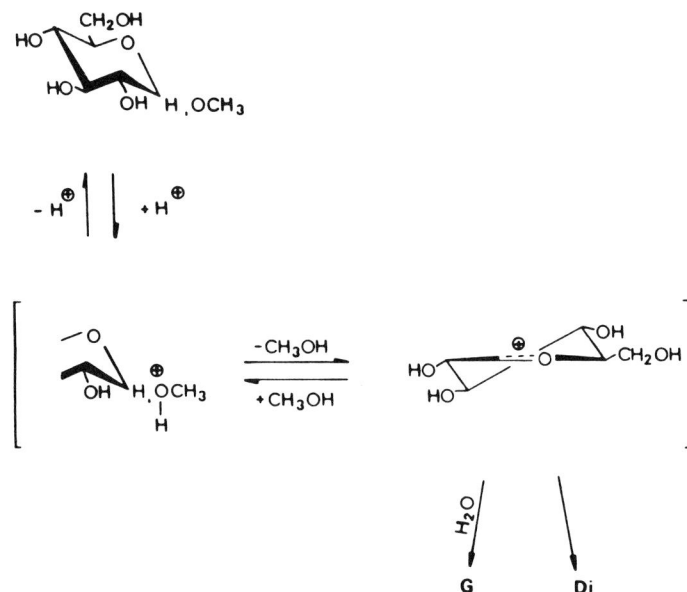

Figure 2. Acid-catalyzed hydrolysis of glucopyranosides. In addition to glucose (G), small amounts of disaccharides (Di) are also formed. (Reproduced with permission from Ref. 18. Copyright 1981, Academic Press.)

LIGNIN. The chemical behavior of lignin in degradative reactions has been studied extensively by subjecting model compounds as well as various lignin preparations to the appropriate conditions. These studies have been useful in interpreting both the structure of lignin and its reactions. Among these investigations hydrolysis of lignin linkages under both acidic and alkaline conditions has received much attention (6).

Even mild hydrolysis with hot water or dilute acetic acid is capable of cleaving the easily hydrolyzable benzyl(α)-aryl ether linkages (C in Figure 1), while leaving the β-aryl ether linkages (A in Figure 1) largely intact (7). A similar cleavage of α-aryl ether bonds in acid sulfite pulping provides the principal fragmentation of lignin in that process. More vigorous acidolysis of lignin as by dioxane–water (9:1) containing 0.2 M HCl does cleave the β-aryl ether linkage, presumably via a benzyllium ion and an enol aryl ether (8).

Cleavage of C–C bonds can also occur during acid hydrolysis, probably by reverse condensation reactions. Formaldehyde has been obtained from β–γ cleavage in yields approaching 4% upon heating lignins with 12% HCl or 28% H_2SO_4 (9). Small amounts of vanillin and vanillic acid encountered in acidolysis could only result from α–β cleavage.

When exposed to alkaline reagents at elevated temperatures lignin undergoes mainly cleavage of the ether linkages between the phenylpropane units with liberation of phenolic hydroxyl groups. Cleavage of C–C bonds and secondary condensation reactions also occur.

Benzyl(α)-aryl ether linkages are cleaved by alkali if there is a free phenolic hydroxyl group *para* to the propyl side chain or an adjacent hydroxyl in the β-position (*10*). Quinone methide intermediates are involved in the former hydrolysis and epoxides in the latter.

Alkaline cleavage of nonphenolic β-ethers occurs through an epoxide if there is an adjacent hydroxyl or carbonyl group in the α or γ-position (*11*). The β-aryl ethers also undergo cleavage through this mechanism, but the cleavage is less favored than formation of an enol ether unless sulfide ions are also present (*12*). Diaryl ether cleavage occurs only to a minor extent. Methoxyl groups in lignin generally are resistant toward alkaline hydrolysis.

Formaldehyde liberation from γ-methylol groups by alkaline cleavage of the β–γ C–C bond has been observed, and the action of hot alkali on lignin to form vanillin by cleavage of the α–β C–C bonds is well known. The simultaneous formation of acetaldehyde in the latter case results from a reverse aldol condensation. Traces of guaiacol found after alkaline hydrolysis of wood may result from cleavage of the C–C bond between the α carbon and the ring.

Oxidative Reactions. Oxidizing reagents affect both the carbohydrate and lignin components of wood. It might be inferred that lignin is more susceptible to oxidation because selective removal of lignin from carbohydrate by oxidation is the basis of an analytical procedure for holocellulose as well as the bleaching of pulps. However, carbohydrate oxidation cannot be neglected, especially when considering the initial processes of chemical attack on solid wood. Cellulose is quite sensitive toward oxidizing reagents.

CARBOHYDRATES. Mild oxidants such as chlorine, bromine, or iodine readily convert the aldehyde end groups in the wood polysaccharides to aldonic acid end groups. Nitrogen dioxide selectively converts the primary hydroxyl groups on C-6 in cellulose to carboxyl groups (*13*). Periodic acid is a specific oxidant for vicinal diols and yields formaldehyde from primary hydroxyl groups and aldehydes from secondary.

Although the oxidation of polysaccharides by halogens is confined largely to the aldehydic end group, and oxidation by periodic acid to glycols and nitrogen dioxide is confined to the primary hydroxyl group, other oxidizing agents are less specific and may affect all these groups as well as either of the secondary hydroxyl groups. Stronger oxidants such as nitric acid, potassium dichromate, and potassium permanganate can cause extensive oxidative degradation to a series of dicarboxylic acids.

An important oxidative degradation of polysaccharides occurs by the action of molecular oxygen in the presence of alkali. This reaction is useful in the controlled depolymerization of alkali cellulose before etherification, but more frequently will have undesirable consequences. The mechanism involves attack by radicals generated by decomposition of hydroperoxides, and is catalyzed by transition metals such as cobalt, iron, or manganese. Peroxide decomposition may be inhibited partly by magnesium salts or complexing agents for the transition metals. Oxidative degradation by free radicals is also involved in exposure of wood carbohydrates to ionizing radiation.

Carbonyl groups are introduced into polysaccharides by the action of chlorine, hypochlorite, and ozone. Under alkaline conditions glycosidic bonds may be cleaved. Hydrogen peroxide and chlorine dioxide react much more slowly with polysaccharides, and consequently are less degrading.

LIGNIN. Studies of the oxidation of lignin have received impetus from both attempts to elucidate the structure of lignin and to understand such technical processes as bleaching of pulp. The reactions may be classified into three categories: degradation of lignin to aromatic carbonyl compounds and carboxylic acids, degradation of aromatic rings, and oxidation of specific functional groups (14).

The first category includes oxidations with nitrobenzene, molecular oxygen, or metal oxides under alkaline conditions. Aromatic ring degradation results from exposure to peracetic acid, nitric acid, chlorine, chlorine dioxide, ozone, and the anions of hypochlorous and chlorous acids. Neutral permanganate can bring about both side chain and ring oxidation. Periodic acid and alkali peroxides oxidize specific functional groups. Lignin oxidation is also involved in the photodegradation of wood and the enzymatic degradation of lignin.

Alkaline oxidation of lignin is the commercial source of vanillin. This procedure converts a significant portion of the lignin to aromatic fragments that are, for the most part, aromatic aldehydes. The mechanism probably involves two steps, hydrolysis of the aryl ether linkages followed by side chain oxidation.

Oxidants such as silver, cobalt, mercuric, and cupric oxides in alkali yield a mixture of aromatic aldehydes or aromatic carboxylic acids or both by acting on the side chains. Silver is the strongest oxidant and yields chiefly acids; cupric oxide is the mildest oxidant and yields chiefly aldehydes. Although molecular oxygen in alkali can completely solubilize lignin by conversion to low molecular weight acids at elevated temperature and pressure, under less stringent conditions the aromatic rings are conserved and vanillin can be obtained in good yield.

Strong oxidants such as permanganate and dichromate in acidic solution degrade lignin completely to carbon dioxide and dibasic acids. In between the side chain oxidants described above and the strong oxidants lies a group of less drastic oxidants including chlorine, nitric acid, chlorine dioxide, sodium hypochlorite, peracetic acid, hydrogen peroxide, and ozone which attack primarily the aromatic nuclei of the lignin. Because they exhibit a relative selectivity in attacking lignin more rapidly than carbohydrate, they have found utility in bleaching pulp. An important mechanism for these lignin oxidations involves formation of quinones with subsequent ring opening to provide derivatives of dicarboxylic acids such as muconic, maleic, and fumaric acids.

Sodium periodate specifically oxidizes guaiacyl groups to quinones. Hydrogen and sodium peroxides in alkali are also somewhat selective in oxidizing lignin and destroy chromophoric groups such as quinones and carbonyl functions while also degrading only aromatic units with free phenolic hydroxyls to dibasic acids.

Other Reactions. The hydrolytic and oxidative reactions described in the preceding sections account for most of the degradation of wood by chemicals. However, there are several other processes significant enough to warrant special mention. Two of these might perhaps have been included under hydrolysis because decrystallization could be considered as representing hydrolysis of intermolecular hydrogen bonds and peeling as a special case of alkaline hydrolysis. The examples to be cited under discoloration demonstrate the influence of extractives on the chemical behavior of wood. Another reaction that involves extractives and affects the utilization of wood is the inhibition of the sulfite pulping of pine heartwood by the stilbene pinosylvin and its methyl ethers (15).

DECRYSTALLIZATION. The crystallinity of cellulose is an inherent property that is an important determinant of its mechanical properties, affinity for water, and accessibility to chemical reagents. Because cellulose comprises almost 50% of the wood, its crystallinity is a determinant of the behavior of the wood as well. Any disruption or change in the crystallinity of the cellulose will cause significant changes in properties and, thus by our definition, degradation.

Decrystallization of cellulose by swelling agents or solvents can be brought about by concentrated sodium hydroxide; amines; metallo-organic complexes of copper, cadmium, and iron; quaternary ammonium bases; concentrated mineral acids (sulfuric, hydrochloric, phosphoric); concentrated salt solutions (beryllium, calcium, lithium, zinc); and a number of recently investigated mixed solvents (16).

An increase in the crystallinity of cellulose from chemical treatment is unusual, but it does occur after acid hydrolysis of the amorphous regions. The initial hydrolysis of amorphous cellulose actually increases crystallinity by permitting the broken chains to have greater freedom to become organized in more highly ordered structures (17).

PEELING. Peeling is the term applied to the stepwise depolymerization of polysaccharides from the reducing end groups under alkaline conditions. In conjunction with the formation of new end groups by alkaline hydrolysis of glycosidic bonds, the degradation of the polysaccharides can be extensive.

The mechanism involves alkali-catalyzed rearrangement of the aldose end group to a 2-ketose. Elimination of the β-alkoxy group from the C-4 position generates a new aldohexose end group and the process continues down the chain. Under alkaline pulping conditions as many as 50 glucose units may be peeled from a single cellulose molecule before the reaction is stopped by a direct β-alkoxy elimination from the C-3 position (18).

DISCOLORATION. Staining or discoloration of wood by chemical processes is a frequently encountered form of degradation. It is often confused with discoloration caused by fungi, but results instead from the conversion of originally colorless or light-colored, naturally occurring extractives into intensely colored products that may impart an objectionable appearance to the wood.

Two mechanisms have been identified. Most of the so-called chemical stains result from oxidation of certain wood extractives by air during air seasoning or kiln drying. Colors observed include shades of brown, blue, green, yellow, and red. Species include both hardwoods (oak, birch, maple, alder, basswood, gum, etc.) and softwoods (eastern and western pines, hemlock).

Wet wood can also discolor by contact with iron or copper when tannins are present to form black iron tannate or reddish copper tannate. In contrast to the chemical stains caused by oxidation, which do not significantly alter the wood other than in color, the prolonged action of iron or copper may catalyze further chemical breakdown of the wood structure by free radical oxidative mechanisms.

Consequences of Chemical Attack

The properties of wood ultimately depend on the interaction of the three polymeric components cellulose, hemicelluloses, and lignin at the molecular level. They are intimately associated, with the linear crystalline cellulose microfibrils embedded in the matrix of the amorphous hemicelluloses and lignin. Intramolecular covalent bonding within the individual polymers is important to the wood structure. But intermolecular bonding between similar molecules and among the three major components is also critical. Hydrogen bonding, dipole–dipole forces, and London forces become very large in the aggregate as the size of polymers increases. Covalent bonding, as between lignin and hemicellulose, is probably also of significance.

Some chemical reactions of the wood components involve functional groups that do not form part of the polymer chain and may have only a slight effect on some wood properties and may enhance some. For example, esterification or etherification of free hydroxyl groups in carbohydrates or lignin may reduce hygroscopicity, increase dimensional stability, and actually increase wood strength by reducing the equilibrium moisture content.

However, any of the multitude of chemical reactions that disrupt the intramolecular and intermolecular bonds within and among the wood components will have deleterious effects on wood properties. The obvious extreme would be the complete depolymerization and dissolution of the wood components with complete destruction of the wood. These processes of depolymerization or solubilization of the wood components do not have to go to completion to bring about marked changes in wood properties or degradation.

Initial attack of the wood by any of the hydrolytic pathways described above reduces the degree of polymerization of the wood polymers involved and may be reflected in such manifestations as serious loss in impact strength before any loss in weight caused by solubilization can be detected. Insofar as many of the reactions exhibit selectivity among the wood components, one property or another may suffer the initial changes.

As a broad generalization the cellulose provides impact resistance and tensile strength, the lignin provides stiffness, and both matrix polymers contribute to hardness and compressive strength. Hydrolytic and oxidative processes that depolymerize and solubilize the wood components will affect the wood properties, often to an extent far greater than might be expected from the limited initial reaction.

Summary

On exposure to chemicals, functional groups and bonds in the component wood polymers undergo reactions that lead to changes in the wood properties. The reactions may affect both carbohydrates and lignin, and are chiefly hydrolytic or oxidative in nature. Hydrolysis of polymer linkages leads to depolymerization and strength loss. Vulnerable to hydrolysis are acetal linkages in carbohydrates and aryl ether linkages in lignin. Oxidative processes also contribute to fragmentation of the wood components. The loss of wood substance through subsequent solubilization of hydrolysis and oxidation products is an important part of the degradation process.

Literature Cited

1. Adler, E. *Wood Sci. Technol.* **1977**, *11*, 169–218.
2. Lora, J. H.; Wayman, M. *Tappi* **1978**, *61*(6), 47–50.
3. Goldstein, I. S. in "Wood and Agricultural Residues: Research on Use for Feed, Fuels and Chemicals"; Soltes, E. J., Ed.; Academic Press: New York, 1983; 315–28.
4. Harris, J. F. *Appl. Polym. Symp.* **1975**, *28*, 131–44.
5. Ferrier, R. J.; Collins, P. M. "Monosaccharide Chemistry"; Penguin Books: Baltimore, 1972; pp. 61–62.
6. Wallis, A. F. A. in "Lignins"; Sarkanen, K. V.; Ludwig, C. H., Eds.; Wiley-Interscience: New York, 1971; 345–69.
7. Nimz, H. *Holzforschung* **1966**, *20*, 105.
8. Adler, E.; Lundquist, K.; Mikshe, G. E. in "Lignin Structure and Reactions," Marton, J., Ed.; ACS ADVANCES IN CHEMISTRY SERIES No. 59; ACS: Washington, D.C., 1966; 22–35.
9. Brauns, F. E. "The Chemistry of Lignin"; Academic Press: New York, 1952; p. 440.
10. Gierer, J.; Noren, I. *Acta Chem. Scand.* **1962**, *16*, 1713.
11. Gierer, J.; Lenz, B.; Noren, I.; Soderberg, S. *Tappi* **1964**, *47*, 233.
12. Adler, E.; Falkehag, I.; Marton, J.; Halvarson, H. *Acta Chem. Scand.* **1964**, *18*, 1313.
13. Yackel, E. C.; Kenyon, W. O. *J. Am. Chem. Soc.* **1942**, *64*, 121.
14. Chang, H-M.; Allan, G. G. in "Lignins"; Sarkanen, K. V.; Ludwig, C. H., Eds.; Wiley-Interscience: New York, 1971; pp. 433–85.
15. Lindstedt, G.; Misiorny, A. *Acta Chem. Scand.* **1951**, *5*, 121.
16. Turbak, A. F. "Solvent Spun Rayon, Modified Cellulose Fibers and Derivatives"; ACS SYMPOSIUM SERIES No. 58; ACS: Washington, D.C., 1977.
17. Hermans, P. H.; Weidinger, A. *J. Polym. Sci.* **1949**, *4*, 317.
18. Sjostrom, E. "Wood Chemistry"; Academic Press: New York, 1981; pp. 43–46.

RECEIVED for review May 19, 1983. ACCEPTED July 27, 1983.

ABBREVIATIONS

a	one-half the thickness of wood	ESR	electron spin resonance
A	empirical sorption isotherm constant	EVAc	ethylene–vinyl acetate
		f	subscript denoting fiber-saturation point
AAC	alkylammonium compounds	f.s.p.	fiber-saturation point
ACA	ammoniacal copper arsenate	FSPL	fiber stress at proportional limit
ACC	acid copper chromate	g	subscript denoting green condition of wood
ASE	antishrink efficiency		
ATR	attenuated total reflectance	G	specific gravity of wood
B	empirical constant relating Q_L and W	ΔG	free energy decrease during sorption of water
BET	Brunauer, Emmett, and Teller equation	ΔG_1 ΔG_2	free energy changes (Dent sorption theory)
c	specific heat of wood		
C	empirical sorption isotherm constant	ΔG_h ΔG_s	free energy changes (Hailwood–Horrobin theory)
c_m	moisture concentration (mass/volume)	GMA	glycidyl methacrylate
CCA	chromated copper arsenate	h	relative vapor pressure
Ci	Curie	H	percent relative humidity
D	moisture diffusion coefficient (desorption isotherm)	H_o	static magnetic field strength
		IB	internal bond
DP	degree of polymerization	J	moisture flux
$\overline{DP}_n$	number average degree of polymerization	k_1, k_2	equilibrium constants (Dent sorption theory)
$\overline{DP}_w$	weighted average degree of polymerization	K	moisture transport coefficient
DSC	differential scanning calorimetry	K_1, K_2	equilibrium constants (Hailwood–Horrobin sorption theory)
DTA	differential thermal analysis		
DTG	differential thermogravimetry	l	longitudinal direction in the wood (also subscript)
EDTA	ethylenediaminetetraacetic acid	m	fractional moisture content (dry weight basis)
EDXA	energy dispersive X-ray analysis	M	percent moisture content (dry weight basis)
EMC	equilibrium moisture content	M_f	apparent fiber-saturation point
ESCA	electron spectroscopy for chemical analysis	M_w	percent moisture content (wet weight basis)

MF	melamine–formaldehyde	Sh	percent shrinkage in wood
ML	middle lamella	Sw	percent swelling in wood
MMA	methyl methacrylate	t	tangential direction in wood (also subscript)
MOE	modulus of elasticity		
MOR	modulus of rupture	t	time
o	subscript denoting dry condition of wood	T_g	glass transition temperature
OD	oven-dry	TBT	tri-n-butyltin
p	vapor pressure of atmospheric water	TBTMA	tri-n-butyltin methacrylate
		TBTO	tri-n-butyltin oxide
p_o	saturated vapor pressure of water	TEA	thermal evolution analysis
		TEM	transmission electron microscopy
P	primary cell wall		
P_m	hydrostatic pressure of water in cell wall	TG	thermogravimetry
		TMA	thermomechanical analysis
PEG	polyethylene glycol	v	specific volume of water vapor
PF	phenol–formaldehyde resins		
		v	subscript denoting volume
PVA	poly(vinyl alcohol)	V	volume of wood
PVAc	poly(vinyl acetate)	UF	urea–formaldehyde resins
PVC	poly(vinyl chloride)	VAc	vinyl acetate
Q_L	enthalpy change when liquid water is sorbed by wood	Vazo	2,2′-azobisisobutyronitrile catalysts
		w	subscript denoting wet condition of wood
Q_v	enthalpy change when water vapor is sorbed by wood	W	weight of wood
		W	heat of wetting
r	radial direction in the wood (also subscript)	WML	work to maximum load
		WPL	work to proportional limit
r	rate of strength increase with decrease in moisture content	WRP	water-repellent preservatives
		x	direction of moisture transport
R	gas constant		
rd	rads	X	moisture expansion coefficient
S	strength of wood at a given moisture content		
		ϵ	swelling strain in Barber equation
ΔS	entropy decrease associated with moisture sorption		
		γ_o	magnetic dipole frequency in NMR
S_1, S_2, S_3	layers in secondary wall of wood cell	Π	osmotic pressure
		ρ	wood density
		Φ	spreading pressure
SEM	scanning electron microscopy		

INDEX

INDEX

A

AAC—*See* Alkylammonium compounds
Abrasion resistance, formaldehyde-treated woods, 189
ACA—*See* Ammoniacal copper arsenate
ACC—*See* Acid copper chromate
Accelerated weathering, 426
  erosion, 417*t*
Acetal linkages
  acid hydrolysis, 580
  hydrolysis, 579
Acetaldehyde, reaction with wood, 189
Acetals, 182
  reactions with wood, 188–90
Acetic acid, 185
  effect on strength, 239
Acetic anhydride
  for chemical modification, 201*t*
  southern pine treatment, 200*t*
Acetyl content, analytical determinations, 72
Acetyl-4-*O*-methylglucuronoxylan, structure, 64*f*
Acetylated wood
  attack resistance, 184
  density, 184–86
  gas permeability, 184
  mechanical properties, 184
  method of acetylation, 185
  as painting surface, 185
  UV radiation effect, 185
Acetylation
  anhydride method, 185
  catalysts, 184
  with ketene gas, 185
  optimum conditions, 184
  by vapor-phase treatment, 185
ACF—*See* Ammoniacal copper fluoride
Acid activation, nonconventional bonding, 390–91
Acid chlorides, for esterification, 186–87
Acid content
  correlation with gel time, 347*t*
  wood, 346*t*
Acid copper chromate
  components, 309
  effects on mechanical properties, 245–48*t*
Acid dehydration–condensation reactions, 388
Acidic solutions, adsorption, 239
Acidity
  acetylation, 185
  wood, 345–47

Acids
  bifunctional, nonconventional bonding, 382
  effect on strength, 238
  surface activation, 372–73
Acrylic latex resins, 437
Acrylonitrile(s)
  chemical modification, 201*t*
  reactions with wood, 191–92
  southern pine treatment, 200*t*
  structure, 262
Activation energy, for moisture diffusion, 168
Activation of wood surface, 349–400
Activators, in nonconventional bonding of polymers, 384
A/D ratio—*See* Adsorption/desorption ratio
Additives
  effect on catalyzed monomer, 276
  effect on pyrolysis products, 551*t*
Adherend and adhesives, molecular forces, 326–27
Adhesion, 323–47
Adhesive(s)
  bond formation, 324–27
  cold-setting resole, 333
  emulsions, 342
  hot-melt, 341–42
  hot-setting phenolic, 333
  isocyanate-based, 337–39
  liquid to solid bond transformation, 324–26
  polymers, 324–26
  tannin-based, 339–40
  wood, 327–45
Adhesive, nature's—*See* Lignin
Adsorption, Douglas-fir, 136*f*
Adsorption isotherms, 137
  wood components, 139*f*
Adsorption/desorption ratio, 137
African pencil cedar, chemical composition, 113*t*
Afterglow, 249
Albizzia board, attack by termites, 466*f*
Algae, deterioration responsible for, 459*t*
Alkali
  component solubility, 62
  oxidation of lignin, 584
  solutions, adsorption, 239
Alkaloids, 458
Alkoxy radicals, in cellulose, 429
Alkyl chlorides, reaction with wood, 190

Alkylammonium compounds
 as biocides, 310
 fixation mechanism, 318
Alkyldimethylammonium chloride, fixation mechanism, 318
Allophanate, formation, 339f
Alodining process, 375
Alpha cellulose, analytical determinations, 71
Aluminum trihydrate, 565
 basis for fire retardants, 563
American Society for Testing and Materials, 214
Amines, bifunctional, nonconventional bonding, 382
Aminoresins, 566–67
Ammoniacal copper arsenate
 components, 308
 effect on mechanical properties, 245–48t
 effect on strength, 243
 mechanism of fixation, 317
Ammoniacal copper fluoride, effect on mechanical properties, 245–48t
Ammonium chloride, 532
Ammonium dihydrogen phosphate
 degree of polymerization studies, 546
 treatment of southern pine, 250f
Ammonium nitrate, 378
Ammonium phosphate, effect on mechanical properties, 251–52t
Ammonium sulfate, 532
 effect on mechanical properties, 251–52t
Ammonium treatment, 260–61
Anatomical structure
 UV light exposure changes, 420
 and weathering, 404
Angiosperm(s)
 See also Hardwoods
 complete lignin removal, 71
 definition, 4
 wood tissue, 6f
Anhydride method, of acetylation, 185
Anisotropy
 cross-sectional distortion, 150f
 definition, 146
 longitudinal, 149
 shrinking, 146–51
 swelling, 146–51
 transverse shrinkage, 149
 theories, 150–51
Anobiid beetle, attack of cuangare, 466f
Antishrink efficiency
 benzaldehyde modification, 190
 calculation, 178
 effect of organotin polymers, 297
 epoxides, 192–93
 tracheid wall splitting, 188
 from water-soaking method, 203t
 wood–polymer composites, 297t
AP—See Ammonium phosphate
Apical meristems, location, 9
 See also Terminal meristems
Araban, weathering effects, 408
Arabinogalactan, in hemicelluloses, 65t
Arabinoglucuronoxylan, in hemicelluloses, 65t

α-L-Arabinose, 63f
Aromatic char formation, effect of inorganic additives, 552 t
Aromatic compounds, oxidation, 509f
ASE—See Antishrink efficiency
Ash, 58
 analysis, 74
 definition, 68
 elemental composition determination, 74
Aspen
 heat of combustion, 523t
 distribution, 524t
 nonconventional bonding, 361
Aspen stem, tension wood, 51f
Aspergillus spp., 459
ASTM Standard D 1102, 74
ASTM Standard D 1104, 69
Attack, chemical, consequences, 586–87
Autolysis, xylem, 11
Axial loading properties, 223–25
Axial shrinkage, 147
2,2'-Azobisisobutyronitrile catalyst—See Vazo

B

Bacteria, deterioration responsible for, 459t
Ball milling procedure, 68
Balsam fir
 elemental composition, 119t
 microfibrils, TEM, 26f
Barber's model, 147–49,149f
Barkas effect, 160
Barrier theories, 542,543
Base-catalyzed reactions, with epoxides, 192
Bases, effect on strength, 238
Basic density, 38
 compression wood, 48
 hardwoods, 40
 and range of variability, 41t
 significance, 39
 softwoods, 40
Basic specific gravity, definition, 39
Basswood
 elemental composition, 120t
 light exposure, erosion, 417t
 physical property enhancements, 280–81t
 strength properties, 284t
 treated with *tert*-butylstyrene, 279f
Basswood–MMA composite
 effect of trimethylol propane trimethacrylate, 274t
 TMPTMA effects, 275f
 vazo catalyst effects, 270t,270f,271f
Basswood–polymer composites
 hardness, 285f
 load deflection curves, 287f
Beech
 modified with diisocyanate, 188
 physical property enhancements, 280–81t
 sorption vs. moisture content, 156f
Bending strengths, formaldehyde-treated woods, 189
Bending test
 data, 282f

Bending test—*Continued*
  derivation of mechanical properties, 221
Benzaldehyde modification, antishrink
  efficiency, 189,190
Benzene dicarboxylic acid, aromatic
  carbon content, 511f
Benzyl(α)-aryl ether linkages, alkaline
  cleavage, 583
BET—*See* Brunauer-Emmett-Teller model
Betula verrucosa Ehrh., polysaccharide
  content, 70t
  See also Birch
Bifunctional acids, nonconventional
  bonding, 382
Bifunctional amines, nonconventional
  bonding, 382
Bifunctional isocyanate
  bonding of preoxidized wood, 362f
  nonconventional bonding, 382-83
Bifunctional molecules, bonding, 361-64
Bioactive wood-polymer composites,
  291-306
Biochemical decomposition
  mechanism, 460
  and wood decays, 462-79
Biocide, tri-*n*-butyltin methacrylate, 302
Biodegradation
  prevention, 177
  resistance
    improvement, 295
    organotin polymers, 300-303
  wood, 292
Biological decomposition, 455-87
  lignin, 460
Biological deterioration, types, 459t
Biological resistance
  chemical modification of wood, 177
  effect on strength, 243
Biotoxicity
  organotin compounds, 292-93
  polymers, 292-93
Birch
  See also Betula verrucosa Ehrh.
  light exposure, erosion, 417t
  lignins, bonding, 581t
  polysaccharide content, 70t
Birds, deterioration responsible for, 459t
Bis(tri-*n*-butyltin)oxide, 439
Bishop pine, properties, H_2O_2 effects, 366t
Black cherry, physical property
  enhancements, 280-81t
Black spruce
  heterocellular ray, 20f
  microscopic studies, 352
  transwall failure, 352
Blackjack oak, heartwood distribution, 42f
Blowing agents, 543
Board density vs. internal bond strength,
  360f
Bond formation, adhesive, 324-27
Bonded chemical, distribution, 204-6
Bonding
  bifunctional molecules, 361-64
  corona discharge treatment, 372
  by covalently attached polymer, 364-72

Bonding—*Continued*
  effect of extractives, 345
  through polymeric chains, 365f
  use of acids, 370-71
  use of oxidants, 364-70
  wood-to-wood, 358-61
    Douglas-fir veneer, 359
    mechanism, 359f
    particle board, 360f
    Pinus ponderosa Laws., 361
Bonding proof
  criteria, 197-204
  increases in wood volume, 197-98
  IR data, 202-4
  resistance to leaching, 198
Borax, 532
  for fire retardants, 564
  water vapor release, 544
Borax treatments, degree of polymerization
  studies, 546
Bordered pit membranes, 29
Bordered pit structures, southern yellow
  pine, 421f
Boric acid, for fire retardants, 564
Boric acid salts, 250
Borneo, woods, chemical composition,
  98-100t
Boron, for fire retardants, 563,564
Botanical origin, commercial woods, 5f
Bound water, 128
  effect on shrinkage, 140
  effect on strength, 218
  enthalpy level, 154
  flow rate, 168
  heat of sorption, 154
  relative energy level, 155f
Brazil, woods, composition, summary, 121t
Brown-rot fungi
  depolymerizing agent, 474
  effect on organotin polymer-
    impregnated wood, 300
  effect on sweetgum, 243f
  effect on wood strength, 471
  features, 468-69t
  glucanase activity, 474
  hemicellulase production, 477
  hydrogen peroxide secretion, 474
  lignin decomposition, 479
  water conduction, 480
Brunauer-Emmett-Teller model
  primary sorption sites, 162
  sorption, 162
  isotherms, 163f
Bulk density, 38
  See also Density
Bulking, by vinyl polymers, 262
Burning, 176
  chemistry, 541
Butyl isocyanate, chemical modification,
  425
Butylene oxide, chemical modification,
  201t,425
Butylene oxide-modified wood
  decay resistance, 195
  termite attack, 196f

Butylene oxide treatment, 193f
tert-Butylstyrene
  properties, 286t
  structure, 262
Butyric anhydrides, 186

C

Calcium carbonate deposition, southern pine, 244f
Cambial zone, definition, 11
Cambodia, woods, chemical composition, 98–100t,121t
Canada, woods, composition, summary, 122t
Capacitance dielectric moisture meter, 133
Capillary water, 128
Carbamic acid, formation, 339f
Carbohydrates, 58–66
  analytical determinations, 69–73
  distribution, 68–69
  enzymatic hydrolysis, 68
  hydrolytic reactions, 579–81
  oxidative reactions, 583
Carbon cycle, simplified, 456
Carboxylic acids, esterification, 187
Carcinogenicity
  β-propiolactone, 191
  of reactants for modification, 182
Catalyzed monomer, additives effect, 276
Cativo, oxide treatment, 198t
CC—*See* Copper chromate
CCA—*See* Chromated copper arsenate
Cell
  degradation, 457
  volumetric swelling, 143f
Cell cavities
  liquid water, enthalpy, 154
  microfibril orientation effect, 144
  water vapor, enthalpy, 154
Cell wall
  amount of polysaccharides, 70t
  apparent density, equation, 141
  bulking, 177
  chemical components, 404f
  chromium distribution, 316
  components, penetration and reactivity, 175–207
  diagram, 144f
  dry, density, 141
  oven-dried, impermeability to vinyl monomers, 279
  polymer migration, 296
  shrinkage
    equation, 142
    maximum, equation, 142
    and swelling, 141–42
  specific gravity, 142
  swelling, 141
Cellobiose
  cleavage, 473
  conversion, *Phanerochaete chrysosporium*, 473
Cellophylic systems, 350
Cellular level, structural performance, 226

Cellulases
  See also Enzymes
  endogenous, 462
  wood digestion, 461
Cellulose, 7,58–62
  acid treatment, 372
  alkoxy radicals, 429
  Arrhenius plots, 493f,497f
  bond scission, 494,497f
  chain length, 60,230
  combustion, 492f
    maximum heat intensity, 560t
  decrystallization, 585
  degree of polymerization, 60,496f
  dehydration reactions, 504–5,507f
  depletion, 463
    by fungi, 470f
  depolymerization, 472f,496t
    by transglycosylation, 500
  disproportionation reactions, 505–8
  elemental composition, 510t
  fission reactions, 505–8
  free radical reactions, 427–29
  furan grafting, 388f
  of 1,4-β-D-glucopyranose, 59f
  heat of combustion, 523t
    distribution, 524t
  hydrogen bonds, 230
  hydroxyl groups, 37
  isothermal degradation, 493f
  low temperature pyrolysis, 492–500
  macrofibrils, TEM, 25f
  matrix substance, 7
  microfibrils, TEM, 25f
  molecular structure, 59f
  oxidation, 377
    with hydrogen peroxide, IR reflectance, 374f
  oxidative depolymerization, 474
  oxidized, IR bands, 377
  oxidized groups, 499f
  phosphoric acid-treated, products, 550
  pyrolysis, 356,492f
    to anhydro sugars, 501f
    products, 502f,503t
    temperature, 503f
  pyrolyzate GLC analysis, 507f
  reaction with copper, 315
  reaction with hydrogen peroxide, 373
  residual, vs. pyrolysis time, 495f
  site of degradation attack, 578
  solubility, 62
  surface activation, 379
  thermal degradation, free radical mechanism, 497
  treated, thermogravimetric analysis, 547f
  treated with acidic salts, 373
  (trimethylsilyl)ated, pyrolysis GLC of tar, 506f
  UV degradation, 176
  volatile products, formation, 490–508
  weathering effects, 407
  xanthate derivative, 60
Cellulose I, 60,61f
Cellulose II, 60

INDEX

Cellulose char
  cross-polarized/magic angle spinning, 514*f*
  differential heat of chemisorption, 519*f*
  FTIR spectra, 516*f*
  IR spectra, 517
  oxygen chemisorption, 518*f*
Cellulose combustion, high temperatures, 500–504
Cellulose content, 58
  tension wood, 50
Cellulose decomposition, 473–77
  hydrolytic and oxidative enzymes, 475*f*
Cellulose degradation, mechanism, 473
Cellulose hydroperoxide, formation and decomposition, 500*f*
Cellulose pyrolysis
  carbon dioxide yields, 498*f*
  carbon monoxide yields, 498*f*
  dehydration products, 504
Cellulose-lignin mixtures
  reaction with hydrogen peroxide, 375
  SEM, 383*f*
Central America, woods, chemical composition, 82–87*t*,88*t*
Ceratocystis spp., 459
CFA—*See* Chromated fluorarsenate
Chain branching, vapor-phase reactions, 544
Char
  aromatic carbon content, 511*f*
  cellulose
    cross-polarized/magic angle spinning, 514*f*
    differential heat of chemisorption, 519*f*
    FTIR spectra, 516*f*
    IR spectra, 517
  chemisorption, 526*f*
  CO and CO_2 production, 527*t*
  elemental composition, 510*t*
  formation, 508–18
  oxidation, 526
  oxidation product gas chromatogram, 510*f*
  oxygen adsorption, 526*f*
  reactivity, 518–20
  residue weight, heat treatment temperature effects, 521*f*
  structure and functionality, 513
Char formation, 176
  aromatic, effect of inorganic additives, 552*t*
  carbon distribution, 515*t*
  heat treatment temperature effect, 509,512*f*
  temperature effects, 515*t*
Char generation, treated wood, 196
Char oxidation, effect of prepyrolysis additives, 528*t*
Checks, formation, plywood, 423
Chemical attack, consequences, 586–87
Chemical catalysts, 264
Chemical components, 58–68
  macroscopic, 226–27
  microscopic, 227
  strength characteristics, 226–36

Chemical composition
  strength relationship, 231–35
  structure relationship, 226–31
  variability, 57
  wood-destroying fungi changes, 463
Chemical composition data, 57–126
  Borneo woods, 98–100*t*
  Cambodia woods, 98–100*t*
  Central America woods, 82–87*t*,88*t*
  Ghana woods, 89–90*t*
  Japanese woods, 91–97*t*
  Kalimantan woods, 98–100*t*
  Mexico woods, 82–87*t*
  Mozambique woods, 89–90*t*
  North American woods, 115–16*t*
  Papua New Guinea woods, 98–100*t*
  Philippine woods, 101–9*t*
  Puerto Rico woods, 82–87*t*
  South America woods, 82–87*t*,88*t*
  Taiwan woods, 110–11*t*
  U.S. woods, 76–81*t*
    southeastern, 117–18*t*
  U.S.S.R. woods, 112–13*t*
Chemical degradation, 577–87
Chemical energy, effects, 402*t*
Chemical impregnation, fire retardants, 532
Chemical modification
  for biological resistance, 177
  conditions, 182–83
  definition, 176
  gas reactants, 182
  penetration, 178–81
  and physical properties, 175–78
  reactants, 181–82
Chemically modified woods, weathering, 425–26
Chemisorption, char, 526*f*
Chemisorption rate, 519
Chemistry of wood strength, 211–56
Chemotaxonomy, definition, 57–58
Chloral, reaction with wood, 190
Chlorates, 381
Chlorite holocellulose, analytical determinations, 71
Chromated copper arsenate
  components, 308–9
  effect on mechanical properties, 245–48*t*
  effect on strength, 243
  fixation diagram, 314*f*
  pH effects, 315
  reactions with wood, 315
  treatment of southern pine, 249*f*
Chromated copper arsenate-treated wood, soft-rot attack, 313
Chromated copper borate, 245–48*t*
Chromated fluorarsenate, 245–48*t*
Chromated zinc chloride
  components, 309
  effect on mechanical properties, 245–48*t*
Chromic acid treatment, 441–42
Chromium, 316
Chromium trioxide
  complex, 316
  as surface treatment, 440
Clear varnish, 437
Clear wood, mechanical properties, 219*t*

Coating interactions, 445–46
Cobalt-60, safety requirements, 263
Coconut shell flour, 333
Coefficient of variation, 215,217t
Cold hardeners, 335
Cold-setting resole adhesives, 333
Color, light exposure effects, 357
Color changes
  artist's rendition, 412f
  cause, 402t
  outdoor weathered wood, 412f,413f
  weathering, 411–14
Combustible volatiles, reduction, mechanism, 546
Combustion, 489–528
  carbon monoxide-to-carbon dioxide ratio, 526
  cellulose, 492f
  definition, 490
  flaming, 491f
    heat of combustion, 522
    heat release, 526
  intensity, equation, 520
  physical transformations, 490
  smoldering, 491f
    driving force, 522
  solid-phase, 541
  vapor-phase, 541
    reactions, 544
  wood, maximum heat intensity, 560t
Commercial woods
  botanical origin, 5f
  effect of bordered pit membranes, 29
Composition
  elemental, 58
  microscopic, 228
  wood, 4
Composition data, tables, 75–122
Compound middle lamella, definition, 26
Compreg, 258–59
Compression parallel to grain, 224
  coefficient of variation, 217t
  specific gravity effects, 216t
Compression perpendicular to grain, 224
  coefficient of variation, 217t
  specific gravity effects, 216t
Compression test data, 283f
Compression wood
  chemistry, 46–47t
  cross-sectional views, 49f
  definition, 48
  Douglas-fir, 45f
  softwoods, 48
  structure, 46–47t
  variability, 48–51
Compressive stress, 213
Conifers, 4
  pit aspiration, 30
Coniferyl alcohol, 66
Coniophora cerebella, 184,188
Coniophora puteana, 301
Coniophora puterana, attack resistance, 184
Conventional bonding, 350
Copper, reactions, 315
Copper chromate, effects on mechanical properties, 245–48t

Copper naphthenate, 309–10,439
Copper-8-quinolinolate, 439
  structure, 310
Corewood, definition, 53
Corona discharge, 372,381
Costa Rica, woods, composition, summary, 121t
Cotton, heat of wetting vs. moisture content, 157f
Cottonwood, thermogravimetry, 524f
p-Coumaryl alcohol, 66
Covalent wood-to-wood bonding, direct, 358–61
CP/MAS—*See* Cross-polarized/magic angle spinning
Creep, definition, 231
Creosote
  chemical composition, 309t
  components, 308
  effects on mechanical properties, 245–48t
Critical oxygen index test, 537–38
Cross and Bevan cellulose, analytical determinations, 71
Cross-field pits, communication, 32
Cross-link density, 325
Cross-linking
  catalysis, 189
  effect on polymerization, 272–74
Cross-linking agent(s)
  bifunctional molecules, 363
  sulfuric acid, 370
Cross-polarized/magic angle spinning, 513
  cellulose char, 514f
Cross-sectional distortion, 150f
Crown of tree, 4
Crushing strength, formaldehyde-treated woods, 189
Cuangare, attack by anobiid beetle, 466f
Cure chemistry, effect of extractives, 345
Curing reaction, initiation, 329
Cyanoethylated wood, impact strength, 192
Cyanoethylation, 191
CZC—*See* Chromated zinc chloride

D

DC resistance moisture meter, temperature calibration curves, 132f
DC resistivity, vs. moisture content, 130f,131f
Decay
  definition, 401
  role of gelatinous sheaths, 471
  types, 462–63
  wood
    cellulose, 473–77
    hemicellulose, 476–77
    lignin, 477–79
    mechanisms, 471–79
Decomposed Douglas-fir, heat of combustion, 523t,524t
Decomposition
  biological, 455–87
  economics, 481
  structural polymers, organism responsible, 459t

INDEX

Decomposition—*Continued*
  wood
    by insects, 464-65t
    by marine borers, 465t
Defibration temperature, effect on lignin content, 353
Degradation
  decrystallization, 585
  definition, 577
  oxidative reactions, 583-85
  sites of initial attack, 578-86
  UV, 408
  wood, 356
Degree of polymerization
  definition, 58
  number average, 60
Dehydration, catalyst, 561
Dehydration products, cellulose pyrolysis, 504
Delignified cell-wall softwood fiber, photomicrograph, 230f
Density, 38-40
  apparent, cell wall, equation, 141
  moisture effects, 152
  relationship to thermal conductivity, 154
  vs. moisture content, 153f
Dent's surface sorption theory, 162-64
Depolymerization, transglycosylation, 500
Desorption
  Douglas-fir, 136f
  isotherm, 137
Deterioration
  biological, types, 459t
  with decomposition, 460
  without decomposition, 458-60
  organism responsible, 459t
  types, 458-60
Dialdehyde reactions, 190
Diammonium phosphate, 250
  reduction of flame spread, 532
Diammonium sulfate, 250
Dibutylphthalate, 342
Dichlorohydrin-treated wood, decay resistance, 195
Dichromate, lignin degradation, 585
Dicyandiamide, 543,544
Didecyldimethylammonium chloride, structure, 310
Dielectric constant
  correlation to swelling, 180
  vs. dry wood specific gravity, 133f
Dielectric moisture meters, 133-34
Diethyl ether, wood solubility, 74
Differential scanning calorimetry, fire retardancy test method, 534
Differential thermal analysis
  cellulose, 554f,555f,556f,558f
  fire retardancy test method, 534
  inorganic fire retardants, 553
  lignin, 554f,555f,556f,558f
  wood, 558f
  in helium, 553
Diffuse-porous, 19f
Diffusion coefficient
  effects, 167

Diffusion coefficient—*Continued*
  moisture content effects, 168
  relationship to moisture transport coefficients, 167t
  vs. wood moisture content, 168f
Difunctional aldehyde reactions, 190
Digestibility
  by cellulases and hemicellulases, 461f
  maximum, 461
Dilution-by-noncombustible-gases theories, 542,544
Dimensional changes, minimizing for wood in use, 140
Dimensional stability
  bulking treatments, 177
  improvement, 295
  wood-polymer composites, 286t
Dipolar forces, 326-27
Dipoles, cause, 326
Direct covalent wood-to-wood bonding, 358-61
Discoloration
  causes, 402t
  form of degradation, 586
  by fungi, 458
  by light, 406
  organisms responsible, 459
Discontinuous rings, 14-15
Dispersion forces, 326
Douglas-fir
  acid content, 346t
  adsorption, 136f
  attack by marine borer, 466f
  bark, heat of combustion, 523t,524t
  basic density, 41t
  butylene oxide treatment, 198t
  color changes, 412f,413f
  compression wood, 45f
  decomposed, heat of combustion, 523t,524t
  distinguishing characteristic, 33
  free radicals vs. moisture content, 433f
  gamma-irradiated, 261
  heat degradation, 241
  initial desorption, 136f
  light exposure, erosion, 417t
  lignin, heat of combustion, 523t,524t
  secondary desorption, 136f
  SEM, 13f
  spiral thickenings, SEM, 33f
  veneer, wood-to-wood bonding, 359
Dry cell wall, density, 141
Dry mass fraction, definition, 129
Dry mass percent, definition, 129
Dry wood
  hydration, 164
  specific gravity vs. dielectric constant, 133f
  specific heat, 153
  water absorption, 278
DSC—*See* Differential scanning calorimetry
DTA—*See* Differential thermal analysis
Durability, 333-34
Duration of load, 218

E

Earlywood
  interfiber pits, 28
  liquid penetration, 181
  weathering, 424
Earlywood cells, role in strength, 226
Earlywood tracheids, cell walls, 181
Earlywood-latewood interaction theory, 151
Earlywood-to-latewood transition, softwood, 18f
Eastern cottonwood, elemental composition, 120t
Eastern hemlock, elemental composition, 120t
Eastern larch, growth rings, 18f
Eastern white pine
  elemental composition, 119t
  growth rings, 18f
Elastic materials, stress and strain, 213
Elastic properties, 213
Elastic strength, 232
Elasticity
  elastomers, 325
  wood, 214
Elastomer, elasticity, 325
Electrical resistance moisture meters, 130–33
Electron spectroscopy for chemical analysis, 378
Electron spin resonance, of wood free radicals, 428f
Electronic absorption spectroscopy, 353–54
Elemental composition, 58
Elementary fibrils, definition, 24
Elm, propylene oxide treatment, 198t
Elovich kinetics, 519
Emulsions, adhesive, 342
End-use applications, 18
Endogenous cellulases, 462
Energy forms, effects, 402t
English oak, chemical composition, 113t
Entropy change, vs. moisture contents, 158f
Environmental effects, strength, 238–43
Enzymatic hydrolysis, carbohydrate, 68
Enzymes
  cellulose decomposition, 473
  lignin decomposition, 478
Epichlorohydrin, polymerization with amines, 388
Epichlorohydrin-treated wood, decay resistance, 195
Epithelial cells, 21
Epoxide(s)
  base-catalyzed reactions, 192
  butylene oxide treatment, 193f
  propylene oxide treatment, 193f
  reactions with wood, 192–97
Epoxide modification, antishrink efficiency, 193f
Equilibrium moisture content, definition, 215
Erosion
  after accelerated weathering, 417t
  reducing, 426
Erosion rate, 415–16
ESCA—*See* Electron spectroscopy for chemical analysis
Ester(s), 182,579
  acetylation, 183–84
  density, 184–86
  hydrolysis, acidic conditions, 580
  phthalylation, 186
  reactions with wood, 183–88
Ester linkages, hydrolysis, 579
Ethanethioic acid, for vapor-phase acetylation, 186
Ethanol/benzene, as extractives solvent, 73–74
Ether(s)
  as extractives solvent, 73
  methylation, 190
  reaction with wood, 190–92
Ether linkages, 182
Ethylene, structure, 262
Ethylene oxide, vapor-phase treatment, 192
Ethylenediamine, as bonding agent, 362
Eucalyptus obliqua, butylene oxide treatment, 198t
European spruce, sorption vs. moisture content, 156f
External plasticizer, 342
External surfaces, 350
Extractives, 68,73–74
  See also Extraneous components
  appropriate solvents, 73
  effect on bonding, 345
  effect on cure chemistry, 345
  redistribution, 354–55
  surface coverage, 353
Extraneous components, 68,73–74
  See also Extractives
Extraneous materials, 58

F

Fatigue, failures, 219
FCAP—*See* Flour chrome arsenate phenol
Ferric chloride, 442
Ferric compounds, 358
Fiber(s), 19
  architecture, 24–28
  definition, 4
  macroscopic level, 234
  strength, 211
  wall architecture, 25f
Fiber cells, degradation, 457
Fiber saturation
  definition, 37
  relationship to volumetric shrinkage data, 144
  swelling of cell wall, 142
Fiber-saturation point
  apparent, obtaining, 138
  definition, 129,278
  measurement, 141–42
Fiber stress at proportional limit, 222
  wood–polymer composites, 286t
Fiber volume, hardwoods, 4
Fibril orientations, of S_1, S_2, and S_3, 144f
Fick's second law, 167

INDEX

FID—*See* Free induction decay
Film-forming finishes, 437–38
  natural wood, 444
Fine grinding, effect, 461
Finishing, properties, 436
Fir, nonconventional bonding, 368t
Fire retardancy, 531–71
  mechanisms, 541–63
  role in vapor-phase reactions, 545
  test methods, 532–41
  theories, 542–63
Fire retardant(s)
  chemical basis, 563
  coating systems, 569
  effect on strength, 250
  halogens, 545
  inorganic, differential thermal analysis, 553
  in monomer solutions, 277
  protection, 532
  reduced smoke and toxicity, 569
  toxicity, test methods, 541
Fire retardant chemicals, effect on oxygen index, 538f
Fire retardant formulations, 563–68
  leaching resistance, 564
Fire retardant treatment, 212
  classification, 532
  for panel products, future research, 568–69
Fire retardant-treated wood, thermogravimetric analysis, 548f
First law of photochemistry, 406
Flame propagation, 544
Flame retardants, aluminum trihydrate, 565
Flame spread
  borax inhibition, 564
  Phosgard, 277
  reduction, 532
Flame spread index, effect of inorganic additives, 537f
Flame spread ratings, determination, 535
Flaming combustion, 491f
  definition, 541
  heat of combustion, 522
  heat release, 526
Flammability
  and oxygen index, 537
  reduction by boron compounds, 565
Flexibility, of polymers, 325
Flexural loading properties, 221–23
Flexural modulus, polymer impregnation effect, 299f
Flexural strength, polymer impregnation effect, 299f
Flexure formula, 222
Flour chrome arsenate phenol, effect on mechanical properties, 251–52t
Foreign materials, deposition, 355
Forest management, scientific, aims, 41
Formaldehyde
  acetylation, 188–89
  reactions with phenol, 330
Formaldehyde-treated wood
  fungi attack, 189
  mechanical properties, 189

Formation, 3–56
Fraction of dry mass, definition, 129
Fractional moisture content, defined, 127
Fracture toughness, 333–34
  cure temperature dependence, 334
Free energy change
  primary water, 164
  secondary water, 164
  with water of hydration, 166
  with water of solution, 166
Free energy loss
  calculation, 157
  vs. moisture contents, 158f
Free induction decay, vs. moisture content, 134,135f
Free radical characteristics, weathered wood, 430–31
Free radical formation, 427
Free radical mechanism, thermal degradation of cellulose, 497
Free radical reactions
  cellulose, 427–29
  hemicellulose, 427–29
  in lignin, 429–30
  weathered wood, 430–31
Free radical trap theories, 542,544
Free radicals
  gamma-radiation as source, 263
  moisture effects, 431
  water effects, 431
Free water, 128
Freezing, of absorbed water, 405
Fresh-cut wood, drying, 458–59
FR-S rating, 535
FSPL—*See* Fiber stress at proportional limit
Fungi
  cause of wood decay, 456
  deterioration responsible for, 459t
Furan, grafting, 388f
Furfuryl alcohol, acid polymerization, 384,385

G

G fibers, definition, 50
Galactoglucomannan, in hemicelluloses, 65t
Gamma-irradiated Douglas-fir, 261
Gamma-radiation, 261
Gamma-ray, ionization, 263f
Gamma-ray-treated wood, 189
Gas chromatography, sugar separation, 72
Gas penetration, 182
Gas reactants, 182
Gel effect, 268–70
Gel times, correlation, 345
Gelatinous fibers, definition, 50
Gelatinous sheaths, role in decay, 471
Ghana, woods, chemical composition, 89–90t,121t
Giomeralla cingulata, decay resistance, 293
Glass transition temperature, in polymers, 325
Glassy state, 325
Gloephyllum trabeum, attack resistance, 184

Glowing, 541
  promotion, 532
Glucanases
  activity, brown-rot fungi, 474
  cellulose decomposition, 473
Glucomannan, in hemicelluloses, 65t
Glucopyranosides, hydrolyses, 581,582f
Glucose, 7
β-Glucosidases, cellulose decomposition, 473
Glucuronoxylan, in hemicelluloses, 65t
Glycosidic bond scission, rates, 498t
Grafted, definition, 294
Grafting experiments, 387
Grain, shrinkage, 146
Grand fir
  impact strengths, 298
  organotin polymers, impregnation, 300
  polymer incorporation, 298
Gravimetric method, 129
Gravimetric moisture measurements, common errors, 129
Graying, 413
  See also Discoloration
Green moisture content, definition, 128
Green volume specific gravity, 143-44
Green wood
  form of water, 128
  moisture content, 128-29
  shrinking and swelling, 140
  volume, definition, 38
Gross wood
  cell wall structure, 142
  shrinking, 142-51
  swelling, 142-51
  volumetric shrinking and swelling, 143-46
Grotthus–Drapper principle, 406
Growth, by zones, 14-17
Growth characteristics, effect of strength, 215
Growth increment, 14
Growth ring(s)
  counting, 15
  eastern larch, 18f
  eastern white pine, 18f
  growth, 15
  hardwoods, 17
  influences, 14
  Sitka spruce, 14f
  six-year-old tree, 16f
  softwood, 18f
Guaiacyl groups, oxidation, 585
Guanidine, 543
Gymnosperm(s)
  See also Softwood(s)
  complete lignin removal, 71
  definition, 4
  wood tissue, 6f

H

Hailwood–Horrobin solution sorption theory, 164-66
Halogens, as fire retardants, 545

Hard maple, oxide treatment, 198t
Hardness, 226
  coefficient of variation, 217t
  specific gravity effects, 216t
Hardwood(s)
  See also Angiosperm(s)
  anisotropic wood cubes, 227f
  basic density, 40,41t
  calcium oxalate crystals in ray parenchyma, 21f
  cell types, 23
  definition, 3
  erosion, 414
  growth rings, 17
  hemicellulose, 63
  intervessel-pit membranes, SEM, 20f
  leaf structure, 4
  lignin, 66
  parenchyma-cell content, 24
  pentoses, 71
  perforation plates, SEM, 36f
  pore patterns, 19f
  reaction wood, 48
  southeastern U.S., chemical composition, 117-18t
  vessel pitting, SEM, 35f
  vessels, 22
  wood anatomy, 22
  wood tissue, 6f
Hardwood fibers, volume occupied, 24
Hardwood xylem, lignin content, 24
Hartig, Robert, 456
Hawaii, termite attack, 312
Heartwood
  definition, 42
  distinguishing characteristics, 43-45
  distribution in blackjack oak, 42f
  extractives, 44
  moisture content, 43
  polyethylene glycol treatment, 259-60
  resistance to decomposition, 458
  variability, 42-45
  western hemlock, aspirated-pit membrane, 43f
  white oak, tyloses, 44f
Heat, and weathering, 405
Heat capacity, measuring, 534
Heat of combustion, 524t
  fire-retardant treatments, 557
  flaming combustion, 522
Heat degradation, 240-41
  southern pine, 242f
Heat release
  determination, 522
  effect of inorganic additives, 562t
  test methods, 539,541
Heat of sorption, 154
  components, 157
  vs. moisture contents, 158f
Heat treatment temperature, effect on char formation, 509,512f
Heat of wetting
  calculation, 154-55
  relationship to sorption, 157
  total, 157

INDEX

Heat of wetting—*Continued*
vs. moisture content, 156
   cotton, 157f
   Hinoki cypress, 157f
Heat-stabilized wood, 258
Heilman-Gulf epoxy monomer, properties, 286t
Hemicellulases
   production, 477
   synthesis, 477
   wood digestion, 461f
Hemicelluloses, 62-66
   acid treatment, 372
   adsorption isotherms, 139f
   components, 65t
   content, 58
   decomposition, 476-77
   definition, 7,230
   depletion, 463
      by fungi, 470f
   free radical reactions, 427-29
   hardwoods, 63
   molecular weight, 62
   monomer components, 63f
   site of degradation attack, 579
   solubility, 62
Heterocellular ray, 20f
Hexamethylenetetramine, structure, 331
1,6-Hexanediamine, as bonding agent, 362
1,6-Hexanediamine-bonded particle board, properties, 363t
Hickory, acid content, 346t
High lignin content, effect on enzymatic degradation, 313
High performance liquid chromatography, 72
Hinoki cypress, heat of wetting vs. moisture content, 157f
Holocellulose
   adsorption isotherms, 139f
   analytical determinations, 69-73
Homocellular ray, white fir, 20f
Hooke's Law, 213
Hot hardeners, 335
Hot water, as extractives solvent, 73
Hot-melt adhesives, 341-42
Hot-setting phenolic adhesives, 333
HPLC—*See* High performance liquid chromatography
Hydration, of polymer, 164
Hydrogen bonding, between polysaccharide chains, 231f,232f,234f
Hydrogen peroxide
   comparison to HNO_3 as oxidant, 367
   secretion in brown-rot fungi, 474
   surface activation, 373-76
Hydrogen peroxide/ferrous sulfate, 360
Hydrogen peroxide-treated lignin, DSC curves, 376f
Hydrolytic enzymes, cellulose decomposition, 475f
Hydrolytic reactions, 579-83
Hydrostatic pressure, vs. specific volume, 161f

Hydroxyl absorption, 357
Hydroxyl groups, 37
   reactions, 177,182
Hydroxyl substitution, calculations, 206
Hydroxyl-terminated polymers, curing, 337
Hydroxyls, conversion, 295
Hygroelastic effect, 160
Hygroelasticity, moisture sorption effects, 160-61
Hygroexpansion, 149
Hygroexpansion coefficients, 145,147
Hygroscopic gels, osmotic pressure equation, 160
Hygroscopicity, 37
   See also Wood moisture content
   effect of waterborne salts, 244
   heating effects, 140
   with high extractive contents, 139
   lignin, 37
   mechanical stress effects, 140
   temperate-zone woods, 144
   tropical woods. 144
Hysteresis, definition, 136

I

IB—*See* Internal bond
Ice, relative energy level, 155f
Impact bending
   coefficient of variation, 217t
   specific gravity effects, 216t
   strength, 463
Impact loading, 219
Impact strength, cyanoethylated wood, 192
Impreg, 258
Impregnation, 265-66,266f
Increased char/reduced volatiles theories, 545-53
   definition, 542
Indian woods, volumetric shrinkage, 144
Inhibitors, removal, 267-68
Inorganic additives
   aromatic char formation effects, 552t
   decomposition temperature effects, 553
   flame-spread index effects, 537f
   heat release effects, 562t
   levoglucosan production effects, 550t
   oxygen index effects, 550t
   thermal evolution analysis effects, 561t
   thermogravimetric analysis effects, 562t
Inorganic elemental composition, 74
Inorganic fire retardants, differential thermal analysis, 553
Inorganic salts, 308-9
   chemical reactions with wood, 314-17
   susceptibility to leaching, 314
Insects
   decomposition of lignin, 462
   deterioration responsible for, 459t
   wood decomposition, 464-65t
   wood digestion, 461
Interaction of preservatives, 307-20
Interfiber bordered pits, 28,29f

Intermolecular attraction, forces responsible, 326
Internal bond
  cross-linking agents, 364
  vs. board density, 360f
Intervessel-pit membranes, hardwoods, SEM, 30f
Intumescent systems, 543
IR data
  as bonding proof, 202–4
  of outdoor exposed wood, 411f
  oxidized, cellulose, 377
  UV-irradiated wood, 409t
Isocyanate(s)
  bifunctional, nonconventional bonding, 382–83
  moisture sensitivity, 188
  reactions, 187–88,339f
  waferboard bonding, 392
Isocyanate groups, terminal, curing, 337
Isocyanate-based adhesives, 337–39
Isocyanurate, formation, 339f

J

Japan, woods, chemical composition, 91–97t,121t
Jarrah, chemical composition, 113t
Juvenile wood
  axial shrinkage, 147
  chemistry, 54–55
  extractives, 54
  location, 52f
  microfibril angle, 54
  physical properties, 53–54
  variability, 51–54

K

Kalimantan, woods, chemical composition, 98–100t,121t
Karl-Fischer titration, 129–30
Kiln-dried wood, 458–59
Klason lignin—*See* Lignin
Knots
  definition, 54
  effect of natural pruning, 54
  loose, definition, 55
  tight, definition, 55
Knotwood
  composition, 55
  location, 52f
  variability, 54
Kürschner cellulose, analytical determinations, 71

L

Lacquers, 438
Langmuir model, sorption, 162
Larch wood, heat of combustion, 523t,524t
Lateral meristems, location, 9
Latewood
  cells, role in strength, 226
  interfiber pits, 28

Latewood—*Continued*
  liquid penetration, 181
  pit aspiration on drying, 181
  tracheids, cell walls, 181
  weathering, 424
Latex paints, 437
Latex stains, 444
Leach-resistant chemicals
  for flame retardancy, 566–68
  future research, 568
Leaching, resistance, 314
  as bonding proof, 198–202
Lead acetate, 187
Lentinus lepideus, attack resistance, 184
Levoglucosan, acid-catalyzed resistance, 550
Levoglucosan inhibition, proposed mechanism, 552f
Levoglucosan production, additives effect, 550t
Levoglucosenone, formation, 507f
Light, and weathering, 405
Light energy, effects, 402t
Light exposure, surface changes, 357
Light penetration and wood surface deterioration, 406–7
Light scattering, determination of degree of polymerization, 60
Lignification, defined, 13
Lignin, 66–68
  alkaline conditions, 410
  alkaline hydrolysis, 583
  alkaline oxidation, 584
  analytical determination, 73
  benzyl (α)-aryl ether linkages, cleavage, 582
  biosynthesis, precursors, 66
  bonded chemical distribution, 205
  C–C bonds, cleavage, 582
  char formation, 176
  characterization by methoxyl content, 68
  crystallinity, 585
  decomposition by insects, 462
  definition, 7
  degradation
    site of attack, 579
    strong oxidants, 585
  distribution, 69
  elemental composition, 510t
  free radical reactions, 429–30
  hydrogen peroxide-treated, DSC curves, 376f
  hydrolytic reactions, 582–83
  hygroscopicity, 37
  isolation, 66,68
  low temperature decomposition, 356
  material definition, 229
  methoxyl group determination, 73
  oxidative degradation, 584–85
  photochemical reactions, 430
  reaction with copper, 315
  role in biological decomposition, 460
  softwood
    partial structure, 67f
    solubility, 73

INDEX 603

Lignin—Continued
  spruce-milled, composition, 68
  as stiffening agent, 211
  treated, thermogravimetric analysis, 549f
  UV absorption, 429f
  UV degradation, 176
  UV light absorption, 408
  water absorption, 7
  weathering effects, 407
  white-rot fungi, 470f
Lignin combustion, maximum heat
    intensity, 560t
Lignin content, effect of defibration
    temperature, 353
Lignin decomposition, 477-79
  by white-rot fungi, 478f
Lignin polymer degradation, prominent
    reactions, 477
Lignin-phenylpropane units, common
    linkages, 580f
Lignocellulosic materials, oxygen:carbon
    ratio, 354t
Lignosulfonates, oxidative cross-linking,
    369,370f
Lignophylic systems, 349
Liquid adhesive to solid bond, 324
Liquid water
  relative energy level, 155f
  specific heat, 153
Liquids, movement, 28,32
Load
  below proportional limit, 232
  beyond proportional limit, 233-35
  duration, 218
  factors, strength, 218-20
  fatigue, 219
Loblolly pine, normal wood, 49f
London forces, 326
Long-term protection, prerequisites, 445
Longhorn beetle, attack of white pine, 466f
Longitudinal shrinkage, 147
  Pinus jeffreyi, 148f
Longitudinal swelling, Barber's model, 149f
Longitudinal tracheid, 19
Loose knots, definition, 55
Low molecular weight polyethylene, 341
Lumen-filling modification, 426
Lumens, 4,11

M

Macrodistribution, of preservative,
    influences, 311
Macrofibrils
  cellulose, TEM, 25f
  definition, 24
Macroscopic level, strength, 236
Major-use preservatives, 307-9
Mammals, deterioration responsible for,
    459t
Maple
  electron spectroscopy for chemical
    analysis, 380f
  light exposure, erosion, 417t
  oxygen:carbon ratio, 354t
  surface oxidation, 379

Maple-MMA composite, moisture effects
    on polymerization, 277f
Marine bacterium, decay resistance, 293
Marine borer(s)
  attack of Douglas-fir, 466f
  attack of yellow pine, 466f
  deterioration responsible for, 459t
  wood decomposition, 465t
  wood digestion, 461
Marine decay, field test data, 303t
Marine organisms, preventing attack, 304
Mature wood
  axial shrinkage, 147
  variability, 51-54
Maximum digestibility, 461
Maximum transverse swelling pressure, 159
Mechanical disintegration, 460
Mechanical energy, effects, 402t
Mechanical properties, 212-13
  acetylated wood, 184
  anisotropy, 212
  axial loading conditions, 223
  coefficient of variation, 217t
  formaldehyde-treated wood, 189
  methylated wood, 190
  moisture changes, 219t
  moisture effects, 152
  propylene oxide-modified wood, 196
  relationship to moisture content, 220
  relationship to stress-strain diagram, 222f
  specific gravity effects, 216t
  strength, 212-26
Mechano-biochemical decomposition,
    460-62
Melamine, 543
  structure, 335
Melamine-formaldehyde resins, 327,334-37
Mercerization, definition, 60
Meristems, 9,11
Methenamine, structure, 331
Methyl iodide, formation, 73
Methyl isocyanate, 188
  for chemical modification, 201t
  southern pine treatment, 200t
Methyl isocyanate-modified southern pine
  IR spectra, 204f
  substitution, 206t
Methyl methacrylate, 426
  properties, 286t
  southern pine treatment, 200t
  structure, 262
  vapor pressure, 266
Methylated wood, mechanical properties,
    190
Methylation, 182
4-*O*-Methylglucuronic acid, 63f
Methylol phenols, 329
Mexico, woods, composition, summary,
    121t
MF—*See* Melamine-formaldehyde resins
Microbial degradation, strength, 241-43
Microdistribution theory, 312
Microfibril angle, 26
  relationship to longitudinal shrinkage, 147
Microfibril orientation
  effect on cell cavity changes, 144

Microfibril orientation—*Continued*
  microscopic, 228-29,228f
Microfibrils
  cellulose, TEM, 25f
  definition, 24,228
Microscopic composition, 228
Microscopic level, strength, 235
Microscopic microfibril orientation, 228-29
Middle lamella
  composition, 7
  definition, 7
Minor-use preservatives, 309-10
MMA—*See* Methyl methacrylate
Modulus of elasticity, 223
  definition, 214
  of wood-polymer composites, 286t
Modulus of rigidity, definition, 214
Modulus of rupture, 221-22
  of wood-polymer composites, 286t
MOE—*See* Modulus of elasticity
Moist wood, specific heat, 153
Moisture
  effect on free radical formation, 431
  effect on polymerization exotherms, 275-76
  effect of strength, 215-18
  and the environment, 127-40
  graying, 413
  and weathering, 404-5
Moisture content
  determination, 74
  entropy change, 158f
  free energy loss, 158f
  of green wood, 128-29
  heat of sorption, 158f
  relationship to mechanical properties, 220
  target, of wood in use, 140
  vs. DC resistivity, 130f, 131f
  vs. density, 153f
  vs. free induction decay, 135f
  vs. heat of wetting, 156
    cotton, 157f
    Hinoki cypress, 157f
  vs. specific gravity, 153f
  vs. wood volume, 145f,146f
  wet basis, 128
Moisture diffusion coefficient, 166-69
Moisture effects, 151
Moisture movement, driving potential, 167
Moisture sorption
  enthalpy changes, 154-57
    measurement, 155
  free energy changes, 157
  free entropy changes, 157
  hygroelastic effects, 160-61
  isotherm, 136-40
  swelling pressure, 158-60
  thermodynamics, 154-61
Moisture transport, 166-70
  coefficients, 167t
Molecular, relationship of strength to chemical composition, 229-31
Molecular level, strength, 235
Monoammonium phosphate, 532
MOR—*See* Modulus of rupture
Mother cells, 11

Mozambique, woods, chemical composition, 89-90t,121t
Multiple ring, 14-15
Multiseriate rays, redwood, 22f
Mushroom-forming fungi, white-rot, 481
Mushrooms, cultivation on solid wood, 483t

N

NaPCP—*See* Sodium pentachlorophenol
Naphthenic acids, structure, 309
National Bureau of Standards smoke density chamber, 538,540f
Natural pruning, effect on knots, 54
Natural wood finishes, 443-45
  film-forming, 444
  opaque stains, 445
  penetrating, 444
Neutron moisture meter, 135-36,135f
New preservatives, 310
Nitrate, surface activation, 376-79
Nitrate salts, thermal decomposition, 377
Nitric acid
  comparison to H_2O_2 as oxidant, 367
  preoxidation, 362
  surface activation, 376-79
Nitrogen, basis for fire retardants, 563
Nitrogen-containing esters, formation, 187
NMR, pulsed, 134
Nonconventional binders, nonpolar, 393
Nonconventional bonding, 349-400
  with acid activation, 390-91
  advantages, 351
  bifunctional acids, 382
  bifunctional amines, 382
  bifunctional isocyanates, 382-83
  with cross-linking mixture, 387
  direct, 381-82
  economics, 351
  methods, 349
  with oxidant activation, 391
  polymers, 383-89
    role of activator, 383-84
  wood properties, 368t
Nonconventionally bonded particle board, advantages, 390
Nonpolar liquids, penetration, 181
Nonpolar nonconventional binders, 393
Nontoxic preservatives, function, 292
Normal wood, 49f
Norrish acceleration, 273
North American woods, chemical composition, 115-16t
Northern red oak, white rot, 467f
Novolak, 330-31
Nylon, structure, 342

O

Oak
  attack by termites, 466f
  wood cell location, 10f
Obeche, chemical composition, 113t
Oil-base paints, 437
Oil-based stains, semitransparent, 444
Oleoresin, 21

INDEX

Opaque stains, natural wood finishes, 445
Organic components, 402
Organic preservatives, chemical reactions with wood, 317–18
Organic solutions, 440
Organisms, wood-decomposing
 control, 479–81
 uses, 481–83
Organolead-treated southern pine, durability, 441
Organophosphorus compounds, for fire retardants, 564
Organotin compounds, biotoxicity, 292–93
Organotin polymers
 biodegradation resistance, 300–303
 dimensional stability, 297
 distribution, 295–97
 effect on antishrink efficiency, 297
 mechanical properties, 298–300
 in wood, 293–304
Osmometry, 60
Osmotic pressure equation, 159
 assumptions, 160
Osmotic pressure theory, Barkas, 160
Outdoor exposed wood, IR spectra, 411f
Outerwood, definition, 53
Oven-dried cell wall, impermeability to vinyl monomers, 279
Oxidant activation, nonconventional bonding, 391
Oxidation, char, 526f
Oxidative cross-linking of lignosulfonates, mechanism, 370f
Oxidative depolymerization, cellulose, 474
Oxidative enzymes, cellulose decomposition, 475f
Oxidative reactions, 583–85
Oxidized surfaces, oxygen content, 410
Oxidizing activators, 381
Oxygen adsorption, char, 526f
Oxygen chemisorption, cellulose, 518f,520f
Oxygen effect, peroxide formation, 434f
Oxygen index
 critical, 537–38
 effect of fire-retardant chemicals, 538f
 effect of inorganic additives, 550t
Oxygen:carbon ratio
 lignocellulosic materials, 354t
 maple wood, 354t
Ozone gas-phase treatment, 260

P

Paint, and fungal growth, 437
Painting surfaces, acetylated wood, 185
Papua New Guinea, woods, chemical composition, 98–100t,121t
Parenchyma cells, 11
Particle board
 bonding preparation, 358
 nonconventionally bonded, advantages, 390
 production, organic emissions, 368
 properties, H_2O_2 effects, 366t
 wood-to-wood bonding, 360f

PCP—See Pentachlorophenol
Pectin membranes, degradation, 459
Peeling, 586
PEG—See Polyethylene glycol
Penetrating finishes, 438–43
 natural wood, 444
Penetration, 178–81
 of gas, 182
Penicillium spp., 459
Pentachlorophenol, 307–8,439
 effects on mechanical properties, 245–48t
 preservatives, analysis by SEM–EDXA, 312
 structure, 308
Pentosan, analytical determinations, 71
Percent of dry mass, definition, 129
Perforation plates
 definition, 35
 in hardwoods, SEM, 36f
 sculpturing, 35
Permanganate, lignin degradation, 585
Peroxide formation
 oxygen effect, 434f
 rose bengal effect, 434f
 triethylamine effect, 435f
Peroxy compounds, as oxidants, 365
Peroxyacetic acid, effect on shear strength, 361
Persimmon, butylene oxide treatment, 198t
Pesticides, as wood preservatives, 480
PF—See Phenol–formaldehyde resins
pH
 changes, effect on strength, 238
 decrease with chromated copper arsenate, 315
 effect on strength properties, 212
Phanerochaete chrysosporium, cellobiose conversion, 473
Phenol, reactions with formaldehyde, 330
Phenol–formaldehyde compressed wood composite, 258–59
Phenol–formaldehyde resin, 327,330f
 applications, 332–33
 durability, 333–34
 fracture toughness, 333–34
Phenol–formaldehyde wood composite, 258
Phenol–formaldehyde-bonded particle board, properties, 363t
Phenolic resin adhesives, 328–34
 hot-setting, 333
Phenols, 458
Phenylpropane units-lignin, common linkages, 580f
Philippine woods, chemical composition, 101–9t,121t
Philippou process, 365–69
 economics, 369
 particle board, properties, 386t
Phloem, primary, 9
Phosgard, 277
Phosphates, as fire retardants, 564
Phosphorus, for fire retardants, 563,564
Phosphorus pentoxide, 129
Phosphorus–nitrogen synergism theories, 557–63

Photochemical degradation, 176
Photochemistry, laws, 406
Photodegradation, mechanisms, 436f
Photooxidation of wood, singlet oxygen, 434
Phthaldehydic acid, reaction with wood, 190
Phthalylation, esters, 186
Physical changes, weathering, 414–16
Physical properties, 37–40,282–88
and chemical modification, 175–78
moisture effects, 151–54
Picea glauca (Moench) Voss, extractives deposition, 355
Pigmented penetrating stain, definition, 439
Pine
See also *Pinus sylvestris* L.
nonconventional bonding, 368t
polysaccharide content, 70t
wood cell location, 10f
Pinus jeffreyi, shrinkage, 148f
Pinus ponderosa Laws., wood-to-wood bonding, 361
Pinus sylvestris L., polysaccharide content, 70t
See also Pine
Pinus taeda
green moisture content, 128
toluene-soluble extractives, 354
Piperidine, for swelling, 179–80
Pit aspiration, 30
Pit membrane, construction, 28
Pitch pine, chemical composition, 113t
Pits
definition, 28
ray cross-field, 30–32
sculpturing, 35
types, 28
Plasma activation, of surfaces, 379–81
Plastic strength, 233–35
Plasticizers, 325
Plexiglas, 267
Plywood
check formation, 423
weathering, 423–25,424f
Polar liquids, penetration, 181
Poly(vinyl acetals), as hot-melt adhesives, 344
Poly(vinyl acetate) emulsions, adhesion characteristics, 342–43
Poly(vinyl butyral), 344
Poly(vinyl formal), 344
Polyamides, 341
Polybor, 565
Polyethylene, low molecular weight, 341
Polyethylene glycol, 259
Polymer adhesives, 324
Polymer grafting, 295
Polymer impregnation, effects, 299f
Polymer loading, 283
Polymeric chains, in bonding, 365f
Polymeric materials, thermal decomposition studies, 533
Polymerization
cross-linking effect, 272–74

Polymerization—*Continued*
exotherm, wood moisture effect, 275–76,276t
vazo catalyst effect, 269–71
of vinyl monomers, 267
Polymers
biotoxicity, 292–93
hydroxyl-terminated, curing, 337
migration into cell walls, 296
nonconventional bonding of, role of activator, 383–84
Polymethyl methacrylate, depolymerization, 276
Polyphosphates, for fire retardants, 564
Polyporus versicolor
attack, resistance, 184
effect on organotin polymer-impregnated wood, 300
Polysaccharide(s), 7
in cell walls, 70t
content, 69
oxidative degradation, 584
peeling, 586
Polysaccharide chains, hydrogen bonding, 231f,232f,234f
Polysaccharide molecule, flexing and elongation, 234f
Polystictus versicolor, beech wood attack, 188
Ponderosa pine
heat of combustion, 523t,524t
propylene oxide treatment, 198t
Poplar wood, heat of combustion, 523t,524t
Pore patterns, hardwoods, 19f
Pores, 4
Poria microspora, attack resistance, 184
Poria monticola
attack resistance, 184
effect on sweetgum, 243f
Poria placenta, molecular weight, 477
Porter equation, 160
Power-loss moisture meter, 133,134
Preconditioning, 293
Preoxidation
increasing, 363
with nitric acid, 362
Preoxidized wood, bonding mechanism, 362f
Preservative(s), 440
chemical reactions with wood, 313–18
distribution, 311–13
effects on mechanical properties, 245–48t
function, 292
inorganic salt, 314–17
interaction, 307–20
macrodistribution, 311
major-use, 307–9
microdistribution, 312–13
minor-use, 309–10
new, 310
organic, chemical reactions with wood, 317–18
Preservative oils, 440
Preservative protection, influences on longevity, 311

INDEX

Preservative-wood interaction, disadvantages, 317
Primary growth, definition, 9
Primary phloem, 9
Primary wall, 144f
 manufacture, 26
Primary water, calculations, 163,164
Primary xylem, 9
Proof of bonding
 criteria, 197-204
 increases in wood volume, 197-98
 IR data, 202-4
 resistance to leaching, 198
β-Propiolactone, 191
Propionic anhydrides, 186
Proportional limit, fiber stress, 222
Propylene, structure, 262
Propylene oxide, treatments, 193f,200,201t
Propylene oxide-modified wood, mechanical properties, 196
Propylene oxide-triethylamine, treatment of southern pine, 194f
Protection
 chemistry, 401-47
 fire retardants, 532
 weathering, 436-43
 wood-based materials, 445
Protoplast, 11
Pseudomonas nigrifaciens, decay resistance, 293
Puerto Rico, woods, chemical composition, 82-87t
Pulsed NMR method, 134
PVA—*See* Poly(vinyl alcohol)
PVAc—*See* Poly(vinyl acetate)
Pyrolysis, 489-528,541
 cellulose, 356,492f,501f
 products, effect of additives, 551t

Q

Quaking aspen, elemental composition, 120t
Quality of wood, 17-19

R

Radial section, UV-light exposure, 419-20
Radial shrinkage, 146
Radial surface, 15f,41
Radial transport of liquids, 32
Radiata pine, propylene oxide treatment, 198t
Radiation catalysis, 261
Ray cross-field pits, 30-32
 softwoods, SEM, 32f
Ray parenchyma, calcium oxalate crystals, hardwood, 21f
Ray restraint theory, 151
Ray tracheids, 21
Rays, 11
 arrangement in softwoods, 20
 location, 12f
 multiseriate, redwood, 22f
 uniseriate, redwood, 22f

Reaction wood
 axial shrinkage, 147
 chemistry, 46-47t
 definition, 45
 in hardwoods, 48
 in softwoods, 48
 structure, 46-47t
 variability, 45-48
Reconstituted panel products, weathering, 425
Red gum, physical property enhancements, 280-81t
Red maple
 elemental composition, 119t
 physical property enhancements, 280-81t
Red oak
 acid content, 346t
 light exposure, erosion, 417t
 northern, white rot, 467f
 propylene oxide treatment, 198t
Red pine
 heterocellular ray, 20f
 physical property enhancements, 280-81t
Red spruce, elemental composition, 119t
Redistribution
 extractives, 354-55
 preservatives, 311
Reduced heat content of volatiles theories, 553-57
 definition, 542
Redwood
 color changes, 412f,413f
 free radicals vs. moisture content, 433f
 light exposure, erosion, 417t
 multiseriate rays, 22f
 uniseriate rays, 22f
Regeneration, definition, 60
Relative humidities, maximum possible, 138f
Resin canals, 21
Resin ducts, 21
Resins
 acrylic latex, 437
 with low monomer content, 329
 resorcinol, 331-32
Resistance, causes, 458
Resistance moisture meters, high humidity, 133
Resoles, 328-29
 cold-setting, 333
 formed from phenol and formaldehyde, 328f
 ortho substitution, 329
Resorcinol, structure, 331
Resorcinol resins, 331-32
Resorcinol-formaldehyde resins, adhesive strength, 185
Ring-porous, 19f
Rose bengal effect, peroxide formation, 434f
Russia woods—*See* U.S.S.R. woods

S

S_1 layer
 composition, 26-27
 fibril orientations, 144f

S_2 layer
  composition, 26-27
  fibril orientations, 144f
S_3 layer
  composition, 27
  fibril orientations, 144f
Sap stream, movement, 28
Sap-stain fungi, 458
Sapwood
  cells, in the living tree, 128
  definition, 42
  polyethylene glycol treatment, 259-60
  resistance to decomposition, 458
  variability, 42-45
Sarcina lutea, decay resistance, 293
Scanning electron micrograph
  cellulose-lignin mixture, 383f
  Douglas-fir, 13f
  softwood, 8f
Scientific forest management, aims, 41
Scotch pine, microfibril orientation, 228f
Sculpturing
  fibers, 28-34
    perforations, 35
Second law of photochemistry, 406
Secondary wall, 144f
  definition, 26
  tension wood, 50
Secondary water, calculations, 163
Secondary xylem, 4
SEM—*See* Scanning electron micrograph
Semi-ring-porous, 19f
Semitransparent oil-based stains, 444
Shear, 225
Shear modulus, definition, 214
Shear parallel to grain, coefficient of variation, 217t
Shear stress, 213
Sheaths, gelatinous, role in decay, 471
Shellacs, 438
Shiitake, 482f
Shock resistance, 226
  moisture effects, 152
Shrinkage, 37,140-51
  anisotropy, 146-51
  bound water loss, 140
  cell wall, 141-42
    maximum, equation, 142
  coefficient of variation, 217t
  fibril angle effect, Barber's model, 147-48
  gross wood, 142-51
  percent, equation, 142
  polymer impregnation effects, 294
  volumetric, equation, 143
Silica, determination, 74
Silviculture, definition, 41
Sinapyl alcohol, 66
Singlet oxygen, photooxidation of wood, 434
Sitka spruce
  growth rings, 14f
  sorption vs. moisture content, 156f
Smoke, promotion, 532

Smoke density chamber, National Bureau of Standards, 538,540f
Smoke production, test methods, 538-39
Smoldering, 541
Smoldering combustion, 491f
  driving force, 522
  inhibition theories, 563
Sodium chlorate, 359
Sodium pentachlorophenol, effects on mechanical properties, 245-48t
Sodium periodate, 585
Sodium silicates, 543
Sodium tetraborate, 250
Soft rot
  attack, in CCA-treated wood, 313
  features, 468-69t
  fungi, preventing attack by, 312
Softwood(s)
  See also Gymnosperm(s)
  ammonium treatment, 260
  anisotropic wood cubes, 227f
  basic density, 40,41t
  cell types, 19
  color changes, 412f
  compression wood, 48
  definition, 3
  deterioration, 416
  earlywood fibers, 17
  earlywood-to-latewood transition, 18f
  erosion, 414
  growth rings, 18f
  heartwood, moisture content, 43
  interfiber pits, 28,29f
  leaf structure, 4
  pentoses, 71
  ray cross-field pits, SEM, 32f
  SEM, 8f
  warty layer, TEM, 34f
  weathered surface, 415f
  wood anatomy, 19-22
  wood tissue, 6f
Softwood fiber, photomicrograph, 229f,230f
Softwood lignin
  partial structure, 67f
  precursor, 66
  solubility, 73
Soil fungus, decay resistance, 293
Solid, burning, 541
Solid-phase combustion, 541
Solubility, wood, 73
Sooty smoke, 277
Sorption, relationship to heat of wetting, 157
Sorption hysteresis, benefits, 137
Sorption isotherms
  Brunauer-Emmett-Teller model, 162,163f
  temperature effects, 138f
Sorption sites, 162,163f
South America, woods, chemical composition, 82-87t,88t
Southeastern U.S., hardwoods, chemical composition, 117-18t

INDEX

Southern pine
  acetic anhydride treatment, 199*t*
  acid content, 346*t*
  calcium carbonate deposition, 244*f*
  chemical treatment, 200*t*
  earlywood, void volume, 181
  heat degradation, 242*f*
  latewood, void volume, 181
  light exposure, erosion, 417*t*
  methyl isocyanate reaction, 338
  methyl isocyanate-modified, IR spectra, 204*f*
  methyl isocyanate-modified, substitution, 206*t*
  ocean air adsorption, 239
  organolead-treated, durability, 441
  propylene oxide treatment, 198*t*,199*t*
  propylene oxide–triethylamine treatment, 194*f*
  sapwood, volumetric swelling coefficients, 179*t*,180*t*
  sodium chloride deposition, 239*f*,240
  treated
    ammonium dihydrogen phosphate, 250*f*
    chromated copper arsenate, 249*f*
  untreated, 249*f*
  weathering effects, 241*f*
Southern red oak, elemental composition, 120*t*
Southern yellow pine
  acetylation, loss, 184
  bordered pit structures, 421*f*
  chromic acid treated, 441*f*,442*f*,443*f*
  color changes, 412*f*,413*f*
  cross section, 418*f*
  decay resistance by organotin polymers, 301
  deterioration, 421*f*
  free radicals vs. moisture content, 433*f*
  radial section, 420*f*
  reaction with β-propiolactone, 191
  UV light exposure, 419*f*
Soviet Union woods—*See* U.S.S.R. woods
Specific gravity, 38–40
  coefficient of variation, 217*t*
  definition, 39
  effect on mechanical properties, 216*t*
  effect of strength, 215
  green volume, 143–44
  moisture effects, 152,153*f*
Specific heat
  dry wood, 153
  liquid water, 153
  moist wood, 153
  water, 153
Specific volumes, vs. hydrostatic pressure, 161*f*
Spiral thickenings, 32–33
  Douglas-fir, SEM, 33*f*
  wood behavior, 36
Spruce
  adsorption–desorption isotherms, 218*f*
  butylene oxide treatment, 198*t*

Spruce—*Continued*
  European, sorption vs. moisture content, 156*f*
  lignin
    bonding, 581*t*
    distribution, 69
  propylene oxide treatment, 198*t*
Spruce-milled wood lignin, composition, 68
Staining, 586
Stains, 439–40
Stark–Einstein principle, 406
Static bending, 216*t*,217*t*
Staybwood, 258
Staypak, 258
Steam bending, alternative, 260
Stearic acid, translocation, 354
Stiffness, 223
Storage, parenchyma cells, 11
Storage cells, degradation, 458
Strain, definition, 213
Strain reduction, 237
Strength
  acetic acid effect, 239
  acids effect, 238
  altered composition effects, 236–53
  bases effect, 238
  bound water effects, 218
  brown-rot fungi effect, 471
  chemical components, 226–36
  chemistry, 211–56
  compression parallel to the grain, equation, 224
  effect on moisture, 215–18
  effect on specific gravity, 215
  environmental effects, 212,238–43
  fire retardation effect, 250
  growth characteristics effect, 215
  heat degradation effects, 240–41
  load factors, 218–20
  loss, 176
  macroscopic level, 236
  mechanical properties, 212–26
  microbial degradation, 241–43
  molecular level, 235
  oven-dry wood, 217
  relationship to chemical composition, 231–35
  relationship to structure, 235–36
  sulfuric acid effect, 239
  swelling solvents effect, 239
  temperature effect, 218,220*f*
  tension parallel to the grain, 225
  tension perpendicular to the grain, 225
  treatment effects, 243
  ultimate, 215
  UV degradation effects, 240
  wood design effects, 220–26
Strength properties
  effect of wood-destroying fungi, 463,471
  S_2 contribution, 26–27
Streptomycete, soft rot, 463
Stress
  definition, 213

Stress—*Continued*
  proportional limit, 215
  reduction, 237
  types, 213f
  unit, 215
Stress-strain curve, mechanical behavior, 231f
Stress-strain diagram, 214
  relationship to mechanical properties, 222f
Structural performance, cellular level, 226
Structure, 3–56
  relationship to chemical composition, 226–31
  relationship to strength, 235–36
Styrene, structure, 262
Sugar maple, physical property enhancements, 280–81t
Sugar maple–polymer composites, hardness, 285f
Sugar pine
  nonconventional bonding, 361
  properties, H_2O_2 effects, 366t
Sugars, wood, chromatographic analysis, 71
Sulfuric acid
  as cross-linking agent, 370
  effect on strength, 239
Sunlight, 427
  and weathering, 405
Surface, calculation of oxygen:carbon ratio, 353
Surface activation, 349–400
  acids, 372–73
  hydrogen peroxide, 373–76
  nitrate, 376–79
  nitric acid, 376–79
  plasma, 379–81
Surface changes, light exposure, 357
Surface composition, prior to activation, 351–58
Surface deterioration, and light penetration, 406–7
Surface emission coefficient, 169–70
Surface formation
  chemical changes, 355–58
  conditions and methods, 351–54
  foreign materials deposition, 355–58
  redistribution of extractives, 354–55
Surface molds, discoloration effects, 459
Surface treatments, 437,440
Surface wood change, artist's rendition, 412f
Sweetgum
  brown-rot fungus attack, 242
  effect of brown-rot fungus, 243f
  effect of *Poria monticola*, 243f
  elemental composition, 119t
Swelling, 37,140–51
  anisotropy, 146–51
  cell, 143f
  cell wall, 141–42
    at fiber saturation, 142
  correlation to dielectric constant, 180
  effect of restraining, 159
  gross wood, 142–51

Swelling—*Continued*
  maximum possible, 141
  mean value, 142
  piperidine, 179–80
  polymer impregnation effects, 294
Swelling percent, of cell wall, 141,142
Swelling pressure
  calculation, 159
  moisture sorption, 158–60
  transverse, maximum, 159
Swelling solvents, 239–40
Synergism
  aluminum trihydrate–boron, 565
  phosphorus–nitrogen, 557–63

T

Taiwan, woods, chemical composition, 110–11t,121t
Tangential section, UV light exposure, 420–22
Tangential shrinkage, 146
  Pinus jeffreyi, 148f
Tangential surface, 15f
Tangential swelling, Barber's model, 149f
Tannin-based adhesives, 339–40
Tannins, source, 339
Tappi Provisional Test Method T 250, 72
Tappi Standard 249, 69
Tappi Standard T 15, 74
Tappi Standard T 223, 71
Tar acids, 308
Tar bases, 308
Target moisture contents, of wood in use, 140
TBT—*See* Tri-*n*-butyltin
TBTMA—*See* Tri-*n*-butyltin methacrylate
TBTMA–GMA—*See* Tri-*n*-butyltin methacrylate–glycidyl methacrylate
TBTO—*See* Tributyltin oxide
Teak
  chemical composition, 113t
  propylene oxide treatment, 198t
TEM—*See* Transmission electron micrograph
Temperate-zone woods, hygroscopicity, 144
Temperature
  defibration, effect on lignin content, 353
  effect on strength, 218,220f
  increasing, 218,357
Tensile stress, 213
Tension parallel to the grain, 224–25
Tension perpendicular to the grain
  coefficient of variation, 217t
  strength, equation, 225
Tension wood
  aspen stem, 51f
  cellulose content, 50
  chemistry, 46–47t
  definition, 48
  fibers, 52f
  formation, 50
  secondary wall, 50
  structure, 46–47t

INDEX

Tension wood—*Continued*
  variability, 48–51
Terminal isocyanate groups, curing, 337
Terminal meristems, 9
  See also Apical meristems
Termites
  attack, 196f
    albizzia board, 466f
    in Hawaii, 312
    oak, 466f
  wood digestion, 462
Terpenes, 458
Thawing, of absorbed water, 405
Thermal conductivity
  moisture effects, 154
  relationship to wood density, 154
Thermal decomposition of nitrate salts, 377
Thermal diffusivity, 154
Thermal energy, effects, 402t
Thermal evolution analysis system, 525f,561t
Thermal properties, moisture effects, 153
Thermal theories, 543–44
  definition, 542
Thermodynamics, of moisture sorption, 154–61
Thermogravimetric analysis
  effect of inorganic additives, 562t
  fire retardancy test method, 533–34
  fire-retardant-treated wood, 548f
  system, 533f
  treated lignin, 549f
Thermogravimetry of cottonwood, 524f
Thermoplastic adhesives, 340–45
Thermoplastic polymers, 324–26
Thermosetting polymers, 324–26
Thermosetting resins, 327
Tight knots, definition, 55
TMPTMA—*See* Trimethylol propane trimethacrylate
p-Toluenesulfonic acid, 189
2,4-Tolylene diisocyanate, reaction with white cedar, 187
Toughness
  coefficient of variation, 217t
  decay effects, 463
  formaldehyde-treated woods, 189
  moisture effects, 152
Toxic preservatives, function, 292
Toxicity
  of reactants for modification, 182
  test methods, 541
Transglycosylation, depolymerization, 500
Transition temperatures, measuring, 534
Transverse plane, 15f
Transverse section, UV-light exposure, 418
Transverse shrinkage
  anisotropy, theories, 150–51
  Barber's model, 147–48
Transverse swelling pressure, maximum, 159
Transwall failure, 352
Treated wood, pyrolysis products, 508t
Treatment effects on strength, 243
Tree(s)
  formation of wood, 9–19

Tree(s)—*Continued*
  sapwood cells, 128
  water formation, 128
Tree crown, 4
Trialkyltin, reactivity, 293
Tri-*n*-butyltin
  decay resistance, 301
  preference, 293
Tri-*n*-butyltin methacrylate
  grafting, 295t
  in situ polymerization, 294
Tri-*n*-butyltin methacrylate-glycidyl methacrylate
  concentration in cell walls, 296f
  decay resistance, 301
Tri-*n*-butyltin methacrylate-methyl methacrylate, role in marine decay, 304
Tributyltin oxide
  condensation reaction with cellulose, 318
  degradation, 318
  structure, 310
Trichloroacetaldehyde, reaction with wood, 190
Triethylamine effect, peroxide formation, 435f
Trimethylol propane trimethacrylate, 273–74
  advantages, 274
  concentration effect, 273f
  effect on basswood-MMA composite, 274t
(Trimethylsilyl)ated cellulose, pyrolysis GLC of tar, 506f
Trisodium phosphate, 250
Trommsdorff effect, 268–70
  See also Gel effect
Tropical woods
  elemental composition, 120t
  hygroscopicity, 144
Tunnel flame-spread test, 534–37
  8-foot, 536f
  25-foot, 535
Tyloses
  definition, 43
  in white oak heartwood, 44f
  wood permeability effects, 43

U

UF—*See* Urea-formaldehyde resins
Ultimate strength, 215
Uniseriate rays, redwood, 22f
United States, woods, chemical composition, 76–81t,122t
Unsteady-state diffusion equation, 167
Urea, 543
  formation, 339f
  noncombustible gas release, 544
  reaction with formaldehyde, 335
Urea-formaldehyde resins, 327,334–37
  adhesive strength, 185
  advantages, 335
  curing, 335
  disadvantages, 335
  unextractable acids, 346
Urethane link, formation, 337

Uronic acids
 analytical determinations, 72
 HPLC, 75
U.S.S.R., woods, chemical composition, 112–13t,122t
UV absorption, lignin, 429f
UV degradation, 240,408
UV irradiation, 408
UV light, penetration, 446
UV-irradiated wood
 IR absorbance, 409t,410t
 lignin content, 410t
 UV absorption spectra, 410f

V

van der Waals forces, 326–27
Vanillin, commercial source, 584
Vapor-phase acetylation, with ethanethioic acid, 186
Vapor-phase combustion, 541
Vapor-phase reactions
 chain branching, 544
 combustion, 544
 role of fire retardancy, 545
Vapor-phase treatment
 acetylation, 185
 ethylene oxide, 192
Variability
 causes, 40–42
 compression wood, 48–51
 consequences, 40
 green moisture content, 128
 heartwood, 42–45
 juvenile wood, 51–54
 knotwood, 54
 major types, 42–55
 mature wood, 51–54
 reaction wood, 45–48
 sapwood, 42–45
 tension wood, 48–51
 along tree radius, 41
Varnishes, 437–38
Vascular bundles, definition, 9
Vascular cambium
 definition, 9
 wood cell production, 11
Vazo catalyst, 264
 effect on polymerization, 269–71
 effects, basswood–MMA composite, 270t,270f
 half-life vs. temperature, 264t
Vessel elements, 4
Vessel pitting, in hardwoods, SEM, 35f
Vessel system, liquid movement, 28
Vessels, in hardwoods, 22
Vinyl chloride, structure, 262
Vinyl composites, 261–88
Vinyl monomers, 267
Vinyl polymers, properties, 261
Vinyltoluene, structure, 262
Viscosity, definition, 214
Visible light, penetration, 446
Void volume, moisture effects, 153
Volume change, of wood–polymer composites, 286t

Volumetric shrinkage
 data, relationship to fiber saturation, 144
 equation, 143
 Indian woods, 144
Volumetric swelling coefficients
 calculation, 177
 determined by water-soaking method, 203t
 southern pine sapwood, 179t,180t
 standardization, 178

W

Waferboard, preparation, 371
Waferboard bonding, isocyanates, 392
Walnut, oxide treatment, 198t
Walnut shell flour, 333
Warts, 33–34
 vessel elements, 36
Warty layer, 33
 softwood, TEM, 34f
Waste treatment, use of wood-decomposing fungi, 483
Water
 absorption, by dry wood, 278
 affinity, 37
 conduction
  by brown-rot fungi, 480
  roots to leaves, 128
 content
  measuring, 129
  primary, calculation, 163
  secondary, calculation, 163
 damage, reducing, 439
 effect on free radical formation, 431
 as extractives solvent, 73
 hot, as extractives solvent, 73
 hydration, free energy change, 166
 liquid, relative energy level, 155f
 in living tree, 128
 primary, free energy change, 164
 relationship to wood, 127–72
 repellents, 438–39
 resistance, in wood bonding, 367
 secondary, free energy change, 164
 solution, 164
  free energy change, 166
 sorption
  decreasing, 150
  theories, 161–66
 specific heat, 153
 vapor, 128
  enthalpy, cell cavities, 154
  relative energy level, 155f
Water-soaking method, 203t
Waterborne preservatives, advantages, 243
Waterborne salts, 440,444
Weathered wood, free radical characteristics, 430–31
Weathering, 176
 accelerated, 426
 erosion, 417t
 and anatomic structure, 404
 chemically modified woods, 425–26
 chemistry, 401–47
 color changes, 411–14

INDEX
613

Weathering—*Continued*
  definition, 401
  earlywood, 424
  effects on southern pine, 241f
  factors responsible, 403
  general aspects, 403-7
  and heat, 405
  latewood, 424
  and light, 405
  and moisture, 404-5
  participation of singlet oxygen, 432-35
  physical changes, 414-16
  plywood, 423-25
  property changes, 407-22
  protection, 436-43
  reactions, chemical aspects, 427-36
  and sunlight, 405
  as surface phenomenon, 240
  wood-based materials, 422-25
Western hemlock
  heartwood, aspirated-pit membrane, 43f
  softwood bordered-pit membranes, 31f
Western redcedar
  brown-rot fungus, 467f
  color changes, 412f,413f
  free radicals vs. moisture content, 433f
  light exposure, erosion, 417t
  soft-rot fungus, 467f
Wet basis moisture content, 128
Wetting, heat, calculation, 154-55,157
White ash, elemental composition, 119t
White birch, elemental composition, 119t
White cedar, reaction with 2,4-tolylene diisocyanate, 187
White fir
  acid content, 346t
  homocellular ray, 20f
  nonconventional bonding, 361
White glue, 342
White oak
  elemental composition, 120t
  heartwood
    tyloses, 44f
    white-pocket rot fungus, 467f
  light exposure, erosion, 417t
White pine, attack by longhorn beetle, 466f
White rot
  classification, 462
  effect on organotin polymer-impregnated wood, 300
  features, 468-69t
  hemicellulase production, 477
  lignin decomposition, 477-79,478f
  northern red oak, 467f
White spruce wood lignin, TEM, 8f
White-pocket rot fungus, white oak heartwood, 467f
WML—*See* Work to maximum load
Wood
  anatomy, 19-24
  behavior, spiral thickenings, 36
  cell production, 11
  decay
    cause, 456
    cellulose decomposition, 473-77

Wood—*Continued*
  decay—*Continued*
    hemicellulose decomposition, 476-77
    lignin decomposition, 477-79
    mechanisms, 471-79
  decomposition
    by insects, 464-65t
    by marine borers, 465t
  degradation, 356
  design, effects on strength, 220-26
  deterioration, types, 458-60
  elasticity, 214
  elemental composition, 510t
  fibers, differentiated, 13f
  finishes, properties, 436
  flour oxidation, by nitrates, 378f
  formation, in trees, 9-19
  major constituents, 8f
  pyrolysis products, 508t
  solubility, 73
  stress-strain diagram, 214f
  thermal conductivity, moisture effects, 154
  treated, thermogravimetric analysis, 548f
  use
    moisture content, 140
    target moisture contents, 140
Wood moisture, and the environment, 127-40
Wood moisture content
  extractives effects, 139-40
  measurement
    by neutron moisture meter, 135-36
    NMR, 134-35
  rate of change, 166
  relative humidity effects, 136-37
  sorption history effects, 136-37
  temperature effects, 137-39
  in use, 140
  vs. diffusion coefficients, 168f
  wood species effects, 139-40
Wood preservatives, from pesticides, 480
Wood properties, nonconventional bonding, 368t
Wood quality, 17-19
Wood rays—*See* Rays
Wood solubility, in 1% NaOH, 73
Wood sources, 3-8
Wood strength, environmental effects, 212
Wood sugars, chromatographic analysis, 71
Wood surface
  activation, 349-400
  in conventional bonding, 350
Wood tissue, construction, 6-7
Wood variability
  See also Variability
  causes, 40-42
  consequences, 40
  major types, 42-55
  tree radius, 41
Wood volume, vs. moisture content, 145f,146f
Wood-based materials
  protection, 445
  weathering, 422-25

Wood–coating interactions, 445–46
Wood-decomposing organisms
  control, 479–81
  uses, 481–83
Wood–polymer composites
  antishrink efficiency, 297t
  bioactive, 291–306
  properties, 286t
Wood–polymer materials, 257–89
  impregnation, 265–66
Wood–water relationships, 127–72
Wood-to-wood bonding, 358–61
  Douglas-fir veneer, 359
  mechanism, 359f
  particle board, 360f
  Pinus ponderosa Laws., 361
Work to maximum load, 223
  reduction, 237
Work to proportional limit, 223
WRP—*See* Water repellents

X

Xanthate derivative, cellulose, 60

Xylan, weathering effects, 408
Xylem
  autolysis, 11
  function, 11
  primary, 9
  secondary, 4

Y

Yellow birch, physical property
  enhancements, 280–81t
Yellow pine
  attack by marine borer, 466f
  brown-rot fungus, 467f
  soft-rot fungus, 467f
Yellow poplar, light exposure, erosion, 417t

Z

Zinc ammonium borate, effect on
  mechanical properties, 251–52t
Zinc chloride, 189, 532
Zinc naphthenate, 439

Production by Paula Bérard
Copyediting and indexing by Deborah Corson
Jacket design by Anne G. Bigler

Typeset by The Sheridan Press, Hanover, Pa.
and Hot Type Ltd., Washington, D.C.
Printed and bound by Maple Press Co., York, Pa.

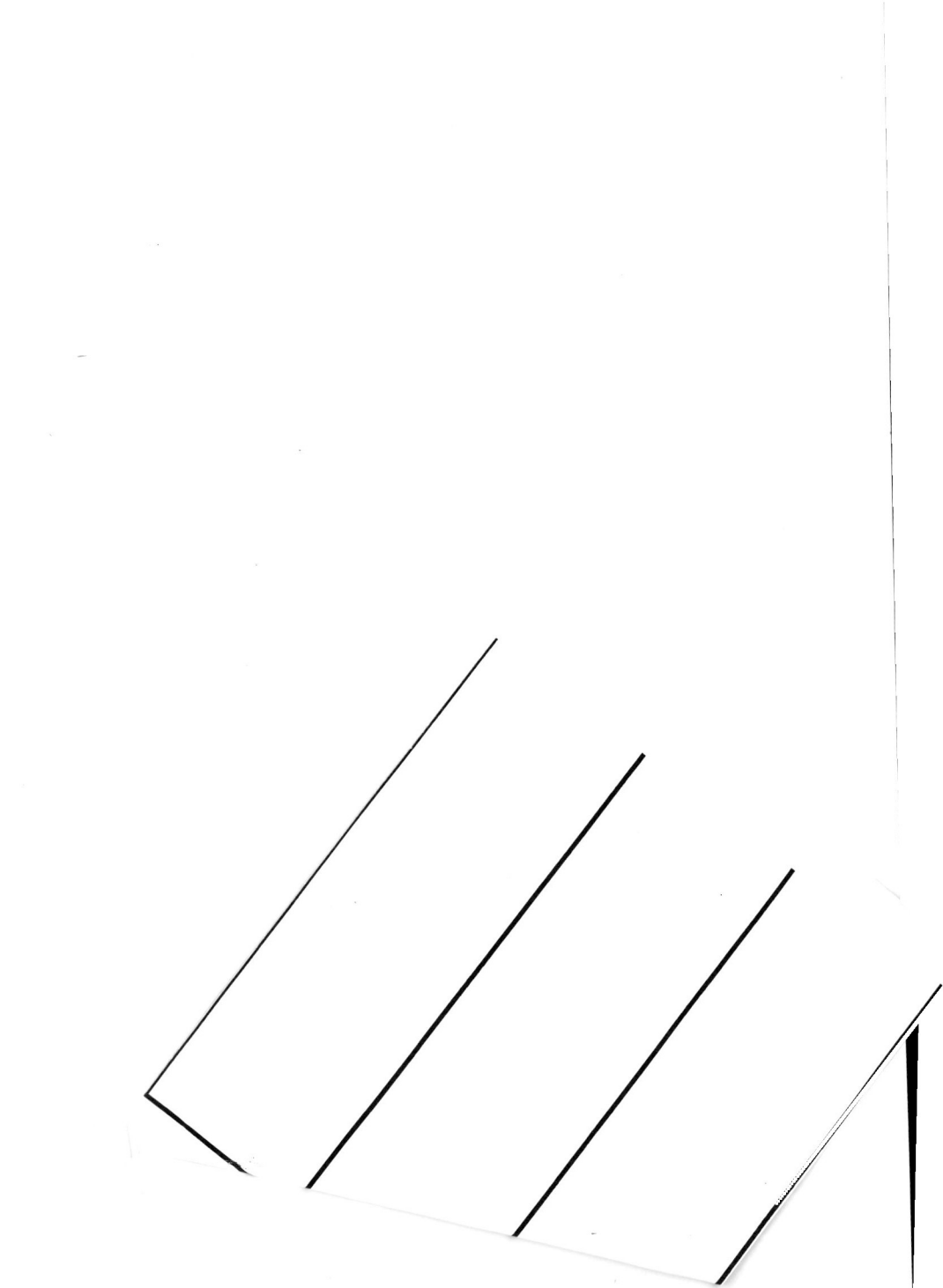